Ion Exchange Membranes

Ion Exchange Membranes

Fundamentals and Applications

Second Edition

Yoshinobu Tanaka
IEM Research, Ibaraki, Japan

ELSEVIER
AMSTERDAM • BOSTON • HEIDELBERG • LONDON • NEW YORK • OXFORD
PARIS • SAN DIEGO • SAN FRANCISCO • SINGAPORE • SYDNEY • TOKYO

Elsevier
Radarweg 29, PO Box 211, 1000 AE Amsterdam, Netherlands
The Boulevard, Langford Lane, Kidlington, Oxford OX5 1GB, UK
225 Wyman Street, Waltham, MA 02451, USA

Notices
Knowledge and best practice in this field are constantly changing. As new research and experience broaden our understanding, changes in research methods, professional practices, or medical treatment may become necessary.

Practitioners and researchers must always rely on their own experience and knowledge in evaluating and using any information, methods, compounds, or experiments described herein. In using such information or methods they should be mindful of their own safety and the safety of others, including parties for whom they have a professional responsibility.

To the fullest extent of the law, neither the Publisher nor the authors, contributors, or editors, assume any liability for any injury and/or damage to persons or property as a matter of products liability, negligence or otherwise, or from any use or operation of any methods, products, instructions, or ideas contained in the material herein.

ISBN: 978-0-444-63319-4

British Library Cataloguing in Publication Data
A catalogue record for this book is available from the British Library

Library of Congress Cataloging-in-Publication Data
A catalog record for this book is available from the Library of Congress

Contents

Author Biography

Yoshinobu Tanaka

Educational Career

Graduated from Tokyo University of Science in 1959

Professional Experience

Joined Japan Monopoly Corporation (now Japan Tobacco Inc.) in 1959.

Investigated ion exchange membranes at Central Research Institute and Odawara Experimental Station. Carried out the following research and development for industrial application of ion exchange membranes.

Preparation of ion exchange membranes
Measurement of membrane characteristics and electrodialysis phenomena
Maintenance and trouble prevention in electrodialyzer operations
Designing and manufacturing of electrodialyzers etc.

Retired from Japan Tobacco Inc. in 1996.

Established IEM Research in 1996 and continued the investigation on ion exchange membranes up to the present.

Degree

1983—Doctor of Engineering; Tokyo University

Activation in Academic Societies

Member of planning board and editorial board

Japan Association of Ion exchange
Society of Sea Water Science Japan
International journal, Membrane Water Treatment

Lead Guest Editor: Special Issue on Ion Exchange Membranes; International Journal of Chemical Engineering (2012)
Tokyo Institute of Technology Lecturer (1994–1995)

Award

1984—Society of Sea Water Science Japan; Science prize
1998—Japan Association of Ion Exchange; Achievement prize

Publication

Books

Electrodialysis. In: Theory and Design of Membrane Separation Process, 1993. Industrial Publishing & Consulting Inc., Tokyo.

Electrodialysis in membrane separation. In: Membrane Separation Technology Handbook, 1990. Industrial Publishing & Consulting Inc., Tokyo.

Fundamental theory and experimental method in ion exchange membrane electrodialysis. In: Advanced Membrane Separation Technology Handbook, 1987. Science Forum Co., Tokyo.

Ion exchange and ion exchange membranes. In: Functions and Applications of Ion Exchange Membranes, 2004. Industrial Publishing & Consulting Inc., Tokyo.

Ion exchange membranes: fundamentals and applications first ed., 2007. In: Membrane Science Technology Series 12, 2010. Elsevier, Amsterdam. The book is translated into Chinese and published from Chemical Engineering Publication, Beijing.

Ion exchange membrane experimental method. In: Membrane Science Experimental Method, 1984. Kitami Shobo Co., Tokyo.

Ion Exchange Membrane Electrodialysis: Fundamentals, Desalination, Separation, 2010. Nova Science Publishers, New York.

Ion exchange membrane salt production. In: Sea Water, its Property and Technology, 1994. Tokai University Press, Tokyo.

Ion exchange membrane separation technology. In: Ion Exchange: Fundamentals of Advanced Separation Technology, 1991. Kodansya Scientific Co., Tokyo.

Mass transport in ion exchange membrane. In: Encyclopedia of Membrane Science Technology, 2013. John Wiley & Sons, Hoboken.

Principles of ion exchange membrane electrodialysis for saline water desalination. In: Ion Exchange Technology, 2012. Springer, New York.

Resources dissolving in seawater. In: Advanced Membrane Treatment Technology and its Applications, 1984. Fuji Techno System Co, Tokyo.

Journals

Acceleration of water dissociation generated in an ion exchange membrane. J. Membr. Sci. 303, 2007, 234–243.

A computer simulation of continuous ion exchange membrane electrodialysis for desalination of saline water. Desalination 249, 2009, 809–821.

A computer simulation of batch ion exchange membrane electrodialysis for desalination of saline water. Desalination 249, 2009, 1039–1047.

A computer simulation of feed and bleed ion exchange membrane electrodialysis for desalination of saline water. Desalination 254, 2010, 99–107.

A computer simulation of ion exchange membrane electrodialysis for concentration of seawater. Membr. Water Treat 1, 2010, 13–37.

Boundary layer phenomena in ion exchange membrane electrodialysis. Chem. Eng. 38, 1993, 342–346.

Concentration polarization and water dissociation in ion exchange membrane electrodialysis. J. Electrochem. Soc. Jpn. 42, 1974, 450–456: 43, 1975, 584–588: 50, 1982, 667–672; 821–824: 51, 1983, 465–470; 267–271.

Changes of membrane electric resistance and water dissociation generation in the treatment giving low permeability for bivalent ions to cation exchange membranes. Bull. Soc. Sea Water Sci. Jpn. 31, 1977, 123–127.

Current density distribution around an insulator in an ion exchange membrane electrodialyzer. Bull. Soc. Sea Water Sci. Jpn. 37, 1984, 295–298.

Concentration polarization and water dissociation in ion-exchange membrane electrodialysis: mechanism of water dissociation. J. Chem. Soc. Faraday Trans. 1 82, 1986, 2065–2077.

Concentration polarization in ion exchange membrane electrodialysis. J. Membr. Sci. 57, 1991, 217–235.

Concentration polarization in a substance layer attached to an ion exchange membrane. Bull. Soc. Sea Water Sci. Jpn. 51, 1997, 228–236.

Concentration polarization, limiting current density and water dissociation in ion exchange membrane electrodialysis. J. Ion. Exch. 8, 1997, 14–28.

Current density distribution and limiting current density in ion-exchange membrane electrodialysis. J. Membr. Sci. 173, 2000, 179–190.

Current density distribution, limiting current density and saturation current density in an ion-exchange membrane electrodialyzer. J. Membr. Sci. 210, 2002, 65–75.

Concentration polarization in ion-exchange membrane electrodialysis – the events arising in a flowing solution in a desalting cell. J. Membr. Sci. 216, 2003, 149–164.

Concentration polarization in ion-exchange membrane electrodialysis, the events arising in an unforced flowing solution in a desalting cell. J. Membr. Sci. 244, 2004, 1–16.

Continuous ion-exchange membrane electrodialysis of mother liquid from a salt-manufacturing plant and transport of Cl^- ions and so_4^{2-} ions. Membr. Water Treat 3, 2012, 63–76.

Distribution of electrodialytic conditions and limiting current density in ion exchange membrane electrodialyzer. J. Electrochem. Soc. Jpn. 45, 1977, 16–21.

Distribution of electrodialytic conditions in an electrodialyzer and limiting current density. J. Membr. Sci. 92, 1994, 217–228.

Development of a computer simulation program of batch ion-exchange membrane electrodialysis for saline water desalination. Desalination 320, 2013, 118–133.

Development of a computer simulation program of feed-and-bleed ion-exchange membrane electrodialysis for saline water desalination. Desalination 342, 2014, 126–138.

Electrodialysis of solutions of high-temperature and high-concentration with ion exchange membranes. Bull. Soc. Sea Water Sci. Jpn. 24, 1970, 104–128: 25, 1971, 189–197: 26, 1972, 74–83.

Effect of electric field in the treatment giving low permeability for bivalent ions to cation exchange membranes. Bull. Soc. Sea Water Sci. Jpn. 32, 1978, 95–99.

Effects of temperature on ion exchange membrane electrodialysis. Bull. Soc. Sea Water Sci. Jpn. 34, 1980, 31–36.

Electrodialysis process in a salt manufacturing plant. Bull. Soc. Sea Water Sci. Jpn. 34, 1980, 61–90.

Electrochemical studies on ion exchange membranes and electrodialysis phenomena. Bull. Soc. Sea Water Sci. Jpn. 38, 1985, 257–282.

Electrochemical properties of ion exchange membranes. Ionics, October 1985, 151–162.

Effect of red tide on limiting current density. Bull. Soc. Sea Water Sci. Jpn. 47, 1993, 4–10.

Electrochemical dissociation of water molecules generated in an interfacial layer on an ion exchange membrane. Bull. Soc. Sea Water Sci. Jpn. 47, 1993, 242–247.

Ionic transport in a boundary layer in ion exchange membrane electrodialysis. J. Ion. Exch. 9, 1998, 2–13.

Ion-exchange membrane electrodialytic salt production using brine discharged from a reverse osmosis seawater desalination plant. J. Membr. Sci. 222, 2003, 71–86.

Irreversible thermodynamics and overall mass transport in ion-exchange membrane electrodialysis. J. Membr. Sci. 281, 2006, 517–531.

Ion-exchange membrane electrodialysis for saline water desalination and its application to seawater concentration. Ind. Eng. Chem. Res. 50, 2011, 7494–7503.

Ion-exchange membrane electrodialysis of saline water and its numerical analysis. Ind. Eng. Chem. Res. 50, 2011, 10765–10777.

Ion-exchange membrane electrodialysis program and its application to multi-stage continuous saline water desalination. Desalination 301, 2012, 10–25.

Limiting current density in the ion exchange membrane electrodialysis. Bull. Soc. Sea Water Sci. Jpn. 29, 1976, 209–217.

Limiting current density of an ion-exchange membrane and of an electrodialyzer. J. Membr. Sci. 266, 2005, 6–17.

Mass transport in an ion exchange membrane electrodialyzer for seawater concentration. J. Electrochem. Soc. Jpn. 45, 1977, 630–635.

Membrane applications in salt production. Special Issue Metals, January 1990, 90–95.

Mass transport and energy consumption in ion-exchange membrane electrodialysis of seawater. J. Membr. Sci. 215, 2003, 265–279.

Mass transport in a boundary layer and in an ion exchange membrane: mechanism of concentration polarization and water dissociation. Russ. J. Electrochem 48, 2012, 735–755.

Measurement of membrane characteristics using the phenomenological equation and overall mass transport equation in ion-exchange membrane electrodialysis of saline water. Int. J. Chem. Eng. 2012, 2012, 1–12. http://dx.doi.org/10.1155/2012/148147. Article ID 148147.

Overall mass transport and solution leakage in an ion-exchange membrane electrodialyzer. J. Membr. Sci. 235, 2004, 15–24.

Problems in ion exchange membrane electrodialysis technology. Bull. Soc. Sea Water Sci. Jpn. 35, 1981, 93–98.

Pressure distribution, hydrodynamics, mass transport and solution leakage in an ion-exchange membrane electrodialyzer. J. Membr. Sci. 234, 2004, 23–39.

Recent advances in ion exchange membrane technology. Bull. Soc. Sea Water Sci. Jpn. 34, 1981, 293–300: 37, 1983, 18–26: 40, 1986, 111–123: 45, 1991, 61–77: 49, 1995, 3–21; 280–305: 52, 1998, 51–73: 55, 2001, 268–288.

Recent trends in ion exchange membrane technology. Water 26, 1984, 21–27.

Regularity in ion-exchange membrane characteristics and concentration of seawater. J. Membr. Sci. 163, 1999, 277–287.

Studies on the treatment giving low permeability for bivalent ions in electrodialytic concentration using ion exchange membranes. J. Electrochem. Soc. Jpn. 42, 1974, 26–31; 192-198; 512-517; 572-577: 49, 1981, 594-598.

Seawater concentration using electrodialysis. Chem. Plants 27, 1983, 32–38.

Simulation of an ion exchange membrane electrodialysis process for continuous saline water desalination. Desalin. Water Treat. 22, 2010, 271–285.

The effect of solution leakage on mass transport in an ion exchange membrane electrodialyzer for seawater concentration. J. Electrochem. Soc. Jpn. 46, 1978, 208–211.

The changes of effects with time in low permeability treatment for bivalent ions of a cation exchange membrane. Bull. Soc. Sea Water Sci. Jpn. 32, 1978, 100–103.

Treatment of ion exchange membranes to decrease divalent ion permeability. J. Membr. Sci. 8, 1981, 115–127.

The effect of solution leakage in an ion-exchange membrane electrodialyzer on mass transport across a membrane pair. J. Membr. Sci. 231, 2004, 13–24.

Water dissociation in ion-exchange membrane electrodialysis. J. Membr. Sci. 203, 2002, 227–244.

Water dissociation reaction generated in an ion exchange membrane. J. Membr. Sci. 350, 2010, 347–360.

Preface

Principles of ion transport across ion exchange membranes had been investigated using biological membranes in old time. Fundamental study and industrial application of ion exchange membranes started over a half century ago. Ion exchange membrane electrodialysis is now one of the basic technologies in saline water desalination industry. It is also applied to many fields such as demineralization and reuse of sewage or industrial waste, refining of amino acid solutions, demineralization of whey and sugar liquor, among others. Further, the membranes are applied to succeeding technologies such as electrodialysis reversal, bipolar membrane electrodialysis, electrodeionization, electrolysis, diffusion dialysis, fuel cell, among others. Through on going research and development, ion exchange membranes contribute to the improvement of our standard of living.

The *Ion Exchange Membrane: Fundamentals and Applications* 2nd edition is edited by revising and developing the contents of the 1st edition published in 2007. It consists of Part 1 *Fundamentals* and Part 2 *Applications* and provides a full and wide explanation on membrane preparation, membrane characterization, membrane phenomena, membrane applications, process design, process operation, process performance, economics, among others. The fundamental studies are the basis of the application studies. At the same time, the application studies induce the development of the fundamental studies. Both studies influence each other and consequently contribute to the technology development.

Many topics are selected and they are explained across 11 chapters in the *Fundamentals* and 13 chapters in the *Applications*. The phenomena are discussed using many figures and tables for easier understanding and equations are supplied to facilitate the explanation. "Introduction" or "Overview of Technology" is described at the beginning of each Chapter. The Computer Simulation Program is one of the topics in this book and it is developed for saline water desalination and water production. Readers can operate the program using the *Companion Sites* attached in Chapter 24 in the *Applications*.

The author started the fundamental research on ion exchange membranes in 1959 at Japan Monopoly Corporation (now Japan Tobacco Inc.). The research was carried out to contribute to the seawater concentration technology conversion from salt field operations to electrodialysis operations for salt production. The author worked to investigate membrane preparation, membrane characterization, membrane phenomena, process design, operation, maintenance, and trouble prevention. The author thanks for many coworkers cooperated in the above investigations. The technology conversion was completed and ion exchange membranes are now incorporated into the salt production system in Japan.

In 1996, the author established IEM Research and started the investigation of saline water desalination and water production. The investigation was carried out on the basis of the seawater concentration technology described above. This book is written based on the above author's experience during about 50 years. Considerable parts are tried to be described based on the author's opinion and are expected to be investigated further.

In writing the book, I am grateful to the late professor Takeo Yamabe and Professor Manabu Seno, Tokyo University, for guidance in the research. Further thanks to the late Mr Reo Ehara and Mr. Shigeru Itoi, IEM Research, for cooperation.

The Computer Simulation Program was attracted by Dr Anita Koch in Elsevier at ICOM 2011 and this event takes an opportunity of publishing this book. The author thanks for her enthusiasm and cooperation. I am grateful to Dr Sarah Jane Watson and Elsevier B.V. for proofreading, checking, and publishing the book. Thanks to Dr. Mohana Priyan Rajendran (Project Manager, Book Production Science and Technology, Elsevier) is expressed. I thank my wife Satsuko for supporting my research life.

Yoshinobu Tanaka

PART 1

Fundamentals

1

Preparation of Ion Exchange Membranes

1.1 Introduction

Ion exchange membranes are classified based on the functions of ion exchange groups fixed in the membrane, as follows.

1. Cation exchange membranes

   Cation exchange groups are fixed. The membranes permeate cations selectively.

2. Anion exchange membranes

   Anion exchange groups are fixed. The membranes permeate anions selectively.

3. Amphoteric membranes

   Cation exchange groups and anion exchange groups are fixed randomly throughout the membranes.

4. Bipolar membranes

   Bilayered membranes in which a cation exchange layer and an anion exchange layer are joined.

5. Mosaic membranes

   Cation exchange domains and anion exchange domains are included in the membranes.

Ion exchange membranes are also classified based on the materials constituting the membrane, as follows.

1. Hydrocarbon membranes.
2. Perfluorocarbon membranes.
3. Inorganic membranes.
4. Composite membranes of inorganic ion exchangers and organic polymers.

The membrane types are classified into the following types according to their functional groups.

1. Strongly acidic (typically sulfonic acid groups).
2. Weakly acids (mostly carboxylic acid groups).

Ion Exchange Membranes. http://dx.doi.org/10.1016/B978-0-444-63319-4.00001-8

3. Strongly basic (typically quaternary amino groups).
4. Weakly basic (mostly primary, secondary, or tertiary amino groups).

The membranes are also classified into two types by their microstructure:

1. Homogeneous membranes.
2. Heterogeneous membranes.

The membranes are required to have the following properties:

1. Low electric resistance.
2. High transport number of counter-ions.
3. Low diffusion coefficient of salt.
4. Low osmotic water and low electro-osmotic water.
5. Permselectivity for specific ions with the same charge.
6. Anti-organic properties.
7. High mechanical strength.
8. Dimensional stability.
9. High chemical stability and durability.
10. Low cost.

1.2 Hydrocarbon Ion Exchange Membranes

In 1950, Ionic, Inc. invented a ion exchange membrane prepared by condensation and polymerization of phenol sulfonic acid and formaldehyde (Juda and MacRae, 1950). This invention attracted a great deal of attention across the world and contributed to the succeeding development of membrane synthesis technology. Various preparation methods of ion exchange membranes have been reported so far (Sata, 2004).

In the recent history of ion exchange membrane preparation, styrene–divinylbenzene copolymers are the most commonly used neutral starting material because of excellent physical (strength) and chemical (anti-oxidation, anti-hydrolysis, and heat resistance) stability. The copolymerization is carried out with vinyl polymerization proceeding with radical reactions between vinyl double bonds (–CH=CH·) in styrene and divinylbenzene. Strongly acidic cation exchange membranes are prepared from the copolymers by sulfonation with chlorosulfonic acid or concentrated sulfuric acid (Fig. 1.1). Strongly basic anion exchange membranes are prepared by copolymerization between chloromethyl styrene and divinylbenzene and then amination using trimethyl amine (Fig. 1.2).

These membranes are now widely available commercially and used industrially. The fundamentals and applications of ion exchange membranes described in this book are mainly based on these membranes. Method of preparing hydrocarbon ion exchange membranes are introduced based on practical examples in patents, and so forth, as described in Sections 1.3–1.5.

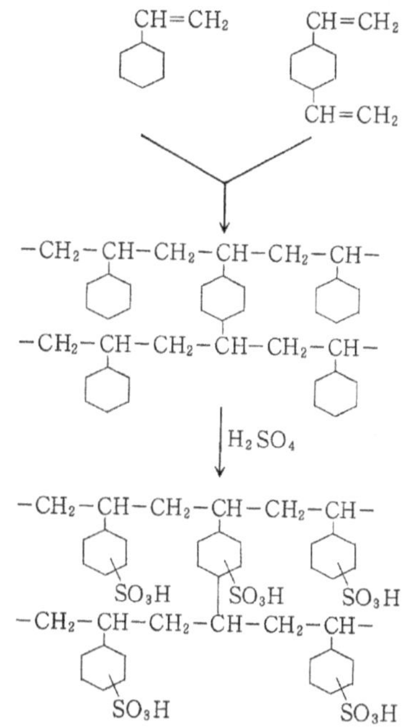

FIGURE 1.1 Styrene–divinylbenzene cation exchange membrane.

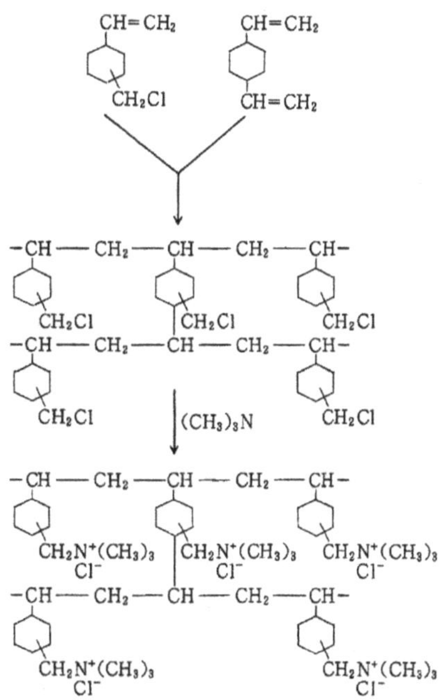

FIGURE 1.2 Chloromethyl styrene–divinylbenzene anion exchange membrane.

1.3 Homogeneous Membranes

1.3.1 Sandwich Method

This method was developed by Ionic, Inc. (MacDonald et al., 1992) and provides an environmentally friendly continuous process for rapidly manufacturing reinforced ion exchange membranes (Nepton membrane). Special features of this invention are as follows:

1. Liquids composed of polymerizable and non-polymerizable components are characterized by having a viscosity less than about 10 P at shear rates in excess of about 1 cm/s and by composing solvents that have a boiling point of less than about 185 °C;
2. Supporting substrates are characterized by being insoluble in this liquid at temperatures and times prevailing during the process.

Steps of the process are:

1. Forming a continuous sandwich of a substrate by juxtaposing films on each face of the substrate.
2. Passing the sandwich of the substrate and the film in a vertical direction downward between a pair of squeeze rolls.
3. Introducing the liquid, including polymerizable and non-polymerizable components, into the pocket formed in the sandwich above the squeeze rolls.
4. Pulling the sandwich and liquid between the squeeze rolls, thereby filling the apertures in the substrate with the liquid.
5. Thereafter, rapidly polymerizing the polymerizable component in the liquid, which results in the formation of a reinforced polymer sheet.

Referring to Figs. 1.3 and 1.4, the number 10 indicates a continuous substrate (woven or nonwoven fabric paper, etc.) that is played out from substrate roll 10A, passes over roll 10B, and continues downward in a vertical direction between horizontal rolls 11B and 11′B. Numbers 11 and 11′ represent pliable films with compositions that are not soluble or swollen in the polymerizable liquid. In addition, they should release relatively easily from the finished polymer sheet. Films 11 and 11′ are played out from film rolls 11A and 11A′, respectively, continuing downward in a vertical direction between the rolls on each side (face) of substrate 10, initially forming a loose sandwich at rolls 11B and 11′B.

Liquid characterized as having a viscosity less than about 10 P at share rates in excess of about 1 cm/s is prepared from the mixture of the following components to form polymer:

54.6% Divinylbenzene	16.6l
Vinyl toluene	5.5l
Dixylyl ethane	18.9l

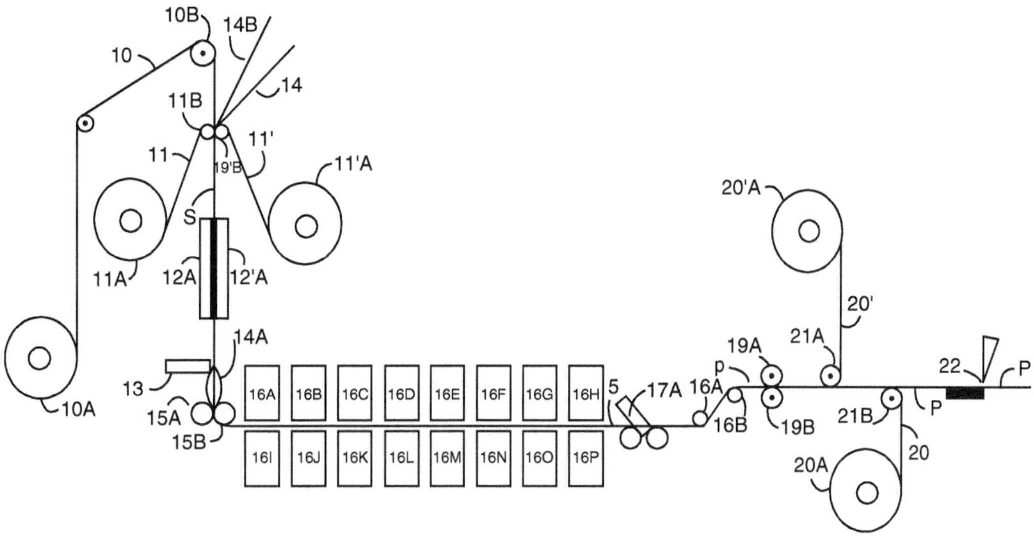

FIGURE 1.3 Sandwich method (Ionics Inc.). *MacDonald et al. (1992), Sheet 1, Fig. 1.*

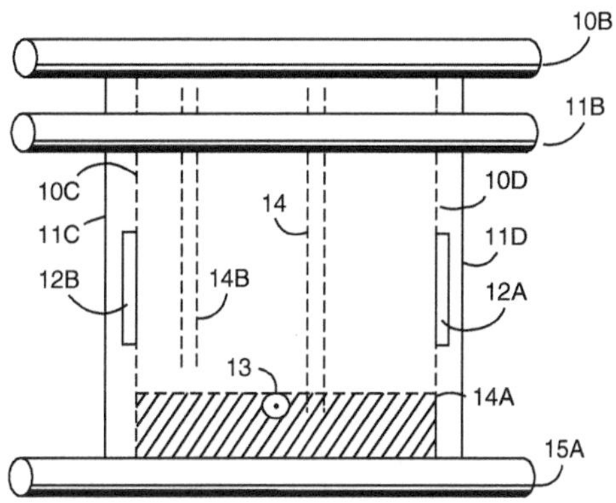

FIGURE 1.4 Sandwich method (Ionics Inc.). *MacDonald et al. (1992), Sheet 2, Fig. 2.*

with the following polymerization initiators:

bis(4-Tertiary-butyl-cyclohexyl) peroxyl dicarbonate	650 g
Dilauroyl peroxide	410 g
Dibenzoyl peroxide	410 g

The mixture is thoroughly deoxygenated by spraying into a vacuum chamber and is fed into an apparatus as shown in Fig. 1.3 through conduit means 14.

Films 11 and 11′ are Mylar (3 mils thick and 24 in wide). Substrate 10 is an acrylic plain weave fabric (20 in wide and 19 mils thick). The linear velocity is adjusted to give a polymerization time of about 25 min. Element 16A is inert and exerts a pressure of about 4 g/cm^2 on sandwich S. Element 16B and D are fans, regulated manually to control the temperature of sandwich S.

Elements 16C and E are inert and exert a pressure of about 0.6 g/cm^2 on sandwich S. Elements 16F, G, H, O, and P are omitted. Elements 16I through N are heated electrically to cause the temperature of sandwich S to increase from about 60 °C at element 16I to about 110 °C at element N. Elements 16A through E and I through R are separated from sandwich S by Kapton adherent film anchored to the apparatus.

Means 12A and B (Fig. 1.4) and 12′A and ′B (Fig. 1.4) are heat welding elements (12′B lies behind 12′A in Fig. 1.3 and behind 12B in Fig. 1.3). Element 13 is a capacitative sensor. Rolls 15A and 19A are covered with 30-durometer rubber to a depth of at least 1 cm. Rolls 15B and 19B are chrome steel. The edges of the sandwich are slit by elements 17A and B and films 20 and 20′ (i.e., 11 and 11′) taken up by rolls 20A and 20′A, respectively. Polymer sheet P is cut by shear 22 into length of about 40 in each. The resulting sheets are leached in methylene dichloride to remove the dixylyl ethane and sulfonated. The resulting cation exchange membranes have an ion exchange capacity of about 2.8 mEq/g of dry membrane and water content of about 50% of wet membrane. The membranes are useful in the electrodialytic production of potable water from brackish water.

1.3.2 Deaeration Method

Asahi Glass Co. developed the following synthetic method of ion exchange membranes (Selemion membrane) (Mineki et al., 1972) as follows:

1. In Fig. 1.5, monomer mixture 11 (consisting of styrene, chloromethyl styrene, divinylbenzene, etc., viscosity 0.1–20 P) is introduced into vessel 10 with a reinforcement (polyvinylchloride net, 100 m long and 100 cm wide) wound around a stainless rod. In this step the reinforcement is immersed in the monomer mixture and consequently base film 13 is formed around base film roller 12.

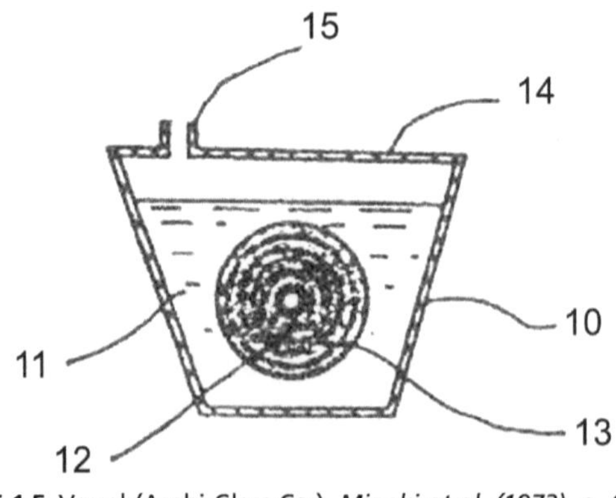

FIGURE 1.5 Vessel (Asahi Glass Co.). *Mineki et al. (1972), p. 6, Fig. 1.*

2. The vessel is sealed with cap 14 and deaerated (1–500 mmHg) from hole 15 connected to a decompression pump (positioned at the outside of the figure). Such a deaeration step is crucial in this process because it prevents pinhole formation in the membrane as a result of air bubbles.
3. Next, vessel 10 is incorporated with the winding unit in Fig. 1.6.
4. The base film in base film roller 12 in Fig. 1.6 is rewound and moves via doctor roller 9 toward windup roller 8.
5. On the other hand, tearing film 2 (Mylar, vinylon, etc.) is played out from tearing film roller 1 and is supplied to windup roller 8 via direction-converting roller 3, 3′, balance roller 4, and expander roller 6.
6. The base film and tearing film are juxtaposed and wound up at the windup roller.
7. The thickness of the base film is controlled by regulating the weight of balance roller 4.
8. When the viscosity of the monomer mixture is low, it is best to use it when the mixture is accumulated in pocket 7.
9. If the viscosity of the monomer mixture is high, the volume of the mixture is regulated by adding pressure with press roller 16.
10. Polymerization is carried out in windup roller 8 by adjusting the temperature (85 °C for 6 h).
11. The polymerized film is taken out by rewinding the windup roller.

For example, an anion exchange membrane is synthesized by adjusting the viscosity of the monomer mixture to 0.1 P, the pressure in the vessel to 20 mmHg for 30 min, and the components of the monomer mixture to:

chloromethyl styrene, 80 parts
divinylbenzene, 10 parts
ethylvinyl benzene, 75 parts
styrene, 3 parts
a mixture of polyvinylchloride (1 vol) and vinyl chloride-vinyl acetic acid mixture (3 vol), 20 parts

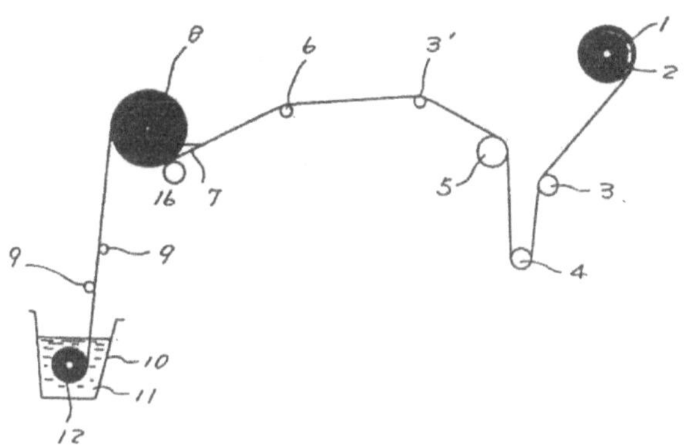

FIGURE 1.6 Winding unit (Asahi Glass Co.). *Mineki et al. (1972), p. 6, Fig. 2.*

dioctyl phthalate (plasticizer), 3 parts
benzoyl peroxide (polymerization initiator), 3 parts.

The polymerized film is aminated in a 15% trimethyl amine methanol solution at room temperature for 40 h and an anion exchange membrane is obtained by washing it in water. Electrochemical properties of the membrane synthesized above are:

Thickness, 175 μ
Area electric resistance, 25 Ωcm^2
Transport number, 0.95.

1.3.3 Paste Method

This is a method for preparing ion exchange membranes (Neosepta membrane) developed by Tokuyama Co. (Mizutani et al., 1964). In this invention, a polyvinylchloride net is passed through a mixture of polymerizable components (styrene, chloromethyl styrene, divinylbenzene, etc.), initiators of polymerization, and plasticizers to obtain a paste mixture. The mixture is continuously coated onto a backing fabric (woven cloth or net) and covered on both sides with a separating film. The membrane obtained in this process is not easily broken and deformed in the succeeding ion exchange group introducing step. The following practical example is found in the patent.

1. A paste solution is prepared by adding polyvinylchloride powder (100 parts) into monomer mixture consisting of styrene (80 parts), divinylbenzene (10 parts), dioctyl phthalate (25 parts), and benzoyl peroxide (1 part).
2. The solution is introduced into vessel 5 (Fig. 1.7).

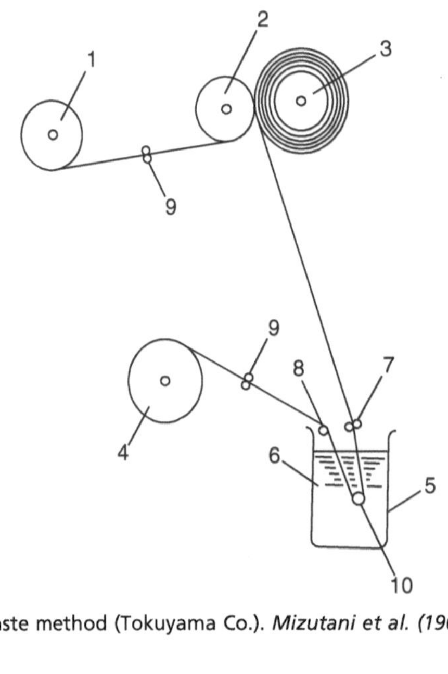

FIGURE 1.7 Paste method (Tokuyama Co.). *Mizutani et al. (1964), p. 6, Fig. 1.*

3. A polyvinylchloride net (reinforcement) wound up in roll 4 is passed through paste solution 6 in vessel 5 and is wound around roll 3, pressing by roll 2 with vinylon film (tearing material) moving from roll 1.
4. After the winding process, roll 3 is dismantled and put into a sealed reaction vessel in which the monomer mixture is polymerized at 130 °C for 30 min and 80 °C for 10 h.
5. In this situation, polyvinylchloride powders expand to include monomer mixtures and form a thin film. A cross-linking reaction between styrene and divinylbenzene is advanced in the films. Accordingly, three-dimensional configurations obtained by the cross-linking reaction are mixed and intertwined uniformly with polyvinylchloride molecular chains. Furthermore, graft polymerization arises between monomer mixtures and polyvinylchloride molecular chains owing to the chain reaction in the polymerization process.
6. After the polymerization reaction, the roll is rewound and the base film is taken out.
7. A cation exchange membrane is obtained by sulfonating the film. An anion exchange membrane is obtained by amination.

There is another method for synthesizing an anion exchange membrane using vinylpyridine instead of styrene to obtain a vinylpyridine–divinylbenzene copolymer (Fig. 1.8).

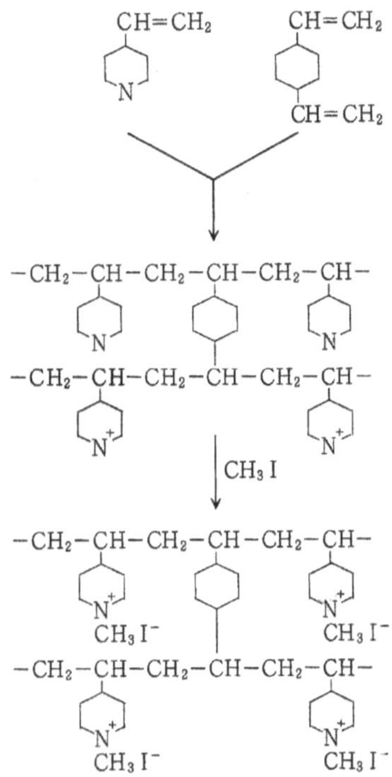

FIGURE 1.8 Styrene–vinylpyridine anion exchange membrane.

1.3.4 Spiral Winding Method

An ion exchange membrane is generally synthesized in an open system by applying a paste solution dissolving polymerizable components into a continuous substrate and promoting polymerization reactions with heating. The process needs to add many kinds of components to prepare a sticky paste solution. To overcome such a problem, Asahi Chemical Co. developed a method for preparing a styrene–divinylbenzene copolymer membrane (Aciplex membrane) (Misumi et al., 1974). The invention makes it possible to use low–boiling point solvents because the reactions are performed in a closed system. The synthetic apparatus is illustrated in Fig. 1.9: (1) vertical cross-section and (2) horizontal cross-section. The synthetic process is carried out as follows in ring-shaped space 10 (diagonal lines) formed between cylindrical drum 1 and drum 2.

1. First, a polypropylene net (reinforcement) is placed on a polyester film (tearing material) and they are wound around the inner drum (drum 1) on which an outer drum (drum 2) is set.
2. A monomer mixture (styrene, 80 parts; divinylbenzene, 20 parts; dioctyl phthalate plasticizer, 80 parts; benzoyl peroxide reaction initiator. 0.5 parts) is injected into the doughnut-shaped space, maintaining a temperature of 10 °C.
3. Circulation water is supplied to adjust the temperature in the inner part of drum 1 from valve 4 to valve 5, and the outer part of drum 2 from valve 7 to valve 8 polymerization is started at 30 °C and proceeds for about 10 days, changing temperature and controlling the reaction states.
4. Next, drums 1 and 2 are separated; the block polymerized material is taken out and rewound, and it is separated into a tearing material and a polymerized membrane film.

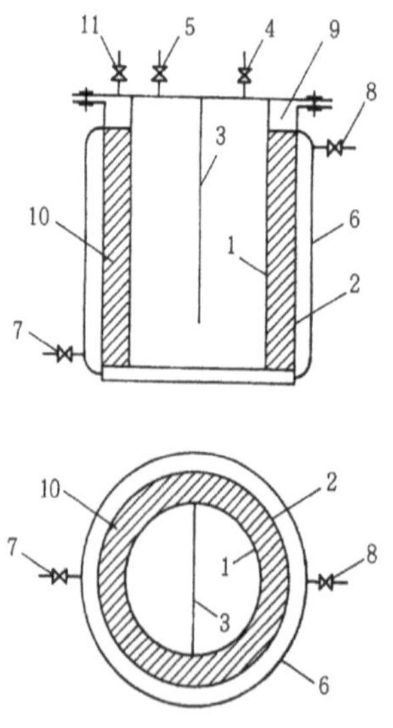

FIGURE 1.9 Membrane synthesizing apparatus (Asahi Chemical Co.). *Misumi et al. (1974), p. 4, Fig. 1.*

5. Dioctyl phthalate in the membrane film is extracted by washing the film with methanol.
6. A cation exchange membrane is obtained by sulfonating the film.

The reinforcement (polypropylene) is pre-irradiated by an electron beam before the winding process around drum 1 (Araki et al., 1981). Accordingly, graft polymerization takes place between polypropylene and styrene, resulting in an increases in membrane strength.

1.4 Heterogeneous Membranes

1.4.1 Synthetic Method

Heterogeneous membranes are durable at temperatures up to 50 °C. In a strong alkali, acid, and oxidizer, membrane properties are stable. These enable us to employ effective hard chemical cleaning to prevent fouling of the membrane (Kishi et al., 1977). Furthermore, the price of the membrane is reasonable and they are apt to store dry and are convenient to transport (Yingqui and Zhongqin, 1990). Early references (Bodamer and Cheltenham, 1954) on heterogeneous ion exchange membranes reveal that they may be made by mechanical incorporation of powdered ion exchange resin into sheets of rubber, polyvinylchloride, acrylonitrile copolymers, or some other extrudable or moldable matrix. Synthetic methods of heterogeneous membranes are classified as follows (Kosaka and Shimizu, 1963):

1. Press forming
   a. A flat plate is formed by applying pressure to a powdered mixture of ion exchange resins and binders.
   b. An ion exchange resin film is formed by applying pressure to the ion exchange resin particles, and then partially polymerized liquid binders are loaded into the gaps between the ion exchange resin particles in the film and are polymerized.
   c. A mixture of binders, plasticizers, and ion exchange resins is kneaded and extruded to form a flat plate.
2. Roll processing
   a. A mixture of ion exchange resins and binders is heated and kneaded on a calender roll.
   b. Precursory ion exchange resin powder is heated with binders and kneaded on a calender roll with cross-linking agents, and then ion exchange groups are introduced.
3. Flat plate forming
   Ion exchange powder is dispersed into a binder solution, which is molded, cross-linked, and insolubilized.
4. Coating
   Ion exchange resin powder is dispersed into a fluid binder solution to form a muddy solution, which is coated on a supporting film to form a membrane film.

5. Dipping

Ion exchange resin powder is dispersed into a binder solution to form a muddy solution, in which a fabric or reinforcement net is dipped and pulled up, squeezing the muddy solution, and then the solvent is evaporated to form the membrane.

Here, we look at representative examples of synthetic procedures of heterogeneous membranes appearing in the literature.

1.4.2 Resin/Binder Ratio and Resin Particle Size

The electric resistance of heterogeneous membranes is influenced by the resin/binder ratio and resin particle size, as follows (Vyas et al., 2001).

Anion exchange resin particles were dried in an oven at 60 °C for 24 h, powdered in a ball mill, and sieved to the desired mesh size. The average particle size of 90% of the resins was 30, 70, and 80 μm, respectively. Finely powdered ion exchange resins were then dispersed into a solution of polyvinylchloride (PVC binder) in tetrahydrofuran. Then membranes were prepared using spray-coating technique. The membranes were dried at ambient temperature (~35 °C) for 30 min and almost-dried membranes were immersed in water. The membranes thus prepared were conditioned by equilibrating in a (~1 N) NaOH solution and subsequently in an NaCl solution.

The experimental results reveal that membranes with higher binder (PVC) content possess good mechanical strength but lack electrochemical properties, whereas membranes with higher resin content possess good electrochemical properties but the mechanical properties are comparatively poor. Both of these mechanical and electrochemical properties depend on the particle size of the resin. Thus, to achieve a better heterogeneous membrane, a compromise is needed between the electrochemical and mechanical properties; i.e., optimization of the resin-binder ratio and particle size is essential. From a thorough analysis of the data, it is concluded that a membrane with a thickness of 0.2 mm prepared from resin with a average particle size (90% of the resins) of 39 μm (−300 + 400BBS mesh) with 60% resin loading has a bursting strength of 1.9 kg/cm^2, areal resistance 8–9 Ωcm^2, capacity 2.14 mEq/g, and transport number 0.90.

1.4.3 Form of Binders

Binders prepared by copolymerization of polyethylene and α-olefin decrease the electric resistance of a heterogeneous membrane as follows (Terada et al., 1997).

1. α-Olefin binders were synthesized by copolymerization of ethylene and 4-methylpentene-1. Ion exchange resins (Mitsubishi Chemical Co., Diaion SK-1B) were dried at 60 °C for 24 h and pulverized with a rotary mill. Ion exchange resin powder was sieved by a stainless-steel mesh to remove particles with sizes exceeding 145 μm. The resin powder with particle sizes of less than 145 μm was mixed with the above α-olefin binders in a mixing ratio of 70/30 (weight ratio) and

the mixture was kneaded with a laboplast mill at 130 °C for 20 min and melt-pressed with a plate press at 160 °C to obtain a 500-μm-thick heterogeneous cation exchange membrane. The membrane was immersed in pure water for 2 days. Characteristics of this membrane were: specific electric resistance of the membrane in a 0.5 M NaCl solution: 110 Ωcm. Tensile strength at the break of the membrane: 2 MPa. Elongation at break: 200%. Bursting strength: 0.12 MPa.

2. The ion exchange resins powder described above was mixed with low-density polyethylene binders in a mixing ratio of 60/40 (weight ratio). Specific electric resistance of this membrane in a 0.5 M NaCl solution was 600 Ωcm (extremely large).

1.5 Graft Copolymerization

Figure 1.10 illustrates radiation-induced graft copolymerization showing that radicals are generated by irradiating an electron beam to a main chain of trunk polymer A (polyethylene, etc.), and monomer B (styrene, chloromethyl styrene, etc.) is bonded with the radicals. The nomenclature "graft polymerization" is derived from monomer B being grafted onto trunk polymer A. In this reaction process, primary radicals are formed on the trunk polymer with high-energy radiation and irradiated polymer is allowed to react with monomer units, which then propagate to form side chain grafts. Radiation-induced graft copolymerization involves the following two main methods (Mohamed and El-Sayed, 2004).

1.5.1 Simultaneous Irradiation

Polymer matrix P irradiated in the presence of monomer M is available in different forms: vapor, liquid, or bulk solution. Irradiation can be carried out in air or an inert

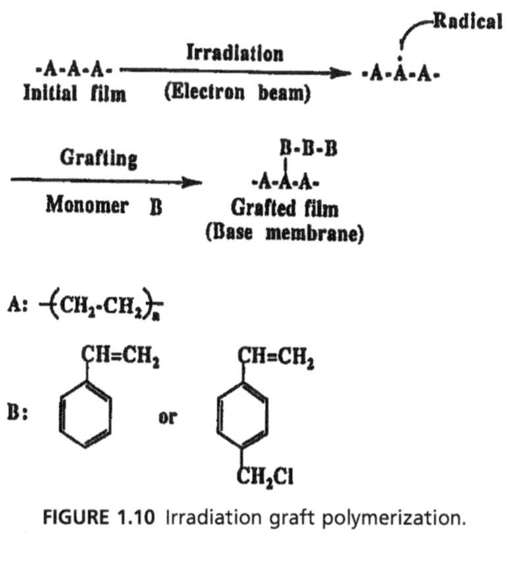

FIGURE 1.10 Irradiation graft polymerization.

atmosphere (e.g., N_2) or under vacuum. The reaction mechanism can be represented as follows.

$$\text{Irradiation}: \ P \xrightarrow{\gamma \text{ rays}} P^* \ \text{(primary radicals)} \tag{1.1}$$
$$\text{Initiation}: \ P^* + M \rightarrow PM^* \ \text{(graft chain)} \tag{1.2}$$
$$\text{Propagation} \ PM^* + nM \rightarrow PM^*_{n+1} \ \text{(graft growing chains)} \tag{1.3}$$
$$\text{Termination} \ PM^*_n + PM^*_m \rightarrow PM_{n+m} \ \text{(graft copolymer)} \tag{1.4}$$

where P is the polymer matrix (–A–A–A– in Fig. 1.10), M is the monomer units (B in Fig. 1.10) and P* and M* are their primary radicals, respectively. PM* is the initiated graft chain. PM^*_n and PM^*_m are the graft growing chains of the copolymer.

In simultaneous irradiation, there is a high possibility of deactivation of the primary radicals by recombination or homopolymerization, represented as follows:

Recombination:

$$P^* + P^* \rightarrow PP \tag{1.5}$$

Homopolymerization:

$$M \xrightarrow{\gamma \text{ rays}} M^* \ \text{(primary radicals)} \tag{1.6}$$
$$M^* + nM \rightarrow M^*_n \tag{1.7}$$
$$M^*_n + M^*_m \rightarrow M_{n+n} \ \text{(homopolymer)} \tag{1.8}$$

1.5.2 Pre-irradiation

Pre-irradiation involves a combination of two steps:

1. Irradiation of the polymer backbone to form active radicals, and
2. Contact of the irradiated polymer backbone with monomer. The overall reaction in graft copolymerization of a monomer onto a hydrocarbon backbone can be represented by Eqns (1.9)–(1.12).

$$\text{Formation of hydroperoxides} \ PH + O_2 \xrightarrow{\gamma \text{ rays}} POOH \tag{1.9}$$
$$\text{Thermal decomposition} \ POOH \xrightarrow{\Delta} PO^* + OH^* \tag{1.10}$$
$$\text{Initiation} \ POM^* + nM \rightarrow POM^*_{n+1} \tag{1.11}$$
$$\text{Propagation} \ POM^* + nM \rightarrow POM^*_{n+1} \tag{1.12}$$

Polyethlene (Tsuneda et al., 1995), poly(ethylene-tetrafluoroethylene) (Momose et al., 1989), FEP (copolymer of tetrafluoroethylene and perfluoropropylene) (Gupta et al., 1996), PFA (copolymer of tetrafluoroethylene and perfluoroalkylvinylether) (Nasef et al., 1999), etc., are irradiated, e.g., by an electron beam to form radicals in the polymers, and then vinyl monomers, styrene, acrylic acid, etc., are grafted onto the film.

1.5.3 Practical Examples

American Machine and Foundry Co. developed a method for forming graft copolymer ion exchange membranes (AMF membranes) as follows (Chen, 1966). A polyethylene film 4 mils thick and 3 in square was immersed in styrene monomer for about 30 min.

Absorption or swelling reached an equilibrium maximum within this time and the whole system was next irradiated by exposure for 5 h to a Co-60 source. The total radiation dose was about 1.5×10^6 R. The film was removed with unpolymerized and homopolymerized styrene. The copolymer film was sulfonated at room temperature for 40 min by immersion into chlorosulfonic acid. It was then immersed in carbon tetrachloride and washed with water to remove excess acid. Finally, it was heated in about 20% caustic soda for about 15 min at 60–70 °C. The membranes were manufactured at an industrial-scale and were on the market for a considerably long period.

Japan Atomic Energy Research Institute developed an irradiation graft copolymerization technology (Machi et al., 1980). A polyethylene film 20 μm thick was irradiated with 20 Mrad in a nitrogen gas atmosphere at a dose of 20 MeV electron beams from a cascade-type accelerator. The irradiated film was immersed in an aqueous acrylic acid solution deaerated by bubbling nitrogen in a glass ampoule. Then, 0.25 wt% of Mohr's salt (ferrous ammonium sulfate) was added into the monomer solution to minimize homopolymerization. The graft polymerization proceeded at 25 °C for 5 h. After grafting, the film was taken out and washed thoroughly with distilled water to remove residual monomer and homopolymer contained in the film. The membrane obtained was treated in a 25% aqueous KCl solution at 90 °C for 120 min. The specific electric resistance of the membrane was 16 Ωcm in 40% aqueous KCl. This technology is applied to produce the diaphragm of a battery (Ishigaki et al., 1981).

Another synthetic method (Tanaka, 1999) of irradiation graft polymerization is as follows. A polyethylene film 100 μm thick was irradiated with a 200-kGy electron beam under nitrogen gas and free radicals formed in the film. The irradiated film was put into an evacuated glass ampoule. A mixed solution of styrene or chloromethyl styrene and benzene (swelling solution) that had been deaerated by nitrogen gas was introduced into the ampoule and graft polymerization was carried out at 50 °C. After the polymerization, the film was washed repeatedly with benzene and methanol and was dried under a vacuum. The styrene-grafted film was put into a 10% chlorosulfonic acid solution of dichloromethane at 0–2 °C for 1 h. After sulfonation, a cation exchange membrane was obtained by washing with methanol and distilled water. The chloromethyl styrene-grafted film was put into a 10% trimethyl amine aqueous solution at 50 °C for 1 h. After quaternarization, an anion exchange membrane was obtained by washing repeatedly with distilled water.

1.6 Bipolar Membranes

The bipolar membrane is a layered membrane constructed so that one surface is a cation exchange layer, whereas the opposite surface is an anion exchange layer, and it must be able to split water into protons and hydroxyl ions at a very fast rate (Applications, Chapter 17). The membranes are prepared using the following several techniques (Wilhelm, 2000).

1. Lamination by pressure at room temperature or at elevated temperature of solid cation or anion exchange resin sheets, one of which contains the catalytic active component.
2. Lamination by pressure at room temperature or at elevated temperature of solid cation or anion exchange resin sheets with an interface layer containing the catalytic active material.
3. Casting a film from a solution containing an anion exchange resin with the catalytic active component onto a solid sheet of the cation exchange resin.
4. Casting a film from a solution containing a cation exchange resin with the catalytic active component onto a solid sheet of the anion exchange resin.
5. Co-extrusion of a cation and anion exchange resin, one of which contains the catalytically active component.

Today, most bipolar membranes are composed of a three-layer structure, i.e., a cation exchange layer, an anion exchange layer, and an interface layer containing a weak acid or base catalyst. This interface layer is a transition region between the two ion exchange membranes where the water dissociation takes place. To have the required water dissociation capacity, it must contain the appropriate concentration of catalytic components. Weakly dissociated ionic groups such as tertiary amines fulfill the requirements for catalytic components. Furthermore, certain heavy metal ion complexes, such as chromium(III) or iron(III) hydroxides with an extremely low solubility constant provide the required catalytic water dissociation effect for a long period. However, their specific electrical resistance is relatively high. To minimize the electrical area resistance, the thickness of the interface or transition region between the oppositely charged layers should be as thin as possible. The most commonly used catalytic components in bipolar membranes are listed in Table 1.1 (Strathmann, 2004).

Following are several examples of the preparation of bipolar membranes.

Instance 1

Bauer et al. (1988) prepared the mono-sheet bipolar membrane using the following steps.

Table 1.1 Component Used as Catalytic Material for the Preparation of Bipolar Membranes

Material	Form of Application	pK_a value	Reference
$Cr(OH)_3$	As salt in the cation exchange layer on the interface		Simons (1993)
$Fe(OH)_3$	As salt in the cation exchange layer or the interface		Hanada et al. (1991)
$-NP_2$	Bond to the matrix of the anion exchange membrane	~9	Bauer et al. (1988)
$R-PO_3H$	Bond to the matrix of the cation exchange membrane	~7	Shel'deshov et al. (1986)

Strathmann (2004), p. 111 Table 3.1.

1. Preparation of a cation selective layer.

One to 1.5 (mEq/dry-g) cation exchange resins consisting of cross-linked sulfonated divinylbenzene–styrene copolymer was dispersed into a 20% (w/w) solution of poly (vinylidene fluoride) in dimethylformamide (DMF). The solution was cast onto a glass plate as a 0.4-mm-thick film and the solvent evaporated at 125 °C, resulting in a 35-μm-thick film.

2. Preparation of an anion selective layer.

The anion selective layer was prepared by dispersing 21–32% (w/w) chloromethylated polystyrene in a solution of 20% (w/w) of poly (vinylidene fluoride) in DMF. Amination was performed by adding 1.0–1.5 mEq amine in 2 g dimethyl sulfoxide. Here, for the amination, the following large amounts of different amines were used, and diamines were used to obtain cross-linking of the polymer. (1) *N,N'*-dimethylpiperazine, (2) bis(dimethylamino)methane, (3) 1,2-bis(dimethylamino) ethane, (4) 1,3-bis(dimethylamino)propane, and (5) 1,4-bis(dimethylamino)butane. The solution was cast as a 0.3-mm-thick film on a glass plate. By evaporating the solvent at 125 °C, a 30-μm-thick anion exchange layer was obtained.

3. Preparation of bipolar membranes.

The mono-sheet bipolar membranes were prepared by casting a cation selective layer on top of an anion exchange layer.

4. Performance of bipolar membranes.

Bipolar membranes, prepared by casting a cation selective layer on an anion selective membrane consisting of chloromethylated polystyrene cross-linked with 1,4-bis(dimethlamino) butane showed good water-splitting capabilities at low potential drop and high current efficiencies owing to their relatively low electrical resistance and high ion selectivity. For the production of concentrated acids and bases, however, the chemical stability, particularly under alkaline conditions, was not yet satisfactory.

Instance 2

Zabolotskii et al. (2014) manufactured heterogeneous bipolar membranes MB-1, MB-2, and MB-3 (JSC Schekinoazot, Russia) by means of the hot pressing method.

1. Preparation of bipolar membranes

In the first step, monopolar heterogeneous membranes preforms are made from a fine powder composition (consisting of a resin and the inert binder). Ion exchange

Table 1.2 Properties of Heterogeneous Bipolar Membranes

Properties	MB-1	MB-2	MB-3
Iongenic Groups			
Cation exchange layer	$-SO_3^-$	$-SO_3^-$	$-PO(OH)_2$
Anion exchange layer	$\equiv N^+$, $=N^+$, $-N^+$	$-N^+(CH_3)_3$	$-N^+(CH_3)_3$
Ion Exchange Capacity (mol-Eq/l)			
Cation exchange layer	1.4	1.4	1.9
Anion exchange layer	4.0	1.4	1.4
Thickness, mm	>0.5	>0.5	>0.5

Zabolotskii et al. (2014), p. 185, Table 1.

groups in the membranes are SO_3^- and $-PO(OH)_2$ for the cation exchange membranes and $\equiv N^+$, $=N^+$, $-N^+$, and $-N^+(CH_3)_3$ for the anion exchange membranes, as shown in Table 1.2. Then, the preforms of the cation exchange membrane and anion exchange membrane are superimposed and subjected to a hot press to obtain a laminated bipolar membrane sheet.

2. Performance of bipolar membranes

The cation exchange layer and anion exchange layer of the bipolar membrane were in contact with a 0.01 M HCl solution and a 0.01 M NaOH solution, respectively. The flow rate and temperature of the solutions were set constant at 2 ml/min and 25 °C, respectively. Figure 1.11 gives current–voltage characteristics. The MB-2 membrane, which has strongly basic and strongly acidic groups, shows significantly higher voltages compared with the other membranes (the water splitting reaction is weak). The MB-3 membrane shows the best results by operating voltage

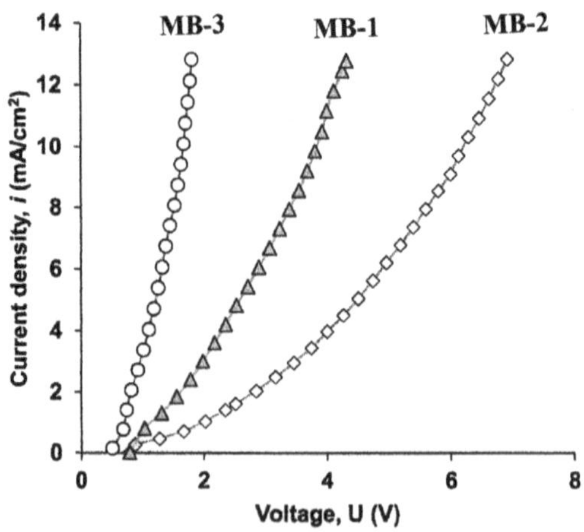

FIGURE 1.11 Current–voltage characteristics of bipolar membranes. *Zabolotskii et al. (2014), p. 187, Fig. 6.*

among the three. This phenomenon is due to the strong water splitting reaction generated in the membrane and it is understandable from larger pK value of $-PO(OH)_2$ groups compared to the value of $-SO_3H$ groups. The water splitting reaction in the MB-1 is stronger than the MB-2 due to smaller pK value of $\equiv N+$, $=N+$ and $-N+$ groups compared to the value of $-N+(CH_3)_3$ groups (Section 7.8.5).

Instance 3

Hanada et al. (1993, 1996) prepared bipolar membranes with cation exchange groups exchanged with heavy metallic ions at the interface between a cation exchange layer and an anion exchange layer.

1. Preparation of bipolar membranes

The surface of a Neosepta CM-1 cation exchange membrane was sandpapered to make the surface uneven, and then the membrane was immersed in a 25 °C aqueous 2 wt% $FeCl_2$ solution for 1 h, washed sufficiently with ion-exchanged pure water, and dried at room temperature. Fe content of the membrane was measured as 98% of its-ion exchange capacity. Poly sulfone was chloromethylated, quaternarized with trimethyl amine and obtained aminated poly sulfonic acid (ion exchange capacity: 0.92 mEq/g), which was dissolved into a methanol-chloroform mixed solution (volume ratio, 1:1, 15 wt%).

Next, the aminated poly sulfonic acid (anion exchange layer) was coated on the sandpapered cation exchange membrane (layer) and dried to prepare a bipolar membrane 90 μm thick.

2. Performance of bipolar membranes

Adhesion intensity between the cation exchange layer and anion exchange layer was 5.1 kg·f/25 mm. Current efficiencies were $\eta_H = 99.2\%$, $\eta_{OH} = 99.2\%$, $\eta_{Cl} = 0.3\%$, and $\eta_{Na} = 0.5\%$. Electrolysis voltage was 1.2 V. The performance of the membrane was unchanged and air bubbles and water bubbles did not form at the interface in the bipolar membrane for 2 months. After that, the adhesion intensity was unchanged (5.2 kg·f/mm) and electron spectroscopy analysis revealed that Fe ions remained at the interface between the cation exchange layer and anion exchange layer.

The experiment described above was repeated without a coating of $FeCl_2$. The performance of this bipolar membrane was: adhesion intensity = 5.3 kg·f/25 mm, current efficiency $\eta_H = 99.3\%$, $\eta_{OH} = 99.3\%$, $\eta_{Cl} = 0.3\%$, $\eta_{Na} = 0.4\%$, and electrolysis voltage = 3.2 V, which was larger than 1.2 V obtained in the previous experiment. This demonstrated that the Fe ions decrease the electric potential to generate a water-splitting reaction. The experiments in this patent show that heavy metallic ions such as Fe(II, III), Ti(IV), Sn(II, IV), Zr(IV), Pa(II), and Ru(III) accelerate the water-splitting reaction and maintain an electrolysis voltage less than 2.0 V or 1.8 V.

Instance 4

1. Practical Example

Umemura et al. (1994, 1995) invented the following perfluorinated bipolar membrane. A sufficiently dried quaternary ammonium type anion exchange membrane (styrene–divinylbenzene type, reinforced with polypropylene, ion exchange capacity: 3.0 mEq/g dry resin, 120 μm thick) was fixed onto a glass plate. On this membrane, a 10 wt% ethanol solution of perfluorinated sulfonic acid type cation exchange copolymers (ion exchange capacity: 1.1 mEq/g dry resin, consisting of $CF_2{=}CF_2$ and $CF_2{=}CFOCF_2CFCF_3CF_2CF_2SO_3H$) was poured. Before the solvent was sufficiently evaporated, the copolymer layer (30 μm thick) was pressed down onto the anion exchange membrane. The bipolar membrane was obtained by drying this membrane in an oven at 60 °C for 30 min.

The bipolar membrane was incorporated with an electrodialyzer; an Na_2SO_4 solution was supplied and electrodialyzed at a current density of 10 A/dm^2. The water dissociation current efficiency and voltage applied to the membrane were 0.95 and 3.4 V, respectively.

2. Comparative Example

The same anion exchange membrane described in the practical example was immersed in a 70 °C, 10 wt% $CrCl_2$ solution for 10 min, immersed in a 5 wt% NaOH solution for 10 min, washed sufficiently with water, and dried. Using the anion exchange membrane described above, the bipolar membrane was prepared in the same way as in the practical example. By means of the same electrodialysis experiment performed in the practical example, the water dissociation current efficiency was measured as 0.95. The voltage applied to the membrane was 1.3 V, which was less than the 3.4 V obtained in the practical example. The experiment seems to show that Cr(II) ions decrease the voltage owing to water dissociation acceleration.

The water-splitting process operates at significantly higher temperature because of heat generation in the bipolar membrane. Accordingly, can be used to develop the technology to prepare the perfluorinated bipolar membrane described in this section.

1.7 Fluorocarbon Membranes

1.7.1 Fuel Cell Membranes

The ion exchange membrane used for a fuel cell diaphragm in NASA space development in the latter half of the 1950s. At first, the membranes incorporated into the fuel cell were inferior in chemical durability because they were based on polystyrene. However, the perfluoronated Nafion membrane by Du Pont Co. exhibited excellent chemical durability and was applied for the fuel cell in Biosatellite in 1969. This membrane is based on a perfluorosulfonyl fluoride monomer copolymerized with tetrafluoroethylene (DuPont, 2002).

The membranes are prepared by the polymerization of monomers, which contain a moiety that can be made either cationic or anionic by further treatment. There is a four-step procedure for the synthesis of a Nafion membrane (Grot, 1973, 1974).

Step 1 Reaction of tetrafluoroethylene with SO_3 to form the sulfone cycle and fluorosulfonylacetylfluoride.

$$CF_2=CF_2+SO_3 \rightarrow \begin{array}{ccc} CF_2 & - & CF_2 \\ | & & | \\ O & - & SO_2 \end{array} \rightarrow FSO_2CF_2COF$$

Step 2 The fluorosulfonylfluoride is heated with hexafluoropropylene oxide and sodium carbonate to form the polyvinylethersulfonylfluoride

$$FSO_2CF_2COF + 2CF_3\overset{\overset{O}{\wedge}}{CF-CF_2} \rightarrow FSO_2CF_2CF_2OCF(CF_3)CF_2OCF(CF_3)COF$$

$$\xrightarrow{Na_2CO_3} FSO_2CF_2CF_2OCF(CF_3)CF_2OCF=CF_2$$

Step 3 The polyvinylethersulfonylfluoride is copolymerized with tetrafluoroethylene to form the insoluble resin.

$$FSO_2CF_2CF_2OCF(CF_3)CF_2OCF=CF_2 + CF_2=CF_2$$

$$\rightarrow (CF_2-CF_2)_m-\underset{\underset{OCF_2-CF(CF_3)OCF_2CF_2SO_2F}{|}}{CF}-CF_2- \text{ (insoluble resin)}$$

Step 4 The insoluble resin is hydrolyzed with NaOH to form a perfluorosulfonic polymer (ion exchange membrane).

$$\text{Insoluble resin} + NaOH \rightarrow (CF_2-CF_2)_m-\underset{\underset{OCF_2-CF(CF_3)OCF_2CF_2SO_3^-Na^+}{|}}{CF}-CF_2- + NaF$$

(ion exchange membrane)

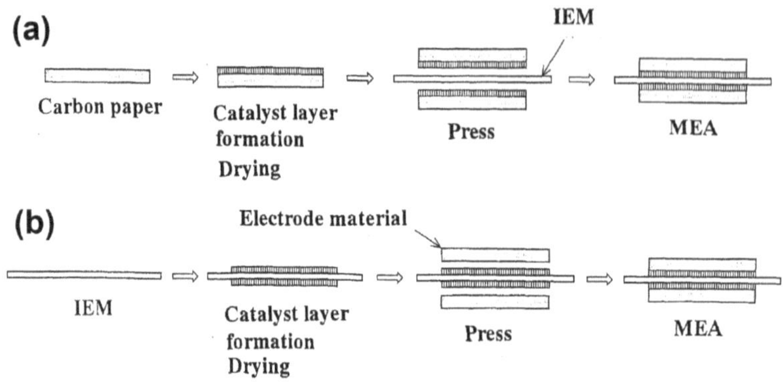

FIGURE 1.12 Formation of catalyst layers in a membrane electrode assembly. *NEDO Research and Development Report, (1999).*

The perfluorosulfonic acid membrane is usually reinforced with a support cloth made of Teflon and is applied in the polymer electrode cell (PEFC) at 80–100 °C (Applications Chapter 22). To reduce the energy consumption of the cell, catalyst (platinum) layers are formed on both surfaces of the ion exchange membrane as shown in Fig. 1.12 (NEDO Research and Development Report, 1999) to form a membrane electrode assembly.

1. The catalyst is applied to a base plate (carbon plate) of both electrodes. Then they are pasted with the membrane.
2. Catalyst layers are formed on both surfaces of the membrane. Then the membrane is pasted with the carbon plates.

The maximum operation temperature of the perfluorosulfonic acid membranes is below 100 °C under highly humidified conditions. In the automotive industry, an operating temperature above 100 °C at ambient pressure is needed to simplify water and temperature management (Wieser, 2004). Polymeric phosphonic acids attracted much attention as proton exchange membranes in fuel cells over the past decade owing to their higher thermal stability (Rusanov et al., 2008).

1.7.2 Chlor-Alkali Membranes

In the chlor-alkali membrane process (Applications Chapter 19), the electric conductivity of the perfluorosulfonic acid membrane is high enough. However, it cannot perfectly prevent back migration of OH^- ions generated at a cathode toward an anode. The perfluorocarboxylic acid membrane effectively prevents the back migration of OH^- ions (Figure 19.4). A double-layer membrane consisted of sulfonic acid and carboxylic acid layers has been developed to prevent such a problem.

An Aciplex membrane developed by Asahi Chemical Co. is a double-layered membrane prepared by coating perfluorocarboxylic acid polymer on perfluorosulfonic acid membrane or laminating of two different polymer films containing carboxylic and

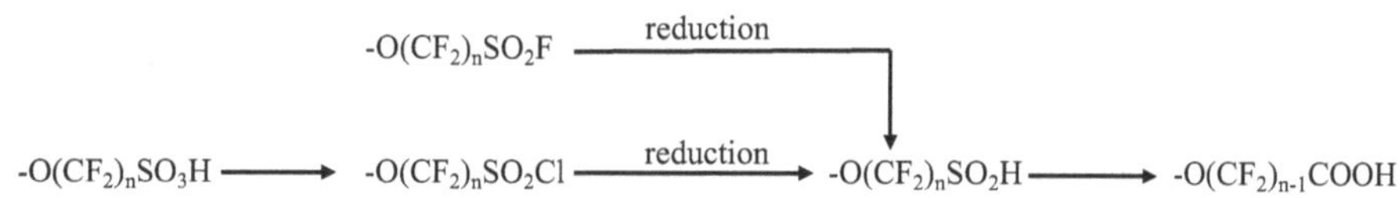

FIGURE 1.13 Chemical treatment to prepare a perfluorocarboxylic acid membrane from a perfluorosulfonic acid membrane or its precursor. *Seko et al. (1982), p. 473, Table 4.*

sulfonic acid, or by chemical conversion of perfluorosulfonic acid to perfluorocarboxylic acid. Figure 1.13 shows the most appropriate method of chemical treatment for converting perfluorosulfonic acid or its precursor to perfluorocarbolic acid membrane (Seko et al., 1979). In this process, sulfonyl halide groups in the membrane is converted to sulfinic acid, which is then converted into perfluorocarboxylic acid groups through a sulfonylation reaction. The perfluorocarboxylic acid pendant has one CF less than the original chain of perfluorosulfonic acid. The extremely high selectivity of this reaction effectively prevents the formation of impure products, which would be impossible to remove from the membrane. The following alternative reactions for conversion perfluorosulfonic acid to perfluorocarboxylic acid have been reported (Grot et al., 1978; Seko et al., 1978; Sata et al., 1978; Onoue et al., 1980; Seko et al., 1979):

1. Conversion through CF_2I.
2. Direct preparation of carboxylic acid group from SO_2Cl or SO_3H groups.
3. Addition reaction of the pendant with CF_2I or SO_2Cl.

Flemion, developed by Asahi Glass Co., is a double-layered membrane and is produced by the lamination of two different polymer films containing carboxylic and sulfonic acid, as follows (Umemura and Shimodaira, 2004):

Step 1

Resin A is synthesized by the copolymerization of $CF_2{=}CF_2$ and $CF_2{=}CFO(CF_2)_3CO_2CH_3$. Resin B is synthesized by the copolymerization of $CF_2{=}CF_2$ and $CF_2{=}CFOCF_2CF(CF_3)OCF_2CF_2SO_2F$. Two-layered film X is produced by affixing resin A layer (25 μm thick) on resin B layer (65 μm). Film Y (20 μm) is produced from resin B.

Step 2

A textile is woven with a polytetrafluoroethylene monofilament and a polyethlene terepthalete multifilament. A reinforcing material (100 μm) is formed by flattening the textile using a roll press.

Step 3

Film Y, reinforcing material, film X, and a polyethylene terepthalate film (tearing material) are piled up in this order and pressed using a roller. A multilayered membrane is obtained by stripping the tearing material.

Step 4

A paste is prepared by mixing zirconium oxide (29.0 wt%), methyl cellulose (1.3 wt%), cyclohexanol (4.6 wt%), cyclohexane (1.5 wt%), and water (63.6 wt%). A gas-removing layer is formed on the surface of film Y of the multilayered membrane by transcribing the paste using a roll press.

Step 5

$-CO_2CH_3$ groups and $-SO_2F$ groups in the multilayered membrane are hydrolyzed to ion-exchange groups by immersing the membrane in a solution dissolving 30 wt% of dimethl sulfoxide and 15 wt% of potassium hydroxide.

References

Araki, K., Daido, H., Sasaki, T., Suzuki, H., Tsushima, S., Misumi, Y., 1981. Production of polymer structure of an ion exchange membrane including polypropylene strings. JP, S56–8857.

Bauer, B., Gerner, F.J., Strathmann, H., 1988. Development of bipolar membranes. Desalination 68, 279–292.

Bodamer, G.W., Cheltenham, 1954. Permselective films of anion-exchange films, US 2,681,319; Permselective films of cation-exchange resins, US 2,681,320.

Chen, W.K.W., 1966. Method of forming graft copolymer ion exchange membrane, US 3,247,133.

DuPont, 2002. Nafion Perfluorinated Membranes. Product information (3.18.02.).

Grot, W.G., 1973. Laminates of support material and fluorinated polymer containing pendant side chains containing sulfonyl groups, US 3,770,567.

Grot, W.G., 1974. Films of fluorinated polymer containing sulfonyl groups with one surface in the sulfonamide or sulfonamide salt form and a process for preparing such, US 3,784,399.

Grot, W.G., 1978. Use of Nafion perfluorosulfonic acid products as separators in electrolytic cells. Chem. Ing. Tech. 50, 299–301.

Grot, W.G., Molnar, C.J., Resnick, P.R., 1978. BE 866 122.

Gupta, B., Buchi, F.N., Scherer, G.G., Chapiro, A., 1996. Cross-linked ion exchange membranes by radiation grafting of stylene/divinylbenzene into FEP films. J. Membr. Sci. 118, 231–238.

Hanada, F., Hirayama, K., Ohmura, N., Tanaka, S., 1991. Bipolar membrane and method for its production, EP 0459820.

Hanada, F., Hirayama, K., Ohmura, N., Tanaka, S., 1993. Bipolar membrane and method for its production, U. S. 5,221,455.

Hanada, F., Hirayama, K., Tanaka, S., Ohmura, N., 1996. Bipolar membrane and method for its production, JP 2,524,012.

Ishigaki, I., Sugo, T., Senoo, K., Takayama, T., Machi, S., Okamoto, J., Okada, T., 1981. Synthesis of ion exchange membrane by radiation grafting of acrylic acid onto polyethylene. Radiat. Phys. Chem. 18 (5–6), 899–905.

Juda, W., McRae, W.A., 1950. Coherent ion-exchange gels and membranes. J. Am. Chem. Soc. 72, 1044.

Kishi, M., Serizawa, S., Nakano, M., 1977. New electrodialyzer system with automatic chemical cleaning. Desalination 23, 203–212.

Kosaka, Y., Shimizu, H., 1963. Ion Exchange Membrane, pp. 28–37. Kyoritsu Publishing Co., Tokyo.

MacDonald, R.J., Hodgdon, R.B., Alexander, S.S., 1992. Process for manufacturing continuous supported ion selective membranes using non-polymerizable high boiling point solvent, U. S. 5,145,618.

Machi, S., Ishigaki, S., Sugo, T., 1980. Production of new ion exchange membranes, JP S55–106231.

Mineki, Y., Gunzima, T., Arai, S., 1972. Production of an ion exchange membrane, JP S47–40868.

Misumi, T., Kawashima, Y., Takeda, K., Kamaya, M., 1974. Production of an ion exchange membrane structure and an apparatus, JP S49–34476.

Mizutani, Y., Yamane, R., Kimura, K., 1964. Production of an ion exchange membrane, JP S39–27861.

Mohamed, M.N., El-Sayed, A.H., 2004. Preparation and applications of ion exchange membranes by radiation-induced graft copolymerization of polar monomers onto non-polar films. Prog. Polym. Sci. 29, 499–561.

Momose, T., Yoshioka, H., Ishigaki, I., Okamoto, J., 1989. Radiation grafting of α, α, β- trifluorostyrene onto poly(ethylene-tetrafluoroethylene)film by preirradiation method. I. Effects of preirradiation dose, monomer concentration, reaction temperature, and film thickness. J. Appl. Polym. Sci. 37, 2817–2826.

Nasef, M.M., Saidi, H., Nor, H.M., Dahlan, K.Z.M., Hashim, K., 1999. Cation exchange membranes by radiation-induced graft copolymerization of styrene onto PFA copolymer films. I. Preparation and characterization of the graft copolymer. J. Appl. Polym. Sci. 73, 2095–2102.

NEDO Research and Development Report, 1999. Development of Polymer Electrolyte Fuel Cell – Development of a High Current Density 10 KW Mobile Power Source System.

Onoue, K., Sata, T., Nakahara, A., Ito, J., 1980. U. S. 4 200 711.

Rusanov, A.L., Kostoglodov, P.V., Abadie, M.J.M., Voytekunas, V.Y., Likhachev, D.Y., 2008. Proton-conducting polymers and membranes carrying phosphonic acid groups. Adv. Polym. Sci. 216, 125–155.

Sata, T., 2004. Ion Exchange Membranes: Preparation, Characterization, Modification and Application. Royal Society of Chemistry, Cambridge.

Sata, T., Nakahara, A., Ito, J., 1978. JP S53–137 888.

Seko, M., Yamakoshi, Y., Miyauchi, H., Fukumoto, M., Kimoto, K., Hane, T., Hamada, M., 1978. JP S53–125 986.

Seko, M., Yamakoshi, Y., Miyauchi, H., Fukumoto, M., Kimoto, K., 1979. JP S54–6887.

Seko, M., Miyauchi, H., Ohmura, J., Kimoto, K., 1982. Ion-exchange membrane for chlor-alkali process. J. Electrochem. Jpn. 50, 470–476.

Simons, R., 1993. Preparation of a high performance bipolar membrane. J. Membr. Sci. 78, 13–23.

Shel'deshov, N.V., Zabolotsky, V.I., Pis'menskaya, N.D., Gunsin, N.P., 1986. Catalysis of water dissociation by phosphoric acid groups and an MB-3 bipolar membrane. Elektrokhimiya 22, 791–795.

Strathmann, H., 2004. Ion-echange Membrane Separation Process. In: Membrane Science Technology Series, vol. 9. Elsevier, Amsterdam, 111.

Tanaka, Y., 1999. Regularity in ion exchange membrane characteristics and concentration of seawater. J. Membr. Sci. 163, 277–287.

Terada, I., Umemura, K., Miyake, H., 1997. Heterogeneous ion exchange membrane and process for its production, JP H9–124805.

Tsuneda, S., Saito, K., Mitsuhara, H., Sugo, T., 1995. Novel ion-exchange membranes for electrodialysis: prepared by radiation-induced graft polymerization. J. Electrochem. Soc. 142, 3659–3663.

Umemura, K., Naganuma, T., Miyake, H., 1994. Production of a bipolar membrane, JP H6–32918.

Umemura, K., Naganuma, T., Miyake, H., 1995. Bipolar membrane, U. S. 5,401,408.

Umemura, K., Shimodaira, T., 2004. Perfluorinated cation-exchange membrane and electrolysis of sodium chloride, JP 2004–188375.

Vyas, P.V., Shah, B.G., Trivedi, G.S., Ray, P., Adhikary, S.K., Rangarajan, R., 2001. Characterization of heterogeneous anion-exchange membrane. J. Membr. Sci. 187, 39–46.

Wieser, C., 2004. Novel polymer electrolyte membranes for automotive applications – requirements and benefits. Fuel Cells 4, 245–250.

Wilhelm, F.W., 2000. Bipolar membrane preparation. In: Kemperman, A.J.B. (Ed.), Bipolar Membrane Technology. Twente University Press, Enschede, The Netherlands.

Yingqui, Y., Zhonnqin, L., 1990. A brief account of the development of ion exchangers in China. Technol. Water Treat. 16, 119–121.

Zabolotskii, V., Sheldeshov, N., Melnikov, S., 2014. Heterogeneous bipolar membranes and their application in electrodialysis. Desalination 342, 183–203.

2

Fundamental Properties of Ion Exchange Membranes

2.1 Introduction

Ion exchange membranes play a leading part in an electrodialyzer, and the performance of an electrodialyzer strongly depend on the characteristics of ion exchange membranes. In order to operate an electrodialyzer effectively, it is necessary to understand the membrane characteristics definitely. This chapter describes the fundamental theories of the membrane characteristics based on the principles of electrochemistry. The origins of the theories are older than the invention of an ion exchange membrane itself, and the reasonability of the theories had been confirmed by experimental observations. With these investigations, the membrane-characteristic measuring methods are established and have been applied widely. In this chapter, the measuring methods are explained clearly, assuming that the membranes are applied to saline water electrodialysis.

2.2 Ionic Transport Across the Membrane

2.2.1 Donnan Equilibrium Theory

A cation and an anion exchange membrane permeate, respectively, cations and anions selectively. This phenomenon is the most fundamental characteristic of the membrane and it is understandable from the following Donnan equilibrium theory (Donnan, 1934):

The electro-chemical potential of component i in a system η_i is defined as:

$$\eta_i = \mu_i^0 + RT \ln a_i + (P - P^0)V_i + z_i F \psi \tag{2.1}$$

where μ_i^0 is the standard chemical potential, R is the gas constant, T is the absolute temperature, a_i is the activity, P is the pressure, P^0 is the standard pressure (1 atm), V_i is the partial molar volume, z_i is the charge number, F is the Faraday constant, and ψ is the electric potential.

Assuming that cation X and anion Y are dissolved into a solvent W, Eqn (2.1) becomes:

$$\begin{aligned} \eta_X &= \mu_X^0 + RT \ln a_X + (P - P^0)V_X + z_X F \psi \\ \eta_Y &= \mu_Y^0 + RT \ln a_Y + (P - P^0)V_Y + z_Y F \psi \\ \eta_W &= \mu_W^0 + RT \ln a_W + (P - P^0)V_W + z_X F \psi \end{aligned} \tag{2.2}$$

Ion Exchange Membranes. http://dx.doi.org/10.1016/B978-0-444-63319-4.00002-X

The Donnan equilibrium state is represented by Eqn (2.3), which shows that the electrochemical potential of component i in a solution phase η_i is equivalent to that in a membrane phase $\overline{\eta}_i$.

$$\eta_i = \overline{\eta}_i \tag{2.3}$$

From Eqns (2.2) and (2.3):

$$RT \ln\frac{a_X}{\overline{a}_X} - (\overline{P} - P)V_X - z_X F(\overline{\psi} - \psi) = 0 \tag{2.4}$$

$$RT \ln\frac{a_Y}{\overline{a}_Y} - (\overline{P} - P)V_Y - z_Y F(\overline{\psi} - \psi) = 0 \tag{2.5}$$

$$RT \ln\frac{a_W}{\overline{a}_W} - (\overline{P} - P)V_W = 0 \tag{2.6}$$

From Eqns (2.4) and (2.5), the Donnan potential difference between both phases E_{Don} is introduced as:

$$E_{Don} = \overline{\psi} - \psi = \left(\frac{1}{z_i F}\right)\left(RT \ln\frac{a_i}{\overline{a}_i} - \pi V_i\right) \tag{2.7}$$

in which π is the pressure difference between both phases and it is equivalent to the osmotic pressure (swelling pressure) introduced from Eqn (2.6) as follows:

$$\pi = \overline{P} - P = \left(\frac{RT}{V_W}\right)\ln\left(\frac{a_W}{\overline{a}_W}\right) \tag{2.8}$$

Assuming 1 mol of electrolytes is dissociated into υ_X moles of cations and υ_Y moles of anions:

$$\upsilon_X z_X + \upsilon_Y z_Y = 0 \tag{2.9}$$

Canceling E_{Don} from Eqn (2.7) for cation X and anion Y, and taking account of Eqn (2.9):

$$RT \ln\left\{\left(\frac{a_X}{\overline{a}_X}\right)^{\upsilon_X}\left(\frac{a_Y}{\overline{a}_Y}\right)^{\upsilon_Y}\right\} = \pi \upsilon_{XY} \tag{2.10}$$

in which υ_{XY} is the partial molar volume of electrolyte XY defined by:

$$\upsilon_{XY} = \upsilon_X V_X + \upsilon_Y V_Y \tag{2.11}$$

Substituting Eqn (2.8) into Eqn (2.10), the following Donnan equilibrium equation is introduced:

$$\ln\left\{\left(\frac{a_X}{\overline{a}_X}\right)^{\upsilon_X}\left(\frac{a_Y}{\overline{a}_Y}\right)\right\} = \frac{\upsilon_{AX}}{\upsilon_W}\ln\frac{a_W}{\overline{a}_W} \tag{2.12}$$

Putting $r = \upsilon_{XY}/\upsilon_W$ in Eqn (2.12):

$$\frac{a_X^{\upsilon_X} a_Y^{\upsilon_Y}}{a_W^r} = \frac{\overline{a}_X^{\upsilon_X}\overline{a}_Y^{\upsilon_Y}}{\overline{a}_W^r} \tag{2.13}$$

If $a_W \fallingdotseq \overline{a}_W$:

$$a_X^{\upsilon_X} a_Y^{\upsilon_Y} = \overline{a}_X^{\upsilon_X}\overline{a}_Y^{\upsilon_Y} \tag{2.14}$$

Putting the activity coefficient of cation X and anion Y to be 1 in Eqn (2.14) leads to the following approximation of the Donnan equilibrium phenomenon:

$$C_X^{\upsilon_X} C_Y^{\upsilon_Y} = \overline{C}_X^{\upsilon_X} \overline{C}_Y^{\upsilon_Y} \tag{2.15}$$

in which C denotes concentration of cation X or anion Y.

When a Na-type cation exchange membrane is immersed in an NaCl solution, Eqn (2.15) becomes:

$$C_{Na} C_{Cl} = \overline{C}_{Na} \overline{C}_{Cl} \tag{2.16}$$

From the electroneutrality in the membrane:

$$\overline{C}_{Na} = \overline{C}_{Cl} + \overline{C}_R \tag{2.17}$$

where $\overline{C}_R$ is the concentration of ion exchange groups R in the membrane.

Combining Eqn (2.16) with Eqn (2.17):

$$\begin{aligned} \overline{C}_{Na} &= \frac{1}{2}\left(\sqrt{\overline{C}_R^2 + 4C_{Na}C_{Cl}} + \overline{C}_R\right) \\ \overline{C}_{Cl} &= \frac{1}{2}\left(\sqrt{\overline{C}_R^2 + 4C_{Na}C_{Cl}} - \overline{C}_R\right) \end{aligned} \tag{2.18}$$

Putting NaCl concentration as C ($=C_{Na} = C_{Cl}$):

$$\frac{\overline{C}_{Cl}}{\overline{C}_{Na}} = \frac{\sqrt{\overline{C}_R^2 + 4C^2} - \overline{C}_R}{\sqrt{\overline{C}_R^2 + 4C^2} + \overline{C}_R} \tag{2.19}$$

$\overline{C}_{Cl}/\overline{C}_{Na}$ is computed by putting $\overline{C}_R = 4$ and 8 mol dm^{-3} in Eqn (2.19), and plotted against C (Fig. 2.1), which shows that $\overline{C}_{Cl}/\overline{C}_{Na}$ increases with C toward 1. However it decreases with C toward 0. This phenomenon is due to the Donnan exclusion of Cl^- ions caused by the repulsive effect of ion exchange groups R fixed in the membrane at lower C. In the same way, the relationship between $\overline{C}_R$ and $\overline{C}_{Cl}/\overline{C}_{Na}$ is computed by putting C as parameter (Fig. 2.2), indicating $\overline{C}_{Cl}/\overline{C}_{Na}$ to approach 1 at infinitely small $\overline{C}_R$. $\overline{C}_{Cl}/\overline{C}_{Na}$ is decreased with the increase of $\overline{C}_R$. This is because the Donnan exclusion is developed at larger $\overline{C}_R$.

2.2.2 Transport Number

An ion exchange membrane consists of a three-dimensional cross-linked polymer network structure combined with many ion exchange groups. The membrane exhibits gelling states in an electrolyte solution due to water absorption. The ion exchange groups are charged and absorb counter-ions selectively. An electric current applied in this system generates the fluxes J_i of counter-ion i across the membrane. J_i is indicated generally by the following extended Nernst–Planck equation presented by adding the convection term (third term) to the Nernst–Planck equation (Planck, 1890):

$$J_i = -\frac{RT}{F}\overline{u}_i\frac{d\overline{C}_i}{dx} - z_i\overline{u}_i\overline{C}_i\frac{d\psi}{dx} + \overline{C}_i v \tag{2.20}$$

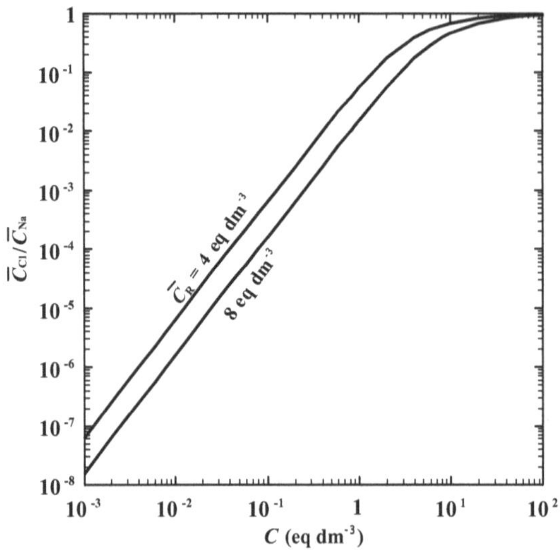

FIGURE 2.1 Relationship between NaCl concentration in a solution and Cl^-/Na^+ ratio in a cation exchange membrane. *Tanaka (2007), p. 40, Fig. 3.1.*

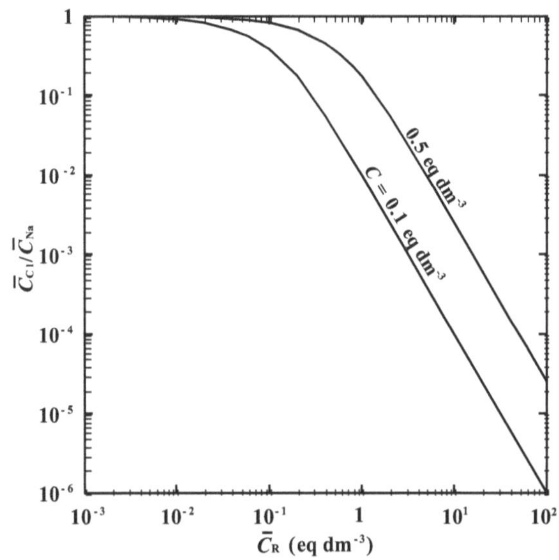

FIGURE 2.2 Relationship between ion exchange group concentration and Cl^-/Na^+ ratio in a cation exchange membrane. *Tanaka (2007), p. 40, Fig. 3.2.*

$\overline{C}_i$ is the concentration, $\overline{u}_i$ is the mobility, z_i is the charge number, $d\overline{C}_i/dx$ is the concentration gradient, and these are the values of counter-ions i in the membrane. $d\psi_i/dx$ is the potential gradient and v is the convection velocity in the membrane. The first, the second, and the third term in Eqn (2.20) correspond to the ionic fluxes of ion i caused by the diffusion, electro-migration, and convection, respectively.

The structure of an ion exchange membrane is generally dense. Furthermore, the ion exchange groups fixed in the membrane exclude the co-ions due to the Donnan exclusion (Figs 2.1 and 2.2), which reduces the diffusion and convection across the membrane (Section 2.4). Accordingly, the influence of diffusion and convection to the flux can be neglected. Thus, Eqn (2.20) is simplified as:

$$J_i = -z_i \overline{u}_i \overline{C}_i \frac{d\psi}{dx} \tag{2.21}$$

Transport number of ion i, $\overline{t}_i$, is defined taking account of Eqn (2.21) as:

$$\overline{t}_i = \frac{z_i J_i}{\sum z_i J_i} = \frac{z_i^2 \overline{u}_i \overline{C}_i}{\sum z_i^2 \overline{u}_i \overline{C}_i} \tag{2.22}$$

Here, one discusses the nature of transport number of an ion exchange membrane integrated in an electrodialyzer. One assumes that a cation exchange membrane is placed in an NaCl solution and an electric current is passed across the membrane through electrodes placed in both sides of the membrane. In this situation, Na^+ ions transfer from the anode side toward cathode side and Cl^- ions transfer from the cathode side toward anode side. The transport number of Na^+ ions $\overline{t}_{Na}$ and that of Cl^- ions $\overline{t}_{Cl}$ passing through the membrane are presented by the following equation taking account of Eqn (2.22):

$$\begin{aligned} \overline{t}_{Na} &= \frac{\overline{u}_{Na} \overline{C}_{Na}}{\overline{u}_{Na} \overline{C}_{Na} + \overline{u}_{Cl} \overline{C}_{Cl}} \\ \overline{t}_{Na} &= \frac{\overline{u}_{Cl} \overline{C}_{Cl}}{\overline{u}_{Na} \overline{C}_{Na} + \overline{u}_{Cl} \overline{C}_{Cl}} \end{aligned} \tag{2.23}$$

The diameter of fine pores in an ion exchange membrane distributes within the range of 0.005–0.1 μm, and the shape of the pores is winding (Mizutani and Nishimura, 1970). Hydrated radius of Na^+ ions and Cl^- ions are 4.5 and 3 Å, respectively (Conway, 1952). The mobility of monovalent counter-ions (excepting H^+ and OH^- ions) and that of divalent counter-ions passing through an ion exchange membrane are about 1/10 and 1/20–1/50 of the values in an aqueous solution, respectively (Yamamoto, 1968). Based on these experimental observations, the ionic mobility in the membrane is estimated to be decreased due to steric hindrance in the pore. Furthermore, sulfonic acid groups fixed in a cation exchange membrane exert an electrical attractive force to Na^+ ions and an electrical repulsive force to Cl^- ions.

The molar conductivity of Na^+ ions and Cl^- ions in water are measured as follows (Convington and Dickinson, 1973):

$$Fu_{Na} = 50.11 \text{ S cm}^2 \text{ mol}^{-1}$$

$$Fu_{Cl} = 76.35 \text{ S cm}^2 \text{ mol}^{-1}$$

$\overline{u}_{Cl}/\overline{u}_{Na}$ in the cation exchange membrane can be defined by the following equation, including the mobility ratio in water u_{Cl}/u_{Na} and the mobility ratio parameter n:

$$\frac{\overline{u}_{Cl}}{\overline{u}_{Na}} = \left(\frac{u_{Cl}}{u_{Na}}\right) n = 1.524n \tag{2.24}$$

where n is less than 1 due to the attractive force between sulfonic acid groups and Na^+ ions and the repulsive force between sufonic acid groups and Cl^- ions. Substituting Eqn (2.24) into Eqn (2.23) (Tanaka, 2007):

$$\bar{t}_{Na} = \frac{1}{1 + 1.524n(\overline{C}_{Cl}/\overline{C}_{Na})} \quad (2.25)$$

$\bar{t}_{Na}$ is calculated using Eqns (2.19) and (2.25), and plotted against C and $\overline{C}_R$ putting n as a parameter (Figs 2.3 and 2.4). In these figures, n is estimated to decrease with the decrease of C and the increase of $\overline{C}_R$ because of Donnan exclusion development and resulting increase of repulsion (attraction) between sulfonic acid groups and Cl^- (Na^+) ions. In this circumstance, $\bar{t}_{Na}$ is increased toward 1 because $\overline{C}_{Cl}/\overline{C}_{Na}$ is decreased toward 0 (Figs 2.1 and 2.2). On the other hand, n is estimated to increase toward 1 with increasing C and decreasing $\overline{C}_R$ because of Donnan exclusion decline and resulting decrease of repulsion (attraction) between sulfonic acid groups and Cl^- (Na^+) ions.

2.2.3 Measurement of Actual Transport Number

1. A sample membrane is incorporated with a two-cell apparatus as shown in Fig. 2.5 (effective membrane area: 4 cm^2).
2. Accurately measured 30 cm^3 of a 0.5-M NaCl solution is put in a desalting cell and 30 cm^3 of a 0.5 M NaCl is put in a concentrating cell.
3. Ag–AgCl electrodes washed sufficiently with water over one night are put in both cells.

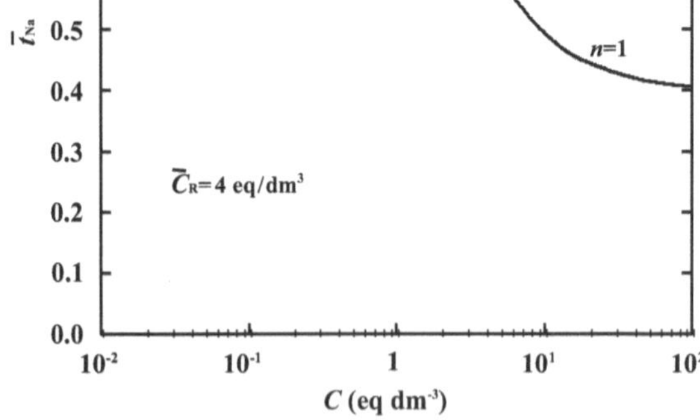

FIGURE 2.3 Relationship between NaCl concentration in a solution and Na^+ ion transport number in a cation exchange membrane. *Tanaka (2007), p. 41, Fig. 3.3.*

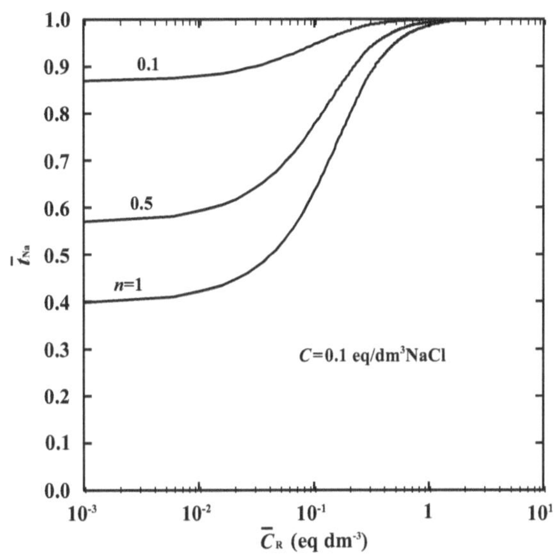

FIGURE 2.4 Relationship between ion exchange group concentration and Na^+ ion transport number in a cation exchange membrane. *Tanaka (2007), p. 42, Fig. 3.4.*

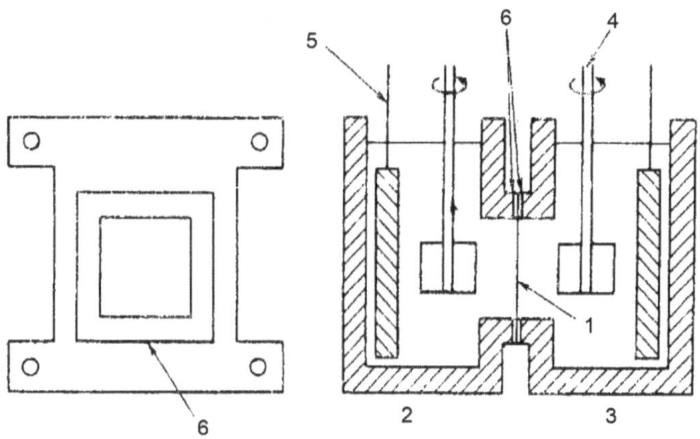

FIGURE 2.5 Actual transport number measuring apparatus. 1. Membrane 2. Desalting cell 3. Concentrating cell 4. Stirrer 5. Ag-AgCl electrode 6. Rubber gasket. *Seno and Tanaka (1984), p. 195, Fig. 3.*

4. The apparatus is set in a 25 °C thermostat and an electric current of 40 mA (1 Adm^{-2}) is passed for 40 min and 10 s (1 mF) through the electrodes, then ions in the desalting cell are transferred to the concentrating cell.
5. The solution in the concentrating cell is discarded and that in the desalting cell is collected into a 250-cm^3 measuring flask.
6. The electrode used in the desalting cell is washed sufficiently with water and immersed in water in a 30-cm^3 beaker. Both washed water and immersed water are collected into the 250-cm^3 measuring flask.

7. The volume of the solution in the measuring flask is adjusted to 250 cm^3 by adding water, and the quantity of Cl^- ions in the solution is analyzed using an $AgNO_3$ standard solution.
8. The transport number ("actual transport number") is calculated using the following equations:

Transport number of a cation exchange membrane, $\bar{t}_+ = a - b$.

Transport number of an anion exchange membrane, $\bar{t}_- = 1 + a - b$, where a is the analytical value of Cl^- ions in the desalting cell before passing an electric current (meq) and b is the analytical value of Cl^- ions in the desalting cell after passing an electric current (meq).

In measuring the transport number of a cation exchange membrane, the NaCl concentration in the desalting cell decreases and that in the concentrating cell increases during electric current passage, resulting in NaCl diffusion across the membrane. In order to reduce the experimental error due to this phenomenon, the NaCl concentration in the desalting cell is adjusted beforehand at higher than 0.5 M, and that in the concentrating cell is at less than 0.5 M. Furthermore, the NaCl concentrations in both cells are devised to be reversed during electric current passage and the experimental error to become a minimum. In case of an anion exchange membrane, the device described above is not necessary.

2.3 Membrane Potential

2.3.1 Nernst Equation

When both surfaces of a cation exchange membrane are in contact with NaCl solutions of different concentrations, Na^+ ions diffuse from a high-concentration side to a low-concentration side; however, the diffusion of Cl^- ions in this system is rather restricted due to the Donnan exclusion. These phenomena develop a positively charged sphere in the low-concentration side and a negatively charged sphere in the high-concentration side, resulting in potential difference (membrane potential) generation that hinders still more diffusion of Na^+ ions. In case of an anion exchange membrane placed in the above mentioned circumstances, a negatively and a positively charged sphere are developed in the low- and high-concentration sides, respectively.

Now, an electric potential difference dE is assumed to be applied to a fine section in a membrane with 1 F of electric current passing. If the changes in this system are reversible, energy consumption is expressed by FdE and it is equivalent to the free energy change dG:

$$dG = -FdE \tag{2.26}$$

The ionic movement generates an electric current, and the free energy change is equivalent to chemical potential change due to the ionic movement. These phenomena are represented by the following equations for ion i:

$$dG = \frac{t_i}{z_i} d\mu_i = \frac{t_i}{z_i} RTd \ln a_i \tag{2.27}$$

$$\mu_i = \mu_i^0 + RT \ln a_i \tag{2.28}$$

where t_i, z_i, a_i, μ_i, and μ_i^0 are the transport number, the charge number, the activity, the chemical potential, and the standard chemical potential of ion i, respectively. From Eqns (2.26) and (2.27), the following Nernst equation (Nernst, 1888, 1889) is introduced:

$$E = -\frac{RT}{F}\int \sum \frac{t_i}{z_i} d \ln a_i \tag{2.29}$$

If a solution 1 and solution 2 dissolving the same kind of electrolytes are partitioned by an ion exchange membrane, Eqn (2.29) leads to:

$$E = -\frac{\bar{t}_+}{z_+}\frac{RT}{F}\ln\frac{(a_+)_2}{(a_+)_1} + \frac{\bar{t}_-}{z_-}\frac{RT}{F}\ln\frac{(a_-)_2}{(a_-)_1} \tag{2.30}$$

$\bar{t}_+$ and $\bar{t}_-$ are the transport number of cations and anions in the membrane, respectively. These parameters are named "apparent transport number" and discriminated against actual transport number, explained in Sections 2.2.2 and 2.2.3.

$(a_+)_2/(a_+)_1$ and $(a_-)_2/(a_-)_1$ are the activity ratio of cations and anions between both solutions, respectively. If the solution dissolves electrolytes, including monovalent cations and monovalent anions, and $(a_+)_2/(a_+)_1 = (a_-)_2/(a_-)_1 = a_2/a_1$ holds, the following equation is introduced:

$$E = -(\bar{t}_+ - \bar{t}_-)\frac{RT}{F}\ln\frac{a_2}{a_1} \tag{2.31}$$

Taking account of $\bar{t}_+ + \bar{t}_- = 1$, the following equations are introduced from Eqn (2.31): For a cation exchange membrane:

$$E = -(2\bar{t}_+ - 1)\frac{RT}{F}\ln\frac{a_2}{a_1} \tag{2.32}$$

For an anion exchange membrane:

$$E = (2\bar{t}_- - 1)\frac{RT}{F}\ln\frac{a_2}{a_1} \tag{2.33}$$

2.3.2 Measurement of Apparent Transport Number

1. The apparent transport number is measured using a two-cell apparatus (Fig. 2.6), in which a sample membrane, immersed in a 1-M NaCl solution over 2 h, is integrated between the cells.
2. A 25 °C NaCl solution and a 2.5-M NaCl solution are poured into the cells, and the apparatus is left about 5 min.
3. Membrane potential (a V) is measured by connecting both cells and calomel electrodes through KCl bridges and letting both solutions flow toward the membrane surface.
4. The apparent transport number is calculated using the following equations introduced from Eqn (2.31):

$$\bar{t} = \frac{a + b}{2ab}$$

$$b(V) = 0.5915 \log\left(\frac{C_1 f_1}{C_2 f_2}\right)$$

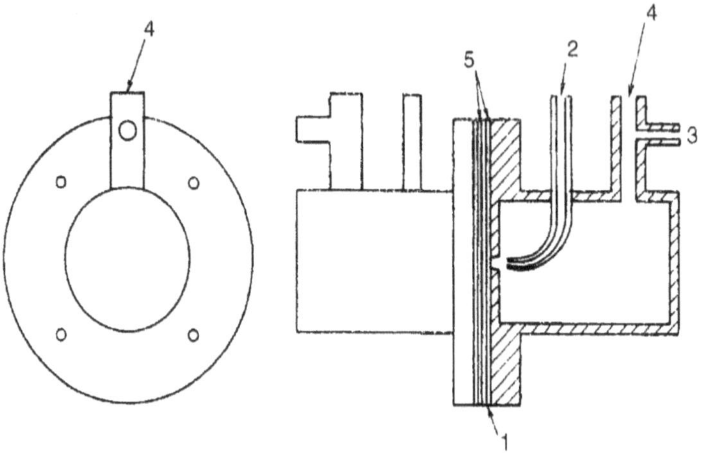

FIGURE 2.6 Apparent transport number measuring apparatus. 1. Membrane 2. Solution inlet 3. Solution outlet 4. Salt bridge insertion hole 5. Rubber gasket. *Seno and Tanaka (1984), p. 195, Fig. 2.*

in which C_1 and C_2 are the NaCl concentration in both cells (M) and f_1 and f_2 the NaCl activity coefficients in both cells.

2.4 Diffusion

2.4.1 Concentration Diffusion

When electrolyte solutions of different concentration are placed on both sides of an ion exchange membrane, the electrolytes diffuse from a high-concentration side to a low-concentration side. The principle of the phenomenon is considered as follows:

Canceling the third term in Eqn (2.20) (extended Nernst–Planck equation) and applying the Nernst–Einstein equation; Eqn (2.34) (Crank, 1957; Ilschner, 1958) lead to Eqn (2.35) (Nernst–Planck equation) expressing the flux of cations J_+ and of anions J_- across the membrane:

$$u_i = \frac{D_i F}{RT} \tag{2.34}$$

$$J_+ = -\overline{D}_+ \frac{d\overline{C}_+}{dx} - z_+ \overline{D}_+ \overline{C}_+ \frac{d\psi}{dx}$$
$$J_- = -\overline{D}_- \frac{d\overline{C}_-}{dx} - z_- \overline{D}_- \overline{C}_- \frac{d\psi}{dx} \tag{2.35}$$

From the electrical neutrality in the membrane:

$$z_+ \overline{C}_+ + z_- \overline{C}_- + \omega \overline{C}_R = 0 \tag{2.36}$$

Differentiating Eqn (2.36):

$$z_+\frac{d\overline{C}_+}{dx}+z_-\frac{d\overline{C}_-}{dx}=0 \tag{2.37}$$

No electric current:

$$z_+J_+ + z_-J_- = 0 \tag{2.38}$$

In a steady state:

$$J_+,\ J_- = \text{constant} \tag{2.39}$$

Canceling dC/dx in Eqn (2.35) and applying Eqns (2.37) and (2.38):

$$\begin{aligned} J_+ &= -\frac{\overline{D}_+\overline{D}_-\left(z_+^2\overline{C}_+ + z_-^2\overline{C}_-\right)}{z_+^2\overline{D}_+\overline{C}_+ + z_-^2\overline{D}_-\overline{C}_-}\frac{d\overline{C}_+}{dx} \\ J_- &= -\frac{\overline{D}_+\overline{D}_-\left(z_+^2\overline{C}_+ + z_-^2\overline{C}_-\right)}{z_+^2\overline{D}_+\overline{C}_+ + z_-^2\overline{D}_-\overline{C}_-}\frac{d\overline{C}_-}{dx} \end{aligned} \tag{2.40}$$

The form of Eqn (2.40) is the same as the Fick's diffusion equation (Eqn (2.43)).

Electric neutrality in the membrane accelerates the movement of ions having lower diffusivity, and deaccelerates the movement of ions having larger diffusivity. This phenomenon induces the acceleration of co-ions and deacceleration of counter-ions in the membrane (Section 2.3.1). The ionic movement described above is due to the Donnan potential developed at the membrane/solution interface, which generates Donnan exclusion against co-ions (Section 2.2.1). When the concentration of co-ions is decreased extremely in the membrane, the diffusion flux of the counter-ions is decreased remarkably, owing to the dominant Donnan exclusion as expressed by the following equation introduced from Eqn (2.40):

For a cation exchange membrane:

$$\lim_{\overline{C}_-\to 0} J_+ = -\overline{D}_-\frac{d\overline{C}_+}{dx} \tag{2.41}$$

For an anion exchange membrane:

$$\lim_{\overline{C}_+\to 0} J_- = -\overline{D}_+\frac{d\overline{C}_-}{dx} \tag{2.42}$$

Equations (2.41) and (2.42) suggest that the diffusion velocity of counter-ions is influenced by a small amount of co-ions remaining in the membrane.

2.4.2 Measurement of Solute Permeability Coefficient

Ionic flux J_S caused by diffusion is given by the following Fick's diffusion equation:

$$J_S = \frac{D_S\Delta C}{d} \tag{2.43}$$

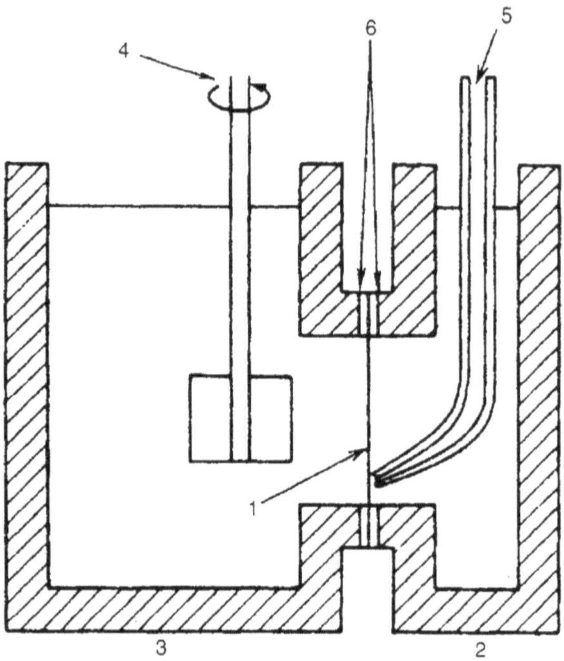

FIGURE 2.7 Solute permeability coefficient measuring apparatus. 1. Membrane 2. Desalting cell 3. Concentrating cell 4. Stirrer 5. Air blowing inlet 6. Rubber gasket. *Seno and Tanaka (1984), p. 196, Fig. 4.*

where d is the thickness of the membrane, ΔC is the ionic concentration difference between both surfaces of the membrane, and D_S is the diffusion constant. D_S/d is defined as the permeability coefficient, which is measured as follows:

1. A sample membrane immersed into a 0.5-M NaCl solution is washed with water, wiped with a filter paper, and integrated in the two-cell apparatus as shown in Fig. 2.7 (effective membrane area: 4 cm^2).
2. An accurately-measured 15 cm^3 of water is put in the desalting cell, and a 4-M NaCl solution is supplied into the concentrating cell so that the solution level in both cells is the same.
3. The apparatus is left in a 25 °C thermostat for 30 min exactly, during which the solution in the desalting cell and that in the concentrating cell are stirred violently by air blowing and stirrer, respectively.
4. The solution in the desalting cell is taken out immediately and the NaCl concentration is measured using a flame spectrochemical analyzer (a M).
5. The solute permeability coefficient is calculated using the following equation:

$$\frac{D_S}{d}(\text{cm s}^{-1}) = \frac{15a}{4 \times 4 \times 1800}$$

2.5 Perm-Selectivity

2.5.1 Perm-Selectivity Between Different Kinds of Ions Charged with the Same Sign

The permeations of different kinds of ions charged with the same sign across the membrane are not equivalent each other. This phenomenon is termed "perm-selectivity between ions charged with the same sign," and is discussed in this section.

When ions A and B charged with the same sign transfer across the membrane, the flux ratio of these ions is introduced from Eqns (2.21) and (2.22) as:

$$\frac{z_B J_B}{z_A J_A} = \frac{\overline{t}_B}{\overline{t}_A} = \frac{z_B^2 \overline{u}_B \overline{C}_B}{z_A^2 \overline{u}_A \overline{C}_A} \tag{2.44}$$

Perm-selectivity coefficient of ion B against ion A, T_A^B is defined using Eqn (2.44) as follows (Yamabe and Seno, 1964):

$$T_A^B = \frac{\left(\frac{z_B J_B}{z_A J_A}\right)}{\left(\frac{z_B C'_B}{z_A C'_A}\right)} = \frac{z_B \overline{u}_B \left(\overline{C}_B / \overline{C}_A\right)}{z_A \overline{u}_A \left(C'_B / C'_A\right)} \tag{2.45}$$

in which C'_A and C'_B are the ionic concentration of ion A and ion B, respectively, in a feeding solution.

Equation (2.45) is practically available to measure the perm-selectivity coefficient. However, in order to understand the mechanism of the perm-selectivity, it is necessary to discuss the influence of concentration polarization and the Donnan equilibrium on T_A^B. That is, C'_B/C'_A in Eqn (2.45) relates to the salt concentration ratio in contact with the membrane; C_B^0/C_A^0. It is necessary to discuss the relationship between C_B^0/C_A^0 and ion concentration ratio in bulk C'_B/C'_A using the extended Nernst–Planck equation, and this phenomenon is the concentration polarization itself. Furthermore, the relationship between C_B^0/C_A^0 and $\overline{C}_B/\overline{C}_A$ is generally understood using the Donnan equation. However, the Donnan equation explains only the overall relationship between monovalent ions and multivalent ions in the solution and in the membrane. So, it is necessary to make clear the relationship experimentally.

2.5.2 Measurement of Perm-Selectivity Coefficient

1. An electrodialysis system for measuring perm-selectivity coefficient (Fig. 2.8) is incorporated with a small-scale electrodialyzer (Fig. 2.9), a circulating tank, a reserve tank, a pump, and a flow meter. The circulation tank is immersed in a thermostat adjusted to a constant temperature.
2. An electrolyte solution dissolving ions A and B (ionic concentration; C'_A/C'_B eq dm^{-3}) is put in the reserve tank and it is supplied into the circulating tank through the flow meter F_1 and further supplied to the electrodialyzer through the flow meter F_2 adjusting the solution velocity in desalting cell to 4–5 cm s^{-1}.

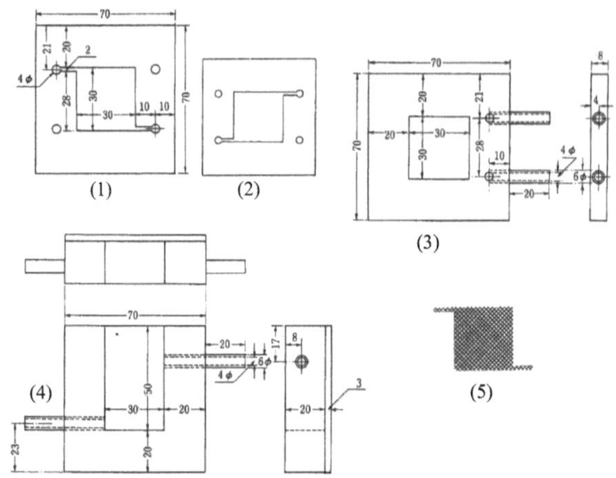

FIGURE 2.8 Components of an electrodialyzer. (1) Desalting cell: Rubber 0.75 mm thick. (2) Concentrating cell: Rubber 0.75 mm thick. (3) Feeding frame: Transparent PVC. (4) Electrode cell: Transparent PVC. (5) Spacer: Polyethylene diagonal net 0.7–0.8 mm thick. *Seno and Tanaka (1984), p. 197, Fig. 7.*

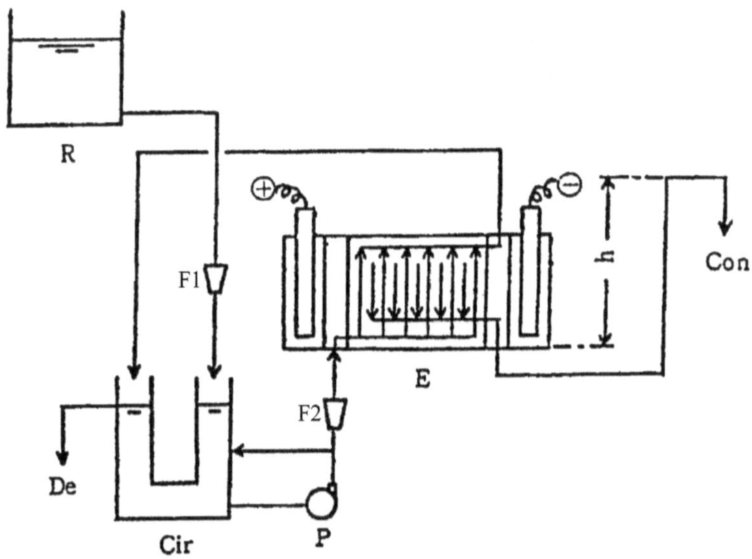

FIGURE 2.9 Electrodialysis system. R: Reserve tank of an electrolyte solution, Cir: Circulating tank, E: Electrodialyzer, P: Pump, F1, F2: Flow meter, Con: Concentrated solution, De: Desalted solution. *Seno and Tanaka (1984), p. 198, Fig. 9.*

The same electrolyte solution is put into the concentrating cell. A 0.5-M NaCl solution is supplied to the electrode cells.
3. An electric current is passed through Ag–AgCl electrodes.
4. An overflow is extracted from the concentrated solution extracting tube and its concentration is observed by a refraction meter.
5. Confirming that the electrolyte concentration of the concentrated solution has become constant, ionic concentration (C''_B/C''_A eq dm^{-3}) is measured.
6. The perm-selectivity coefficient T_A^B is calculated by the following equation:

$$T_A^B = \frac{(C''_B/C''_A)}{(C'_B/C'_A)} \tag{2.46}$$

2.5.3 Monovalent Cation Selective Cation Exchange Membranes

In the electrodialysis process for concentrating seawater, the formation of a polycation layer on the desalting surface of a cation exchange membrane results in the reduction of Ca^{2+} ion permeability and the prevention of $CaSO_4$ precipitation in concentrating cells. In order to accomplish this purpose, many attempts on the membrane treatment to decrease divalent ion permeation had been made over some period of time (Mizutani et al., 1971a,b; Mihara et al., 1970, 1972; Hani et al., 1961).

The polycation layer fixed on the membrane surface is illustrated schematically in Fig. 2.10. In the polycation layer, positively charged groups pack closely together, and the Coulomb and Born repulsive force exerts on surrounding cations. The potential

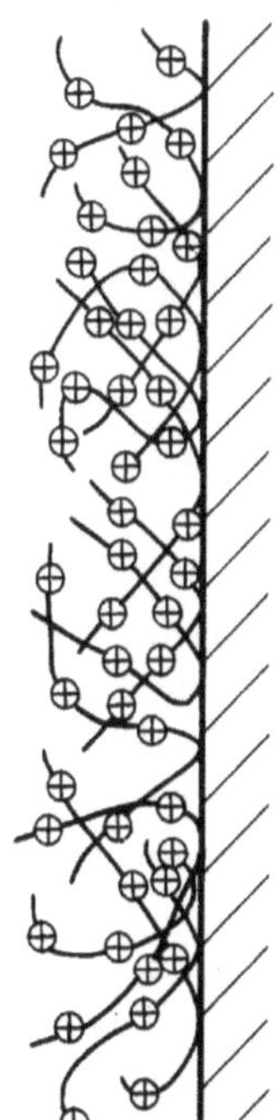

FIGURE 2.10 Polycation layer formed on the cation exchange membrane. *Tanaka and Seno (1981), p. 121. Fig. 8.*

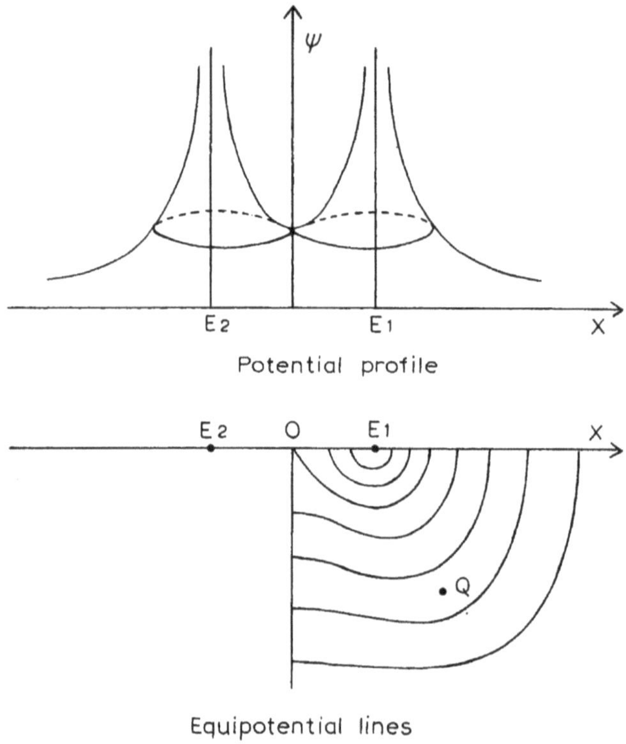

FIGURE 2.11 Potential distribution in the polycation layer. *Tanaka and Seno (1981), p. 121. Fig. 9.*

produced by two adjacent polycation groups is illustrated in Fig. 2.11; under the influence of an electric field, a mobile cation Q move across the lowest potential barrier between two polycation groups, E_1 and E_2. Any mobile cation must pass over the potential barrier of the polycation layer in order to permeate the membrane. Under these circumstances, the transport of divalent ions across the membrane becomes more difficult than for monovalent ions, because greater energy is required for doubly charged cations to pass over the barrier.

The mechanism of monovalent ion perm-selectivity increase (divalent ion permeability decrease) is discussed by applying the Boltzmann equation (Debye and Huckel, 1923) to the potential profile illustrated in Fig. 2.12 (Tanaka and Seno, 1981). In this system, the feed solution contains monovalent cation A and divalent cation B with common anions.

When no polycation layer is formed on the membrane surface, the potential profile is depicted schematically by the dotted line in Fig. 2.12. Under an applied electric current, the concentration polarization is assumed to be neglected because the boundary layer is diminished due to forced flowing solutions in desalting cells. Thus, the relationship between the ion concentration in the membrane and in the desalting cell is

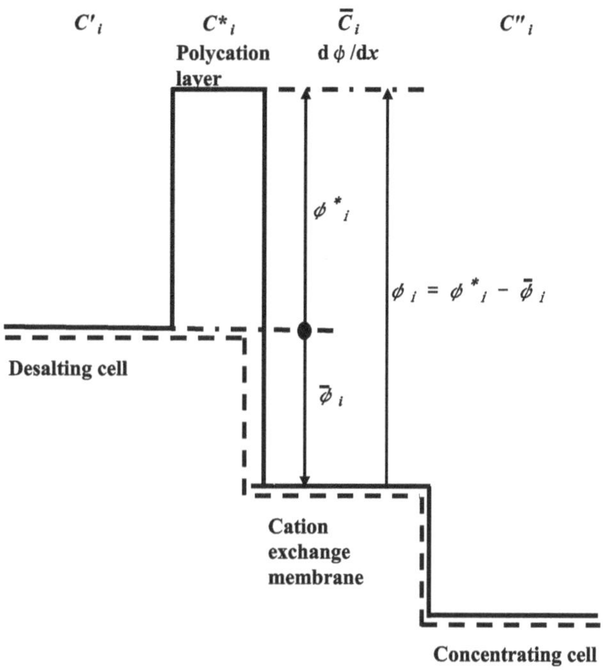

FIGURE 2.12 Potential profile across a polycation layer and a cation exchange membrane. Dotted line shows the profile with no polycation layer. *Tanaka and Seno (1981), p. 122, Fig. 10.*

approximately expressed as follows, applying the Boltzmann equation and taking the potential height in the desalted solution phase as standard:

$$\overline{C}_A = C'_A \exp\left(-\frac{\overline{\psi}_A}{RT}\right)$$
$$\overline{C}_B = C'_B \exp\left(-\frac{\overline{\psi}_B}{RT}\right) \tag{2.47}$$

where $\overline{C}_i$ and C'_i are the concentration of ion i in the membrane and in the desalting cell, respectively. $\overline{\psi}_i$ is the potential height in the membrane for ion i measured from the desalted solution phase. Equation (2.47) shows that the existing probability of ion i in the membrane is governed by the potential barrier for ion i in the membrane.

The ion fluxes are introduced from Eqn (2.47) and the potential gradient term in the Nernst–Planck equation as:

$$J_A^0 = -z_A\overline{u}_A\overline{C}_A\frac{d\psi}{dx} = -z_A\overline{u}_A C'_A\frac{d\psi}{dx}\exp\left(-\frac{\overline{\psi}_A}{RT}\right)$$
$$J_B^0 = -z_B\overline{u}_B\overline{C}_B\frac{d\psi}{dx} = -z_B\overline{u}_B C'_B\frac{d\psi}{dx}\exp\left(-\frac{\overline{\psi}_B}{RT}\right) \tag{2.48}$$

where J_i^0, z_i, and u_i are the flux, the valence, and the mobility of ion i. $d\psi/dx$ is the potential gradient in the membrane. The perm-selectivity coefficient of ion B against ion A; $(T_A^B)^0$ is expressed using Eqns (2.45) and (2.48) as follows:

$$\left(T_A^B\right)^0 = \frac{\left(\frac{z_B J_B^0}{z_A J_A^0}\right)}{\left(\frac{z_B C_B'}{z_A C_A'}\right)} = \frac{z_B \overline{u}_B}{z_A \overline{u}_A} \exp\left(-\frac{\overline{\psi}_B - \overline{\psi}_A}{RT}\right) \tag{2.49}$$

With the polycation layer formed on the membrane surface, the potential profile is formed as indicated by the continuous line in Fig. 2.12. Concentration (existing probability) of ion i in the polycation layer is expressed by:

$$C_A^* = C_A' \exp\left(-\frac{\psi_A^*}{RT}\right)$$
$$C_B^* = C_B' \exp\left(-\frac{\psi_B^*}{RT}\right) \tag{2.50}$$

where C_i^* is the concentration in the polycation layer for ion i. ψ_i^* is the potential height in the polycation layer for ion i measured from the desalted solution phase.

When the ionic mobility in the membrane phase is rather high, ion concentration in the membrane is primarily controlled by the ratio of uptake of those ions from the solution phase. This situation is observed for the carrier-mediated transport of ions through liquid membranes (Bloch, 1970). Since the polycation layer serves as a barrier to ion transport across the membrane system, and the permeation step through this barrier is rate-determining, it is reasonable to suppose that a kinetic-controlled mechanism holds in the present case. Under these conditions, it is safely considered that the concentration ratio in the membrane, $\overline{C}_A/\overline{C}_B$, is primarily determined by the concentration ratio in the polycation layer, C_A^*/C_B^*. If $\overline{C}_i$ is tentatively assumed to be proportional to C_i^*, then the ion fluxes across the membrane will be given by Eqn (2.51):

$$J_A = k z_A \overline{u}_A C_A' \frac{d\psi}{dx} \exp\left(-\frac{\psi_A^*}{RT}\right) = k J_A^0 \exp\left(-\frac{\psi_A^* - \overline{\psi}_A}{RT}\right)$$
$$J_B = k z_B \overline{u}_B C_B' \frac{d\psi}{dx} \exp\left(-\frac{\psi_B^*}{RT}\right) = k J_A^0 \exp\left(-\frac{\psi_B^* - \overline{\psi}_B}{RT}\right) \tag{2.51}$$

k is the constant. T_A^B is derived from Eqn (2.51) as follows:

$$T_A^B = \left(\frac{z_B J_B}{z_A J_A}\right)\left(\frac{z_A C_A'}{z_B C_B'}\right) = \frac{z_B \overline{u}_B}{z_A \overline{u}_A} \exp\left(-\frac{\psi_B^* - \psi_A^*}{RT}\right) = \left(T_A^B\right)^0 \exp\left(-\frac{\psi_B - \psi_A}{RT}\right) \tag{2.52}$$

The potential $\psi_i = \psi_i^* - \overline{\psi}_i$ is that of the polycation layer relative to that in the membrane phase. The potential difference $\psi_B - \psi_A$ provides a mechanism for the separation of ion A from ion B, and can be estimated from the electrodialysis experiment as follows:

$$\psi_B - \psi_A = RT \ln\left[\left(\frac{J_B^0}{J_A^0}\right) \Big/ \left(\frac{J_B}{J_A}\right)\right] = RT \ln\left[\left(T_A^B\right)^0 \Big/ T_A^B\right] \tag{2.53}$$

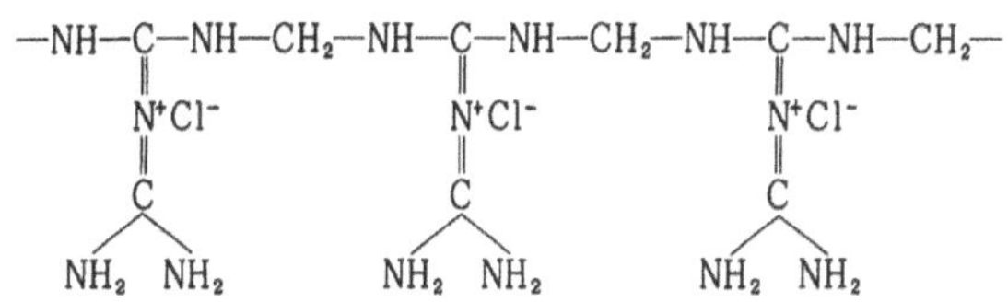

(I)

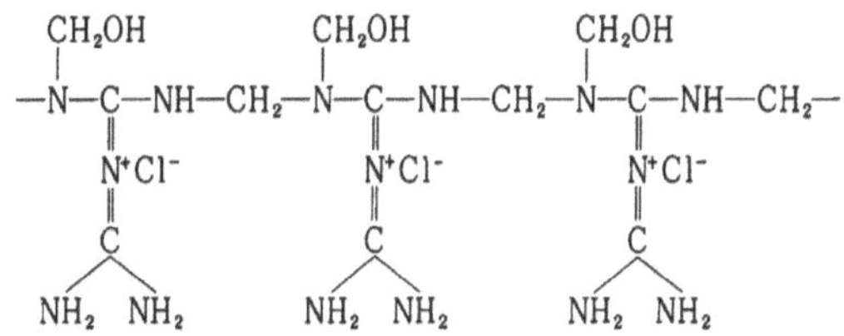

(II)

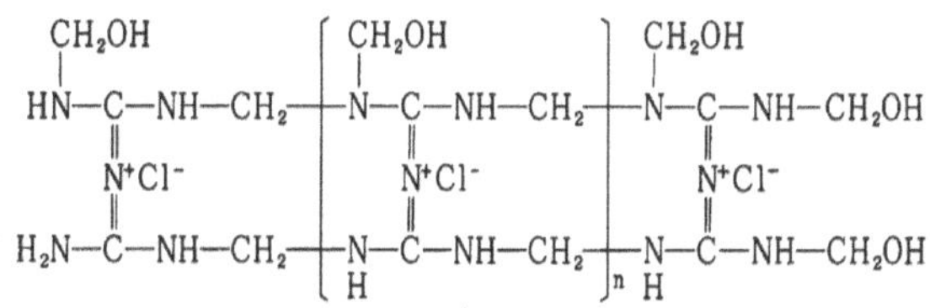

(III)

FIGURE 2.13 Dicyandiamide formaldehyde condensate. *Tanaka and Seno (1981), p. 118, Fig. 3.*

The following is the electrodialysis experiment carried out by injecting polycations into the feeding seawater in order to evaluate $\psi_B - \psi_A$ in Eqn (2.53) (Tanaka, 1974). The polycations are the di-cyandiamide formaldehyde condensates, which include many quaternary ammonium positive groups in the molecule as shown in Fig. 2.13. The electrodialysis experiment was carried out to form the polycation layer on the cation exchange membrane under the following conditions:

Electrodialyzer specifications; Membrane area; 5 dm^2. Space between the membranes; 0.75 mm. Cell pair number; 20. Ion exchange membrane; Selemion CMV-4/AST-4 (Asahi Glass Co.), Spacer; diagonal net.

Operating conditions; Current density; 4 A/dm^2. Seawater temperature; 26.4 °C. Linear velocity in desalting cells; 5 cm/s. Concentration of the polycations in feeding seawater C_X was changed from 0.0462 to 1.929 mg/dm^3. The concentrated solution is extracted to the outside of the process.

T_A^B is calculated using Eqn (2.46) and shown in Fig. 2.14. Figure 2.15 shows $\psi_B - \psi_A$ calculated from T_A^B in Fig. 2.14 using Eqn (2.53). This phenomenon gives the monovalent cation selectivity to the cation exchange membrane. Figure 2.16 shows monovalent ion

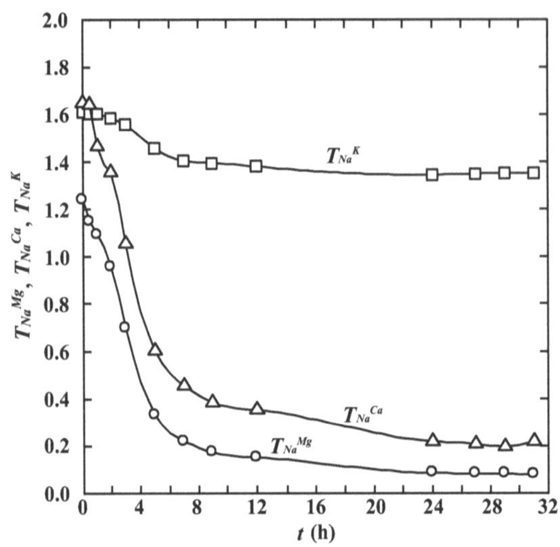

FIGURE 2.14 Changes of T_A^B with time. $C_X = 0.508$ mg/dm^3.

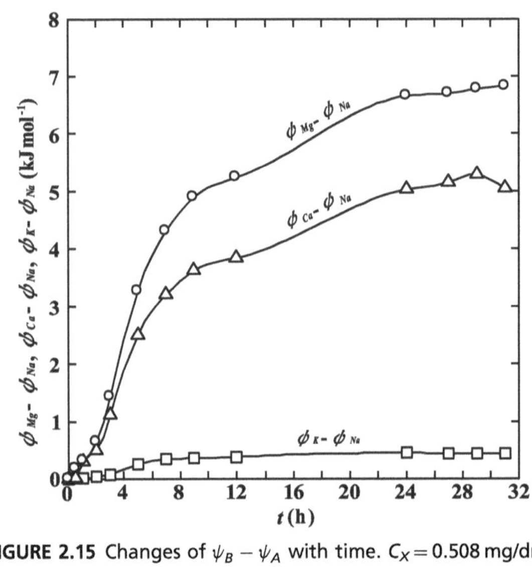

FIGURE 2.15 Changes of $\psi_B - \psi_A$ with time. $C_X = 0.508$ mg/dm^3.

concentration ratio ε $(=(C_{Na} + C_K)/(C_{Na} + C_K + C_{Mg} + C_{Ca}))$ versus cell voltage V_{cell} and energy consumption E. V_{cell} and E are extremely increased when ε is increased beyond 0.99. This phenomenon is caused by salt concentration decrease at the interface between the polycation layer and the cation exchange membrane due to the bipolar type structure formation.

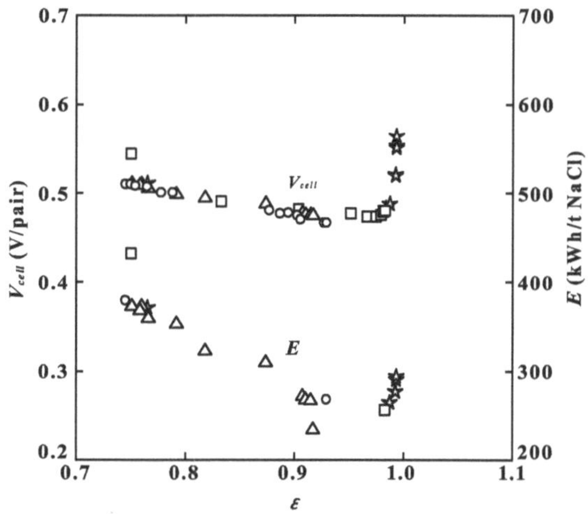

FIGURE 2.16 Monovalent ion ratio in concentrate versus cell voltage and energy consumption. $C_X = 0.0492$ (○), 0.0917 (△), 0.987 (□), 1.939 (☆) mg/dm^3.

2.6 Electric Conductivity

2.6.1 Principles of Electric Conductivity

The convection velocity v in Eqn (2.20) (extended Nernst–Planck equation) is expressed as (Schlogel, 1955):

$$v = \omega \overline{u}_0 \frac{d\psi}{dx} \quad \left(\overline{u}_0 = \frac{F\overline{C}_R}{\rho_0 \varepsilon} \right) \tag{2.54}$$

ω is the charge number of ion exchange groups, being -1 for a cation exchange membrane and $+1$ for an anion exchange membrane; $\overline{u}_0$ ε and ρ_0 are, respectively, the mobility, the volume ratio, and the specific flow resistance of the pore liquid; the liquid in the pores formed in the membrane (not including the matrix volume). The first term (diffusion term) in Eqn (2.20) is generally negligible in an ion exchange membrane due to the Donnan exclusion of co-ions (Section 2.2.1), so the following equation is introduced taking account of Eqn (2.54):

$$J_i = -\overline{C}_i(z_i \overline{u}_i - \omega \overline{u}_0)\frac{d\psi}{dx} \tag{2.55}$$

Equation (2.55) indicates that the velocity of counter-ions is increased and that of co-ions is reduced by $\overline{u}_0$ in the membrane.

Here, the electric current density i is defined by the following equation:

$$\frac{i}{F} = \sum z_i J_i \tag{2.56}$$

The electroneutrality is given by the following equation:

$$\sum z_i \overline{C}_i + \omega \overline{C}_R = 0 \tag{2.57}$$

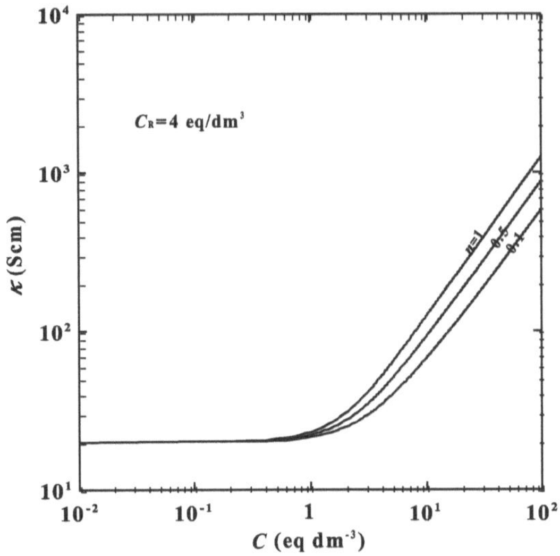

FIGURE 2.17 Relationship between NaCl concentration in a solution and specific conductivity of a cation exchange membrane. *Tanaka (2007), p. 45, Fig. 3.5.*

The electric current density is introduced from Eqns (2.55)–(2.57) as follows:

$$i = -F\left(\sum z_i^2 \overline{u}_i \overline{C}_i + \overline{u}_0 \overline{C}_R\right) \frac{d\psi}{dx} \tag{2.58}$$

Accordingly, the specific electric conductivity of the membrane κ is:

$$\kappa = -\frac{i}{d\psi/dx} = F\left(\sum z_i^2 \overline{u}_i \overline{C}_i + \overline{u}_0 \overline{C}_R\right) \tag{2.59}$$

In an ion exchange membrane, the specific flow resistance of the pore liquid ρ_0 is generally large enough, so $\overline{u}_0$ is relatively small (Eqn (2.54)) because of dense structure of the membrane. In this situation, the second term in Eqns (2.58) and (2.59) is allowed to delete and the conductivity of a cation exchange membrane immersed in a NaCl solution may be presented by the following equation:

$$\kappa = F\left(\overline{u}_{Na} \overline{C}_{Na} + \overline{u}_{Cl} \overline{C}_{Cl}\right) = F\overline{u}_{Na}\left(\overline{C}_{Na} + \frac{\overline{u}_{Cl}}{\overline{u}_{Na}} \overline{C}_{Cl}\right) \tag{2.60}$$

κ is calculated using Eqns (2.18), (2.23), (2.24) and (2.60), assuming Na^+ ion mobility in the membrane is one-tenth of the value in the solution; $F\overline{u}_{Na} = 50.11/10 = 5.011\ \mathrm{S\ cm^2\ mol^{-1}}$ (Section 2.2.2). κ is plotted against C and $\overline{C}_R$ putting n as parameter (Figs 2.17 and 2.18). In these figures, n increases toward 1 with the increase of C and the decrease of $\overline{C}_R$.

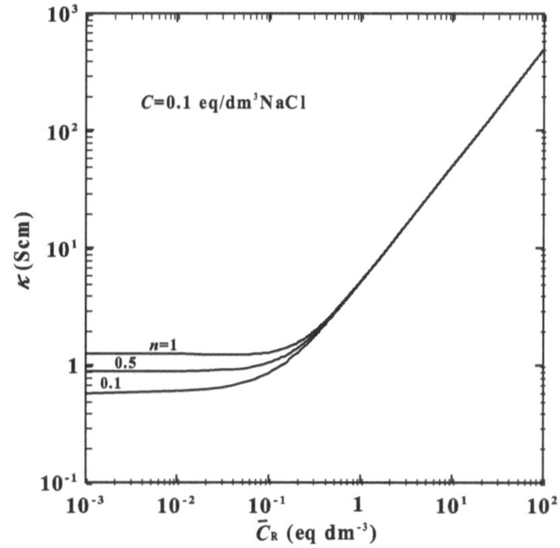

FIGURE 2.18 Relationship between ion exchange group concentration and specific conductivity of a cation exchange membrane. *Tanaka (2007), p. 45, Fig. 3.6.*

2.6.2 Measurement of Electric Resistance

Electric resistance of an ion exchange membrane R (Ω) is defined by the following equation:

$$R = \frac{\nu d}{S} = \frac{d}{\kappa S} \tag{2.61}$$

in which, S (cm^2) is area and d (cm) is thickness of the membrane. ν (Ω cm) is the specific electric resistance of the membrane and it is equivalent R at $d = 1$ cm, and $S = 1\ cm^2$. κ ($S\ cm^{-1}$) is the specific electric conductivity defined in Eqn (2.59). Area electric resistance $r = \nu d$ ($\Omega\ cm^2$) is equivalent to R at $S = 1\ cm^2$.

The electric resistance of the membrane is usually measured using a two-compartment apparatus, which adjusts the effective membrane area to $S = 1\ cm^2$ and is incorporated with platinum black electrodes (Fig. 2.19). The measuring procedure is:

1. A sample membrane is equilibrated with a 0.5-M NaCl solution for about 2 h, and the membrane surface is wiped with a filter paper.
2. The membrane is incorporated with the apparatus, a 0.5-M NaCl solution is supplied into the compartment, and the cell is left in a 25 °C thermostat.
3. Electric resistance (R_1) is measured with an alternating current bridge (frequency: 1000 Hz).
4. The sample membrane is taken away, the apparatus is reintegrated without a membrane, and electric resistance (R_2) is measured again under the same conditions.
5. ν is calculated substituting $R = R_1 - R_2$, $S = 1\ cm^2$ and d measured with a micrometer into Eqn (2.61).

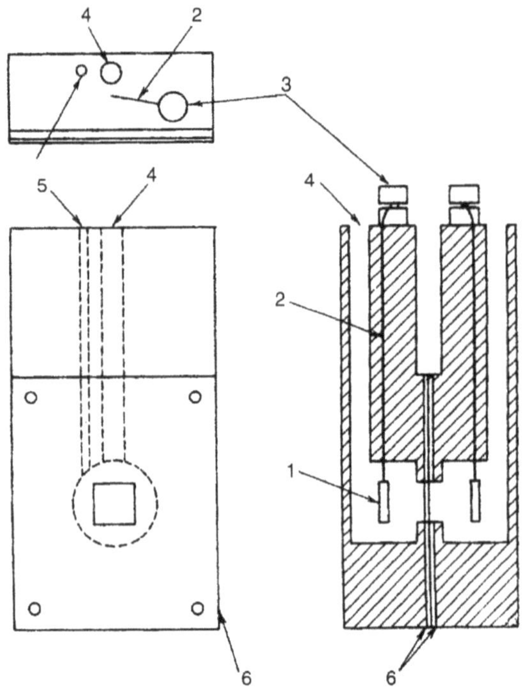

FIGURE 2.19 Electric resistance measuring apparatus. (1) Pt electrode. (2) Pt wire. (3) Terminal. (4) Solution injection hole; Thermometer inserting hole. (5) Air extracting hole. (6) Rubber gasket. *Seno and Tanaka (1984), p. 193, Fig. 1.*

2.6.3 Direct Current Electric Resistance

Ion exchange membranes usually work under a direct electric current. The direct current electric resistance of the membrane is influenced by the salt concentration of the solutions placed on both sides of the membrane. Thus, the direct current electric resistance of the membrane working in an electrodialyzer is not the same as the alternating current electric resistance described in Sections 2.6.1 and 2.6.2.

For measuring the direct current electric resistance of an ion exchange membrane, the membrane is set in a two-cell apparatus and a low-concentration NaCl solution (specific conductivity, κ' (S/cm)) is supplied into both cells. Direct current electric resistance of the membrane r^*_{dire} ($\Omega\,cm^2$) is measured at 25 °C by passing a direct current. The relationship between κ' and r^*_{dire}/r_{alter} is expressed by the following empirical equation (Tanaka et al., 2003):

$$\log\left(\frac{r^*_{dire}}{r_{alter}}\right) = 0.3380 + 0.6386 \log \kappa' + 0.2961(\log \kappa')^2 \tag{2.62}$$

A low-concentration NaCl solution (specific conductivity, κ') is supplied to the desalting side, and a high-concentration NaCl solution (specific conductivity, κ'') is supplied to the concentrating side of the above apparatus. The direct current electric resistance of the membrane r_{dire} is measured at 25 °C by passing a direct current and subtracting the effect of membrane potential. The empirical relationship between κ''/κ' and r_{dire}/r^*_{dire} is:

$$\frac{r_{dire}}{r^*_{dire}} = 1.000 - 0.1359 \log\left(\frac{\kappa''}{\kappa'}\right) \tag{2.63}$$

The direct electric current of a membrane pair r_{memb} is estimated by multiplying Eqn (2.62) by Eqn (2.63):

$$r_{memb} = r_K + r_A = \left(\frac{r^*_{dire}}{r_{alter}}\right) \times \left(\frac{r_{dire}}{r^*_{dire}}\right) \times r_{alter} = r_{dire} \tag{2.64}$$

2.7 Osmosis

2.7.1 Anomalous Osmotic Behavior

When a membrane is in contact with electrolyte solutions differing in concentration (Fig. 2.20), the following osmotic pressures (swelling pressure) π' and π'' generate on both sides of the membrane surface (cf. Eqn (2.8)).

Low-concentration (C') side:

$$\pi' = \left(\frac{RT}{V_W}\right) \ln\left(\frac{a'_W}{\bar{a}'_W}\right) \tag{2.65}$$

High-concentration (C'') side:

$$\pi'' = \left(\frac{RT}{V_W}\right) \ln\left(\frac{a''_W}{\bar{a}''_W}\right) \tag{2.66}$$

In Fig. 2.20, $C' < C''$, $\pi' > \pi''$ and $a'_W/\bar{a}'_W > a''_W/\bar{a}''_W$ hold, and the solvent flux, $J_{V,osmotic}$, due to osmotic pressure is given as:

$$J_{V,osmotic} = k(\pi' - \pi'') \tag{2.67}$$

k is the proportional constant. However, the osmotic behavior described above is generally observed with a nonionic or a completely impermeable membrane. With an

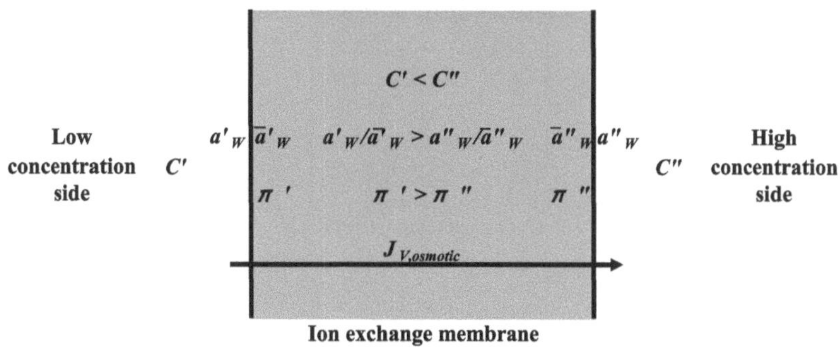

FIGURE 2.20 Solvent flux caused by osmotic pressure.

ion exchange membrane, accelerated (anomalous positive) or depressed (negative) osmotic behaviors are observed (Grim and Sollner, 1957; Tasaka et al., 1969). The mechanism of both anomalous positive and negative osmotic behavior was investigated by Schlogel (1955) and explained to be caused by diffusion of solutes across the membrane as shown in Fig. 2.21, exemplifying the solvent flux across a cation exchange membrane. In this figure, counter-ions (cations) and co-ions (anion) diffuse from a high-concentration (C'') side to a low-concentration (C') side. The mechanism of anomalous positive and negative osmotic behavior is understandable conceptually as follows:

The anomalous positive osmosis is observed when the mobility of counter-ions $\overline{u}_+$ in the cation exchange membrane is larger than that of anions $\overline{u}_-(\overline{u}_+ > \overline{u}_-)$. In this situation, cation flux due to diffusion $J_{+,diffusion}$ becomes larger than anion flux $J_{-,diffusion}$ $(J_{+,diffusion} > J_{-,diffusion})$, and consequently the membrane surface is charged positively in the low-concentration side and charged negatively in the high-concentration side; this phenomenon generates the electric potential gradient across the membrane. An ion exchange membrane consists of a three-dimensional cross-linked structure, including

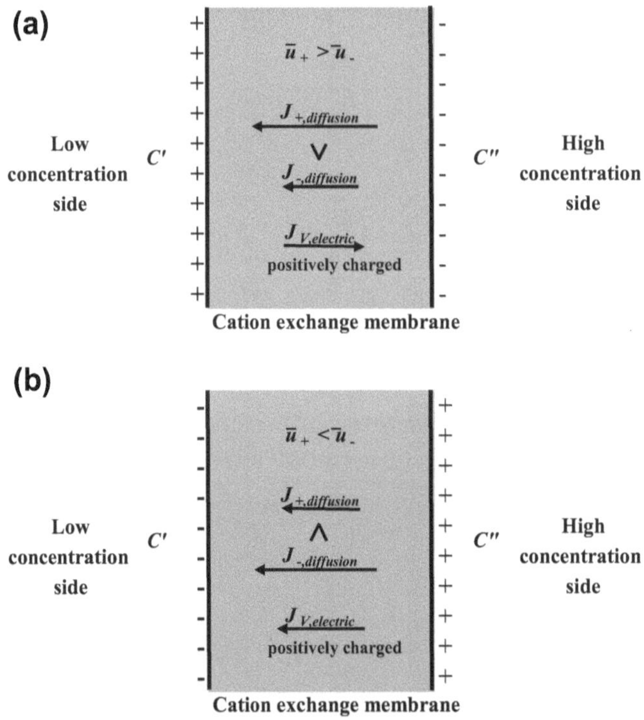

FIGURE 2.21 Anomalous positive (a) solvent flux and negative solvent flux (b).

many micro-pores in which water molecules are absorbed. Water molecules in the pores dissolve counter-ions and carry the same signs as those of the counter-ions (opposite to the signs of the exchange groups). In the cation exchange membrane, the pore liquid (absorbed water molecules) carries plus signs, so that electric potential gradient established in the membrane generates the flux of pore liquid $J_{V,electric}$ from the low-concentration side to high-concentration side. Consequently, the solvent flux J_V in the anomalous positive osmosis is expressed by adding $J_{V,osmotic}$ and $J_{V,electric}$ as follows:

$$J_V = J_{V,osmotic} + J_{V,electric} \tag{2.68}$$

The negative osmosis occurs when the mobility of counter-ions $\overline{u}_+$ in the cation exchange membrane is less than that of anions $\overline{u}_-(\overline{u}_+ > \overline{u}_-)$, and its mechanism is explained in contrast to that of anomalous positive osmosis. Namely, cation flux due to diffusion $J_{+,diffusion}$ becomes smaller than anion flux $J_{-,diffusion}$ $(J_{+,diffusion} < J_{-,diffusion})$, and consequently the membrane surface is charged negatively in the low-concentration side and charged positively in the high-concentration side, so that electric potential gradient established in the membrane generates the flux of pore liquid $J_{V,electric}$ from the high-concentration side to low-concentration side. Consequently, the solvent flux J_V in the negative osmosis is expressed by subtracting $J_{V,electric}$ from $J_{V,osmotic}$ as follows:

$$J_V = J_{V,osmotic} - J_{V,electric} \tag{2.69}$$

Negative osmosis occurs when $J_{V,electric} > J_{V,osmotic}$ holds.

2.7.2 Measurement of Osmosis

When an electrolyte concentration difference ΔC is established between both membrane surfaces, water permeates across the membrane according to the mechanism described in Section 2.7.1. Water flux J_W is expressed by the following equation:

$$J_W = \frac{D_W \Delta C}{d} \tag{2.70}$$

where d is the thickness of the membrane and D_W the osmotic water transport coefficient. D_W/d is defined as water permeability coefficient, which is measured as follows:

A sample membrane immersed into a 0.5-M NaCl solution is washed with water, wiped with a filter paper, and integrated in the two-cell apparatus (Fig. 2.22). Then, 25 °C water is put in the measuring cell, and a 4-M NaCl solution is put in the other cell so that the solution level in both cell is the same. The apparatus is left in a 25 °C thermostat and the solution in both cells are stirred violently using magnetic stirrers. After 10 min, a decrease of the volume in the measuring cell is observed for 60 min with a measuring pipette (a cm^3). The water permeation coefficient is calculated using the following equation:

$$\frac{D_W}{d}\left(\text{cm}^4\ \text{s}^{-1}\ \text{mol}^{-1}\right) = \frac{a \times 10^3}{4 \times 3600S}$$

in which S is effective membrane area (cm^2).

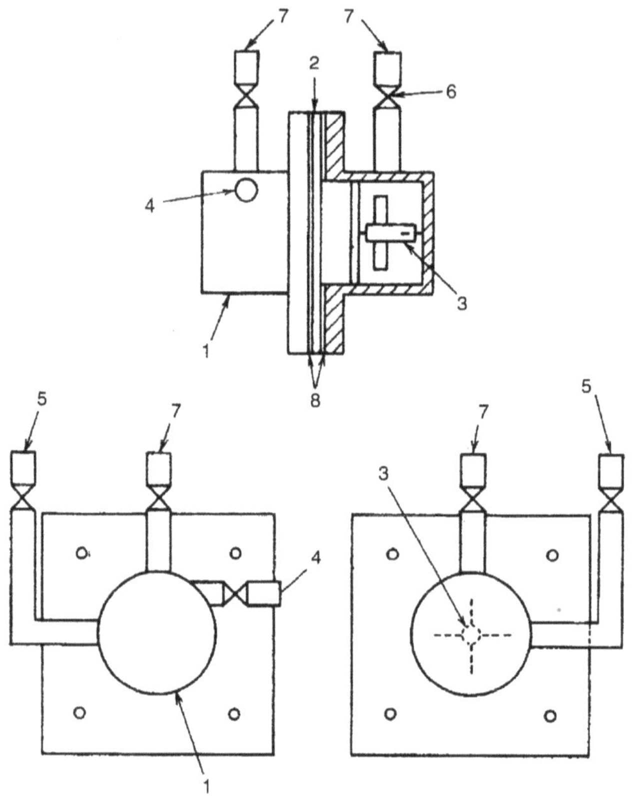

FIGURE 2.22 Water permeation coefficient measuring apparatus. (1) Measuring cell. (2) Membrane holding plate. (3) Magnetic stirrer. (4) Pipette holding plate. (5) Solution injecting hole. (6) Cock. (7) Air extracting hole. (8) Rubber gasket. *Seno and Tanaka (1984), p. 197, Fig. 6.*

2.8 Electro-Osmosis

2.8.1 Electro-Osmotic Permeability

When an electric current is passed across an ion exchange membrane, counter-ions are transferred through the membrane with water molecules (electro-osmotic water). Because the liquid in the membrane is charged the same sign with the counter-ions, the liquid moves with the counter-ions toward the same direction. The flux of counter-ions J_i (mol cm^{-2} s^{-1}) is given by the following equation:

$$J_i = C_i v \tag{2.71}$$

where v (cm s^{-1}) is the velocity of the solution in the membrane. It is defined by Eqn (2.54) described in Section 2.6.1.

$$v = \omega \overline{u}_0 \frac{d\psi}{dx} \quad \left(\overline{u}_0 = \frac{F\overline{C}_R}{\rho_0 \varepsilon} \right) \tag{2.72}$$

Flux of a solution due to electro-osmosis J_V is accordingly expressed by the following equation:

$$J_V = \varepsilon\, v = \omega\varepsilon\overline{u}_0\frac{d\psi}{dx} = \frac{\omega F\overline{C}_R}{\rho_0}\frac{d\psi}{dx} \tag{2.73}$$

On the other hand, current density i is defined by Eqns (2.58) and (2.59) introduced in Section 2.6.1.

$$i = -F\left(\sum z_i^2\overline{u}_i\overline{C}_i + \overline{u}_0\overline{C}_R\right)\frac{d\psi}{dx} = -\kappa\frac{d\psi}{dx} \tag{2.74}$$

The electro-osmosis permeability β is introduced from Eqns (2.73) and (2.74) as:

$$\beta = \frac{J_V}{i} = \frac{\omega\varepsilon\overline{u}_0}{F\left(\sum z_i^2\overline{u}_i\overline{C}_i + \overline{u}_0\overline{C}_R\right)} = \frac{\omega F\overline{C}_R}{\rho_0\kappa} \tag{2.75}$$

2.8.2 Measurement of Electro-osmotic Permeability

Solution flux across the membrane J_V is given by the function of β as follows (cf. Eqn (2.75)):

$$J_v = \beta i \tag{2.76}$$

In order to measure β in Eqn (2.76):

1. A sample membrane is immersed for a whole day and night in a 0.5-M NaCl solution. The membrane is integrated in a two-cell apparatus incorporated with Ag-AgCl electrodes (Fig. 2.23), and the apparatus is put in a 25 °C thermostat.
2. β of a cation exchange membrane is measured putting a 25 °C, 0.5-M NaCl solution in both cells and an electrode (cathode) in the measuring cell, and then current density of 1 A dm^{-2} is passed. In this case, NaCl concentration in the measuring cell increases, that in the other cell decreases during electric current passage, and it gives rise to solution permeation across the membrane due to a NaCl concentration difference generated in the cells.
3. In order to minimize the experimental error due to the phenomenon described above, the NaCl concentration in the measuring cell is adjusted beforehand less than 0.5 M, and that in the other cell is higher than 0.5 M. Furthermore, the NaCl concentrations in both cells are devised to be reversed during electric current passage.
4. In case of an anion exchange membrane, the device described above is not necessary.
5. The solution volume increase in the measuring cell is observed accurately for 60 min after 10 min from the beginning of electric current passage with measuring pipette (a cm^3).

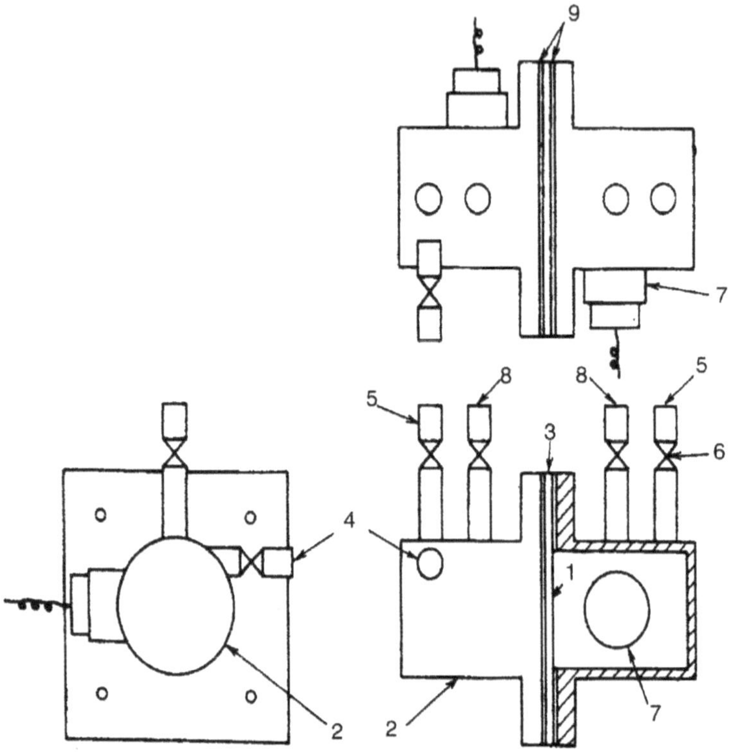

FIGURE 2.23 Electro-osmotic coefficient measuring apparatus. (1) Membrane. (2) Measuring cell. (3) Membrane holding plate. (4) Pipette holding hole. (5) Solution injecting hole. (6) Cock. (7) Electrode inserting hole. (8) Air extracting hole. (9) Rubber gasket. *Seno and Tanaka (1984), p. 196, Fig. 5.*

6. β is calculated using the following equation:

$$\beta(\mathrm{cm^3\,A^{-1}\,s^{-1}}) = \frac{(10^2/3600)a}{S}$$

in which S is the effective membrane area (cm^2).

7. Temperature changes and bubbles generated from a cathode during electric current passage cause experimental errors, so the temperature must be adjusted accurately and the cathode should be prepared perfectly in an AgCl type.

2.9 Ion Exchange Capacity and Water Content

2.9.1 Ion Exchange Group Concentration in an Ion Exchange Membrane

The relationship between the ion exchange group concentration C_W (eqiv./cm^3 water) and ion exchange capacity C_R (equiv./g dry membrane) is defined as:

$$C_W = C_R/W \tag{2.77}$$

W (cm^3 water/g dry membrane) is the water content. Various ion exchange membrane properties are strongly influenced by C_W.

2.9.2 Measurement of Ion Exchange Capacity and Water Content

1. Cation exchange membrane
   a. A sample membrane is put in a 1-M HCl solution over a 6-h changing, and the solution and the exchange groups are converted to an H type. Then, it is washed with water sufficiently.
   b. The membrane is immersed in a 2 M NaCl solution, which is changed several times and further washed with water sufficiently.
   c. The immersed solutions and washed water are collected and H^+ ions in the collected solution are analyzed with a 0.1-M NaOH (a meq).
2. Anion exchange membrane
   a. A sample membrane is put in a 2-M $NaNO_3$ solution for 30 min, then it is immersed in a 2-M NaCl solution over 6 h, during which the solution is changed and the ion exchange groups are converted to a Cl^- type. The membrane is washed with water sufficiently.
   b. The membrane is immersed in a 2-M $NaNO_3$ solution for over 6 h and washed sufficiently with water.
   c. The immersed solutions and washed water are collected and Cl^- ions in the collected solution are analyzed with a 0.1-M $AgNO_3$ (a meq).
3. Weight
   a. The moisture adhered on the sample membrane is wiped with a filter paper and the membrane is weighed (b g).
   b. The membrane is dried in a 65 °C thermostat until the weight becomes constant (c g).
4. Calculation

   The ion exchange capacity A_R and water content W are calculated as:
   AR (meqiv./g dry membrane) = a/c.

$$W(\%) = 100(b - c)/c$$

2.10 Swelling Ratio

The following is the measuring procedure:

1. A sample membrane is immersed in a 0.5-M NaCl solution for 24 h, during which the solution is changed more than three times.
2. A vertical (*a* mm) and a horizontal (b mm) length of the membrane are measured with a ruler, and a thickness (c mm) is measured with a micrometer.
3. The sample membrane is immersed in a 3.5-M NaCl solution for 24 h, during which the solution is changed more than three times.

4. A vertical (a′ mm) and horizontal (b′ mm) length and a thickness (c′ mm) is measured as described in 2.
5. The swelling ratio is calculated as follows:

$$\text{Vertical swelling ratio (\%)} = 100(a - a')/a$$

$$\text{Horizontal swelling ratio (\%)} = 100(b - b')/b$$

$$\text{Thickness swelling ratio (\%)} = 100(c - c')/c$$

2.11 Mechanical Strength

2.11.1 Mechanical Characteristics

A sample membrane is immersed in a 0.5-M NaCl solution for 24 h, during which the solution is substituted more than three times. Then, the bursting strength and tensile strength are measured as follows.

2.11.2 Bursting Strengths

A sample membrane sheet (about over 6- × 6-cm width) is fixed into a Mullen tester and inflated injecting glycerol. The maximum pressure when the membrane bursts is the bursting strength.

2.11.3 Tensile Strength

A sample membrane sheet (about over a 10-cm length × 1-cm width) is fixed to a Schopper's tension tester. The tensile strength is maximum strength when the sample is cut off. The tensile strength value is influenced by the thread (reinforcement) put in the membrane. So it is reasonable to cut off the sample membrane, for example, as an angle of the thread to the sample length is 45°.

2.12 Characteristics of Commercially Available Ion Exchange Membranes

The characteristics of hydrocarbon ion exchange membranes are listed in Table 2 1. The data are quoted from maker's catalogs and technical reports.

Table 2.1 Commercially Available Ion Exchange Membranes

Company/ Product Name	Name	Type	Thickness (mm)	Electric resistance (Ω cm^2)	Transport Number	Burst Strength (Kg/cm^2)	Remarks
Ionics (USA)				0.01 MNaCl			
(GE Infrastructure)	CR61-CMP	Cation memb	0.56	10		22.5	Food/Pharmaceutical
Nepton	CR64-LMP	Cation memb	0.51	11		22.5	General purpose
	CR64-LMRH	Cation memb	1.2	20		23.2	Interstage/Electrode
	CR67-HME	Cation memb	0.30	9		18.3	Water, extra thin, low resistance
	CR67-HMR	Cation memb	0.56	10		10.5	Water
	CR69-HMP	Cation memb	0.56	10		22.5	Special applications
	AR103-QDP	Anion memb	0.51	11		22.5	Food/Pharmaceutical
	AR204-SZRA	Anion memb	0.51	7		10.5	Fouling resistant
	AR204-UZRA	Anion memb	0.56	7		10.5	High efficiency
Membrane International (USA)	CMI-7000S	Cation memb	0.45 ± 0.025	<30	0.94	>80	Gel polystyrene cross-linked with divinyl benzene
	AMI-7001S	Anion memb	0.45 ± 0.025	<40	0.90	>80	Gel polystyrene cross-linked with divinyl benzene
	AMI-7001CR	Anion memb	0.50 ± 0.025	<40	0.90	>80	Gel polystyrene cross-linked with divinyl benzene
Sybron Chemicals (USA)				0.1 MNaCl	0.5/1.0 MNaCl		
Ionac	MC-3470	Cation memb (Na^+)	15 mil	25	0.96	10.3	Heterogeneous membrane
	MA-3475	Anion memb (Cl^-)	16 mil	50	0.99	10.3	Heterogeneous membrane
	MA-7500	Anion memb (Cl^-)	18 mil	30	0.9*9	10.3	Heterogeneous membrane
AGC Engineering (Japan)		(); counter ions		0.5 MNaCl	0.5/1.0 MNaCl		
Selemion	CMV	Cation memb (Na^+)	0.12	3.0	>0.96	2.0	Standard
	CMD	Cation memb (Na^+)	0.38	17	>0.94	9.0	Strong (thick)
	HSF	Cation memb (H^+)	0.15	19		2.0	H^+ selective
	CSC	Cation memb (Na^+)	0.10	2.3	>0.97	1.5	Monovalent ion selective
	CMF	Cation memb (H^+)	0.44	2.5	>0.95	10.0	High durability
	AMV	Anion memb (Cl^-)	0.12	2.8	>0.96	2.0	Standard
	AMT	Anion memb (Cl^-)	0.20	6.0	>0.96	7.0	Strong (thick)

(Continued)

Table 2.1 Commercially Available Ion Exchange Membranes—cont'd

Company/ Product Name	Name	Type	Thickness (mm)	Electric resistance (Ω cm^2)	Transport Number	Burst Strength (Kg/cm^2)	Remarks
	DSV	Anion memb (Cl^-)	0.10	1.1		1.5	Low resistance
	AAV	Anion memb (SO_4^{2-})	0.12	6.4	>0.95	3.0	Low proton leakage
	ASV	Anion memb (Cl^-)	0.12	3.7	>0.97	2.0	Monovalent ion selective
	AHO	Anion memb (Br^-)	0.30	20	>0.95	12.0	High temp. & Oxidant proof
	AP64	Anion memb (SO_4^{2-})	0.15	0.70		2.0	Oxidant proof
Astom (Japan)		(); counter ions		0.5 MNaCl			
Neosepta	CMX	Cation memb (Na^+)	0.17	3.0		≧4.0	Purification of organics; concentration of inorganics etc.
	CIMS	Cation memb (Na^+)	0.15	1.8		≧1.0	Deacidification of metal solution
	CMB	Cation memb (Na^+)	0.21	4.5		≧4.0	Alkali removal
	AMX	Anion memb (Cl^-)	0.14	2.4		≧2.5	Purification of organics; concentration of inorganics etc.
	AHA	Anion memb (Cl^-)	0.22	4.1		≧9.0	Alkali removal
	ACM	Anion memb (Cl^-)	0.11	2.6		≧1.5	Acid concentration
	ACS	Anion memb (Cl^-)	0.13	3.8		≧1.5	Purification of pharmaceutical products etc.
	AFN	Anion memb (Cl^-)	0.16	0.5		≧2.5	Diffusion dialysis of acid; Demineralization of organics etc.
	ACM	Anion memb (Cl^-)	0.17	1.0		≧2.5	Diffusion dialysis
	BP-1	Bipolar membrane	0.20–0.35			4–7	Organic/inorganic acid production
Fumatech GmbH (Germany)				0.5 MNaCl	0.1/0.5 MKCl		
Fumasep	FKS	Cation memb	0.11–0.13	<8	>0.96		Standard demineralization
	FKB	Cation memb	0.08–0.10	<4	>0.98		Withstand high caustic concentration
	FKL	Cation memb	0.11–0.12	<4	>0.92		High chemical and mechanical stability
	FKE	Cation memb	0.005–0.07	<3	>0.98		Excellent chemical and mechanical stability
	FAB	Anion memb	0.10–0.13	<1	>0.96		Low H^+ transport

	FAP	Anion memb	0.06–0.07	<1	>0.92		Plain film, resistant against chlorine
	FAP	Anion memb	0.09–0.10	<1.5	>0.92		PEEK reinforced, resistant against chlorine
	FAD	Anion memb	0.08–0.10	<0.80	>0.91		Diffusion dialysis
PCA GmbH (Germany)					0.1/0.5 MKCl		
	PC-SK	Cation memb	0.09–0.13			4–5	Standard
	PC-AS	Anion memb	0.09–0.13			4–5	Standard
	PC-SA/HD	Anion memb	0.31–0.35			Ca 10	Standard
	PC acid 60	Anion memb	0.08–0.10		>0.95	4–5	Production of mineral acids
	PC acid 100	Anion memb	0.08–0.10		>0.88	4–5	Production of mineral acids
	PC 100D	Anion memb	0.08–0.10	Ca 5Ω		4–5	Production of organic acids
	PC 200D	Anion memb	0.08–0.10	Ca 2Ω (1 MKCl)	Gluconate > 0.9	4–5	Production of organic acids
Mega (Czech Republic)					0.5/0.1 M KCl		
RALEX	CM(H) (PAD)	Cation memb	<0.65	<8.5	>0.95		Ployamide textil fiber, Ed, EDI
	CM(H) (PES)	Cation memb	<0.70	<10	>0.95		Polyester textil fiber, Ed, EDI
	CMH5E-PM	Cation memb	<1.1	<19.0	>0.95		Polyamide textil fiber, Anaphoresis
	AM(H) (PAD)	Anion memb	<0.70	<7	>0.95		Polyamide textil fiber, Ed, EDI
	AM(H) (PES)	Anion memb	<0.75	<7.5	>0.95		Polyester textil fiber, Ed, EDI
	AMH5E-HD	Anion memb	<1.25	<19	>0.94		Polyester textil fiber, Cataphoresis
Hangzhou Iontech Environmental Technology (China)	CN standard	Cation memb	0.42 ± 0.04	≦15	≧0.92	≧6	Heterogeneous membrane
IONSEP	MC	Cation memb	0.42	6–10	0.91	6	Heterogeneous membrane
	HC	Cation memb	0.42	10–12	0.92	6	Heterogeneous membrane
	HC8	Cation memb	0.42	11–13	0.93	6	Heterogeneous membrane
	CN standard	Anion memb	0.42 ± 0.04	≦20	≧0.92	≧6	Heterogeneous membrane
	MC	Anion memb	0.42	8–10	0.9	6	Heterogeneous membrane
	HC	Anion memb	0.42	10–13	0.92	6	Heterogeneous membrane
	HC8	Anion memb	0.42	11–14	0.93	6	Heterogeneous membrane

References

Bloch, R., 1970. Hydrometallurgical separation by solvent membranes. In: Flinn, J.E. (Ed.), Membrane Science and Technology. Plenum Press, New York, p. 171.

Convington, A.K., Dickinson, T. (Eds.), 1973. Physical Chemistry of Organic Solvent System. Plenum Press, New York, p. 525.

Conway, B.E., 1952. Electrochemical Data. Elsevier Publishing Co., London.

Crank, J., 1957. Diffusion coefficients in solids. Their measurement and significance. Discuss. Faraday Soc. 23, 99–104.

Debye, P., Huckel, E., 1923. Zur Theorie der electrolyte. I. Gefrierpunktserniedrigung und verwandte erscheinungen. Phys. Z. 24, 185–206.

Donnan, F.G., 1934. Die genaue thermodynamik der membranegleichgewichte. Z. Phys. Chem. A 168 (5/6), 369–380.

Grim, E., Sollner, K., 1957. The contributions of normal and abnormal osmosis to the osmotic effects arising across charged membranes with solutions of electrolytes. J. Gen. Physiol. 40, 887.

Hani, H., Nishihara, H., Oda, Y., 1961. Anion-exchange membrane having permselectivity between anions, JP S36–15258.

Ilschner, B., 1958. The adaptability and relationship of Nernst–Einstein and of Onsager. Z. Electrochem. 62, 989–992.

Mihara, K., Misumi, T., Miyauchi, H., Ishida, Y., 1970. Anion-exchange membrane having excellent specific permeability between anions, JP S45–19980.

Mihara, K., Misumi, T., Miyauchi, H., Ishida, Y., 1972. Production of a cation-exchange membrane having excellent specific permeability between cations, JP S47–3081.

Mizutani, Y., Nishimura, M., 1970. Studies on ion-exchange membranes. XXXII. Heterogeneity in ion-exchange membranes. J. Appl. Polym. Sci. 14, 1847–1856.

Mizutani, Y., Yamane, R., Sata, T., 1971a. Electrodialysis for transporting selectively smaller charged cations, JP S46–23607.

Mizutani, Y., Yamane, R., Sata, T., Izuo, T., 1971b. Permselectivity treatment of a cation-exchange membrane, JP S46–42083.

Nernst, W., 1888. Zur kintik der losung befindlichen korper. Z. Phys. Chem. 2, 613–637.

Nernst, W., 1889. Die elektromotorische wirksamkeit der ionen. Z. Phys. Chem. 4, 129–181.

Planck, M., 1890. Ueber die potentialdifferenz zwischen zwei verdunnten losungen binarer electrolyte. Ann. Der. Phys. Chem. 40, 561–576.

Schlogel, R., 1955. Zur Theorie der anomalen osmose. Z. Phys. Chem. (Frankfurt) 3, 73–102.

Seno, M., Tanaka, Y., 1984. Ion exchange membrane experimental method. In: Nakagaki, M. (Ed.), Membrane Science Experimental Method. Kitami Shobo Co., Tokyo, pp. 191–207.

Tanaka, Y., 1974. Effect of treatment conditions by a reagent on low permeability of cation exchange membranes for bivalent ions. J. Electrochm. Jpn. 42, 192–198.

Tanaka, Y., Ehara, R., Itoi, S., Goto, T., 2003. Ion-exchange membrane electrodialytic salt production using brine discharged from a reverse osmosis seawater desalination plant. J. Membr. Sci. 222, 71–86.

Tanaka, Y., Seno, M., 1981. Treatment of ion exchange membranes to decrease divalent ion permeability. J.Membr. Sci. 8, 115–127.

Tanaka, Y., 2007. Ion Exchange Membrane: Fundamentals and Applications. In: Membrane Science and Technology Series, vol. 12. Elsevier, Amsterdam.

Tasaka, M., Kondo, Y., Nagasawa, M., 1969. Anomalous osmosis through charged membranes. J. Phys. Chem. 73, 3181.

Yamabe, T., Seno, M., 1964. Ion Exchange Resin Membrane. Gihodo Co., Tokyo, p. 139.

Yamamoto, H., 1968. Mobility of counter ions in an ion exchange membrane. Bull. Soc. Sea Water Sci. Jpn. 22, 323–326.

3

Theory of Teorell, Meyer, and Sievers (TMS Theory)

3.1 Introduction

Teorell (Teorell, 1935) and Meyer and Sievers (Meyer and Sievers, 1936) discussed membrane phenomena in an aqueous electrolyte solution and revealed the mechanism of fundamental characteristics such as the membrane potential, diffusion coefficient, electric conductivity, transport number, etc. In this theory (TMS theory), the ionic mobility and activity coefficient are assumed to be constant. Strictly speaking, these assumptions do not hold in an ion exchange membrane because these parameters are influenced by the electric charge density. Furthermore, the theory is not concerned with the ohmic potential gradient as a driving force operating in electrodialysis. This assumption corresponds to zero current density or electric current interruption in the electrodialysis system. Despite these restrictions, the theory is generally applicable to an understanding of the mechanisms of transport phenomena (Hanai, 1981). The TMS theory is based on the Donnan equilibrium theory (Donnan, 1934) and the Nernst–Planck equation (Planck, 1890). Accordingly, the results introduced from the TMS theory are fundamentally equivalent to those introduced in Sections 2.2, 2.3 and 2.4. However, the concrete expressions of both results are not always the same.

3.2 Membrane Potential

A cation exchange membrane is assumed to be immersed in a solution dissolving monovalent cations and monovalent anions, as shown in Figure 3.1. Ions are transferred by the diffusion from the high concentration side to the low concentration side across the membrane. The mobility of ions in the membrane is so small that the Donnan equilibrium theory is approximately applicable at the interface between the membrane and the solution. The ion flux in the membrane is expressed by the Nernst–Planck equation and the electric neutrality is held in the membrane. In the TMS theory, the membrane potential $\Delta\psi$ generated between the high concentration side (C_1) and the low concentration side (C_2) is written by summing up the Donnan potential $\Delta\psi_{\mathrm{Don}}$ and the diffusion potential $\Delta\psi_{\mathrm{Diff}}$ as follows:

$$\Delta\psi = \left(-\Delta\psi_{\mathrm{Don1}} + \Delta\psi_{\mathrm{Don2}}\right) + \Delta\psi_{\mathrm{Diff}}$$

Ion Exchange Membranes. http://dx.doi.org/10.1016/B978-0-444-63319-4.00003-1

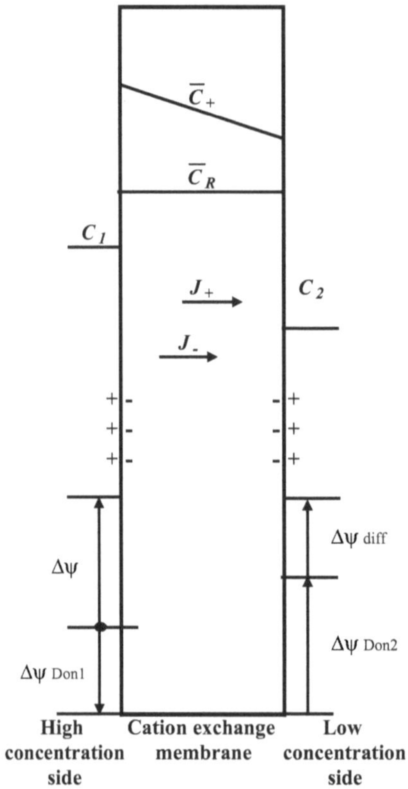

FIGURE 3.1 Ionic concentration and electric potential in an ion exchange membrane. *Tanaka, 2007, p.60, Fig. 4.1.*

$$= -\frac{RT}{F}\ln\frac{C_2}{C_1}\frac{\sqrt{\overline{C}_R^2+4C_1^2}+\overline{C}_R}{\sqrt{\overline{C}_R^2+4C_2^2}+\overline{C}_R} - \frac{RT}{F}\overline{u}\ln\frac{\sqrt{\overline{C}_R^2+4C_2^2}+\overline{u}\overline{C}_R}{\sqrt{\overline{C}_R^2+4C_1^2}+\overline{u}\overline{C}_R} \tag{3.1}$$

where

$$\overline{u} = \frac{\overline{u}_+ - \overline{u}_-}{\overline{u}_+ + \overline{u}_-} \tag{3.2}$$

$\overline{C}_R$ is the concentration of ion exchange groups in the membrane. $\overline{u}_+$ and $\overline{u}_-$ are the mobility of cations and anions in the membrane, respectively. In Eqn (3.1), ionic activity coefficients in the membrane phase and in the solution phase and the ionic distribution coefficient between both phases are assumed to be 1.

Figure 3.2 shows the relationship between $\overline{C}_R/C_1$ and $\Delta\psi$, calculated by substituting $C_2/C_1 = 0.1$ into Eqn (3.1) and putting in $\overline{u}$ as a parameter. Extrapolating $\overline{C}_R/C_1$ to the infinite and zero in Eqn (3.1) leads to:

$$\lim_{\overline{C}_R/C_1 \to \infty} \Delta\psi = -\frac{RT}{F}\ln\frac{C_2}{C_1} \tag{3.3}$$

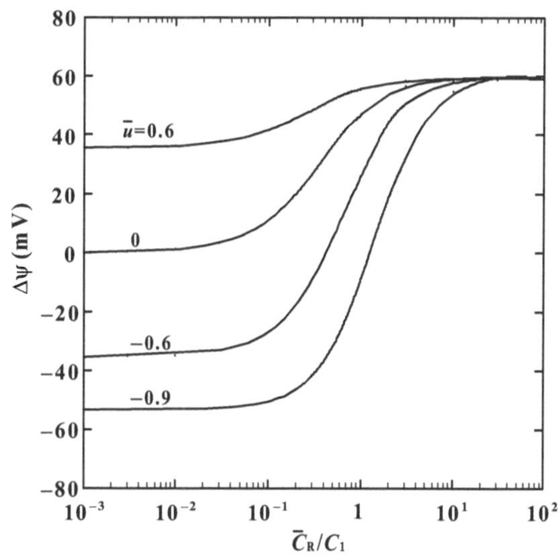

FIGURE 3.2 Ion exchange group concentration versus membrane potential. $C_2/C_1 = 0.125$. *Hanai, 1981, p. 241, Fig. 6.20.*

$$\lim_{\overline{C}_R/C_1 \to 0} \Delta\psi = -\frac{RT}{F}\overline{u}\ln\frac{C_2}{C_1} \tag{3.4}$$

Equations (3.3) and (3.4) show that the membrane potential is governed by the Donnan potential (the first term in Eqn (3.1)) at $\overline{C}_R/C_1 \to \infty$, and is governed by the diffusion potential (the second term in Eqn (3.1)) at $\overline{C}_R/C_1 \to 0$.

Substituting the transport number of cations and anions defined by Eqn (3.5) into Eqn (3.4), the Nernst equation Eqn (3.6) (Eqn (2.31)) (Nernst, 1888, 1889) is introduced.

$$t_+ = \overline{u}_+/(\overline{u}_+ + \overline{u}_-) \tag{3.5}$$

$$t_- = \overline{u}_-/(\overline{u}_+ + \overline{u}_-)$$

$$E = -(t_+ - t_-)\frac{RT}{F}\ln\frac{C_2}{C_1} \tag{3.6}$$

t_+ and t_- correspond to the transport number in an uncharged ion exchange membrane (cf. Figure. 3.5). These discussions mean that the Nernst equation shows the diffusion potential and does not include the influence of the Donnan potential.

3.3 Diffusion Coefficient

Fluxes of cations J_+ and anions J_- across the membrane in Figure 3.1 are:

$$J_+ = J_- = -\frac{\overline{u}_+\overline{u}_- RT}{(\overline{u}_+ + \overline{u}_-)d}\left\{\left(\sqrt{\overline{C}_R^2 + 4C_2^2} - \sqrt{\overline{C}_R^2 + 4C_1^2}\right) - \overline{u}\ln\frac{\sqrt{\overline{C}_R^2 + 4C_2^2} + \overline{u}\overline{C}_R}{\sqrt{\overline{C}_R^2 + 4C_1^2} + \overline{u}\overline{C}_R}\right\} \tag{3.7}$$

where d is the thickness of the membrane.

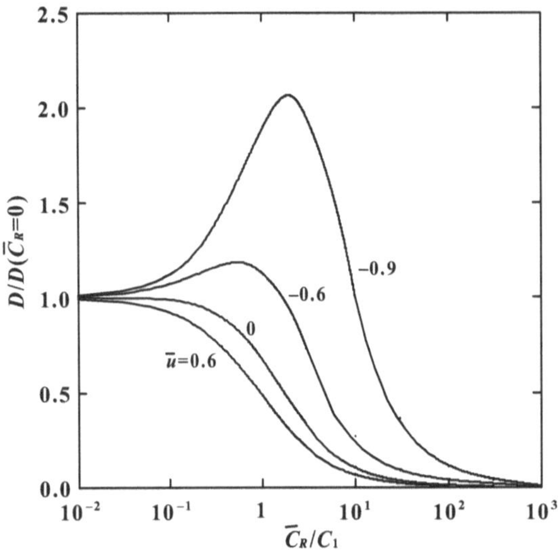

FIGURE 3.3 Ion exchange group concentration versus diffusion constant ratio. $C_2/C_1 = 0.125$. *Hanai, 1981, p. 242, Fig. 6.21.*

Ion flux is:

$$J_+ = J_- = -D(C_2 - C_1)/d \tag{3.8}$$

From Eqns (3.7) and (3.8), diffusion constant D is introduced as:

$$D = \frac{\overline{u}_+ \overline{u}_- RT}{(\overline{u}_+ + \overline{u}_-)\,(C_2 - C_1)} \left(\sqrt{\overline{C}_R^2 + 4C_2^2} - \sqrt{\overline{C}_R^2 + 4C_1^2} - \overline{u}\,\overline{C} \ln \frac{\sqrt{\overline{C}_R^2 + 4C_2^2} + \overline{u}\overline{C}_R}{\sqrt{\overline{C}_R^2 + 4C_1^2} + \overline{u}\overline{C}_R} \right) \tag{3.9}$$

Using $D(\overline{C}_R = 0)$ obtained by putting $\overline{C}_R = 0$ into Eqn (3.9), $D/D(\overline{C}_R = 0)$ is indicated as:

$$\frac{D}{D(\overline{C}_R = 0)} = \frac{1}{2(C_1 - C_2)} \left(\sqrt{\overline{C}_R^2 + 4C_1^2} - \sqrt{\overline{C}_R^2 + 4C_2^2} + \overline{u}\,\overline{C}_R \ln \frac{\sqrt{\overline{C}_R^2 + 4C_2^2} + \overline{u}\overline{C}_R}{\sqrt{\overline{C}_R^2 + 4C_1^2} + \overline{u}\overline{C}_R} \right) \tag{3.10}$$

Putting $C_2/C_1 = 0.1$ into Eqn (3.10) and plotting $D/D(\overline{C}_R = 0)$ against $\overline{C}_R/C_1$ creates Figure 3.3, which indicates that:

1. For $\overline{u} \geqq 0$ $(\overline{u}_+ \geqq \overline{u}_-)$; $D/D(\overline{C}_R = 0)$ decreases with the increase in $\overline{C}_R$ and reaches 0.
2. For $\overline{u} < 0$ $(\overline{u}_- > \overline{u}_+)$; $D/D(\overline{C}_R = 0)$ increases with $\overline{C}_R$ and passes through the maximum and reaches 0.

$$\lim_{\overline{C}_R/C_1 \to \infty} \frac{D}{D(\overline{C}_R = 0)} = 0 \tag{3.11}$$

The above phenomena are because:

1. The diffusion velocity of counter-ions (cations) is reduced by the co-ions (anions).
2. The co-ion concentration is decreased owing to the Donnan exclusion (when co-ion concentration becomes 0, counter-ions are not transferred).

The maximum of $D/D(\overline{C}_R = 0)$ is caused by the acceleration of counter-ions because of the greater mobility of co-ions: $\overline{u}_- > \overline{u}_+ (\overline{u} < 0$ in Eqn (3.2)).

3.4 Electric Conductivity

When a low electric current is passed across the membrane in a salt solution of concentration C, the electric conductivity κ of the membrane is introduced as:

$$\kappa = F^2 \frac{\overline{u}_+ + \overline{u}_-}{2} \left(\sqrt{\overline{C}_R^2 + 4C^2} + \overline{u}\,\overline{C}_R \right) \tag{3.12}$$

From Eqn (3.12), $\kappa/\kappa(\overline{C}_R = 0)$ is:

$$\frac{\kappa}{\kappa(\overline{C}_R = 0)} = \left(\frac{1}{2C}\right) \left(\sqrt{\overline{C}_R^2 + 4C^2} + \overline{u}\,\overline{C}_R \right) \tag{3.13}$$

$\kappa/\kappa(\overline{C}_R = 0)$ is plotted against $\overline{C}_R/C$, setting u as a parameter, and is shown in Figure 3.4, which is comparable to Figures 2.17 and 2.18.

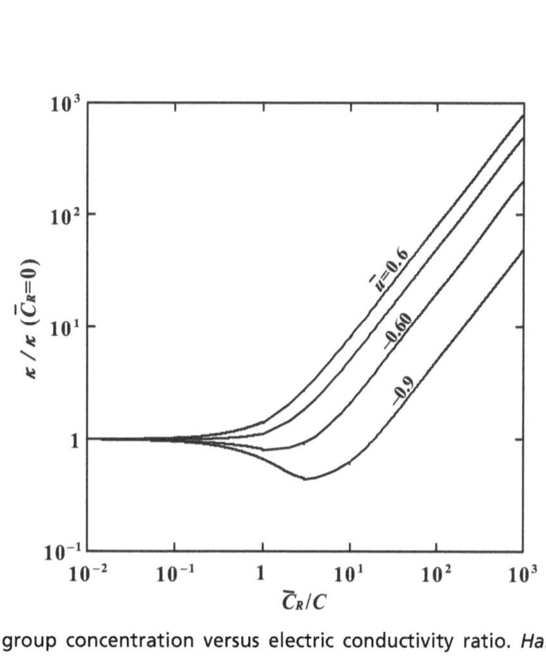

FIGURE 3.4 Ion exchange group concentration versus electric conductivity ratio. *Hanai, 1981, p. 243, Fig. 6.22.*

3.5 Transport Number

The transport number of a cation exchange membrane $\overline{t}_+$ immersed in an electrolyte solution of concentration C is:

$$\overline{t}_+ = \left(\frac{\overline{u}_+}{\overline{u}_+ + \overline{u}_-}\right)\left(\frac{\sqrt{\overline{C}_R^2 + 4C^2} + \overline{C}_R}{\sqrt{\overline{C}_R^2 + 4C^2} + \overline{u}\overline{C}_R}\right) \tag{3.14}$$

The relationship between $\overline{C}_R/C$ and $\overline{t}_+$ is computed using Eqn (3.14), putting $\overline{u}$ and t_+ (transport number in an uncharged cation exchange membrane) as parameters and is shown in Figure 3.5, which is comparable to Figures 2.3 and 2.4. $\overline{t}_+$ converges at $\overline{C}_R/C \rightarrow \infty$ and $\overline{C}_R/C \rightarrow 0$ as:

$$\lim_{\overline{C}_R/C \rightarrow \infty} \overline{t}_+ = 1 \tag{3.15}$$

$$\lim_{\overline{C}_R/C \rightarrow 0} \overline{t}_+ = \frac{\overline{u}+1}{2} = t_+ \tag{3.16}$$

Equation (3.16) means that the transport number in an ion exchange membrane becomes that of an uncharged membrane at $\overline{C}_R = 0$.

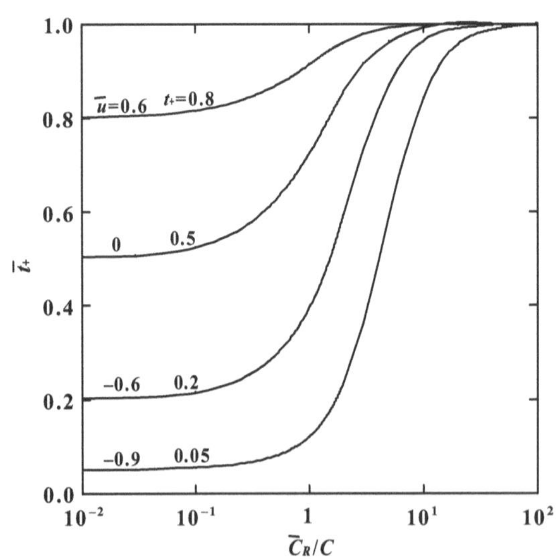

FIGURE 3.5 Ion exchange group concentration versus transport number. *Hanai, 1981, p. 244, Fig. 6.23.*

References

Donnan, F.G., 1934. Die genaue thermodynamik der membrangleichgewichte. Z. Phys. Chem. A168, 369–380.

Hanai, T., 1981. Membrane and Ions Theory and Calculation of Mass Transport. Kagaku Dozin Co., Kyoto, Japan, pp. 221–244.

Meyer, K.H., Sievers, J.F., 1936a. La permeabilite des membranes I, Theorie de la permeabilite ionique. Helv. Chim. Acta 19, 649–664.

Meyer, K.H., Sievers, J.F., 1936b. La permeabilite des membranes II, Essis avec des membranes selectives artifielles. Helv. Chim. Acta 19, 665–677.

Nernst, W., 1888. Zur Kinetik der in losung befindlichen korper. Z. Phys. Chem. 2, 613–679.

Nernst, W., 1889. Die elektromotorische wirksamkeit der ionen. Z. Phys. Chem. 4, 129–181.

Plank, M., 1890. Ueber die potentialdifferenz zwischen zwei verdunnten losungen binarer electrolyte. Ann. d. Phys. u. Chem. 40, 561–576.

Teorell, T., 1935. An attempt to formulate a quantitative theory of membrane permeability. Proc. Soc. Exptl. Biol. 33, 282–385.

Tanaka, Y., 2007. Ion Exchange Membranes: Fundamentals and Applications. In: Membrane Technology Science Series 12. Elsevier, Amsterdam.

4 Irreversible Thermodynamics

4.1 Introduction

Mass transport across the membrane must be discussed fundamentally on the basis of the thermodynamics because the thermodynamics describes the rule of energy inevitably generating in the mass transport. The classical thermodynamics have introduced a lot of information on the basis of the first and the second law. However, we have to notice that the classical thermodynamics discusses only reversible phenomena and is not concerned with the transport rate; in other words, it does not include the parameter of "time." Many phenomena (including membrane phenomena) generating in the natural world proceed with "time" and they are far away from reversible states, so the reversible thermodynamics is ineffective to discuss their mechanism.

The irreversible thermodynamics came to succeed in discussing the transport rate by introducing the concept of "time" in its system (Prigogine, 1961; de Groot et al., 1962; Fitts, 1962; Glansdorff and Prigogine, 1971). Basic theory of the irreversible thermodynamics is established on the assumption of "micro irreversibility." (Onsager, 1931; Casimir, 1945). This assumption means that the thermodynamics functions hold in a fine irreversible space, or more strictly speaking, the functions hold in the circumstance of being more close to equilibrium states. The actual electrodialysis process is considerably dynamic and it is hardly conceivable to proceed close to the equilibrium state. Accordingly, it might be reasonable to assume that the irreversible thermodynamics is merely applicable approximately in the electrodialysis process. The irreversible thermodynamics is not general theory applicable to analyze every reaction in the natural world, but it is believed to be applicable in the circumstances, being apart to some extent from equilibrium states (Dunlop, 1957; Dunlop and Gosting, 1959). Furthermore, it should be noticed that the applicable limit of the irreversible thermodynamics must be determined through experiments. In this chapter, the irreversible thermodynamics is discussed keeping the electrodialysis in mind, and the mechanism of saline water electrodialysis is conceptually explained in Sections 4.5 and 4.6.

4.2 Phenomenological Equation and Phenomenological Coefficient

Mass transport in a solution is often discussed on the basis of the Nernst–Planck equation (Planck, 1890):

$$J_i = -C_i u_i \frac{d\psi}{dx} \tag{4.1}$$

Ion Exchange Membranes. http://dx.doi.org/10.1016/B978-0-444-63319-4.00004-3

in which J_i is the ionic flux, C_i is the ionic concentration, u_i is the ionic mobility, and ψ is the electrochemical potential.

Equation (4.1) is expressed as:

$$J = LX \tag{4.2}$$

where X is the driving force and L is the proportionality constant.

When two kinds of ions are transported with solvent in a solution, their fluxes influence each other, because the fluxes and driving forces are not independent and coupled together. These interactions are not taken into account in the Nernst–Planck equation. However, the interactions are presented in the irreversible thermodynamics as shown in the following phenomenological equation (Kedem and Katchalsky, 1961):

$$J_i = L_{i1}X_1 + L_{i2}X_2 + \ldots + L_{in}X_n = \sum_{k=1}^{n} L_{ik}X_k \quad (i = 1, 2, \ldots n) \tag{4.3}$$

in which, X_i is the force and J_i is the flux. L_{ik} is the phenomenological coefficients, in which i and k are components. The following Onsager's reciprocal theorem (1931) suggests that the matrix composed of the phenomenological coefficients is symmetrical:

$$L_{ik} = L_{ki} \quad (i, k = 1, 2, \ldots n) \tag{4.4}$$

In Fig. 4.1 illustrating a two-cell apparatus, electrolyte solutions are partitioned by an ion exchange membrane. In this system, the following assumptions are set up:

1. The ion exchange membrane forms a continuous boundary interface between cell I and cell II.
2. The system is closed, and there is no mass transfer between the closed system and the surroundings.
3. There is no appearance and disappearance of material caused by a chemical reaction.
4. The system is isothermal.

Staverman (1952) represented the entropy production σ in this system:

$$\sigma = \frac{dS}{dt} = \sum_i J_i X_i \tag{4.5}$$

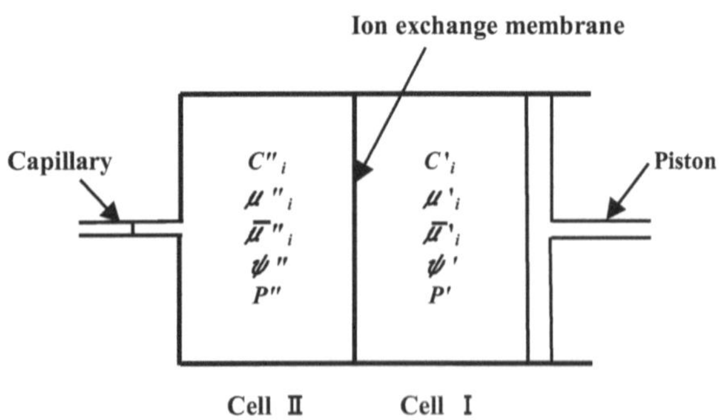

FIGURE 4.1 Two-cell membrane system in which electrolyte solutions are partitioned by an ion exchange membrane. *Tanaka (2007), p. 68, Fig. 5.1.*

in which J_i and X_i are, respectively, the force and flux of component i, and they are indicated as:

$$J_i = -\frac{dM_i^{\mathrm{I}}}{dt} = \frac{dM_i^{\mathrm{II}}}{dt} \tag{4.6}$$

$$X_i = \frac{z_i\Delta\psi + v_i\Delta P + \Delta\mu_i}{T} = \frac{\Delta\overline{\mu}_i}{T} \tag{4.7}$$

where dM_i^{I} and dM_i^{II} are the quantity changes of component i during dt in cells I and II, respectively. $\Delta\overline{\mu}_i(=\overline{\mu}_i' - \overline{\mu}_i'')$ and $\Delta\mu_i(=\mu_i' - \mu_i'')$ are an electrochemical potential difference and a chemical potential difference between the cells. ΔP $(=P' - P'')$ and $\Delta\psi$ $(=\psi' - \psi'')$ are the pressure difference and potential difference between the cells. z_i and v_i are the electric charge number and partial molar volume of component i.

From Eqns (4.5)–(4.7), the following dissipation function is introduced:

$$T\sigma = \sum_i J_i\Delta\overline{\mu}_i = \sum_i J_i(z_i\Delta\psi + v_i\Delta P + \Delta\mu_i) = I\Delta\psi + J\Delta P + \sum_i J_i\Delta\mu_i \tag{4.8}$$

where I, J, and J_i are an electric current, volume flow of a solution, and a mass flux of component i, respectively.

The entropy change dS in a closed system is expressed by the following Carnot–Clausius equation, based on the second law of thermodynamics:

$$dS = d_eS + d_iS = \frac{dQ}{T} + d_iS \tag{4.9}$$

where dQ is quantity of heat supplied to the system from the surroundings. d_eS is the entropy change in the system caused by the heat supply indicating the reversible process. d_iS corresponds to the irreversible change arising in the system and it is equivalent to the entropy production σ (Eqn (4.5)). Td_iS is equivalent to the dissipation function $T\sigma$ (Eqn (4.8)). Accordingly, Eqn (4.8) expresses that the entropy production in the system is generated by the electric current due to electromotive force, the mechanical flow due to pressure difference, and the diffusion due to chemical potential difference.

Based on the irreversible thermodynamics, the phenomenological equation in the ion exchange membrane system is introduced starting from Eqn (4.8) as follows (Kedem and Katchalsky, 1963):

$$I = L_E\Delta\psi + L_{EP}\Delta P + \sum_i L_{Ei}\Delta\mu_i \tag{4.10}$$

$$J = L_{EP}\Delta\psi + L_P\Delta P + \sum_i L_{Pi}\Delta\mu_i \tag{4.11}$$

$$J_i = L_{iE}\Delta\psi + L_{iP}\Delta P + \sum_k L_{ik}\Delta\mu_k \tag{4.12}$$

Furthermore, the phenomenological equation is also introduced as follows from Eqn (4.8):

$$J_i = \sum_k L_{ik}X_k = \sum_k (z_kL_{ik}\Delta\psi + v_kL_{ik}\Delta P + L_{ik}\Delta\mu_{ik}) \tag{4.13}$$

where z_k and v_k are the electric charge number and partial volume of component k.

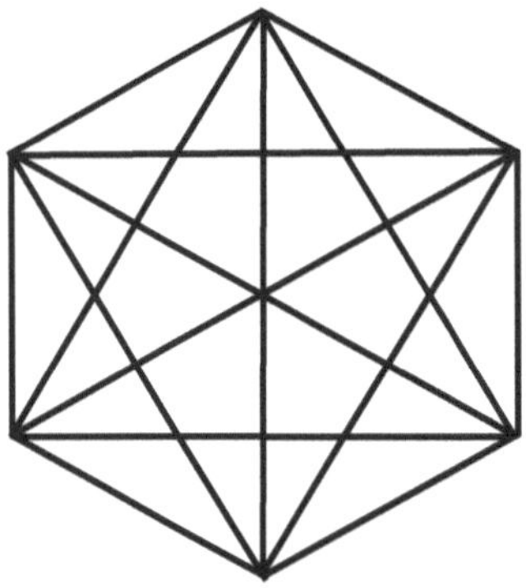

FIGURE 4.2 Fluxes, driving forces, and membrane phenomena. *Kimizuka (1988), p. 162, Fig. 8.2.*

The phenomenological equation introduced above is equivalent to Eqn (4.3) and the Onsager's reciprocal theorem (Eqn (4.4)) is held between each phenomenological coefficient. In the above ion exchange membrane system including cations i_+ and anions i_-, the coupling of ionic fluxes presented in the phenomenological equation (Eqns (4.10)–(4.13)) is depicted in the model of Fig. 4.2 (Kimizuka, 1988).

4.3 Membrane Phenomena

The membrane phenomena are expressed based on the phenomenological equations described in Section 4.2 (Staverman, 1952; Sakai and Seiyama, 1956; Yamabe and Senō, 1964).

4.3.1 Permeability

1. Electric conductivity (permeability)

Putting $\Delta P = 0$ and $\Delta\mu = 0$ in Eqns (4.10) and (4.13):

$$L_E = \left(\frac{I}{\Delta\psi}\right)_{\Delta P=0,\Delta\mu=0} = \frac{1}{E}\sum z_i J_i = \sum_i \sum_k L_{i,k} z_i z_k \tag{4.14}$$

2. Hydraulic (mechanical) permeability

Putting $\Delta\psi = 0$ and $\Delta\mu_k = 0$ in Eqns (4.11) and (4.13):

$$L_P = \left(\frac{J}{\Delta P}\right)_{\Delta\psi=0,\Delta\mu=0} = \frac{1}{P}\sum v_i J_i = \sum_i \sum_k L_{i,k} v_i v_k \tag{4.15}$$

3. Electric transport number of component *i*

Putting $\Delta P = 0$ and $\Delta\mu = 0$ in Eqns (4.10) and (4.13):

$$t_i = \left(\frac{z_i J_i}{\sum z_i J_i}\right)_{\Delta P=0,\Delta\mu=0} = \frac{z_i J_i}{I} = \frac{z_i \sum_k L_{ik} z_k}{L_E} = \frac{z_i \sum_k L_{ik} z_k}{\sum_i \sum_k L_{ik} z_i z_k} \tag{4.16}$$

4. Mechanical (hydraulic) transport number of component *i*

Putting $\Delta\psi = 0$ and $\Delta\mu = 0$ in Eqns (4.11) and (4.13):

$$\tau_i = \left(\frac{v_i J_i}{\sum v_i J_i}\right)_{\Delta\psi=0,\Delta\mu=0} = \frac{v_i J_i}{J} = \frac{v_i \sum_k L_{i,k} v_k}{L_P} = \frac{v_i \sum_k L_{ik} v_k}{\sum_i \sum_k L_{ik} v_i v_k} \tag{4.17}$$

4.3.2 Electrokinetic Phenomena

When the concentrations on both sides of the membrane are the same ($\Delta\mu_i = 0$), the following equations are introduced from Eqns (4.10) and (4.11):

$$I = L_E \Delta\psi + L_{EP} \Delta P \tag{4.18}$$

$$J = L_{PE} \Delta\psi + L_P \Delta P \tag{4.19}$$

From Eqns (4.18) and (4.19), the electrokinetic phenomena are expressed as follows:

1. Electroosmosis

$$\left(\frac{J}{I}\right)_{\Delta P-0} = \frac{L_{PE}}{L_E} \tag{4.20}$$

2. Streaming potential

$$\left(\frac{\Delta\psi}{\Delta P}\right)_{I=0} = -\frac{L_{EP}}{L_E} \tag{4.21}$$

Equation (4.20) shows a solution flux caused by an electric current, and Eqn (4.21) shows a potential difference caused by a pressure difference. Both equations are different at a first glance, but these equations show the same phenomenon from the opposite view points. This suggestion is demonstrated by the Onsager's reciprocal theorem, $L_{EP} = L_{PE}$, accordingly:

$$\left(\frac{\Delta\psi}{\Delta P}\right)_{I=0} = -\left(\frac{J}{I}\right)_{\Delta P=0} = -\frac{\sum_i \sum_k L_{ik} z_i v_k}{\sum_i \sum_k L_{ik} z_i z_k} \tag{4.22}$$

3. Electroosmotic pressure

$$\left(\frac{\Delta P}{\Delta \psi}\right)_{J=0} = -\frac{L_{PE}}{L_P} \tag{4.23}$$

4. Streaming current

$$\left(\frac{I}{J}\right)_{\Delta\psi=0} = \frac{L_{EP}}{L_P} \tag{4.24}$$

Equation (4.23) shows a pressure difference caused by a potential difference, and Eqn (4.24) shows an electric current caused by a solution flux. Both equations show the same phenomenon because of the similar reason described above, accordingly:

$$\left(\frac{\Delta P}{\Delta \psi}\right)_{J=0} = -\left(\frac{I}{J}\right)_{\Delta\psi=0} = -\frac{\sum_i \sum_k L_{ik} z_i v_k}{\sum_i \sum_k L_{ik} v_i v_k} \tag{4.25}$$

4.3.3 Diffusion Potential

Putting $I = 0$ and $\Delta P = 0$ in Eqn (4.10) and expressing the diffusion potential as $\Delta\psi = \Delta\psi_{Diff}$:

$$I = L_E \Delta\psi_{Diff} + \sum_i L_{Ei} \Delta\mu_i = 0 \tag{4.26}$$

Accordingly, taking account of Eqn (4.16):

$$\Delta\psi_{Diff} = -\frac{1}{L_E}\sum_i L_{Ei}\Delta\mu_i = -\frac{1}{L_E}\sum_i \sum_k L_{ik} z_k \Delta\mu_i = -\sum_i \left(\frac{t_i}{z_i}\right)\Delta\mu_i \tag{4.27}$$

Equation (4.27) is a general expression of the Nernst equation (Nernst, 1888, 1889).

4.4 Reflection Coefficient

One assumes in Fig. 4.1 that an electrolyte solution is placed in cells I and II ($C_i' = C_i'' = C_i$, $\Delta C_i = 0$) and the solution in cell I is pressurized through a piston, then the water and solute i in cell I are transferred toward cell II. The fluxes of water and solutes through the membrane in this system are given by J_{water} and J_{solute}. If the membrane in this system does not permeate the solute i at all ($J_{solute} = 0$) and permeates only water, the solute concentration in cell II becomes $C_i'' - \Delta C_i'' = (C_i''V'')(V'' + \Delta V'')$ in which $\Delta C_i''$ and $\Delta V''$ are respectively solute concentration decrease and volume increase in cell II. However, if

J_{water} is equivalent to J_{solute}, C_i'' does not change ($\Delta C_i'' = 0$). The perm-selectivity of solutes against water in this system is defined by the following reflection coefficient σ_i in the reversible thermodynamics (Schultz, 1980).

$$\sigma_i = 1 - \frac{C_i^{filtrate}}{C_i^{filtrand}} \tag{4.28}$$

Phenomenological meaning of σ_i is understandable from the perm-selectivity of solutes against water T_{water}^{solute} defined by the following equation, which is commonly applied concept in electrodialysis:

$$T_{water}^{solute} = \frac{J_{solute}/J_{water}}{C'_{solute}/C'_{water}} \tag{4.29}$$

The relationship between J_{water}, J_{solute}, σ_i and T_{water}^{solute} are given as:

$$\begin{aligned} &J_{solute} = J_{water}, && \sigma_i = 0, && T_{water}^{solute} = 1 \\ &J_{solute} < J_{water}, && 0 < \sigma_i < 1, && 1 > T_{water}^{solute} > 0 \\ &J_{solute} = 0 && \sigma_i = 1, && T_{water}^{solute} = 0 \end{aligned} \tag{4.30}$$

The osmotic pressure π developed between cell I and cell II in Fig. 4.1 is presented by the following Van't Hoff equation:

$$\pi = RT(C_i' - C_i'') = RT\Delta C_i \tag{4.31}$$

Equation (4.31) presents π at $J_{solute} = 0$, $\sigma_i = 1$ and $T_{water}^{solute} = 0$ in Eqn (4.30). Staverman (1951) defined the following effective osmotic pressure π_{eff}:

$$\pi_{eff} = \sigma_i \pi = \sigma_i RT\Delta C_i \tag{4.32}$$

4.5 Electrodialysis Phenomena

Hause (1974) discussed an electrokinetic phenomenon based on the approaches of Kedem and Katchalsky (1963). Referring to this suggestion, Schultz (1980) discussed ionic transport in an electrodialysis process as follows:

Figure 4.3 illustrates a two-cell electrodialysis system consisting of cells I and II filled with solutions dissolving monovalent cations and anions and partitioned by a cation exchange membrane. The concentrations of electrolyte i in cell I are adjusted to C'_+ for cations and C'_- for anions, and those in cell II are adjusted to C''_+ for cations and C''_- for anions. Reversible electrodes are placed in both cells, and an electric current is passed across the membrane through the electrodes setting electric potentials on both sides of the membrane to ψ' and ψ'' and keeping pressures in both cells to P' and P'' with a piston.

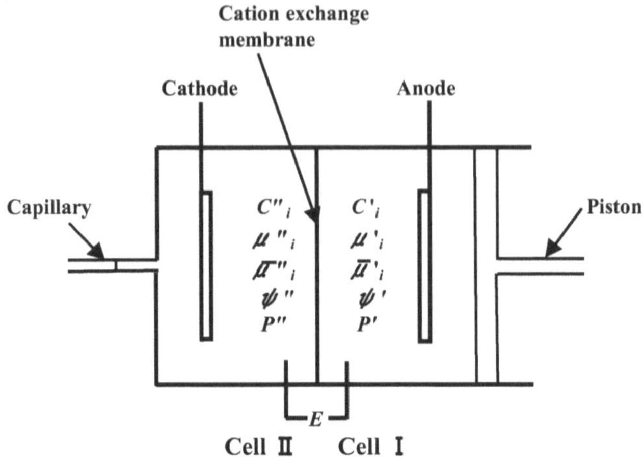

FIGURE 4.3 Two-cell electrodialysis system in which electrolyte solutions are partitioned by a cation exchange membrane. *Tanaka (2007), p. 74, Fig. 5.3.*

The differences of electric potential $\Delta\psi$, electrolyte concentration ΔC_i, pressure ΔP, and electrolyte chemical potential $\Delta\mu_i$ between both cells are defined as:

$$\begin{aligned}\Delta\phi &= \phi'' - \phi' \\ \Delta C_i &= C_i'' - C_i' \\ \Delta P &= P'' - P' \\ \Delta\mu_i &= \mu_i'' - \mu_i'\end{aligned} \tag{4.33}$$

Here, the following dissipation function is set up in the tertiary system consisting of three components, three flows, and three driving forces:

$$\frac{Td_iS}{dt} = J_+\Delta\overline{\mu}_+ + J_-\Delta\overline{\mu}_- + J_W\Delta\mu_W \tag{4.34}$$

$$\Delta\overline{\mu}_+ = \overline{v}_+\Delta P + \frac{RT\Delta C_+}{C_+^*} + F\Delta\psi \tag{4.35}$$

$$\Delta\overline{\mu}_- = \overline{v}_-\Delta P + \frac{RT\Delta C_-}{C_-^*} - F\Delta C \tag{4.36}$$

in which, the subscripts +, – and W denote monovalent cations, monovalent anions, and water (solvent), respectively. $\Delta\overline{\mu}_+$ and $\Delta\overline{\mu}_-$ are electrochemical potential difference of + ions and – ions. $\overline{v}_+$ and $\overline{v}_-$ are the partial molar volume of these ions. C_+^* and C_-^* are the logarithmic mean concentration Eqn (4.50) of these ions.

Electroneutrality of the system is:

$$\begin{aligned}C_+' &= C_-' = C_i' \\ C_+'' &= C_-'' = C_i''\end{aligned} \tag{4.37}$$

The chemical potential difference of electrolytes i, $\Delta\mu_i$ is introduced from Eqns (4.34)–(4.37) as:

$$\Delta\mu_i = \Delta\mu_+ + \Delta\mu_- = \overline{v}_i\Delta P + \frac{2RT\Delta C_i}{C_i^*} \tag{4.38}$$

$$\overline{v}_i = \overline{v}_+ + \overline{v}_- \tag{4.39}$$

The relationship between electromotive force E (volt) and potential difference $\Delta\psi$ is given by the following equation (Katchalsky and Curran, 1965):

$$E = -\frac{\Delta\mu_-}{F} = -\frac{RT}{F}\Delta \ln C_- + \Delta\psi \tag{4.40}$$

The transport number of a cation exchange membrane t_+ incorporated with the apparatus (Fig. 4.3) is given by the following equation because the flux of cations J_+ is equivalent to the flux of electrolytes J_i.

$$t_+ = \frac{FJ_+}{I} = \frac{FJ_i}{I} \tag{4.41}$$

$$I = F(J_+ - J_-) \tag{4.42}$$

Substituting Eqns (4.38) and (4.40) into Eqn (4.34) introduces:

$$\frac{Td_iS}{dt} = J_W\mu_W + J_i\Delta\mu_i + IE \tag{4.43}$$

The following phenomenological equations are introduced from Eqn (4.43):

$$J_W = L_{WW}\Delta\mu_W + L_{Wi}\Delta\mu_i + L_{WI}E \tag{4.44}$$

$$J_i = L_{iW}\Delta\mu_W + L_{ii}\Delta\mu_i + L_{iI}E \tag{4.45}$$

$$I = L_{IW}\Delta\mu_W + L_{Ii}\Delta\mu_i + L_{II}E \tag{4.46}$$

Equations (4.44)–(4.46) include nine coefficients (L). According to the Onsager's reciprocal theorem, $L_{Wi} = L_{iW}$, $L_{WI} = L_{IW}$, and $L_{iI} = L_{Ii}$, so the coefficients reduce to six. Equation (4.44) indicates that J_W consists of the hydrodynamic term ($L_{WW}\Delta\mu_W$), the osmotic pressure term ($L_{Wi}\Delta\mu_i$), and the electroosmosis term ($L_{WI}E$). Equation (4.45) indicates that J_i consists of ultra-filtration term ($L_{iW}\Delta\mu_W$), diffusion term ($L_{ii}\Delta\mu_i$), and electrophoresis term ($L_{iI}E$). Equation (4.46) indicates that I consists of streaming current term ($L_{IW}\Delta\mu_W$), diffusion stream term ($L_{Ii}\Delta\mu_i$), and electromotive force term ($L_{II}E$).

The phenomenological equation described above involves mol flow J_W (Eqn (4.44), the unit, mol/cm^{2}s) which is difficult to measure. So, this parameter is changed to volume flow J_V (cm^3/cm^{2}s) applying $\Delta\mu_W$ (Eqn (4.47)) and $\Delta\mu_i$ (Eqn (4.48)) to Eqn (4.43).

$$\Delta\mu_W = \overline{v}_W(P'' - P') - \overline{v}_W RT(C_i'' - C_i') = \overline{v}_W(\Delta P - \pi) \tag{4.47}$$

$$\Delta\mu_i = \overline{v}_i\Delta P + \frac{RT\Delta C_i}{C_i^*} = \overline{v}_i\Delta P + \frac{\pi}{C_i^*} \tag{4.48}$$

where:

$$\text{Osmotic pressure,}\quad \pi = RT\Delta C_i \tag{4.49}$$

$$\text{Logarithmic mean concentration}\quad C_i^* = \frac{\Delta C_i}{\ln(C_i''/C_i')} \tag{4.50}$$

Consequently, Eqn (4.43) is converted to:

$$\frac{Td_iS}{dt} = J_V(\Delta P - RT\Delta C_i) + \frac{J_i(1 - C_i^*\overline{v})RT\Delta C_i}{C_i^*} + IE \tag{4.51}$$

In a diluted solution:

$$C_i^* << 1 \tag{4.52}$$

Taking account of Eqn (4.52) in Eqn (4.51), the dissipation function is introduced as:

$$\frac{Td_iS}{dt} = J_V(\Delta P - RT\Delta C_i) + \frac{J_iRT\Delta C_i}{C_i^*} + IE \tag{4.53}$$

Furthermore, the following phenomenological equation is introduced from Eqn (4.53):

$$J_i = L_P(1-\sigma_i)C_i^*(\Delta P - RT\Delta C_i) + \omega_i RT\Delta C_i + \frac{t_+ I}{F} \tag{4.54}$$

$$I = \beta G(\Delta P - RT\Delta C_i) + \frac{Gt_+RT\Delta C_i}{FC_i^*} + GE \tag{4.55}$$

$$J_V = L_P\Delta P - \sigma_i L_P RT\Delta C_i + \beta I \tag{4.56}$$

The coefficients in Eqns (4.54)–(4.56) are termed as follows: L_P, the hydraulic conductivity; σ_i, the reflection coefficient; β, the electroosmotic permeability; ω_i, the solute permeability; t_+, the transport number; and G, the electric conductance.

4.6 Separation of Salt and Water by Electrodialysis

4.6.1 Two-Cell Electrodialysis System

A salt solution dissolving monovalent cations and monovalent anions is assumed to be partitioned by a cation exchange membrane in a two-cell electrodialysis system (Fig. 4.3). The concentration of cations+ and anions− in a one-side compartment (cell I) and that in the another side compartment (cell II) are assumed to be maintained to $C_+^{\mathrm{I}} = C_-^{\mathrm{I}} = C_S^{\mathrm{I}}$ and $C_+^{\mathrm{II}} = C_-^{\mathrm{II}} = C_S^{\mathrm{II}}$, respectively. Concentration difference of electrolytes in cell I and II is expressed as $\Delta C_+ = \Delta C_- = C_S^{\mathrm{I}} - C_S^{\mathrm{II}} = \Delta C_S$. Applying an electric current density i through reversible electrodes immersed in both compartment across the membrane, cations in cell I are assumed to be transported toward cell II. In an electrodialysis system, the pressure difference between cell I and cell II is negligible ($P^{\mathrm{I}} - P^{\mathrm{II}} = \Delta P = 0$). In this system, the dissipation function is introduced from Eqn (4.53) putting $\Delta P = 0$ and replacing J_i, J_V, C_i^*, ΔC_i, and I with $J_{S,K}$, $J_{V,K}$, C_S^*, ΔC_S, and i, respectively, as follows:

$$\frac{Td_iS}{dt} = \left(\frac{J_{S,K}}{C_S^*} - J_{V,K}\right)RT\Delta C_S + iE \tag{4.57}$$

$J_{S,K}$ and $J_{V,K}$ are introduced similarly from Eqns (4.54) and (4.56) as follows:

$$J_{S,K} = t_{+K}\frac{i}{F} + \{\omega_K - L_{P,K}\sigma_K(1-\sigma_K)C_S^*\}RT\Delta C_S \tag{4.58}$$

$$J_{V,K} = \beta_K i - \sigma_K L_{P,K} RT\Delta C_S \tag{4.59}$$

4.6.2 Three-Cell Electrodialysis System

The principle of separating salt from water based on the theory mentioned above is introduced using a three-cell electrodialysis system (Fig. 4.4). The system consists of a

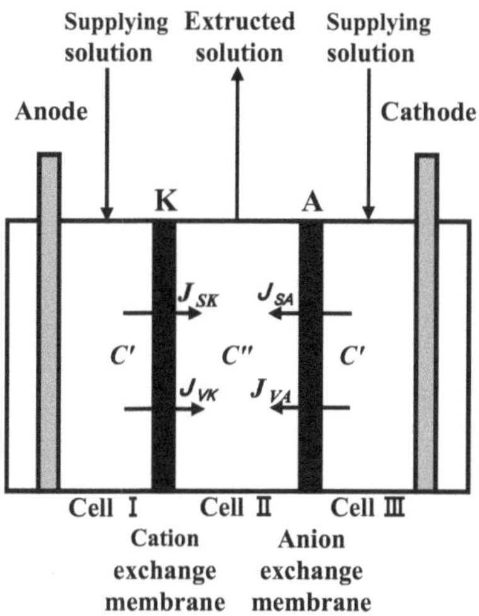

FIGURE 4.4 Three-cell electrodialysis system in which electrolyte solutions are partitioned by a cation and an anion exchange membrane, and are transferred across the membranes with an applied electric current.

central cell (cell II) and electrode cell (cell I and cell III) placed on both outsides of cell II. A cation exchange membrane (K) is placed between cell I and cell II, and an anion exchange membrane (A) is placed between cell II and cell III. Supplying a salt solution into cells I and III, constant current density i is applied and a salt solution being collected in cell II is taken out from the system until the electrolyte concentration in cell II reaches steady constant. Salt accumulation $J_{S,K}+J_{S,A}$ and solution accumulation $J_{V,K}+J_{V,A}$ in cell II in the steady state is given by the following equation introduced from Eqns (4.58) and (4.59) (Tanaka, 2006):

$$J_{S,K}+J_{S,A}=(t_K+t_A-1)\frac{i}{F}-RT\left[(\omega_K+\omega_A)-\{L_{P,K}\sigma_K(1-\sigma_K)+L_{P,A}\sigma_A(1-\sigma_A)\}C_S^*\right]\Delta C \quad (4.60)$$

$$J_{V,K}+J_{V,A}=(\beta_K+\beta_A)i+RT(\sigma_K L_{P,K}+\sigma_A L_{P,A})\Delta C \quad (4.61)$$

where $\Delta C=C''-C'=-\Delta C_S$, $t_+^K+t_-^A=1$, $t_+^K=t_K$ and $t_-^A=t_A$. Subscripts K and A denote a cation exchange membrane and an anion exchange membrane, respectively. Superscripts, single quotation marks, and double quotation marks denote a desalting side (cells I and III) and concentrating side (cell II). C_S^* (logarithmic mean concentration; Eqn (4. 50)) is given by the following equation:

$$C_S^*=C^*=\frac{C''-C'}{\ln(C''/C')} \quad (4.62)$$

References

Casimir, H.B.G., 1945. On Onsager's principle of microscopic reversibility, 1931 Rev. Mod. Phys. 17, 343–350.

Dunlop, P.J., 1957. A study of interacting flows in diffusion of the system raffinose-KCl-H_2O at 25 °C. J. Phys. Chem. 61, 994–1000.

Dunlop, P.J., Gosting, L.J., 1959. Use of diffusion and thermodynamics data to test the Onsager reciprocal relation for isothermal diffusion in the system NaCl-KCl-H2O at 25 °C. J. Phys. Chem. 63, 86–93.

Fitts, D.D., 1962. Nonequilibrium Thermodynamics. Mc-Graw Hill, New York.

Glansdorff, P., Prigogine, I., 1971. Thermodynamic Theory, Structure, Stability and Fluctuations. Wily-Interscience, London.

de Groot, S.R., Mazur, P., 1962. Nonequilibrium Thermodynamics. North-Holland, New York.

House, C.R., 1974. Water Transport in Cells and Tissues. Edward Arnold, London.

Katchalsky, A., Curran, P.F., 1965. Nonequilibrium Thermodynamics in Biophysics. Harvard Univ. Press, Cambridge.

Kedem, O., Katchalsky, A., 1961. A physical interaction of the phenomenological coefficients of membrane permeability. J. Gen. Phys. 45, 143–179.

Kedem, O., Katchalsky, A., 1963. Permeability of composite membranes. Part 1. Electric current, volume flow and flow of solute through membranes. Trans. Faraday Soc. 59, 1918–1930.

Kimizuka, H., 1988. Membrane Transport of Ions. Kyoritsu Shuppan Co., Tokyo.

Nernst, W., 1888. Zur Kinetik der in losung befindlichen korper. Z. Phys. Chem. 2, 613–679.

Nernst, W., 1889. Die elektromotorische wirksamkeit der ionnen. Z. Phys. Chem. 4, 129–181.

Onsager, L., 1931. Reciprocal relations in irreversible process. Phys. Rev. 37, 405–426, 38, 2265–2279.

Planck, M., 1890. Ueber die potentialdifferenz awischen zwei verdunnten losungen binarer electrolyte. Ann. Phys. Chem. 40, 561–576.

Prigogine, P., 1961. Introduction to Thermodynamics of Irreversible Process, second ed. John Wiley, New York.

Sakai, W., Seiyama, T., 1956. Electrochemical studies on ion exchanger. Application of non-equilibrium thermodynamics to ionic membrane system. J. Electr. Chem. Jpn. 24, 274–279.

Schultz, S.G., 1980. Basic Principles of Membrane Transport. Cambridge University Press, Cambridge.

Staverman, A.J., 1951. The theory of measurement of osmotic pressure. Rec. Trav. Chim. 70, 344–352.

Staverman, A.J., 1952. Non-equilibrium thermodynamics of membrane process. Trans. Faraday Soc. 48, 176–185.

Tanaka, Y., 2006. Irreversible thermodynamics and overall mass transport in ion-exchange membrane electrodialysis. J. Membr. Sci. 281, 517–531.

Tanaka, Y., 2007. Ion Exchange Membranes: Fundamentals and Applications, Membrane Science and Technology Series 12. Elsevier, Amsterdam.

Yamabe, T., Senō, M., 1964. Ion Exchange Resin Membrane. Giho Do Co., Tokyo.

5

Overall Mass Transport

5.1 Introduction

Irreversible thermodynamics described in Chapter 4 is a classical and basic theory of transport phenomena across an ion exchange membrane. However, it does not elucidate the mechanism of phenomena occurring in a practical-scale electrodialyzer because the phenomena generated in the electrodialyzer include too many parameters not considered in irreversible thermodynamics. The overall mass transport equation is the empirical equations established from seawater electrodialysis experiments. The performance of an electrodialyzer can be discussed using the computer simulation program (Applications Chapters 13–15) in which the overall mass transport equation functions to express mass transport across the membrane pair. The basis of the equation is supported by irreversible thermodynamics.

5.2 Electrodialysis of Saline Water and Overall Mass Transport Equation

An electrodialysis system (Figure 5.1) was set up with an electrodialyzer ED and commercially available ion exchange membranes (Aciplex CK-2/CA-3, K-172/A-172, Asahi Chemical Co., Selemion CSV-2/AST, CMR/ASR, Asahi Grass Co., and Neosepta CL-25T/AVS-4T, CIMS/ACS-3, Tokuyama Co.). Diagonal net spacers are integrated into desalting cells and concentrating cells. Saline water (the constituent is the same as that of seawater) was supplied into a desalting solution tank DT and then supplied to desalting cells De. An electric current was passed through Ti/Pt-stainless electrodes.

After salt concentration of solutions in concentrating cells Con reaches constant, the fluxes of ions J_S and a solution J_V across an ion exchange membrane pair and electrolyte concentration in desalting cells $C'(=(C'_{in}+C'_{out})/2)$ and in concentrating cells C'' were measured. The salt concentration C was measured by the normality (equiv./dm^3) of main negatively charged ions (Cl^- and SO_4^{2-} ions) dissolving in a solution, applying the definition $C = C_{Cl} + C_{SO_4}$. Changing current density i incrementally, electrodialysis was repeated. Plotting J_S/i and J_V/i against $\Delta C/i$ produces linear lines, as exemplified in Figure 5.2. These lines are expressed by the following overall mass transport equation (Tanaka, 2006):

$$J_S = C''\ J_V = \lambda i - \mu(C'' - C') = \lambda i - \mu\Delta C = \eta i/F \tag{5.1}$$

$$J_V = \phi i + \rho(C'' - C') = \phi i + \rho\Delta C \tag{5.2}$$

Ion Exchange Membranes. http://dx.doi.org/10.1016/B978-0-444-63319-4.00005-5

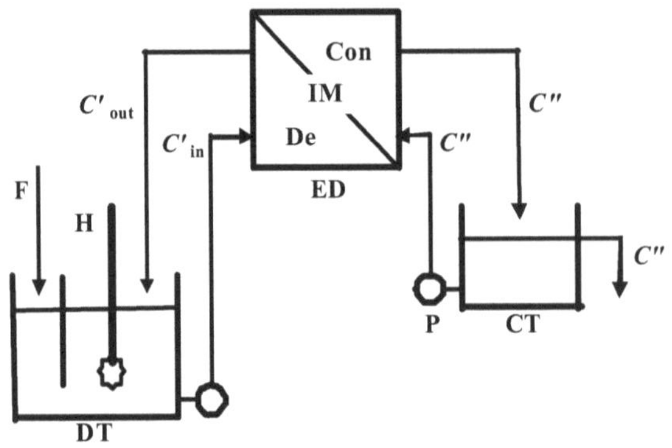

FIGURE 5.1 Electrodialysis system. *Tanaka (2006), p. 520, Fig. 1.*

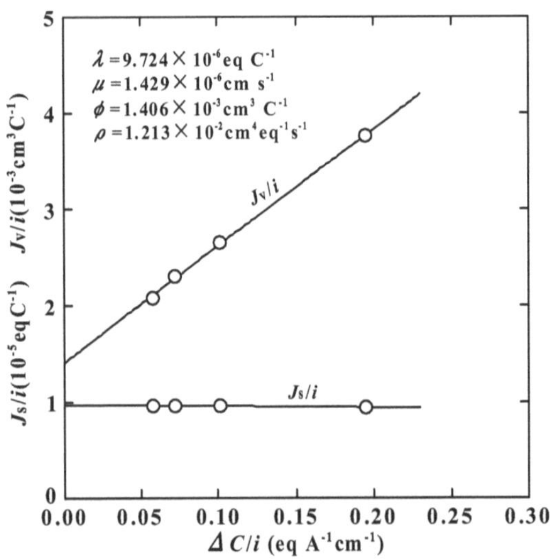

FIGURE 5.2 J_S/i versus $\Delta C/i$ plot and J_V/i versus $\Delta C/i$ plot. Aciplex K172/A172, 25 °C, Seawater. *Tanaka (2006), p. 521, Fig. 3.*

where λ(eq C^{-1}) is the overall transport number, μ(cm s^{-1}) is the overall solute permeability, ϕ (cm^3 C^{-1}) is the overall electro-osmotic permeability, and ρ(cm^4 eq^{-1} s^{-1}) is the overall volume osmotic permeability. η is current efficiency and F (96,485 As/equiv.) is the Faraday constant. The term "overall" means that the coefficients express the contributions of a cation and an anion exchange membrane (membrane pair). It also means that the coefficients express the contributions of many types of ions dissolved in a salt solution. The terms λi and $\mu(C''-C')$ in Eqn (5.1) stand for he migration and diffusion of ions, respectively. The terms ϕi and $\rho(C''-C')$ in Eqn (5.2) correspond to electro-osmosis and volume osmosis, respectively. The direction from a desalting cell to concentrating cell is taken as positive for each term. However, the second term (diffusion term) in Eqn (5.1) is in the direction from a concentrating cell to a desalting cell, so this direction is taken as positive and minus signs are supplied.

5.3 Ion Exchange Membrane Pair Characteristics

5.3.1 ρ versus λ, μ, and ϕ

Based on the saline water electrodialysis experiment described in Section 5.2, λ, μ, and ϕ are plotted against ρ. The plots are shown in Figure 5.3 and by the following empirical equations:

$$\lambda = 9.208 \times 10^{-6} + 1.914 \times 10^{-5}\rho \quad (5.3)$$

$$\mu = 2.005 \times 10^{-4}\rho \quad (5.4)$$

$$\phi = 3.768 \times 10^{-3}\rho^{0.2} - 1.019 \times 10^{-2}\rho \quad (5.5)$$

Figure 5.3 shows the characteristics of commercially available ion exchange membranes. The fluctuations seen in the plotting are caused by the experimental relative errors with the fluctuations of membrane properties. Despite these problems, Figure 5.3 is widely available to discuss the phenomena in the electrodialysis process because it represents the membrane characteristics in extremely general forms. Eqns (5.3)–(5.5) are included in the overall mass transport equation.

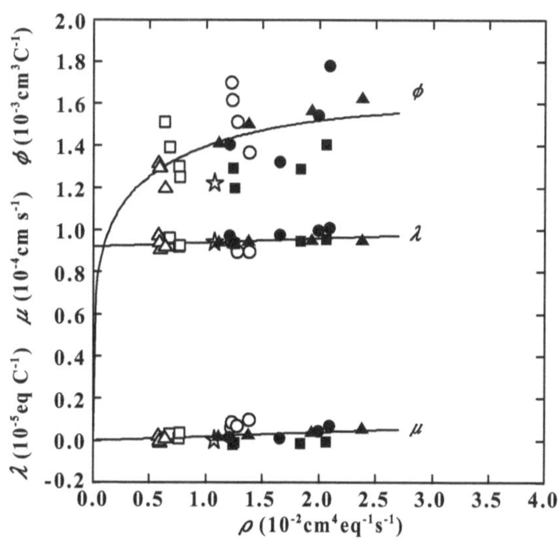

FIGURE 5.3 ρ versus λ, μ, and ϕ. Aciplex: ○ CK-2/CA-3, ● K-172/A-172, Selemion: △ CSV-2/AST, ▲ CMR/ASR, Neosepta: □ CL-2.5T/ASV-4T, ■ CIMS/ACS-3. *Tanaka (2006), p. 521, Fig. 4.*

5.3.2 ρ versus Electric Resistance

The overall volume osmotic permeability ρ of commercially available ion exchange membrane pairs was measured as described in Section 5.2. Then, alternating current electric resistance of an Na^+ type cation exchange membrane $r_{alter,K}$ and a Cl^- type anion exchange membrane $r_{alter,A}$ was measured in a 25 °C 0.5 M NaCl solution using an alternating current bridge at 1000 cycles. Membrane pair electric resistance r_{alter} was calculated by summing up $r_{alter,K}$ and $r_{alter,A}$: $r_{alter} = r_{alter,K} + r_{alter,A}$. Figure 5.4 shows ρ versus r_{alter}, which is represented by the following equations (Tanaka, 2011a).

$$r_{alter} = 1.2323\rho^{-1/3} \tag{5.6}$$

Fluctuations seen in Figure 5.4 are also attributed to the experimental relative errors and fluctuations of membrane properties. Eqn (5.6) is included in the overall mass transport equation.

5.3.3 ρ versus Temperature

Figure 5.5 shows ρ versus temperature T (°C) which is presented by the following equation (Tanaka, 2011b)

$$\rho = 3.421 \times 10^{-3} + 3.333 \times 10^{-4} T \tag{5.7}$$

These equations mean that ρ represents all of the overall membrane characteristics and that λ, μ, ϕ, and r_{alter} are determined by setting T. ρ is termed the "leading parameter." Eqn (5.7) is included in the overall mass transport equation.

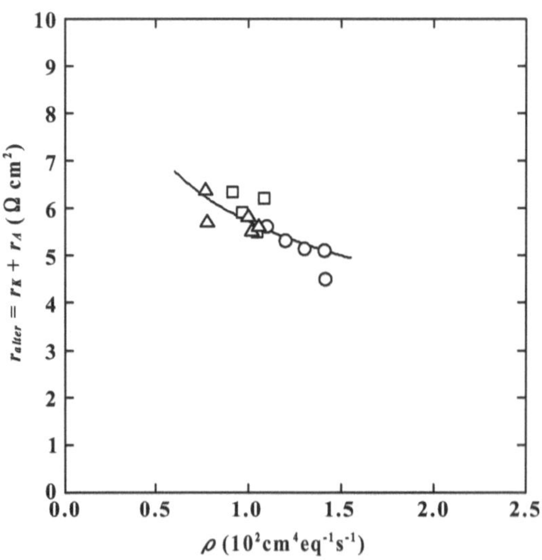

FIGURE 5.4 ρ versus r_{alter}. ○ Aciplex K102/A102, △ Selemion CMV/ASV, □ Neosepta CH-45T/AFS-4T.

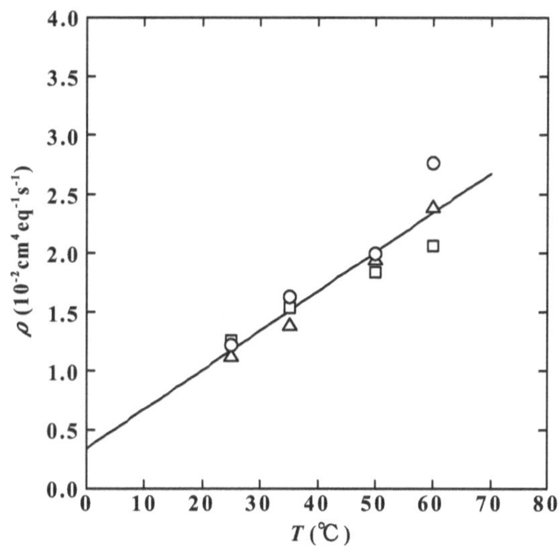

FIGURE 5.5 T versus ρ. ○ Aciplex K172/A172, △ Selemion CMR/ASR, □ Neosepta CIMS/ACS3. *Tanaka (2011b), p. 7496, Fig. 2.*

5.3.4 Influences of Electrolyte Concentration and Feeding Solution Velocity on Membrane Pair Characteristics

ρ is plotted against electrolyte concentration C' of a feeding solution and is shown in Figure 5.6 (Tanaka, 2006). ρ is barely influenced by C'. Figure 5.7 shows J_S/i and J_V/i plotted against $\Delta C/i$, maintaining the linear velocity in desalting cells to 2 and 4 cm/s (Tanaka, 1999). The figure indicates that λ, μ, ϕ, and ρ are independent of the linear velocity. Thus, the membrane pair characteristics are assumed to be little influenced by the concentration polarization.

5.4 Overall Mass Transport and Electrodialysis of Seawater

Seawater is assumed to be supplied to desalting cells in an electrodialyzer to concentrate seawater. Passing a constant electric current between electrodes, concentrated seawater accumulated in concentrating cells (salt concentration; C″) is collected and extracted to the outside of the electrodialyzer. The electrodialyzer is operated until C'' reaches a steady value. Electrodialyzer specifications and operating conditions are as follows:

Thickness of a desalting cell $a = 0.05$ cm.

Flow-pass width in a desalting cell $b = 20$ cm.

Flow-pass length in a desalting cell $l = 25$ cm.

Seawater concentration at the inlets of desalting cell $C'_{in} = 0.6$ eq./dm^3 (34,722 mg/dm^3).

Linear velocity in a desalting cell $u' = 5$ cm/s.

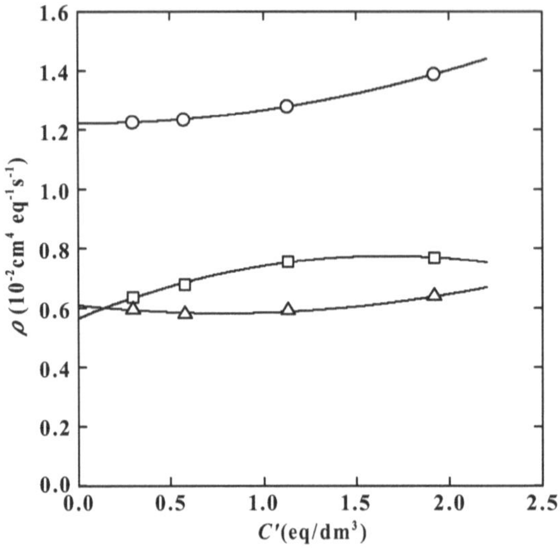

FIGURE 5.6 C' versus ρ. ○ Aciplex CK-2/CA-3, △ Selemion CSV-2/AST, □ Neosepta CL-25T/AVS-4T. *Tanaka (2006), p. 523, Fig. 9.*

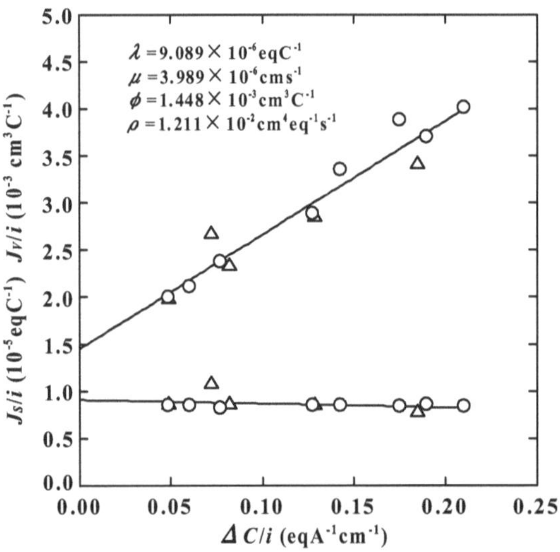

FIGURE 5.7 Influence of solution velocity in desalting cells on the membrane pair characteristics. Aciplex K-102/A/102, 25 °C, ○4 cm/s △ 2 cm/s. *Tanaka (1999), p. 281, Fig. 7.*

Temperature $T = 25$ °C.

The computation is carried out using Eqns (5.1)–(5.5), (5.7) and Eqns (5.8)–(5.12).

Salt concentration at the outlet of a desalting cell C'_{out}:

$$C'_{out} = C'_{in} - \left(\frac{\eta l}{aF}\right)\left(\frac{i}{u'}\right) \tag{5.8}$$

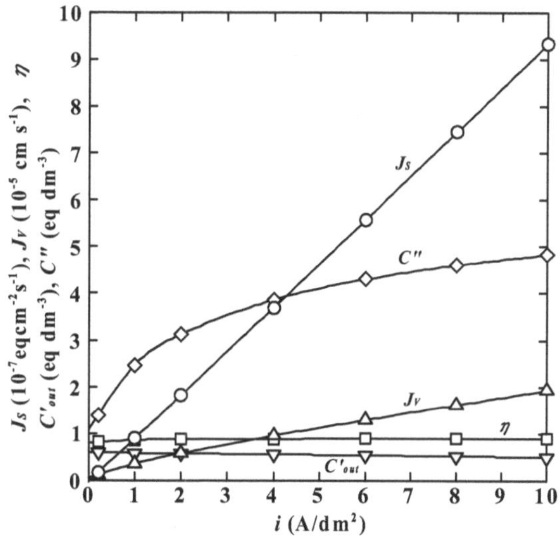

FIGURE 5.8 *i* versus J_S, J_V, η, C'_{out}, and C''.

Salt concentration in a desalting cell C':

$$C' = \frac{C'_{in} + C'_{out}}{2} \tag{5.9}$$

Salt concentration in a concentrating cell C'':

$$C'' = \frac{1}{2\rho}\left(\sqrt{A^2 + 4\rho B} - A\right) \tag{5.10}$$

$$A = \phi i + \mu - \rho C' \tag{5.11}$$

$$B = \lambda i + \mu C' \tag{5.12}$$

Figure 5.8 gives the computed relationship between current density i and J_S, J_V, η, C'_{out}, and C''.

5.5 Overall Mass Transport Equation and Phenomenological Equation

The phenomenological equation introduced in irreversible thermodynamics (Eqns (4.60) and (4.61)) describes the mass transport in a one-pair membrane system. The overall mass transport equation, Eqns (5.1) and (5.2) also represents the mass transport across one pair of membranes. Both equations are substantially identical, so the following equations hold:

$$J_{S,K} + J_{S,A} = J_S \tag{5.13}$$

$$J_{V,K} + J_{V,A} = J_V \tag{5.14}$$

From Eqns (5.13) and (5.14), the overall membrane characteristics are represented by the functions of phenomenological coefficients as follows:

$$\lambda = \frac{t_K + t_A - 1}{F} = \left(\frac{J_S}{i}\right)_{\Delta C=0} \tag{5.15}$$

$$\begin{aligned} \mu &= RT\left[(\omega_K + \omega_A) - \left\{L_{P,K}\sigma_K(1-\sigma_K) + L_{P,A}\sigma_A(1-\sigma_A)\right\}C_S^*\right] \\ &= RT\left[(\omega_K + \omega_A) - \sigma(1-\sigma)(L_{P,K} + L_{P,A})C_S^*\right] = \left(\frac{J_S}{\Delta C}\right)_{i=0} \end{aligned} \tag{5.16}$$

$$\phi = \beta_K + \beta_A = \left(\frac{J_V}{i}\right)_{\Delta C=0} \tag{5.17}$$

$$\begin{aligned} \rho &= RT(\sigma_K L_{P,K} + \sigma_A L_{P,A}) \\ &= RT\sigma\left(L_{P,K} + L_{P,A}\right) = \left(\frac{J_V}{\Delta C}\right)_{i=0} \end{aligned} \tag{5.18}$$

where σ_K and σ_A are reflection coefficients of a cation and an anion exchange membrane, respectively. σ expresses the contributions of a cation and an anion exchange membrane, so it is termed the overall reflection coefficient and is introduced as follows.

From Eqn (5.16):

$$\sigma(1-\sigma) = \frac{L_{P,K}\sigma_K(1-\sigma_K) + L_{P,A}\sigma_A(1-\sigma_A)}{L_{P,K} + L_{P,A}} \tag{5.19}$$

From Eqn (5.18):

$$\sigma = \frac{\sigma_K L_{P,K} + \sigma_A L_{P,A}}{L_{P,K} + L_{P,A}} \tag{5.20}$$

From Eqns (5.19) and (5.20):

$$\sigma = \frac{L_{P,K}\sigma_K^2 + L_{P,A}\sigma_A^2}{L_{P,K}\sigma_K + L_{P,A}\sigma_A} \tag{5.21}$$

when $\sigma_K = 0$, $\sigma = \sigma_A$; $\sigma_A = 0$, $\sigma = \sigma_K$ and $\sigma_K = \sigma_A = 1$, $\sigma = 1$

Yamauchi and Tanaka (1993) measured the reflection coefficient of a cation exchange membrane σ_K (Neosepta CL-25T) by means of pressure-driven dialysis of a KCl solution, and concluded σ_K to be unity. This phenomenon is estimated to be result from the Donnan exclusion of co-ions (Cl^- ions) and the dense structure in the cation exchange membrane. Thus, counter-ions (K^+ ions) are not transported across the cation exchange membrane in the pressure-driven dialysis. From this experiment, the reflection coefficient σ of an ion exchange membrane is generally assumed to be 1. Thus, the membrane pair characteristics defined in the phenomenological equations are introduced from Eqns (5.15)–(5.18) putting $\sigma_K = \sigma_A = 1$ as follows:

$$t_K + t_A = \lambda F + 1 \tag{5.22}$$

$$\omega_K + \omega_A = \frac{\mu}{RT} \tag{5.23}$$

$$\beta_K + \beta_A = \phi \tag{5.24}$$

$$L_{P,K} + L_{P,A} = \frac{\rho}{RT} \tag{5.25}$$

5.6 Reflection Coefficient of Ion Exchange Membranes

Equations (5.15)–(5.18) show that σ_K and σ_A appear in μ and ρ and disappear in λ and ϕ. These phenomena and Eqns (5.1) and (5.2) suggest that σ_K and σ_A do not exert an influence on electric current-driven mass transport, i.e., migration and electro-osmosis. Reflection coefficient is a parameter indicative of the permselectivity between solutes and solvent passing through the membrane. It is essentially the concept developed in pressure-driven dialysis. This phenomenon is conceivable in electrodialysis, but the concept developed in pressure dialysis is not directly applicable to the phenomenon in electrodialysis. To understand the behavior of the reflection coefficient in an electrodialysis process, it is necessary to establish a concept of zero current density. In other words, it is reasonable to image electric current interruption (switch off) for a moment in the electrodialysis process operating under a constant electric current, and assume the disappearance of the migration and electro-osmosis in this moment. Here, one assumes further that solute diffusion and volume osmosis remain as they are just after the electric current interruption (Tanaka, 2006, 2012).

To discuss the behavior of the reflection coefficient in an electrodialysis process, the volume flow J_V and exchange flow J_D in an ion exchange membrane pair are given by the following equations introduced by Schloegel (1964):

$$J_V = (L_{P,K} + L_{P,A})\Delta P + (L_{PD,K} + L_{PD,A})RT\Delta C = \left(\frac{J_S}{C_S^*}\right) + \left(\frac{J_W}{C_W^*}\right) \tag{5.26}$$

$$J_D = (L_{DP,K} + L_{DP,A})\Delta P + (L_{D,K} + L_{D,A})RT\Delta C = \left(\frac{J_S}{C_S^*}\right) - \left(\frac{J_W}{C_W^*}\right) \tag{5.27}$$

in which ΔP and ΔC are the pressure difference and concentration difference across the membrane, respectively. L_P is the hydraulic conductivity and L_D is the exchange flow parameter. L_{PD} is the osmotic volume flow coefficient and L_{DP} is the ultrafiltration coefficient. J_W is the flux of water molecules. C_S^* and C_W^* are, respectively, logarithmic mean concentration (Eqn (4.50)) of solutes and water (solvent) between a desalting cell and a concentrating cell.

It should be noted that Eqns (5.26) and (5.27) are originally defined in the pressure-driven transport (pressure dialysis) of neutral species with no electric current. Eqn (5.26) represents the sum of the solute flux J_S and solvent flux J_W, whereas Eqn (5.27) shows the difference between J_S and J_W. It should be added that Eqn (5.2) expresses the solution flux J_V and not the solvent flux J_W.

In a pressure-driven process, putting $\Delta C = 0$ into Eqns (5.26) and (5.27) introduces the following equation applicable to pressure dialysis:

$$(J_V)_{\Delta C=0} = (L_{P,K} + L_{P,A})\Delta P \tag{5.28}$$

$$(J_D)_{\Delta C=0} = (L_{DP,K} + L_{DP,A})\Delta P \tag{5.29}$$

σ_K and σ_A included in Eqns (5.16) and (5.18) are defined by Staverman (1951) and Kedem-Katchalsky (1958) as follows:

$$\sigma_K = -\left(\frac{J_{D,K}}{J_{V,K}}\right)_{\Delta C=0} = -\left(\frac{L_{DP,K}}{L_{P,K}}\right) \tag{5.30}$$

$$\sigma_A = -\left(\frac{J_{D,A}}{J_{V,A}}\right)_{\Delta C=0} = -\left(\frac{L_{DP,A}}{L_{P,A}}\right) \tag{5.31}$$

The overall reflection coefficient σ is defined using Eqns (5.28) and (5.29).

$$\sigma = -\left(\frac{J_D}{J_V}\right)_{\Delta C=0} = -\frac{L_{DP,K} + L_{DP,A}}{L_{P,K} + L_{P,A}} \tag{5.32}$$

σ given in Eqn (5.32) is the reflection coefficient defined in the pressure-driven dialysis. Here, σ is termed the "overall pressure reflection coefficient" because it is the membrane pair characteristic and reflects pressure difference ΔP-driven phenomena.

σ given in Eqns (5.30)–(5.32) defines the permselectivity between water (solvent) and solutes and it is originally established in pressure dialysis. To understand the permselectivity between water and solutes (ions) in electrodialysis, σ is fundamentally inapplicable. The permselectivity between water and ions in electrodialysis must be explained on the basis of σ^* (overall concentration reflection coefficient) described below. In the electrodialysis process, ΔP is relatively low and negligible, and just after an electric current interruption (switch off), ΔC remains as it is. Putting $\Delta P = 0$ into Eqns (5.26) and (5.27) introduces the following equations applicable to electrodialysis:

$$(J_V)_{\Delta P=0} = (L_{PD,K} + L_{PD,A})RT\Delta C \tag{5.33}$$

$$(J_D)_{\Delta P=0} = (L_{D,K} + L_{D,A})RT\Delta C \tag{5.34}$$

Another reflection coefficient σ^* is introduced from Eqns (5.33) and (5.34) just after an electric current interruption, as follows:

$$\sigma^* = -\left(\frac{J_D}{J_V}\right)_{\Delta P=0} = -\frac{L_{D,K} + L_{D,A}}{L_{PD,K} + L_{PD,A}} \tag{5.35}$$

σ^* (Eqn (5.35)) is the reflection coefficient defined in electrodialysis and is termed the "overall concentration reflection coefficient" because it reflects a concentration difference ΔC-driven phenomenon.

Koter (1993) suggested that the Onsager reciprocity (1931) is not satisfied and the relationship between L_{PD} and L_{DP} depends on the concentration difference across the membrane. One example for Nafion 417 in NaCl/H_2O at $C''/C'=2$ is $L_{DP}/L_{PD}=0.40$. Thus, the reciprocal equation is represented as follows:

$$L_{DP,K}+L_{DP,A}=k(L_{PD,K}+L_{PD,A}) \quad k\neq 1 \tag{5.36}$$

in which k is defined as the Onsager reciprocity coefficient. From Eqns (5.32), (5.35) and (5.36):

$$\sigma\sigma^{*}=k\frac{L_{D,K}+L_{D,A}}{L_{P,K}+L_{P,A}} \tag{5.37}$$

Cancelling J_W/C_W^{*} in Eqns (5.26) and (5.27):

$$J_D=2\frac{J_S}{C_S^{*}}-J_V \tag{5.38}$$

From Eqns (5.35) and (5.38):

$$\sigma^{*}=1-2\left(\frac{1}{C_S^{*}}\right)\left(\frac{J_S}{J_V}\right)_{\Delta P=0} \tag{5.39}$$

σ^{*} presented in Eqn (5.39) gives the permselectivity between ions and water molecules just after an electric current interruption. J_S and J_V are expressed as the following equations by putting $i=0$ into Eqns (5.1) and (5.2):

$$J_S=-\mu(C''-C') \tag{5.40}$$

$$J_V=\rho(C''-C') \tag{5.41}$$

Substituting Eqns (5.40) and (5.41) into Eqn (5.39):

$$\sigma^{*}=1+2\left(\frac{\mu}{\rho}\right)\left(\frac{1}{C_S^{*}}\right) \tag{5.42}$$

The experimental results of seawater electrodialysis (Section 5.2) show that σ^{*} is not influenced by current density. Thus, the observed σ^{*} is averaged for each current density and plotted against temperature T (Figure 5.9). σ^{*} is generally larger than 1; however, it is less than 1 for the Neosepta CIMS/ACS3 membrane in the figure (Tanaka, 2012).

The physical rationale of σ^{*} is validated by the following relationships in an electrodialysis system (Figure 5.10) incorporated with a cation and an anion exchange membrane and cells I, II, and III, in which electrolyte solutions are filled.

$$J_{solute}>0 \quad \sigma^{*}<1 \quad \mu<0 \tag{5.43}$$

$$J_{solute}=0 \quad \sigma^{*}=1 \quad \mu=0 \tag{5.44}$$

$$J_{solute}<0 \quad \sigma^{*}>1 \quad \mu>0 \tag{5.45}$$

In Figure 5.10, an electric current is passed through reversible electrodes and electrolyte concentration in cells I and III is assumed to be decreased to C' and that in cell II it is increased to C'' ($C'<C''$). The x axis is drawn across each ion exchange membrane. The direction of water flux J_{water} is positive ($J_{water}>0$); J_V is positive ($J_V>0$) in every situation. Solute flux across the membrane is generally negative ($J_{solute}<0$) and it moves

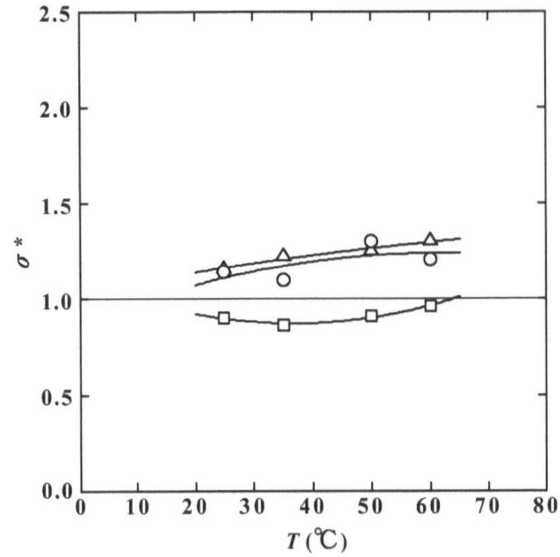

FIGURE 5.9 Relationship between temperature and overall concentration. Reflection coefficient. ○ Aciplex K172/A172 △ Selemion CMR/ASR, □ Neosepta CIMS/ACS3. *Tanaka (2012), p. 10, Fig. 13.*

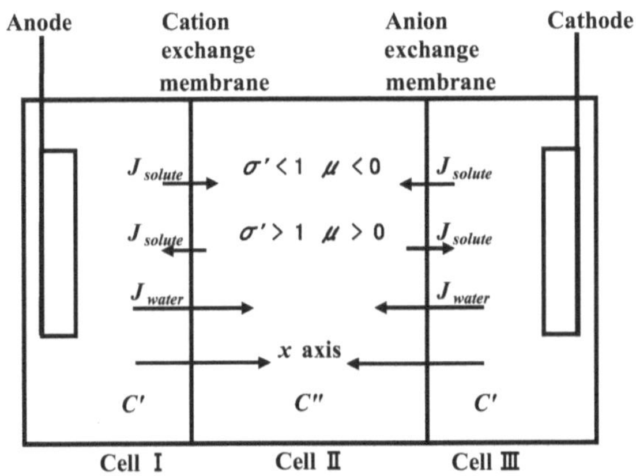

FIGURE 5.10 Electrodialysis system. *Tanaka (2012), p. 10, Fig. 15.*

from the concentrating cell (cell II) toward desalting cells (cells I and III) owing to solute diffusion just after an electric current interruption. In this situation, the concentration reflection coefficient is larger than 1 (Eqn (5.45), $\sigma^* > 1$, Figure 5.9). This phenomenon is ordinary in electrodialysis but extraordinary in pressure dialysis because the direction of the solute resulting from solute diffusion is against water flux owing to volume osmosis. In this case, the overall solute permeability becomes positive, i.e., $\mu > 0$ and $\omega_K + \omega_A > 0$. In Neosepta CIMS/ACS3 membranes, solutes move from the desalting cells (cells I and III) toward concentrating cell (cell II) ($J_{solute} > 0$) just after an electric current interruption. In this situation, the concentration reflection coefficient is less than 1 (Eqn (5.43),

$\sigma^* < 1$, Figure 5.9). This phenomenon is ordinary in pressure dialysis because the direction of the solute is the same as the water flux because of volume osmosis. The overall solute permeability in this situation becomes negative ($\mu < 0$). The negative μ indicates coupled transport between water and solute, which is validated also by negative $\omega_K + \omega_A$. However, this phenomenon is extraordinary in electrodialysis because it means uphill transport of solutes against the concentration difference ($C''-C'$).

References

Kedem, O., Katchalsky, A., 1958. Thermodynamic analysis of the permeability of biological membranes to non-electrolysis. Biochim. Biophys. Acta 27, 229–246.

Koter, S., 1993. Transport of electrolytes across cation-exchange membranes. Test of Onsager reciprocity in zero-current process. J. Membr. Sci. 78, 155–162.

Onsager, L., 1931. Reciprocal relations in irreversible process. Phys. Rev. 37, 405–426.

Schlogel, R.Z., 1964. Fortschritte der physikalischen Chemie. Band 9.

Staverman, A.J., 1951. The theory of measurement of osmotic pressure. Rec. Trav. Chim. 70, 344–352.

Tanaka, Y., 1999. Regularity in ion-exchange membrane characteristics and concentration of sea water. J. Membr. Sci. 163, 277–287.

Tanaka, Y., 2006. Irreversible thermodynamics and overall mass transport in ion-exchange membrane electrodialysis. J. Membr. Sci. 281, 517–531.

Tanaka, Y., 2011a. Ion-exchange membrane electrodialysis of saline water and its numerical analysis. Ind. Eng. Chem. Res. 50, 10765–10777.

Tanaka, Y., 2011b. Ion-exchange membrane electrodialysis for saline water desalination and its application to seawater concentration. Ind. Eng. Chem. Res. 50, 7494–7503.

Tanaka, Y., 2012. Measurement of membrane characteristics using the phenomenological equation and the overall mass transport equation in ion-exchange membrane electrodialysis of saline water. Int. J. Chem. Eng. Article ID 148147 doi:10.1155/2012/148147.

Yamauchi, A., Tanaka Yasuko, 1993. Salt transport phenomena across charged membrane driven by pressure difference. In: Paterson, R. (Ed.), Effective Membrane Process-new Perspectives. BHR Group Limited, Information Press Ltd, Oxford England, pp. 179–185.

6 Concentration Polarization

6.1 Introduction

Ion exchange membrane electrodialysis is a process for separating or extracting ionic species in a solution. In this process, a boundary layer is formed on a desalting surface of the membrane and the salt concentration in the layer is depleted owing to the concentration polarization, which increases voltage drop and energy consumption. Concentration polarization is an inevitable phenomenon that exerts an influence on the performance of an electrodialyzer. The current density at which the ion concentration on the membrane surface becomes zero as a result of the concentration polarization is the limiting current density. The concentration polarization is observed from the current–voltage relationship (Peer, 1956; Cowan and Brown, 1959; Kooistra, 1967; Spiegler, 1971; Valerdi-Prez and Ibannez-Mengual, 2001). At the limiting current density, water dissociation occurs and generates H^+ ions and OH^- ions. At the overlimiting current density, a charged region is developed on the membrane surface (Rubinstein and Shtilman, 1979; Rubinstein, 1981) and ion transport in the boundary layer is promoted by gravitational convection (Frilette, 1957; Gavish and Lifson, 1979; Zabolotsky et al., 1996) or electroconvection (Rubinstein et al., 1997; Balster et al., 2007; Nikonenko et al., 2010), which accompanies vortex flow (Rubinstein et al., 2008; Shaposhnik et al., 2008; Chang et al., 2012), voltage fluctuation (Green and Yafuso, 1968; Kolyubin et al., 1996; Krol et al., 1999), current fluctuation (Maletzki et al., 1992), and light scattering fluctuation (Lifson et al., 1978; Li et al., 1983).

6.2 Current–Voltage (*I–V*) Relationship

Concentration polarization is generally observed by current–voltage (*I–V*) curves. Figure 6.1 shows a representative *I–V* curve for a cation exchange membrane (Permutit C-10) seen in the pioneering report by Peer (1956). The curve has the S-type shape including region I, exhibiting ohmic behavior, region II, in which the current varies slowly with voltage and exhibits a plateau presenting the limiting current density, and region III, corresponding to the overlimiting current sphere in which the current increases gradually.

Spiegler (1971) discussed *I–V* curves for an anion exchange membrane in a 1:1 valent electrolyte solution applying the Nernst–Planck equation and discussed the mechanism

Ion Exchange Membranes. http://dx.doi.org/10.1016/B978-0-444-63319-4.00006-7

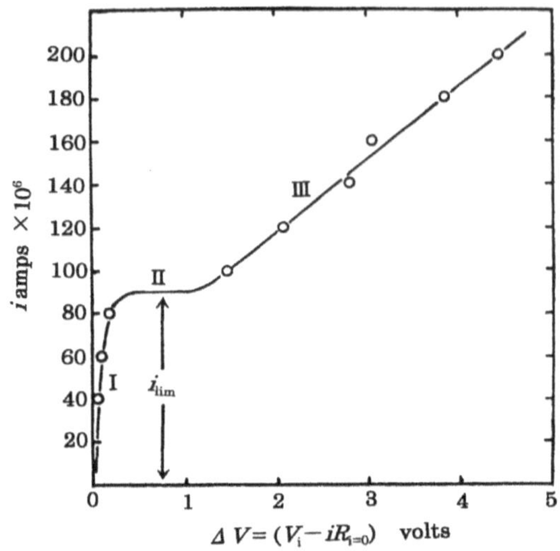

FIGURE 6.1 Current–voltage curve. *Peer (1956), p. 124, Fig. 1.*

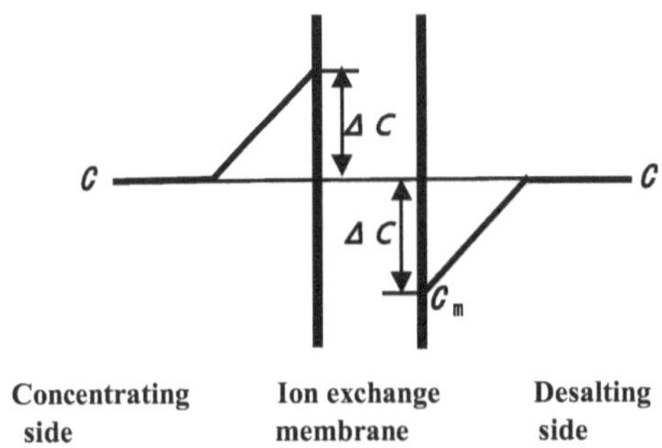

FIGURE 6.2 Concentration polarization.

of the phenomenon in Figure 6.1. He introduced the potential drop across membrane ΔE in the electrodialysis system in Figure 6.2 as:

$$\Delta E = -(2R_S + R_m)i + \left\{\frac{\alpha}{\lambda} + \beta - \gamma\right\}\ln\frac{1 + (i/i_{\text{lim}})}{1 - (i/i_{\text{lim}})} \tag{6.1}$$

where

$$\alpha = -\frac{FD}{\bar{t}_+ - t_+} = \frac{FD}{\bar{t}_- - t_-}$$

$$\beta = \left(\frac{RT}{F}\right)(\bar{t}_- - \bar{t}_+)$$

$$\gamma = \left(\frac{RT}{F}\right)(t_- - t_+)$$

where R_S is the electric resistance of the solution, R_m is the electric resistance of the membrane, λ is the equivalent conductance of the solution, $\bar{t}_-$ and $\bar{t}_+$ are the transport number of anions and cations in the membrane, and t_- and t_+ are the transport number of anions and cations in the solution. ΔE is plotted against i and shown in Figure 6.3. The first and the second terms in the figure correspond to the terms in Eqn (6.1). Figure 6.3 shows that upon increasing the voltage across the membrane, the current plateau (region II in Figure 6.1) is reached at the limiting current density i_{lim} at $C_m = 0$ in Figure 6.2. However, the overlimiting region (region III in Figure 6.1) is not found. This is because Eqn (6.1) is introduced from the Nernst–Planck equation, which is not applicable to the overlimiting region.

Cowan and Brown (1959) expressed the current–voltage relationship (Cowan and Brown plot) by the following equation:

$$\frac{V}{I} = R + \frac{V_e + V_c + V_p}{I} \tag{6.2}$$

where V_e is the electrode potential, V_c is the concentration potential, V_p is the polarization potential, and IR is the ohmic voltage. Equation (6.2) indicates that the intercept of the plot of V/I against I^{-1} represents the resistance of the cell and the slope of the plot represents the cell voltage plus their derivatives. Polarization voltage manifest as a rapid change in the slope, as shown in Figure 6.4, obtained for an anion exchange membrane. The pH of a diluting stream begins to change at a current near the value at which the resistance slope changes and continues to decrease as current density increases. The pH change is attributed to the occurrence of water dissociation on the desalting surface of the anion exchange membrane. The point at which the negative slope cuts the positive slope shows the limiting current i_{lim}.

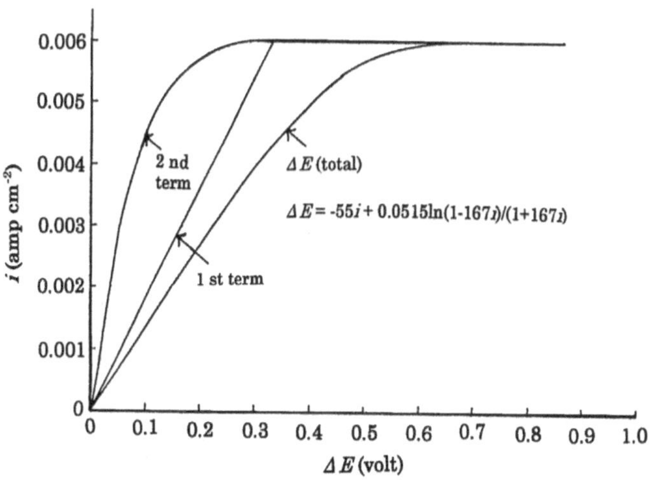

FIGURE 6.3 Current–voltage curve (calculation). *Spiegler (1971), p. 379, Fig. 3.*

FIGURE 6.4 V/I versus 1/I curve. *Cowan and Brown (1959), p. 1445, Fig. 1.*

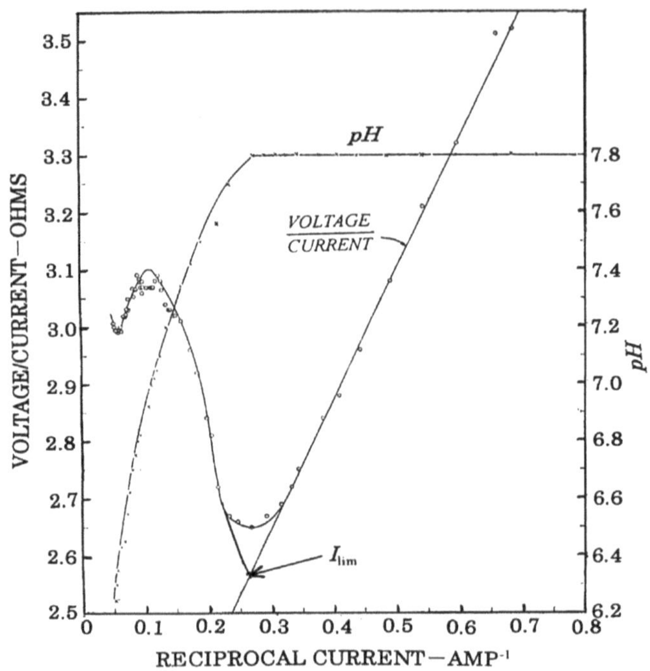

6.3 Concentration Changes in a Boundary Layer

When a ray of light shines on the interface between two solutions that have different refractive indexes, the light is refracted and the refractive index increases with the electrolyte concentration of the solutions. This phenomenon is the principle of the Schlieren-diagonal method, which was applied by Takemoto (1972) to measure the concentration polarization on the surface of an ion exchange membrane. In this experiment, an anion exchange membrane (Aciplex CA-2, Asahi Chemical Co.) and a cation exchange membrane (Aciplex CK-2) were integrated in the three-compartment optical glass cell. An NaCl solution (0.1 or 0.05 M) was put into the central desalting cell without forced flowing. The electrode cells were placed on both sides of the central cell and an electric current was passed through Ag-AgCl electrodes. The NaCl concentration changes in the boundary layer were observed by the Schlieren-diagonal method.

Figures 6.5 and 6.6 show the concentration distribution in the boundary layer formed on the anion and cation exchange membrane, respectively. In the 0.1 M NaCl solution, the limiting current density for the anion exchange membrane (3.99 A/dm^2) was larger than that for the cation exchange membrane (2.93 A/dm^2). The boundary layer thickness δ was developed extensively because of an unforced flowing solution in bulk. δ on the anion exchange membrane (0.0402 cm) was larger than that on the cation exchange membrane (0.0366 cm). In the 0.05 M NaCl solution, the limiting current density of the

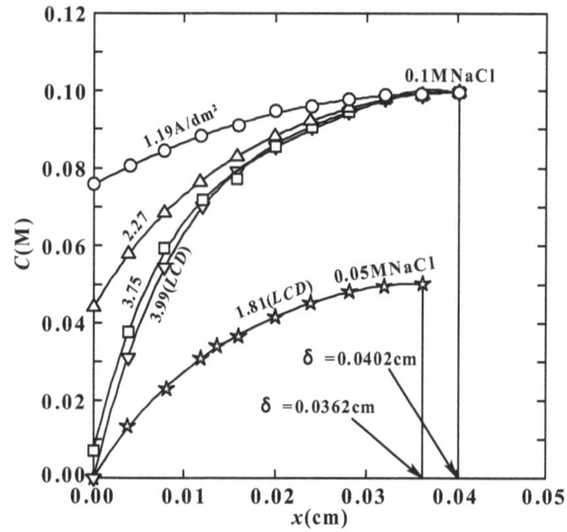

FIGURE 6.5 NaCl concentration distribution in a boundary layer. Anion exchange membrane (Aciplex CA-2). *Takemoto (1972), p. 2065, Fig. 3.*

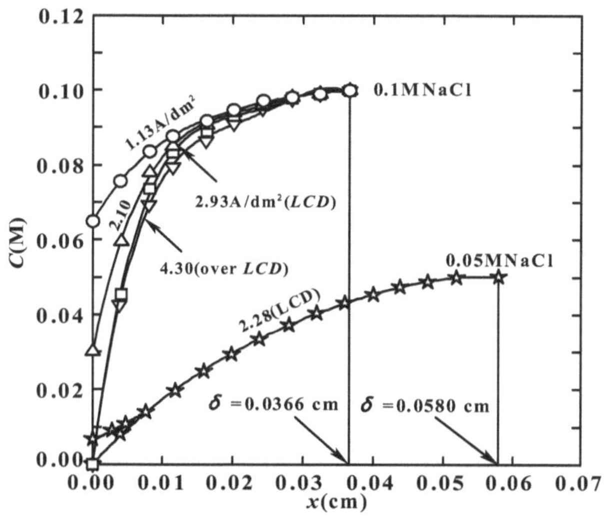

FIGURE 6.6 NaCl concentration distribution in a boundary layer. Cation exchange membrane (Aciplex CK-2). *Takemoto (1972),p. 2055, Fig. 3.*

cation exchange membrane (2.28 A/dm^2) was larger than that for the anion exchange membrane (1.81 A/dm^2). The thickness of the boundary layer on the cation exchange membrane (0.0580 cm) (Figure 6.6) was greatly increased compared with that on the anion exchange membrane (0.0362 cm) (Figure 6.5). Furthermore, the NaCl concentration fluctuated on the cation exchange membrane within the limits of $0 < x < 0.077$ cm distant from the membrane surface (Figure 6.6). The concentration fluctuation propagated from the membrane surface toward the inside of the solution adjacent to the membrane. The velocity of the propagation was unstable but it was estimated to be about 0.1 cm/s. Thus, the frequency of fluctuation was estimated to be 10 times/s.

6.4 Mass Transport in a Boundary Layer

Mass transport is a fundamental phenomenon in ion exchange membrane electrodialysis. Every phenomenon in the electrodialysis process is related to the mass transport and its driving force is divided into diffusion, migration, and convection. Here, the mechanism of the concentration polarization is discussed by evaluating the ion transport across the boundary layer illustrated in Figure 6.7, assuming that an ion exchange membrane is immersed in an unforced flowing solution. The mass transport in Figure 6.7 is given by the extended Nernst–Planck equation; Eqns (6.3) and (6.4) include diffusion J_{diff}, migration J_{migr}, and convection J_{conv} in an NaCl solution.

$$J_{Na} = J_{Na,diff} + J_{Na,migr} + J_{Na,conv} = -D_{Na}\frac{dC_{Na}}{dx} - \frac{FD_{Na}C_{Na}}{RT}\frac{d\psi}{dx} + C_{Na}v \tag{6.3}$$

$$J_{Cl} = J_{Cl,diff} + J_{Cl,migr} + J_{Cl,conv} = -D_{Cl}\frac{dC_{Cl}}{dx} + \frac{FD_{Cl}C_{Cl}}{RT}\frac{d\psi}{dx} + C_{Cl}v \tag{6.4}$$

where J_{Na} and J_{Cl} are the flux of Na^+ ions and Cl^- ions on x; x is the axis drawn in the boundary layer perpendicular to the membrane surface; C_{Na} and C_{Cl} are the

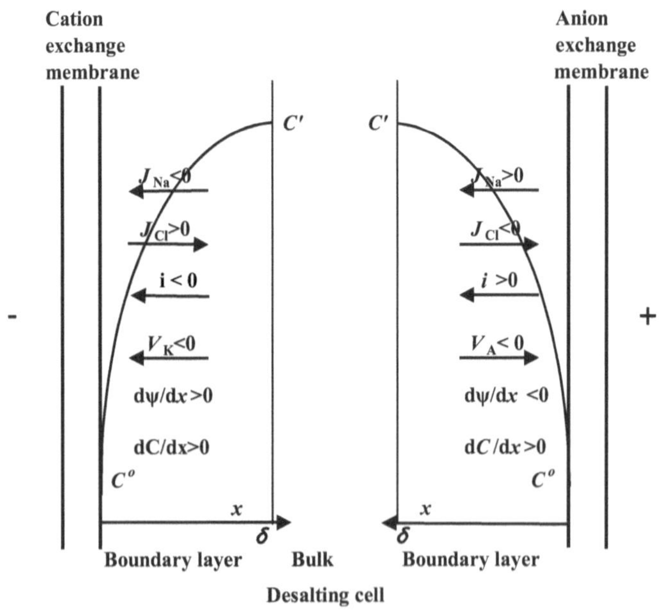

FIGURE 6.7 Boundary layer formed on a desalting surface of an ion exchange membrane. *Tanaka (2004), p. 3, Fig. 2.*

concentration of Na^+ ions and Cl^- ions on x; ψ is the electrical potential on x; D_{Na} and D_{Cl} are the diffusion constants of Na^+ ions and Cl^- ions on x; F is the Faraday constant; R is the gas constant; and T is the absolute temperature. Although the membrane is immersed in an unforced flowing solution, the solution in the boundary layer is disturbed by the convection.

v in Eqns (6.3) and (6.4) is defined as the x component of the solution velocity toward the membrane surface as a result of the convection. The convection is a key phenomenon in the boundary layer. The contribution of the convection to ion transport is described altogether in the convection term ($J_{Na,conv}$ and $J_{Cl,conv}$) in Eqns (6.3) and (6.4). v corresponds to v_A for the anion exchange membrane, and v_K for the cation exchange membrane in Figure 6.7, and takes a minus value on x.

Applying the extended Nernst–Planck equation, Eqns (6.3) and (6.4), to the concentration distribution in the boundary layer in Figures 6.5 and 6.6, ion fluxes at the limiting current density were calculated (Tanaka, 2004) and are shown in Figure 6.8 (anion exchange membrane, 3.99 A/dm^2, 0.1 M NaCl) and Figure 6.9 (cation exchange membrane, 2.93 A/dm^2, 0.1 M NaCl). Near the anion exchange membrane (Figure 6.8) and the cation exchange membrane (Figure 6.9), the absolute value of J_{conv} is decreased. This phenomenon is due to the decrease in concentration C on the membrane surface (Figures 6.5 and 6.6) and the resultant decreases in the convection term J_{conv} in Eqns (6.3) and (6.4). Here, J_{conv} does not carry an electric current because $J_{Na,conv} = J_{Cl,conv}$ holds. Ion flux changes near the anion exchange membrane in Figure 6.8 are interpreted because J_{conv}, which does not carry an electric current, converts to J_{migr} and J_{diff}, which carry an electric current. Here, the increase in absolute value of $J_{Cl,diff}$ (diffusion flux of counter-ions) is larger than that of $J_{Na,diff}$ (diffusion flux of co-ions) because the mobility of

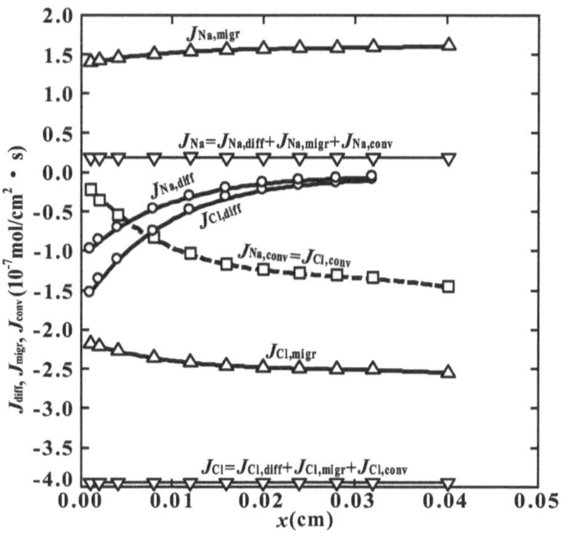

FIGURE 6.8 Ion flux in a boundary layer. Anion exchange membrane 3.99 A/dm^2(LCD) 0.1 M NaCl *Tanaka (2004), p. 7, Fig. 10.*

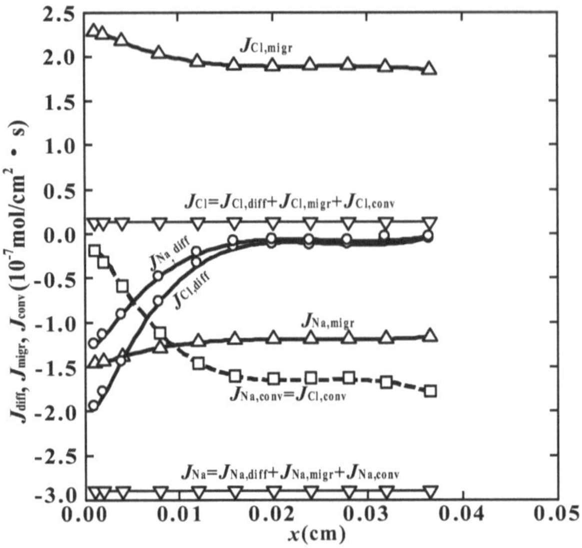

FIGURE 6.9 Ion flux in a boundary layer. Cation exchange membrane 2.93 A/dm^2(LCD) 0.1 M NaCl. *Tanaka (2004), p. 7, Fig. 8.*

counter-ions, Cl^- ions (7.9×10^{-4} cm^2/V s), is larger than that of co-ions, Na^+ ions (5.2×10^{-4} cm^2/V s), on the anion exchange membrane. Namely, the diffusion flux supports current density increase near the anion exchange membrane. On the other hand, the ion flux changes near the cation exchange membrane in Figure 6.9 are interpreted because J_{conv} converts to J_{migr}. Here, the increase in the absolute values of $J_{Na,diff}$ (diffusion flux of counter-ions) is less than that of $J_{Cl,diff}$ (diffusion flux of co-ions) because the mobility of counter ions (Na^+ ions) is less than that of co-ions (Cl^- ions). Accordingly, the diffusion flux lowers current density near the membrane. In addition, the counter-ions (H^+ ions) are almost not supplied from the water dissociation reaction because the reaction is strongly suppressed in the cation exchange membrane (Section 7.7). Thus, J_{conv} is obliged to be changed to J_{migr} to increase an electric current on the cation exchange membrane. In the 0.1 M NaCl solution, the boundary layer thickness δ on the cation exchange membrane (Figure 6.6) is less than that of the anion exchange membrane (Figure 6.5). However, in the 0.05 M NaCl solution, δ on the cation exchange membrane is drastically increased (Figure 6.6). This phenomenon presumably causes an increase in the convection J_{conv} on the cation exchange membrane. The NaCl concentration near the cation exchange membrane placed in the 0.05 M NaCl solution fluctuates at the limiting current density (Figure 6.6). This phenomenon causes the increase in NaCl concentration and J_{conv} and it involves vortex flow, voltage fluctuation, and light scattering fluctuation and accelerates ion transport toward the membrane surface (Section 6.8).

6.5 Space Charge

Rubinstein and Shtilman computed an ion concentration profile at a desalting surface of a cation exchange membrane at an overlimiting current based on the following Nernst–Planck–Poisson equation for 1:1 valent electrolytes (Rubinstein and Shtilman, 1979; Rubinstein, 1981).

$$\frac{\partial p}{\partial t} = div(grad\, p + p\, grad\psi) \tag{6.5}$$

$$\frac{\partial n}{\partial t} = div(grad\, n - n\, grad\psi) \tag{6.6}$$

$$\varepsilon \Delta \psi = n - p \tag{6.7}$$

p and n are the dimensionless counter-ion concentration and dimensionless co-ion concentration, respectively. ε is a square ratio of the bulk Debye radius to the thickness of the unstirred layer. t is dimensionless time. ψ is the dimensionless electric potential.

The computed ion concentration profile in the unstirred layer on the cation exchange membrane is illustrated in Figure 6.10, which is distinguished by three distinct regions, regions I, II, and III. Solid lines and dotted lines show cation concentration and anion concentration, respectively. In region I, the local electro-neutrality is preserved with high accuracy. Region II is a charged region developed in the course of the concentration polarization. Its dimensions can exceed by an order of magnitude the equilibrium Debye radius. In this region the counter-ion (cation) concentration is much higher than that of co-ions (anion), which is low. The electric field at the end of region II is high. Ion transport in this region is dominated by migration as opposed to diffusion. Region III is

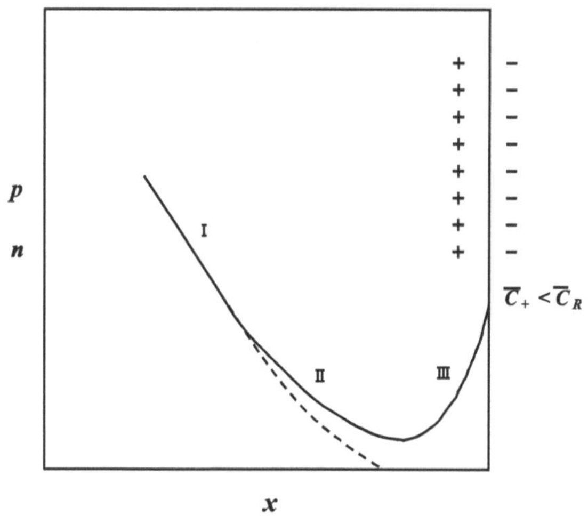

FIGURE 6.10 Concentration profiles of cations and anions in an unstirred layer formed on the desalting surface of a cation exchange membrane. *Rubinstein (1981), p. 1596-1597, Fig. 2 (b, c, d, e, f, g).*

comparable in thickness (although smaller) to the equilibrium Debye radius. If one assumes that the interface concentration of exchange groups within the cation exchange membrane in Figure 6.10 is $\overline{C}_R$, to maintain electric neutrality in a total system, $\overline{C}_R > \overline{C}_+$ must hold and the potential difference must be generated between the solution and the membrane.

6.6 Gravitational Convection

When an electric current is passed across the ion exchange membrane, ionic concentration and solution density in the boundary layer decrease on the desalting side and increase on the concentration side. These phenomena involve upward solution flow on the desalting surface and downward solution flow on the concentration surface owing to gravitational convection, as illustrated in Figure 6.11.

Frilette (1957) observed gravitational convection using an electrodialysis cell (Figure 6.12) incorporated with Permutit cation exchange membranes and equilibrated with a 0.0996 M NaCl solution. A constant electric current was passed and was turned off after the predetermined time had elapsed by keeping total coulombs passed as constant. After completion of the experiment, the solution in the chamber was drained off slowly and collected in two equal portions; the portion first collected corresponded to the solution in the lower half of the central chamber. The two portions were analyzed for chloride ions. In all experiments, a net movement of NaCl into the lower half of the chamber had occurred in which there was no significant change in the total electrolyte content of the chamber. The results of this experiment are shown in Table 6.1, which is characterized by a relatively large transfer of salt from the upper half into the lower half of the chamber.

To determine the ascending flow velocity v_y caused by the gravitational convection on the desalting surface of an ion exchange membrane, fine cellulose fibers were suspended in a 0.5 M NaCl solution in a transparent polyethylene desalting cell incorporated with

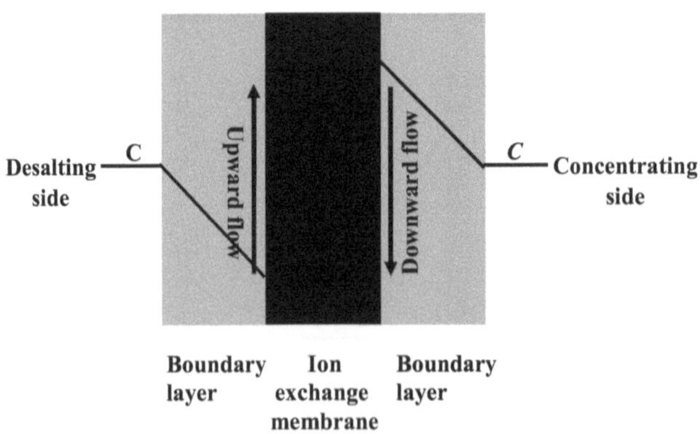

FIGURE 6.11 Gravitational movement of solutions on membrane surfaces.

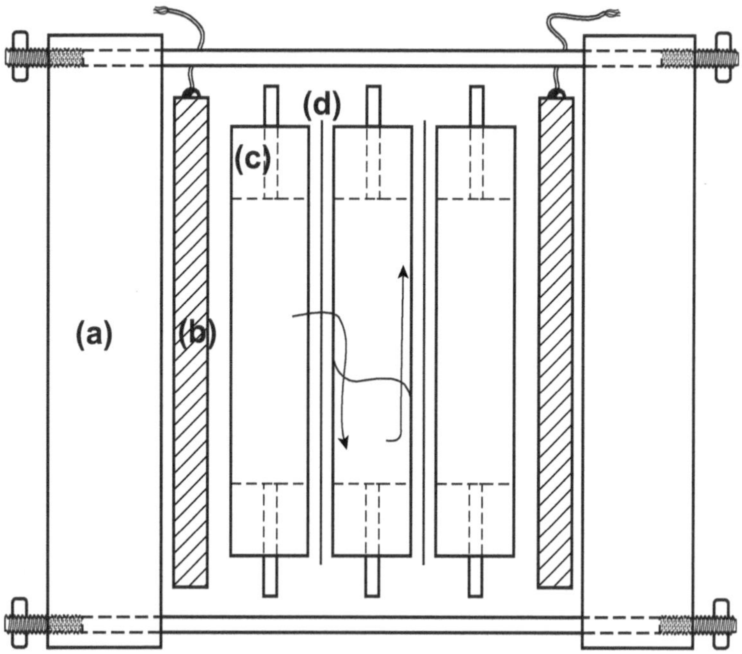

FIGURE 6.12 Apparatus for measuring gravitational convection. (a) Lucite end block, (b) graphite plate, (c) spacer, (d) membrane. *Frilette (1957), p.169, Fig. 1.*

Table 6.1 Transfer of Electrolytes in a Cell as a Result of Gravitational Convection: Constant Current × Time[a]

Current	Current Density	Time	Lost, Upper Layer	Gained, Lower Layer	Transferred Upper Layer	Transferred Lower Layer
mA	A/cm^2	min	mEq		eq./Faraday	
46	1.87	40	0.303	0.310	−0.264	+0.270
61.3	2.49	30	0.345	0.310	−0.301	+0.270
92	3.74	20	0.367	0.365	−0.320	+0.318
138	5.61	$13_{1/3}$	0.465	0.457	−0.406	+0.400
184	7.48	10	0.483	0.488	−0.422	+0.426
230	9.35	8	0.506	0.493	−0.442	+0.430
368	14.96	5	0.504	0.511	−0.440	+0.446

[a]Total of 1.840 ampare minutes for all runs or 1.147 × 10^{-3} Faraday.

Frilette (1957), p. 170, Table 1.

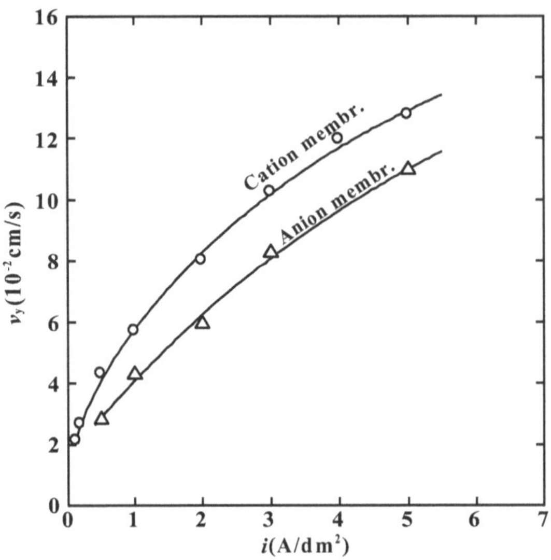

FIGURE 6.13 Ascending flow velocity in a boundary layer. 0.1 M NaCl Aciplex CK-2/CA-3. *Tanaka (2004), p. 5, Fig. 5.*

Aciplex CK-2 cation exchange membrane and Aciplex CA-2 anion exchange membrane (Asahi Chemical Co.) (Tanaka, 2004). The effective membrane area was adjusted to 4 cm^2 (2 cm height and 2 cm width). The movement of the cellulose fibers was observed with a microscope while applying an electric current through Ag-AgCl electrodes. The flow velocity was plotted against the current density and is presented in Figure 6.13, which shows that v_y on the cation exchange membrane is larger than v_y on the anion exchange membrane. This phenomenon means that concentration polarization occurs more drastically on the cation exchange membrane than on the anion exchange membrane. The ascending flow supposedly induces the overlimiting current.

In that experiment, a 0.1 M NaCl solution was put into a concentrating cell and an electric current was passed; then, the solution flow was observed with a microscope on the concentrating surface of the membrane. In this experiment, a downward moving solution was observed as something like a waterfall. This phenomenon is estimated to be caused by mixing a high-density solution passed through the membrane with a low-density solution put in the concentrating cell. The solution mixing was visualized because of the reflection index difference between the low- density and high-density NaCl solutions.

6.7 Electroconvection

When an overlimiting electric current is applied across an ion exchange membrane, a microvortex instability owing to the electroconvection (second kind of electroosmosis) occurs on the membrane surface (Duskin, 1991; Mishchuk et al., 2001). The driving force

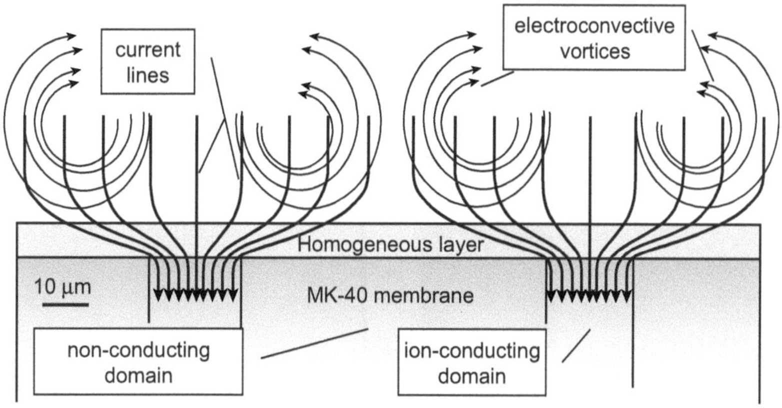

FIGURE 6.14 Scheme of electroconvective vortices. *Nikonenko et al. (2010), p. 119, Fig. 16.*

of the electroconvection is the excess pressure generated in the space charged region (Section 6.5) as the result of a high- pressure Maxwell spot at the membrane surface (Chang et al., 2012). The high-pressure spot drives liquid radially away from it on the membrane surface and produces vortices by which counter-ions are enhanced to enter into the membrane. Figure 6.14 illustrates the electroconvective vortices generated at the entrance of the membrane nanopore with a charged wall (Nikonenko et al., 2010).

Yossifon and Chang (2010) observed the vortices on the nanoslot (equivalent to the membrane) immersed a 0.1 mM weak electrolyte solution applying an electric current. Figure 6.15 shows colloid dynamics for different applied voltages in the range of (1) region I, (2) region II, and (3) region III in Figure 6.1. The dynamics in (1) exhibits linear colloid translation within the colloid band, whereas those in (2) and (3) exhibit complex torroidal movement. Furthermore, the tangential motion of colloids trapped just at the nanoslot entrance becomes more pronounced with increased voltage. At low voltage (1), a nanocolloid suspension at the depleted entrance of a nanoslot is captured by the horizontal vortices. At higher voltage (2), (3), they are captured by the vertical vortex array as a result of instability, producing a complex toroidal pattern at the end.

6.8 Fluctuation

The occurrence of fluctuation of electric currents through ion exchange membranes is attributed to salt depletion, an increase in electric resistance, Joule heating, buoyant force generation, and turbulent convection created on the desalting surface of the membrane. These phenomena are related to the gravitational convection and electro-convection. The turbulence mixes the depletion layer so that the electric resistance fluctuates, and consequently the current flickers.

Lifson et al. (1978) observed the flicker noise on a dual-beam oscilloscope. A 0.01 N KCl solution dispersing latex particles (0.2 μm in diameter) is supplied to the conducting

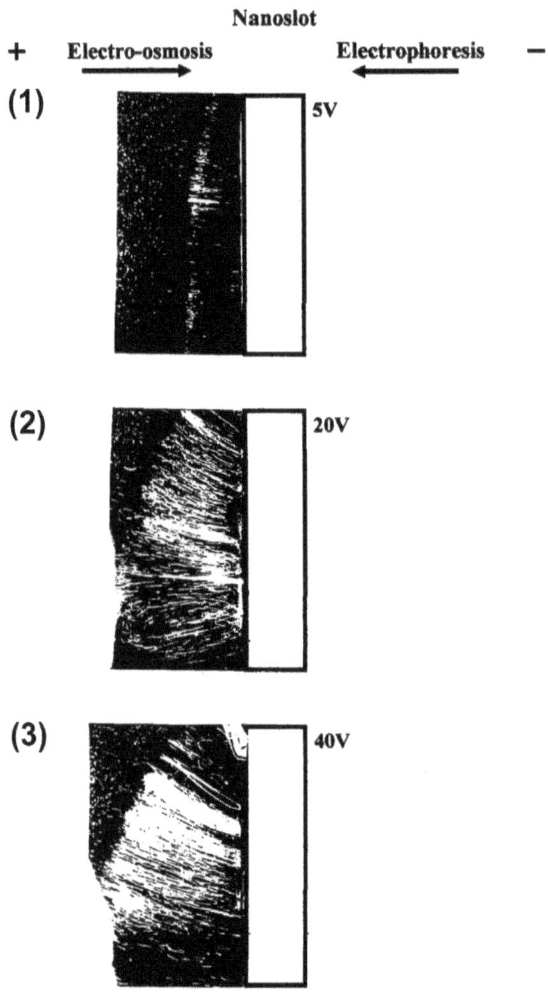

FIGURE 6.15 Colloid dynamics for convective vortices. *Chang et al. (2012), p. 422, Fig. 16.*

cell incorporated with the Selemion membrane (Asahi Glass Co.). A narrow beam of light (~ 0.01 cm in diameter) is focused on the depleted or enriched side of the membrane. The light intensity is detected by a photodiode after passing through a narrow slit and a red filter, and its fluctuations (light noise) are displayed on the dual-beam oscilloscope, in parallel with the current noise. Figure 6.16 shows the light noise and current noise on the depleted side. The experimental results are summarized as follows.

1. Noise coincides with an increase in electric resistance by the depletion process.
2. When the current density decreases, it reaches a critical value below which the convection changes from turbulent to laminar and the fluctuation disappears.
3. Fluctuation reduces with temperature because the expansion coefficient of water decreases with temperature and its viscosity increases.

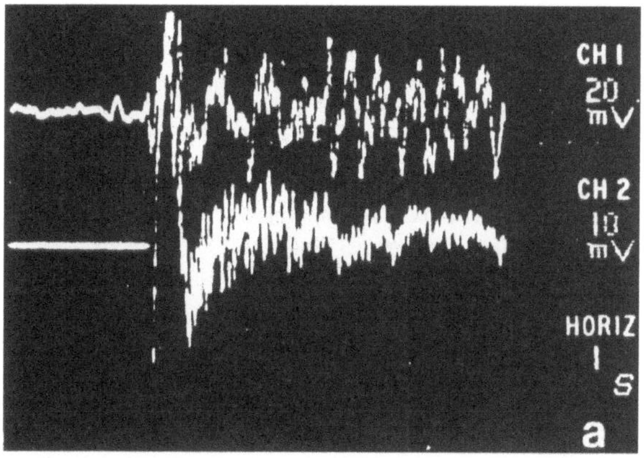

FIGURE 6.16 Current noise and optical noise. *Lifson et al. (1978), p. 62, Fig. 6.*

4. A nonionic water-soluble polymer added to the compartment on the side of the depletion layer reduces the fluctuation by increasing the bulk viscosity of the solution.
5. Fluctuation depends on the membrane's orientation in the gravitational field.
6. The convection in the depletion layer can be observed directly using a laser beam, by adding latex particles that create optical noise as they drift with the convection across the beam. The optical noise is observed only coincidently with the current noise.

Tanaka (2004) observed the voltage fluctuation and water dissociation at an over-limiting current using an electrodialysis cell integrated with Aciplex K-172 cation and Aciplex A-172 anion exchange membrane (Asahi Chemical Co.). The effective membrane area was adjusted to 0.383 cm^2 and a 5 mM NaCl solution was supplied to the desalting cell. An electric current was passed through Ag-AgCl electrodes and the voltage fluctuation was measured using a voltmeter. Current efficiencies for the water dissociation arising in the cation exchange membrane η_H and those arising in the anion exchange membrane η_{OH} were calculated from the changes in pH and solution volume in the concentrating cell. Voltage fluctuation on the membranes observed by setting the current density to 0.131 A cm^{-2} is shown in Figure 6.17, which indicates that voltage fluctuates on the cation exchange membrane but not on the anion exchange membrane. η_H and η_{OH} were calculated as 0.041 and 0.336 for the cation and anion exchange membrane, respectively. The experimental results indicate that water dissociation is generally restricted in the ion exchange membrane (η_H and η_{OH} are considerably less than 1) and it is extremely restricted in the cation exchange membrane (η_H is extremely less than 1) (Section 7.7). These phenomena are understandable as follows:

1. To make up for insufficient water dissociation strength in the ion exchange membrane, gravitational convection or electroconvection is generated in the boundary layer formed on the desalting surface of the membrane and it supplies Na^+ and Cl^- ions from bulk toward the membrane.

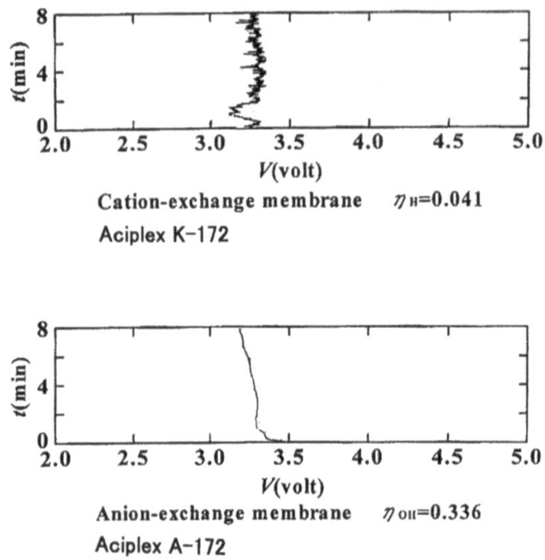

FIGURE 6.17 Voltage fluctuation at overlimiting current density 5 mM NaCl 0.131 A/cm^2. *Tanaka (2004), p. 5, Fig. 6.*

2. To make up for extremely insufficient water dissociation in the cation exchange membrane, solution velocity, ionic flux, ionic concentration, and current density fluctuate in the boundary layer to promote ionic transport and maintain material transport.

Krol et al. (1999) studied overlimiting ion transport through a Neosepta CMX cation and AMX anion exchange membrane (Tokuyama Co.). This technique is used to characterize fluctuations in membrane voltage drop observed in the overlimiting current region. Above the limiting current, measurements show a large voltage drop fluctuation in time indicating hydrodynamic instability. The amplitude of this fluctuation increases with increasingly applied current density. The fluctuation occurs when a setup is used in which there is no forced convection and the depleted boundary layer is stabilized by gravitation. Li et al. (1983) measured light scattering spectra from polystyrene latex near an Ionics, Inc., #61-CZL-386 cation exchange membrane in a 0.020 M HCl and 0.020 M NaCl solution. For HCl, spectra on the depleted side showed evidence for turbulent flow at overlimiting current density. Spectra on the concentrate side showed similar spectra above approximately twice the limiting current density, although of low intensity. For NaCl, depleted side spectra began at currents approximately twice the limiting value. On the concentrated side, at least six to seven times current density was needed for nonzero difference spectra. These experimental results suggest that the electrical noise spectra correspond principally to fluctuation induced by turbulent flow on the depleted side of the membrane.

At an overlimiting current density (region III in Figure 6.1), the salt concentration at the interface between the boundary layer and the membrane is decreased drastically.

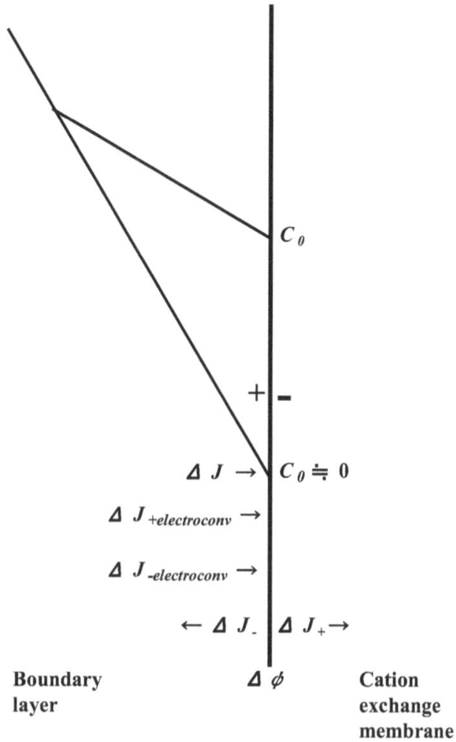

FIGURE 6.18 Fluctuation.

However, the salt concentration on the membrane surface never becomes zero. This is because the salt concentration fluctuates at the interface between the boundary layer and the membrane, as illustrated in Figure 6.18, indicating that the NaCl concentration fluctuates on the cation exchange membrane within the limits of $0 < x < 0.077$ cm distant from the membrane surface (Figure 6.6).

The transport phenomena in Figure 6.18 are generated in an irreversible state and governed by irreversible thermodynamics. Thus, the solvent flux is given by the following phenomenological equation (Kedem and Katchalsky, 1963):

$$J = L_{EP}\Delta\psi + L_P\Delta P + \sum_i L_{Pi}\Delta\mu_i \tag{6.8}$$

At an overlimiting current density, the water dissociation reaction is extremely restricted in the cation exchange membrane (Section 7.7). In this circumstance, the electric neutrality is disturbed owing to the charged region generation and the potential difference $\Delta\psi$ formation at the solution–membrane interface (Section 6.5). $\Delta\psi$ induces the solvent flux ΔJ and the relationship between $\Delta\psi$ and ΔJ is introduced from Eqn (6.8) as follows:

$$\Delta J = L_{EP}\Delta\psi \tag{6.9}$$

ΔJ is equivalent to the solvent flux caused by electroconvection; $J_{electroconv}$ (Section 6.7) and $\Delta\psi$ functions as the driving force of ΔJ. Ions are transferred with $\Delta J = J_{electroconv} = \Delta J_{+electroconv} = \Delta J_{-electroconv}$, which corresponds to electroconvection of positive and negative ions dissolved in the solution. ΔJ does not carry an electric current because $J_{+electroconv} = J_{-electroconv}$ holds (Section 6.4). Ion flux changes near the cation exchange membrane in Figure 6.18 are understood as $\Delta J_{+electroconv}$ and $\Delta J_{-electroconv}$, which do not carry an electric current conversion to ΔJ_+ and ΔJ_-, which corresponds to the migration of positive and negative ions that carry an electric current. Here the contribution of the diffusion is neglected.

Based on this discussion, fluctuations generated on the membrane surface (electric current noise, optical noise, etc.) at the overlimiting current density in Figure 6.18 are explained as follows:

1. Salt concentration at the membrane surface C_0 is extremely decreased toward $C_0 \fallingdotseq 0$.
2. The space charge at the solution–membrane interface is generated and the potential difference between the solution and membrane $\Delta\psi$ is increased (Section 6.5).
3. ΔJ is increased (Eqn (6.9)).
4. Electroconvection of ions, $\Delta J_{+electroconv}$ and $\Delta J_{-electroconv}$, are increased (Section 6.7).
5. $\Delta J_{+electroconv}$ and $\Delta J_{-electroconv}$ are converted to electromigration J_+ and J_- and carry an electric current (Section 6.4).
6. C_0 is increased.
7. $\Delta\psi$ is decreased.
8. ΔJ is decreased.
9. $\Delta J_{+electroconv}$ and $\Delta J_{-electroconv}$ are decreased.
10. J_+ and J_- are decreased with an electric current.
11. Return to 1.

In conclusion, the mechanism of the fluctuation generated at the overlimiting current density is understandable based on the phenomenological equation; Eqns (6.8) and (6.9) held at the space charged region generated on the membrane surface.

6.9 Concentration Polarization Generated on a Concentrating Surface of an Ion Exchange Membrane

Concentration polarization is usually discussed on the desalting surface of the ion exchange membrane as described in the preceding sections. In this section, however, the phenomenon on the concentrating surface of the membrane is discussed.

Takemoto (1969) observed the NaCl concentration distribution in a boundary layer formed on the concentrating surface of an ion exchange membrane (Aciplex CK-2/CA-2, Asahi Chemical Co.) integrated into a three-compartment optical glass cell incorporated with a central concentrating cell and desalting (electrode) cells put on both outsides of the central cell. The effective membrane area was adjusted to 1.5 cm^2 (1.5 cm height and 1 cm width). The distance between the membranes in the central concentrating cell was maintained 2.6–2.8 mm. The 25 °C 0.5 eq/dm^3 NaCl solutions were put into the desalting cells and concentrating cell, and an electric current density of 2 A/dm^2 was applied through Ag-AgCl electrodes. The changes in NaCl concentration distribution in the boundary layer formed on the concentrating surface of the membrane were observed by the Schlieren-diagonal method (Section 6.3).

The concentration changes in the boundary layer observed in this experiment are understandable on the basis of Figure 6.19. In this experiment, the ion concentration in the cation exchange membrane is assumed to be larger than that in the anion exchange membrane. The NaCl concentration in the concentrating cell is 0.5 eq/dm^3 before passing electric current, as shown by 0 in the figure. Applying an electric current, the concentrated counter-ions in the membrane flow out into the concentrating cell and flow down with concentrated co-ions along the membrane surface owing to the electrogravitational movement (Section 6.6). Successively, the concentration distribution is changed as 0 → 1 → 2 → 3, and finally attains 4 in the steady state. Figure 6.20 shows the concentration profile corresponding to 4 in the steady state observed by Takemoto (1969), indicating that the NaCl concentration on the cation exchange membrane

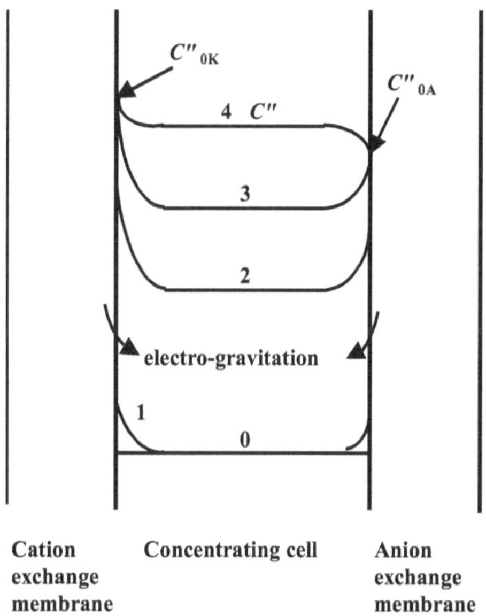

0: before current passing, 1: just after current passing,
2,3: at a transitional state, 4: at a steady state

FIGURE 6.19 Electrolyte concentration changes in boundary layers formed on the concentrating surface of ion exchange membranes. *Tanaka (2006), p. 525, Fig. 12.*

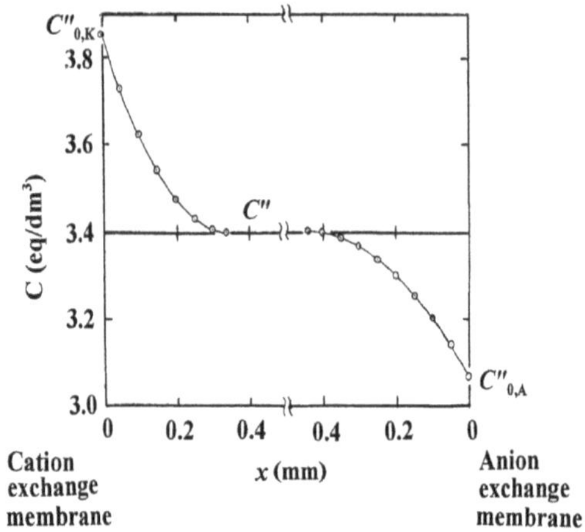

FIGURE 6.20 NaCl concentration distribution on a concentrating surface of a cation and an anion exchange membrane in a steady state owing to concentration polarization. *Takemoto (1969), p.56, Fig. 2.*

becomes larger than that on the anion exchange membrane. From the experiment described above, the concentration polarization generated on the concentrating surface of the membrane is assumed not to be the same in principle as the phenomenon generated on the desalting surface of the membrane.

References

Balster, J., Yildirim, M.H., Stamatialis, D.F., Ibanez, R., Lammertink, R.G.H., Jordan, V., Wessling, M., 2007. Morphology and microtopology of cation-exchange polymers and the origin of the overlimiting current. J. Phys. Chem. B 111, 2152–2165.

Chang, H.C., Yossifon, G., Demekhin, E.A., 2012. Nanoscale electrokinetics and microvortics: how microdynamics affects nanofluidic ion flux. Annu. Rev. Fluid Mech. 44, 401–426.

Cowan, D.A., Brown, J.H., 1959. Effect of turbulence on limiting current in electrodialysis cells. Ind. Eng. Chem. 51, 1445–1448.

Duskin, S.S., 1991. Electrokinetic phenomena of the second kind and their applications. Adv. Colloid Interface Sci. 35, 173–196.

Frilette, V.J., 1957. Electrogravitational transport at synthetic ion exchange membrane surface. J. Phys. Chem. 61, 168–174.

Gavish, B., Lifson, S., 1979. Membrane polarization at high current densities. J. Chem. Soc. Faraday Trans. I 75, 463–472.

Green, M.E., Yafuso, M.A., 1968. Study of the noise generated during ion transport across membranes. J. Phys. Chem. 72, 4072–4078.

Kedem, O., Katchalsky, A., 1963. Permeability of composite membranes. Part 1. Electric current, volume flow and flow of solute through membranes. Trans. Faraday Soc. 59, 1918–1930.

Kooistra, W., 1967. Characterization of ion exchange membranes by polarization curves. Desalination 2, 139–147.

Kolyubin, A.V., Maksimychev, A.V., Timashev, S.F., 1996. The use of flicker-noise spectroscopy for studying the mechanism of the overlimiting current in a system with cation-exchanhe membrane. Russ. J. Electrochem. 32, 206–213.

Krol, J.J., Wessling, M., Strathmann, H., 1999. Chronopotentiometry and overlimiting ion transport through monopolar ion exchange membranes. J. Membr. Sci. 162, 155–164.

Li, Q., Fang, Y., Green, M.E., 1983. Turbulent light scattering fluctuation spectra near a cation electrodialysis membrane. J. Colloid Interface Sci. 91, 412–417.

Lifson, S., Gavish, B., Reich, S., 1978. Flicker noise of ion-exchange membranes and turbulent convection in the depleted layer. Biophys. Struct. Mech. 4, 53–65.

Maletzki, F., Roster, H.W., Staude, E., 1992. Ion transfer across electrodialysis membranes in the overlimiting current range: stationary voltage current characteristics and current noise power spectra under different conditions of free convection. J. Membr. Sci. 71, 105–115.

Mishchuk, N., Gonzalez-Caballero, F., Takhistov, P., 2001. Electroosmosis of the second kind and current through curved interface. Colloids and Surfaces 181, 131–141.

Nikonenko, V.V., Pismenskaya, N.D., Belova, E.I., Sistat, P., Huguet, P., Pourcelly, G., Larchet, C., 2010. Adv. Colloid Int. Sci. 160, 101–123.

Peer, A.M., 1956. Communication in the membrane phenomena. Special Issue, Discus Faraday Soc. 21, 124.

Rubinstein, I., Shtilman, L., 1979. Voltage against current curves of cation-exchange membranes. J. Chem. Soc. Faraday Trans. II 75, 231–246.

Rubinstein, I., 1981. Mechanism for an electrodiffusion instability in concentration polarization. J. Chem. Soc. Faraday Trans. II 77, 1595–1609.

Rubinstein, I., Zaltzman, B., Kedem, O., 1997. Electric fields in and around ion-exchange membranes. J. Membr. Sci. 125, 17–21.

Rubinstein, S.M., Manukyan, G., Staicu, A., Rubinstein, I., Zaltzman, B., Lammertink, R.G.H., Mugele, F., Wessling, M., 2008. Direct observation of a nonequilibrium electro-osmotic instability. Phys. Rev. Lett. PRL 101, 236101.

Shaposhnik, V.A., Vasil'eva, V.I., Grigorchuk, O.V., 2008. The interferometric investigations of electromembrane processes. Adv. Colloid Interface Sci. 139, 74–82.

Spiegler, K.S., 1971. Polarization at ion exchange membrane-solution interface. Desalination 9, 365–385.

Takemoto, N., 1969. Concentration distribution in concentrating surface of membranes. Bull. Soc. Sea Water Sci. Jpn. 23, 54–59.

Takemoto, N., 1972. Concentration distribution in the interfacial layer at desalting side in ion exchange membrane electrodialysis. J. Chem. Soc. Jpn. 1972, 2053–2058.

Tanaka, Y., 2004. Concentration polarization in ion-exchange membrane electrodialysis. The events arising in an unforced flowing solution in a desalting cell. J. Membr. Sci. 244, 1–16.

Tanaka, Y., 2006. Irreversible thermodynamics and overall mass transport in ion-exchange membrane electrodialysis. J. Membr. Sci. 281, 525.

Valerdi-Prez, R., Ibannez-Mengual, J.A., 2001. Current-voltage curves for an electrodialysis reversal pilot plant: determination of limiting currents. Desalination 141, 23–37.

Yossifon, G., Chang, H.C., 2010. Changing nanoslot ion flux with a dynamic nanocolloid ion-selective filter: secondary overlimiting currents due to nanocolloid-nanoslot interaction. Phys. Rev. E81, 066317.

Zabolotsky, V.I., Nikonenko, V.V., Pismenskaya, N.D., 1996. On the role of gravitational convection in the transfer enhancement of salt ions in the course of dilute solution electrodialysis. J. Membr. Sci. 119, 171–181.

7

Water Dissociation

7.1 Introduction

The water dissociation (water splitting) reaction occurs at overlimiting current density (Kressmann and Tye, 1956; Frilette, 1956; Rosenberg and Tirrel, 1957). It dissociates H_2O molecules into H^+ and OH^- ions, which carry an electric current instead of electrolyte counter-ions decreased due to the concentration polarization. The mechanism the water dissociation reaction is investigated extensively from the view points of the diffusional model (Rubinstein, 1977), repulsion zone (Patel et al., 1977), Wien effect (Wien, 1928; Onsager, 1934), protonation and deprotonation reaction (Simons, 1984, 1985), etc. The water dissociation is an equilibrium reaction and its intensity is governed by the forward reaction rate constant k_a and the reverse reaction rate constant k_b. With an applied electric potential field, k_a increases substantially while k_b remains unchanged, and the water dissociation reaction is accelerated. In spite of the considerable increase of k_a, current efficiency for the water dissociation reaction η is generally not so high. η for the anion exchange membrane is larger than that for the cation exchange membrane. η for the cation exchange membrane is strongly suppressed. However, η is drastically increased when metallic hydroxides are precipitated on the cation exchange membrane. The above phenomena are important in understanding the mechanism of the water dissociation reaction.

7.2 Current–pH Relationship

An investigation based on the current–pH relationship was proceeded by Rosenberg and Tirrel (1957) using seven-cell assemblies indicated in Fig. 7.1. Cell 1 was bounded by an anode and an anion exchange membrane, and cell 7 by a cathode and a cation exchange membrane; all other cells were bounded by the membranes as shown. Water dissociation was studied in the central cell (cell 4), which was fed with 0.005- to 0.05-M NaCl solutions. Other cells were fed with a salt 3 to 10 times more concentrated. At a series of increasing applied voltages, an electric current, influent and effluent concentrations in cell 4, and effluent pH values in cells 3, 4, and 5 were determined. The cation exchange membranes were sulfonated divinylbenzene polystyrene types, and anion exchange membranes were quaternarized divinyl benzene pyridine types. A typical history of pH in cell 3, 4, and 5 is given in Fig. 7.2. As the current increases, the interface between the anion exchange membrane and a cell 4 solution becomes depleted sodium chloride. The transfer of hydroxyl ions from cell 4 to cell 3 causes the pH of the effluent to decrease in

Ion Exchange Membranes. http://dx.doi.org/10.1016/B978-0-444-63319-4.00007-9

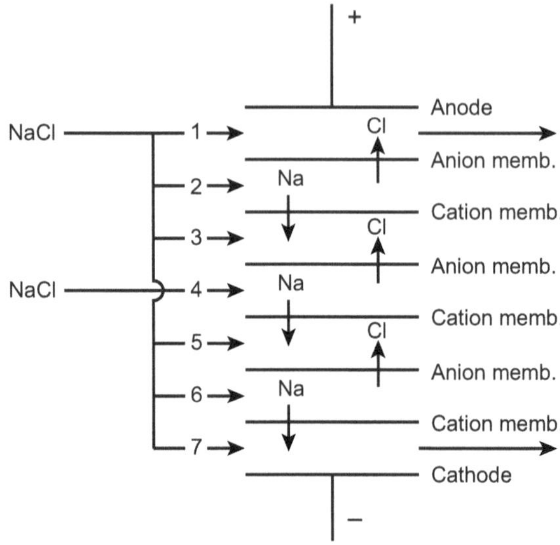

FIGURE 7.1 Cell assembly for studying polarization. *Rosenberg and Tirrel (1957), p. 782, Fig. 4.*

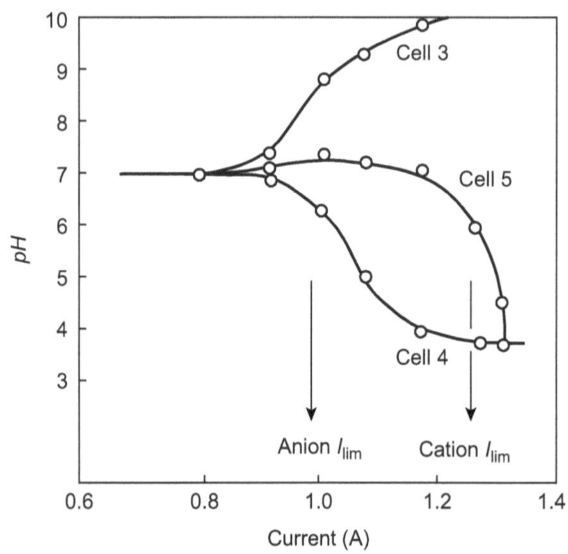

FIGURE 7.2 Electric current–solution pH relationship. *Rosenberg and Tirrel (1957), p. 783, Fig. 6.*

cell 4 and increase in cell 3. At this current, there is no indication of cation exchange membrane polarization. At the current is increased, however, the cation exchange membrane interface becomes depleted; hydrogen ion transfer from cell 4 into cell 5 occurs, and the pH in cell 5 falls. However, the transfer of hydroxyl ions through the anion exchange membrane interface increases at a faster absolute rate, and the pH in cell 4 continues to fall.

The mobility of Na^+ ions is less than that of Cl^- ions in an NaCl solution. Because of this fact, concentration polarization is generally recognized to occur more easily on the cation exchange membrane than the anion exchange membrane. Because of this reason, it is expected that the water dissociation occurs more easily on the cation exchange membrane than on the anion exchange membrane. However, Fig. 7.2 shows that the water dissociation occurs more easily on the anion exchange membrane than on the cation exchange membrane. This phenomenon has been recognized in many investigations showing that the water dissociation is strongly suppressed on the cation exchange membrane. Accordingly, the pH in cell 5 might not decrease due to the water dissociation of the cation exchange membrane as described in this study, but decrease due to the water dissociation of the anion membrane accompanying by hydrogen ion transfer from cell 4 to cell 5. Whether it is true or not, the mechanism of water dissociation has not yet been cleared. Many investigations introduced in this chapter will concern with this phenomenon.

7.3 Diffusional Model

An ion exchange membrane is assumed to be placed in a strong 1, 1 valent electrolyte solution. When an overlimiting current is passed across the membrane, an unstirred boundary layer is formed on the desalting surface of the membrane and the following water dissociation is generated in the boundary layer:

$$H_2O \underset{k_2}{\overset{k_1}{\Leftrightarrow}} H^+ + OH^- \tag{7.1}$$

k_1 and k_2 are the forward and reverse equilibrium constant, respectively. Rubinstein (1977) discussed the steady ion transfer based on the diffusional model by the following mass conservation with the electrical neutrality using subscripts 1, 2, 3, and 4 to represent, respectively, salt cations, salt anions, H^+ ions, and OH^- ions.

Mass conservation:

$$\frac{dJ_i}{dx} = 0 \quad or \quad J_i = constant \quad (i = 1, 2) \tag{7.2}$$

$$\frac{dJ_i}{dx} = R_W \quad (i = 3, 4) \tag{7.3}$$

R_W is the generation rate of H^+ and OH^- ions in Eqn (7.1). The ion fluxes J_i are given by the Nernst–Planck equations:

$$J_i = -D_i\frac{dC_i}{dx} - D_iC_i\frac{d\psi}{dx} \quad (i = 1, 3) \tag{7.4}$$

$$J_i = -D_i\frac{dC_i}{dx} + D_iC_i\frac{d\psi}{dx} \quad (i = 2, 4) \tag{7.5}$$

where $\psi = |z| F\psi/RT$ is the dimensionless electric potential.

Electroneutrality:

$$C_1 + C_3 = C_2 + C_4 \tag{7.6}$$

Equations (7.2)–(7.6) were solved by supplementing the boundary conditions in which nonpenetrability for co-ions of an ideal ion exchange membrane is provided. The computation introduced the current–voltage relationship, including the overlimiting current regime, and the profiles of electric potential with the profiles of H^+ and OH^- ion concentration in the unstirred layer.

The conclusion introduced in this research does not explain the difference of water dissociation phenomena of the cation exchange membrane and that of the anion exchange membrane.

7.4 Repulsion Zone

Patel et al. (1977) developed the theoretical model incorporated with "repulsion zones" in the solutions adjacent to the membrane surface. Here, the membrane is assumed to be ideally perm-selective, allowing the passage only of cations. In the repulsion zones, it is assumed that the repulsive forces exerted by the fixed membrane charges upon co-ions dominate the situation. Outside the repulsion zones, the rest of the solution boundary layers are present, and outside the boundary layers is the bulk solution phase (Fig. 7.3).

The mass conservation Eqns (7.2)–(7.5) are applied to the steady state electrodialysis system. The relationship between the space charge distribution and the electric potential gradient in the repulsion zones is given by the following Poisson's equation:

$$\frac{d^2\psi}{dx^2} = -\frac{4\pi RT}{zF\varepsilon}(C_1 + C_3 - C_2 - C_4) \tag{7.7}$$

where ε is the dielectric constant. However, away from the membrane surface, large changes in the electric potential are not expected so that the assumption of electoneutrality Eqn (7.6) is satisfied instead of Eqn (7.7). The generation rate of H^+ and OH^- ions is presented by Eqn (7.1).

Using the above mentioned equations and the Wien effect theory (Section 7.5), the thickness of the repulsion zone r_C, in which the recombination rate of H^+ and OH^- ions

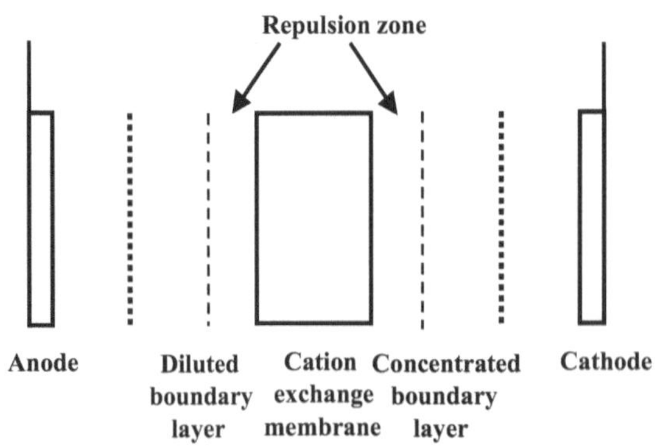

FIGURE 7.3 Formation of repulsion zones on the membrane surfaces. *Patel et al. (1977), p. 341, Fig. 1.*

is strongly affected by the local electric field, was computed. The results show that for cation exchange membranes, very high concentration of H^+ ions are built up at the membrane surface during polarization. The decrease in the recombination rate of H^+ and OH^- ions near the membrane is responsible for this.

The feature of this research is that it takes into account the Wien effect. However, the Wien effect theory does explain the difference between the phenomenon on the cation exchange membrane and that on the anion exchange membrane.

7.5 Wien Effect

The Wien effect (The second Wien effect) is the phenomenon generated in an electrolyte solution under applied, very strong electric potential fields of 10^6–10^7 V/m. In this circumstance, the ionic mobility is increased and Ohm's law has only a limited range of validity (Wien, 1928). For the case of strong electrolytes, this effect has been successfully interpreted as a destruction of the "ionic atmosphere." The initial effect is proportional to the square of the field intensity, and for very strong fields, the equivalent conductance approaches a limiting value, which is not greater than the limiting equivalent conductance for a small concentration. Weak electrolytes, on the other hand, show a much enhanced deviation from the Ohm's law; the conductance increase linearly over a considerable range of the field intensity, and the limit of the increase, if any, corresponds to complete dissociation of the total amount of electrolyte present (Wien, 1931). It was considered an open question whether the prevailing theory for the electrostatic interaction of the ions could account for this increased dissociation of weak electrolytes.

Onsager (1934) presented a result that was computed on the basis of the interionic attraction, and discussed its significance based on the theory of the Wien effect. The agreement with the available measurements of conductance in strong fields was considered satisfactory. In addition, the theory allowed some predictions concerning the rates of dissociation and recombination of the ions, and the considerations involved by this question were extended with equal ease to the case of high field intensities. The computed relative increase of the dissociation rate constant is given by the formula:

$$\frac{k_1}{k_0} = 1 + b + \frac{b^2}{3} + \frac{b^3}{18} + \frac{b^4}{180} + \frac{b^5}{2700} + \frac{b^6}{56700} + \cdots \tag{7.8}$$

with

$$b = 0.09636 \frac{E}{\varepsilon_r T^2} \tag{7.9}$$

where k_1 and k_0 are the forward equilibrium constant in Eqn (7.1) under the influence of an electric field and that without an electrical field, respectively. E is the electric field density (V/m), ε_r is the relative dielectric constant, and T is the absolute temperature. In the case of high field intensities ($E > 10^8$ V/m), Eqn (7.10) is introduced from Eqn (7.8), being used to calculate the effect of the electric field on the dissociation rate constant.

$$\frac{k_1}{k_0} = \left(\frac{2}{\pi}\right)^{1/2} (8b)^{-3/4} e^{(8b)^{1/2}} \tag{7.10}$$

The discussion mentioned above is applicable to analyze the mechanism of the water dissociation generated in the boundary layer adjacent to an ion exchange membrane.

7.6 Protonation and Deprotonation Reaction

Simons (1984, 1985) suggested that with anion exchange membranes, the water dissociation is caused by a following catalytic reversible protonation and a deprotonation reaction of weakly basic groups.

$$B + H_2O \underset{k_{-1}}{\overset{k_1}{\Leftrightarrow}} BH^+ + OH^- \tag{7.11}$$

$$BH^+ + H_2O \underset{k_{-2}}{\overset{k_2}{\Leftrightarrow}} B + H_3O^+ \tag{7.12}$$

where B is a neutral base such as tertiary or secondary amine groups.

It has been known that when the applied electric field is high enough, Ohm's law is no longer valid and the conductance of electrolytes increases rapidly with the field (Onsager, 1934). For weak electrolytes such as H_2O, this phenomenon is known as "the second Wien effect," and the mechanism of catalytic water dissociation reaction described above is presented in two steps as follows (Simons, 1979):

$$\underset{}{\overset{a}{B + H_2O}} \Leftrightarrow \overset{b}{BH^+ ... OH^-} \Leftrightarrow \overset{c}{BH^+ + OH^-} \tag{7.13}$$

$$BH^+ + H_2O \Leftrightarrow B \cdots H_3O^+ \Leftrightarrow B + H_3O^+ \tag{7.14}$$

where $BH^+...OH^-$ and $B...H_3O^+$ are encounter pairs in the same solvent cage. a–b is a chemical transformation process, and b–c is a separation and encounter process. Applying the steady state approximation to the encounter pair, the expressions for the overall forward reaction rate constant $\overrightarrow{k}$ and the overall reverse reaction rate constant $\overleftarrow{k}$ are:

$$\overrightarrow{k} = \frac{k_{ab}k_{bc}}{k_{ba} + k_{bc}}, \quad \overleftarrow{k} = \frac{k_{cb}k_{ba}}{k_{ba} + k_{bc}} \tag{7.15}$$

k_{ab} and k_{ba} are, respectively, the forward and reverse reaction rate constants for the chemical transformations. k_{bc} and k_{cb} are, respectively, the forward and reverse rate constants for the separation and encounter processes. The chemical transformation is so fast that the diffusion-controlled process (separation and encounter process) is the rate-determining step, so that one has $k_{ba} >> k_{bc}$ ($k_{ba} > 10^{12}\ s^{-1}$, while k_{bc} is usually between 10^{10} and $10^{11}\ s^{-1}$). Thus, the overall reaction rate constants defined by Eqn (7.15) is simplified as:

$$\overrightarrow{k} = \frac{k_{ab}k_{bc}}{k_{ba}}, \quad \overleftarrow{k} = k_{cb} \tag{7.16}$$

Onsager suggested that only k_{bc} increases with the electric field, whereas k_{ab}, k_{ba}, and the recombination step k_{cb} are independent of the field (Onsager, 1934; Eigen and Maeyer, 1959). One understands from this suggestion that $\overrightarrow{k}$ increases with the electric field, but $\overleftarrow{k}$ is not affected by the electric field.

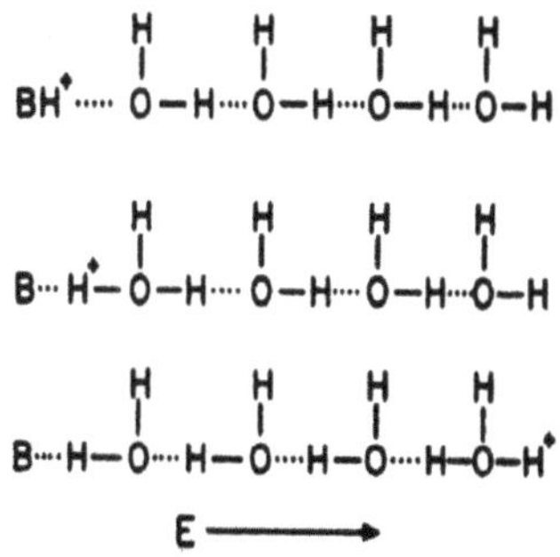

FIGURE 7.4 Deprotonation of amino group and proton conduction along a chain of oriented water molecules. *Simons (1979), p. 825, Fig. 3.*

We shall consider the following two ways by which the water dissociation reaction rate might be increased by the strong electric field (Simons, 1979):

1. In reaction (7.11), k_{bc} depends on the strength of an external field. In reaction (7.12), the site in B, for recombination with H_3O^+, is a lone pair so that the charge product is effectively negative when the proton is sufficiently close to the reaction site. So, k_{bc} would also increase with the electric field intensity for reaction (7.12).
2. Actual proton transfer is very fast due to the Grotthuss protonic conduction (de Grotthuss, 1806; Markovitch and Agmon, 2007) as illustrated in Fig. 7.4. The rate–determining step seems to be the necessary reorientation of water molecules between successive transfers (structural diffusion). When reaction (7.12) occurs, it is conceivable that there is a bridge of water molecules in the reaction layer, already favorably oriented for proton transfer by the strong electric field, leading to the increase of k_{bc} ($\overrightarrow{k}$).

Simons also suggested that quaternary ammonium groups do not bring about the reaction in Eqns (7.11) and (7.12). However, the conversion of the quaternary amino groups into the tertiary form creates the reaction of Eqns (7.11) and (7.12), and accelerates the water dissociation. Water dissociation at cation exchange membranes is generated when the membrane contains weakly acidic groups, such as carboxylic acid, according to the following catalytic reaction:

$$A^- + H_2O \underset{k_{-3}}{\overset{k_3}{\Leftrightarrow}} AH + OH^- \tag{7.17}$$

$$AH + H_2O \underset{k_{-4}}{\overset{k_4}{\Leftrightarrow}} A^- + H_3O^+ \tag{7.18}$$

where AH is a neutral acid groups such as carboxylic acid.

7.7 Experimental Research on the Water Dissociation Reactions

7.7.1 Current–pH Relationship

The water dissociation reaction was studied by current–pH curves as follows (Tanaka, 1974): The apparatus in Fig. 7.5 was assembled with strong base (quaternary ammonium or pyridinium groups) and strong acid (sulfonic acid groups) ion exchange membranes: Selemion ASG/CSG (Asahi Glass Co.), Aciplex CA-2/CK-2 (Asahi Chemical Co.), and Neosepta AVS-4T/CL-25T (Tokuyama Co.). The membrane was immersed in a 0.05- or a 0.1-M NaCl solution, and an electric current was passed. The observed current–pH curves (Fig. 7.6) show that the water dissociation reaction in the cation exchange membranes is strongly suppressed compared to the phenomenon in the anion exchange membranes. Figure 7.6 suggests that the intensity of the water dissociation reaction is strongly influenced by the functional groups fixed in the membrane. From this phenomenon, the water dissociation reaction is estimated to occur in the membrane. If the reaction is generated at the outside of the membrane (in the solution phase), the reaction is presumably the same for both membranes.

The apparatus in Fig. 7.7 was integrated with commercially available membranes. It consisted of a central concentrating cell (Con) partitioned into cells 1–4 by filter papers, desalting cells (De), and electrode cells bounded by the desalting cells. The effective membrane area was reduced to 0.367 cm^2 by using gaskets, and a 0.5-M NaCl solution was put in the electrode cells. Furthermore, 0.1-M NaCl solutions and 0.05-M NaCl solutions were put in the concentrating and desalting cells, respectively. After an electric current was passed through Ag–AgCl electrodes for 30 min, the solutions in cells 1–4 were taken out to measure pH values. The experiment was repeated, changing current density incrementally (Tanaka et al., 1982).

Table 7.1 is the results for Neosepta CH-60T/AFS-4T. The pH in cell 1 is declined to alkaline at first due to the water dissociation of the anion exchange membrane (Neosepta

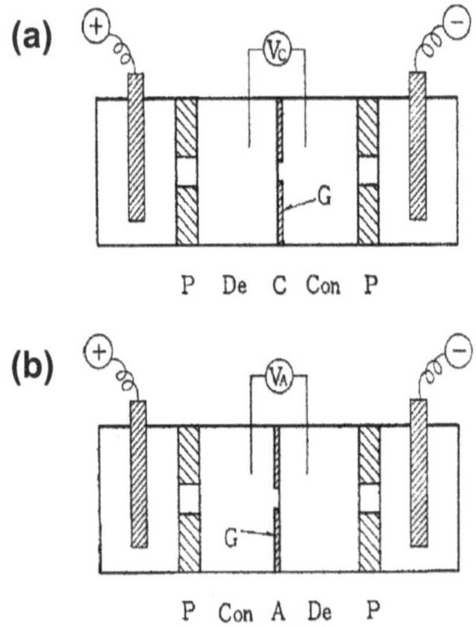

FIGURE 7.5 Cell arrangement. A, anion exchange membrane; C, cation exchange membrane; Con, concentrating cell; De, desalting cell; G, gasket; P, partition (agar). *Tanaka (1974), p. 451, Fig. 1.*

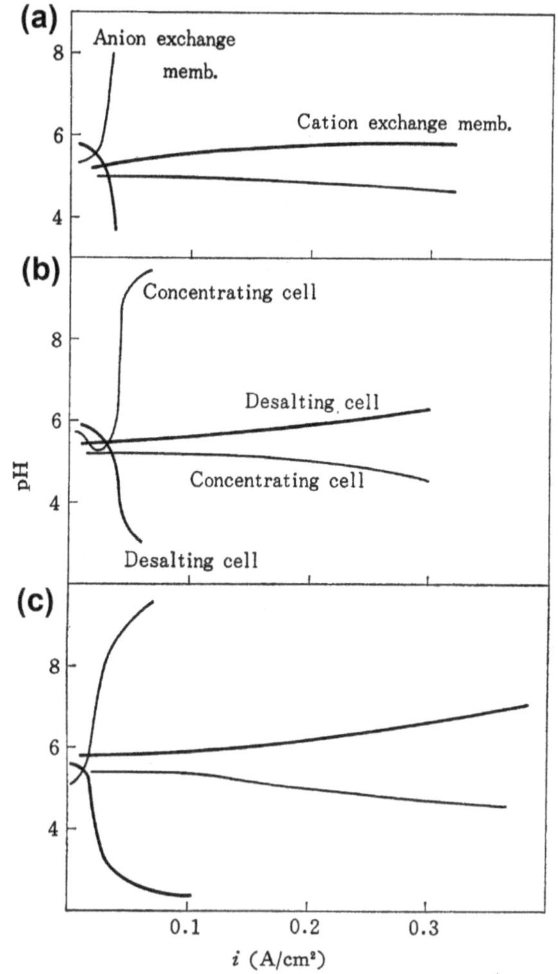

FIGURE 7.6 Electric current–solution pH relationship. (a) 0.05-M NaCl Selemion CMG/ASG, (b) 0.1-M NaCl Aciplex CK-2/CA-2, (c) 0.05-M NaCl Neosepta CL-25T/AVS-4T. *Tanaka (1974), p. 452, Fig. 4.*

AFS-4T). The pH in cells 2 and 3 is also declined to alkaline with increasing current density. The pH in cell 4 is decreased slightly at first due to the water dissociation of the cation exchange membrane (Neosepta CH-60T), but it turns to alkaline with current density increase due to the water dissociation of the anion exchange membrane. The water dissociation reaction is accelerated by the quaternary ammonium groups in the anion exchange membrane.

Next, in the above experiment, 0.1-M $MgCl_2$ solutions and 0.05-M $MgCl_2$ solutions were put in the concentrating and desalting cells, respectively, and pH shifts were measured in cells 1–4 (Table 7.2). In this experiment, the pH in cell 4 declined to acidic

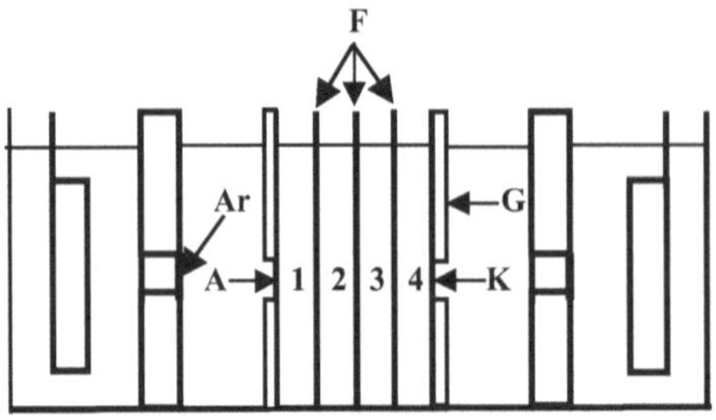

FIGURE 7.7 Electrodialysis apparatus for measuring water dissociation. 1–4, cells 1–4; A, anion exchange membrane; Ar, agar; Con, concentrating cell; De, desalting cell; F, filter; G, gasket; K, cation exchange membrane; P, partition. *Tanaka et al. (1982), p. 667, Fig. 1.*

due to the water dissociation of the cation exchange membrane, and the pH in cells 2 and 3 is also declined to acidic with increasing current density. The pH in cell 1 seems slightly decline to alkaline because of the water dissociation of the anion exchange membrane. However, it turns acidic due to the water dissociation of the cation exchange membrane. The water dissociation observed on the cation exchange membrane is caused by catalytic effects of $Mg(OH)_2$ precipitated on the desalting surface of the membrane. Such an accelerated water dissociation is seen in the bipolar membrane electrodialysis due to the catalytic effect of inorganic substances introduced into the interface layer in the bipolar membrane.

Finally, in the above experiment, the ion exchange membranes were replaced with Selemion CMV/AST. Seawater and 10 times as much as diluted seawater were put in the concentrating and desalting cells, respectively. The pH changes in cells 1–4 (Table 7.3) show that the pH in cells 1 and 2 declines to alkaline caused by water dissociation of the

Table 7.1 pH Changes of NaCl Solutions in Each Cell (Neosepta CH-60T/AFS-4T)

Cell No., i (A/cm^2)	1	2	3	4
0.02	6.66	6.72	6.62	6.79
0.04	6.89	6.71	6.64	6.81
0.06	10.84	8.08	6.67	6.68
0.08	11.31	10.40	6.91	6.71
0.10	11.63	10.85	7.10	6.71
0.12	11.91	11.22	9.01	6.65
0.14	12.02	11.68	10.33	6.91
0.16	12.20	11.71	10.79	7.08
0.18	12.30	11.89	11.01	8.22
0.20	12.29	11.86	11.11	8.21
0.30	12.39	12.19	11.63	10.91
0.40	12.41	12.21	11.92	11.59

Tanaka et al. (1982), p. 669, Table 1.

Table 7.2 pH Changes of $MgCl_2$ Solutions in Each Cell (Neosepta CH-60T/AFS-4T)

Cell No., i (A/cm^2)	1	2	3	4
0.008	5.32	5.93	6.03	6.20
0.020	5.41	6.00	6.11	6.12
0.032	5.90	6.12	6.13	5.91
0.040	6.08	6.00	5.92	5.50
0.052	6.21	5.98	5.82	5.15
0.060	6.18	5.97	5.65	4.00
0.072	6.00	4.00	2.50	1.98
0.080	6.80	3.19	2.09	1.61
0.100	6.70	2.69	1.86	1.49
0.160	3.08	1.92	1.60	1.31
0.240	2.08	1.68	1.48	1.21

Tanaka et al. (1982), p. 669, Table 2.

Table 7.3 pH Changes of Diluted Seawater in Each Cell (Selemion CMV/ASV)

Cell No., i (A/cm^2)	1	2	3	4
0.01	7.75	7.73	7.67	7.52
0.02	8.36	7.86	7.78	7.23
0.036	8.46	7.88	7.86	7.32
0.05	9.88	7.91	7.74	7.33
0.07	10.47	8.37	7.67	7.14
0.08	10.08	9.36	7.59	7.13
0.09	10.14	8.95	7.46	6.85
0.10	10.51	9.86	7.50	6.67
0.11	10.15	9.54	7.86	7.11
0.12	10.62	9.88	8.05	7.46
0.13	12.02	9.73	7.86	6.83
0.15	12.25	9.94	7.95	7.00
0.17	12.43	10.30	8.20	6.96
0.20	12.65	10.35	8.22	6.65
0.21	12.55	9.95	9.18	7.42
0.22	12.63	10.10	9.33	7.16
0.24	12.94	10.23	9.74	7.47
0.25	12.78	9.77	1.15	0.84
0.26	12.03	11.00	9.86	6.23
0.27	12.93	11.23	9.75	7.36
0.28	12.24	10.05	1.65	1.25
0.30	12.73	10.06	1.16	0.95

Tanaka (1975), p. 585, Table 1.

anion exchange membrane (Selemion AST). The pH in cell 4 is decreased with increasing current density due to the water dissociation of the cation exchange membrane (Selemion CMV). The pH in cell 3 is at first increased due to the water dissociation of the anion exchange membrane. However, it is decreased with the increase of current density due to the water dissociation of the cation exchange membrane. The water dissociation generated on the cation exchange membrane is caused by the strong catalytic effect of $Mg(OH)_2$ precipitated on the cation exchange membrane.

7.7.2 Influence of Ionic Electrolytes in a Solution on the Water Dissociation

The setup in Fig. 7.8(a) and (b) was assembled for measuring the water dissociation of the test strong base anion exchange membrane A* (Aciplex A-102, Asahi Chemical Co.) and the test strong acid cation exchange membrane K* (Aciplex K-102). The effective area of A* and K* was reduced to 0.264 cm^2 by a gasket, so that the water dissociation occurs more easily in A* and K* than the other membranes. D and C are the desalting cells and concentrating cells, respectively. A 0.1 eq/dm^3 solution of several chlorides was supplied to D by a pump at the rate of 0.1 cm^3/s. The 1 eq/dm^3 solution of the same electrolytes, which were supplied to D, was added into D′. A 1 eq/dm^3 NaCl solution was added into C (5 cm^3) and D″ as well. After these preparations, an electric current was passed for 10 min between Ag and AgCl electrodes. When water dissociation takes place in A* or K*, H^+ or OH^- ions generated are transferred into C. Solution pH in C was

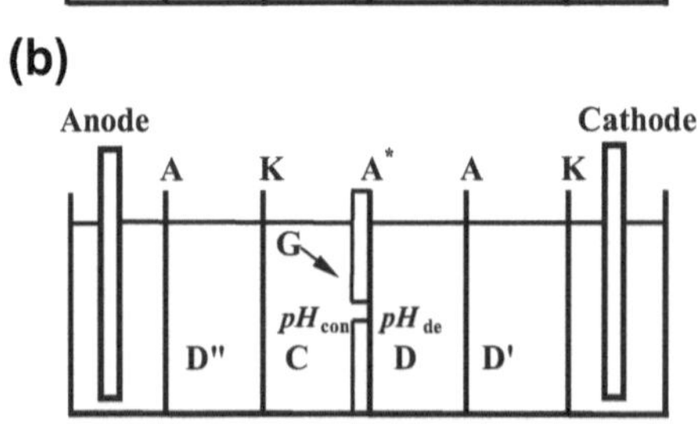

FIGURE 7.8 Electrodialysis apparatus for measuring water dissociation. A, anion exchange membrane; C, concentrating cell; D, desalting cell; G, gasket; K, cation exchange membrane, * supplied to K and A denotes the test membrane. *Tanaka et al. (1982), p. 668, Fig. 2.*

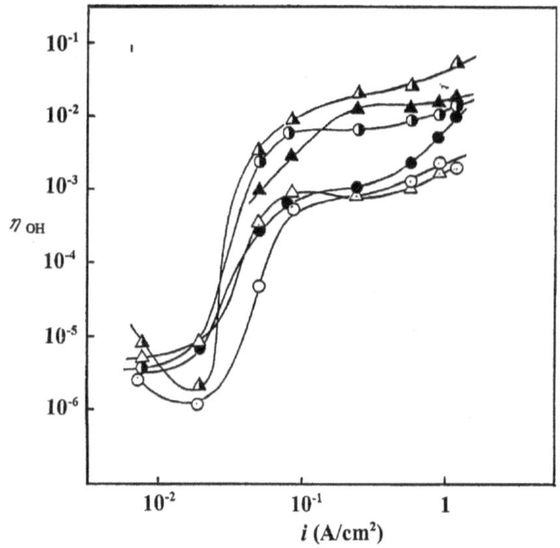

FIGURE 7.9 Current density versus current efficiency of OH^- ions. Anion exchange membrane, Aciplex A-102. ○: NaCl, ◑: $CaCl_2$, ●: $MgCl_2$, △: KCl, **△**: $BaCl_2$, ▲: $NiCl_2$. *Tanaka et al. (1982), p. 670, Fig. 7.*

confirmed to be almost steady 10 min after applying the electric current. The electrodialysis experiment was repeated, changing current densities incrementally. Current efficiencies for OH^- ions η_{OH} and for H^+ ions η_H were calculated from the pH changes in the solution in C (Tanaka et al., 1982).

Experimental results are shown in Fig. 7.9 (anion exchange membrane) and Fig. 7.10 (cation exchange membrane) and are understood as follows:

1. When the anion exchange membrane was immersed in a chloride solution, η_{OH} was in the range of 10^{-4} to 10^{-1}. The water dissociation reaction was moderately suppressed (Fig. 7.9).
2. When the cation exchange membrane was immersed in an $MgCl_2$ or an $NiCl_2$ solution, η_H was increased drastically at overlimiting current density. The violent water dissociation recognized here is caused by the $Mg(OH)_2$ or $Ni(OH)_2$ precipitated on the cation exchange membrane and its mechanism is discussed in Section 7.8.6. When the cation exchange membrane was immersed in the other chloride solutions, η_H was in the range of 10^{-5} to 10^{-3}, and water dissociation was confirmed to be suppressed extraordinarily (Fig. 7.10).

7.7.3 Influence of an Electric Potential Field on the Water Dissociation Reaction

An electrodialysis unit with cells I and II was assembled as shown in Fig. 7.11. A small central hole (diameter of 1 mm, length of 1.2 mm) was bored into the poly vinyl chloride

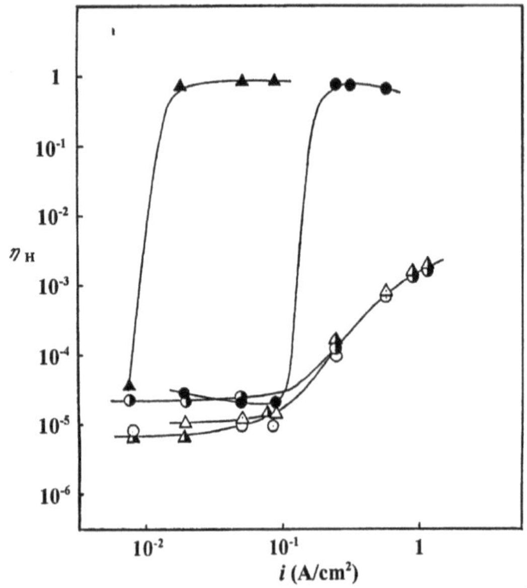

FIGURE 7.10 Current density versus current efficiency of H^+ ions. Cation exchange membrane, Aciplex K-102. ○: NaCl, ◑: $CaCl_2$, ●: $MgCl_2$, △: KCl, ▲: $BaCl_2$, ▲: $NiCl_2$. *Tanaka et al. (1982), p. 670, Fig. 5.*

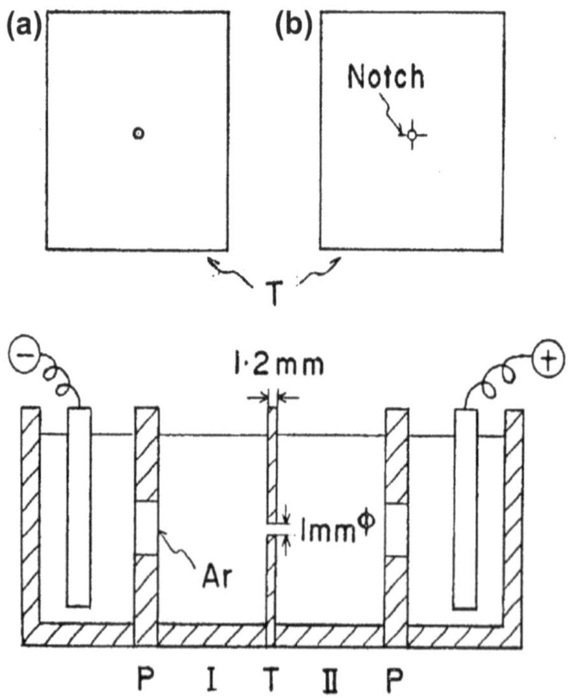

FIGURE 7.11 Experimental apparatus for measuring water dissociation. T: Test plate, I: Cell I, II: Cell II, P: Partition, Ar: Agar, ⊕: Anode, ⊙: Cathode. *Tanaka and Seno (1983b), p. 466, Fig. 1.*

test plate T (Fig. 7.11(a)). Agar including a 0.1 eq/dm^3 electrolyte solution was supplied into the central hole bored in poly vinyl chloride partition P (thickness 12 mm), 5 cm^3 of diluted electrolyte solutions (10^{-2}, 10^{-1}, 1, and 10 meq/dm^3) were put in cells I and II with one or two droplets of pH indicators (methyl red and bromo thymol blue), and 1 eq/dm^3 NaCl solutions were supplied in the electrode cells. An electric current was passed applying 500 V to both ends of the small hole in the test plate. The potential gradient in the hole was about 4×10^5 V/m (Tanaka and Seno, 1983b). The above small central hole in the test plate images the cation exchange membrane/solution interface applied an overlimiting current. The water dissociation occurred in the hole is observed from the color changes of the pH indicators. From the color changes, intensities of water dissociation were classified and listed in Table 7.4.

When a 0.1 meq/dm^3 NaCl solution was supplied in cells I and II with methyl red, the color of the bottom of cell I turned to light pink after 2 min of electric current passing.

Table 7.4 Intensities of Water Dissociation

Concentration (meq/dm^3)	10^{-2}	10^{-1}	1	10
Electrolyte				
$MnCl_2$	□	○	○	○
$CoCl_2$	□	○	○	□
$MgCl_2$[a]	□	○	○	△
$NiCl_2$	□	□	○	○
$MgCl_2$	□	□	△	×
$AlCl_3$	□	□	△	×
$Cr(NO_3)_2$	□	□	△	×
$CuCl_2$	□	□	△	×
$ZnCl_2$	□	□	△	×
NaBr	□	□	△	×
NaI	□	□	△	×
$NaNO_3$	□	□	△	×
Na_2SO_4	□	□	△	×
LiCl	□	□	×	×
NaCl	□	□	×	×
$CaCl_2$	□	□	×	×
$BaCl_2$	□	□	×	×
$Na_2S_2O_3$	□	□	×	×
$Fe(NO_3)_3$	□	△	△	×
$CH_3 \cdot NH_2 \cdot HCl$	□	△	△	×
$CH_3C_6H_4SO_3Na$	□	△	△	×
KCl	□	△	×	×
Na_2SO_3	□	△	×	×

○, Violent; □, mild; △, extremely mild; ×, not occurred.

[a]Before the experimental run, $Mg(OH)_2$ was deposited on the surface in the hole of the test plate.

Tanaka and Seno (1983b), p. 467, Table 1.

This is because H^+ ions generated in the hole due to the mild water dissociation transfer into cell II. When the NaCl solution with bromo thymol blue was electrodialyzed in the same way, the bottom of cell II turned light blue, indicating OH^- ion transfer due to the mild water dissociation.

When a 0.1 meq/dm^3 $MnCl_2$, $CoCl_2$, or $NiCl_2$ solution was added in cells I and II with methyl red, the bottom of the cell I turned deep red just after the electric current passing. When the same electrolyte solution was electrodialyzed with bromo thymol blue, the bottom of cell II turned to deep blue. It is estimated to occur mild water dissociation at first and precipitate $Mn(OH)_2$, $Co(OH)_2$, or $Ni(OH)_2$ crystals in the hole, then these hydroxides generate the catalytic violent water dissociation.

The phenomenon observed in the $MgCl_2$ solution electrodialysis was slightly complex. Namely, the water dissociation was moderate when the inner surface of the hole in the test plate was washed with an aqueous HCl solution ($MgCl_2$ in Table 7.4). However, when $Mg(OH)_2$ was precipitated on the inner surface, the violent water dissociation was observed ($MgCl_2^a$ in Table 7.4).

In the electrodialysis of other electrolyte solutions, violent water dissociation was not observed, but mild or extremely mild water dissociation was generated. Such a water dissociation reaction is assumed to be caused by the Wien effect generated under a strong electric potential gradient of 10^6–10^7 V/m (Section 7.5) (Onsager, 1934). However, the potential gradient 4×10^5 V/m applied in the hole in this experiment was rather small compared to the potential gradient suggested by Onsager. Thus, that the intensity of the water dissociation reaction is generally weak and the reaction is substantially not occurred in the solution.

The catalytic violent water dissociation generated in $MnCl_2$, $CoCl_2$, and $MgCl_2$ solutions is due to precipitation of the metallic hydroxides. The experimental results validate the violent water dissociation due to metallic hydroxides precipitated on the cation exchange membrane.

It is noticed that the primary amine $CH_3NH_2 \cdot$ HCl did not cause the violent water dissociation. Simons (1984, 1985) suggested that with anion exchange membranes the water dissociation is caused by the catalytic protonation and deprotonation reaction due to weakly basic groups such as tertiary or secondary amine groups in the membrane. It is estimated that the primary amine did not cause the water dissociation because it is dissolved in the solution.

In order to make sure the mechanism of the violent water dissociation, the inner surface of the test plate was notched in four points (Fig. 7.11(b)). Various insoluble hydroxides prepared as shown in Table 7.5 were inserted in the notches; 1 meq/dm^3 NaCl solutions were put in cells I and II and an electric current was passed similarly. The intensities of the water dissociation are classified and listed in Table 7.5, showing that $Mg(OH)_2$, $Ni(OH)_2$, $Co(OH)_2$, $Mn(OH)_2$, $Cu(OH)_2$, $Fe(OH)_3$, and $Al(OH)_3$ generate the violent water dissociation. $MnCO_3$ is estimated to convert to $Mn(OH)_2$, which arises the catalytic violent dissociation in the notches.

Table 7.5 Effect of Insoluble Salts on Water Dissociation

Insoluble Salts	Synthetic Method	Intensities of Water Dissociation
$Mg(OH)_2$	Commercial	○
$Ni(OH)_2$	$NiCl_2 + 2NaOH$	○
$Co(OH)_2$	$CoCl_2 + 2NaOH$	○
$Mn(OH)_2$	$MnSO_4 + 2NaOH$	○
$Cu(OH)_2$	$CuCl_2 + 2NaOH$	○
$Fe(OH)_3$	$Fe(NO_3)_3 + 3NaOH$	○
$Al(OH)_3$	$AlCl_3 + 3NaOH$	○
$MnCO_3$	$MnSO_4 + NaHCO_3$	○
$Ca(OH)_2$	Commercial	×
$Zn(OH)_2$	$ZnCl_2 + 2NaOH$	×
$CaSO_4$	Commercial	×
$BaSO_4$	$BaCl_2 + 2NaOH$	×
$CaCO_3$	$CaCl_2 + NaHCO_3$	×

○, Violent; ×, not occurred.
Tanaka and Seno (1983b), p. 468. Table 2.

7.7.4 Precipitation of Insoluble Metallic Hydroxides on the Membrane Surface and Generation of the Water Dissociation

In seawater electrodialysis, the electrodialyzer operations were some times interrupted by $Mg(OH)_2$ precipitation. The following experiments are carried out to find the cause of the troubles and to operate electrodialyzers stably.

$Mg(OH)_2$ was precipitated on the desalting surface of the cation exchange membrane K* (Selemion CMV) in the apparatus illustrated in Fig. 7.8(a). Supplying a 0.1-M NaCl solution into D, an electric current was passed (Tanaka and Seno, 1983b; Tanaka, 2007a). Current density i versus H^+ ion current efficiency η_H exhibits stronger water dissociation on the cation exchange membrane as shown in Fig. 7.12. Next, $Mg(OH)_2$ was precipitated on the desalting surface of the anion exchange membrane A* (Selemion ASV). Electrodialysis in this case resulted in the generation of weaker water dissociation and the dissolution of $Mg(OH)_2$. The mechanism of these phenomena is understandable from the illustration in Fig. 7.13, which displays that $Mg(OH)_2$ layer on the cation exchange membrane is stable because OH^- ions generated in the membrane pass through the $Mg(OH)_2$ layer. However, $Mg(OH)_2$ layer formed on the anion exchange membrane is dissolved because H^+ ions pass through the $Mg(OH)_2$ layer.

In order to confirm the influence of inorganic substances on the water dissociation on the cation exchange membrane (Aciplex K-102), the 0.02 M NaCl solutions suspending 0.1% (W/V) of $Mg(OH)_2$, $Ca(OH)_2$, $Fe(OH)_3$ $Al(OH)_3$, $MgCO_3$, $CaCO_3$, $CaSO_4$, or SiO_2 was fed into D in Fig. 7.8(a). Passing an electric current for 10 min, η_H was measured. The results show that inorganic hydroxides; $Mg(OH)_2$, $Ca(OH)_2$, $Fe(OH)_2$, and $Al(OH)_3$, generate catalytic strong water dissociation; however, $CaCO_3$, $CaSO_4$, and SiO_2 did not cause strong water dissociation. η_{OH} was measured for the anion exchange membrane

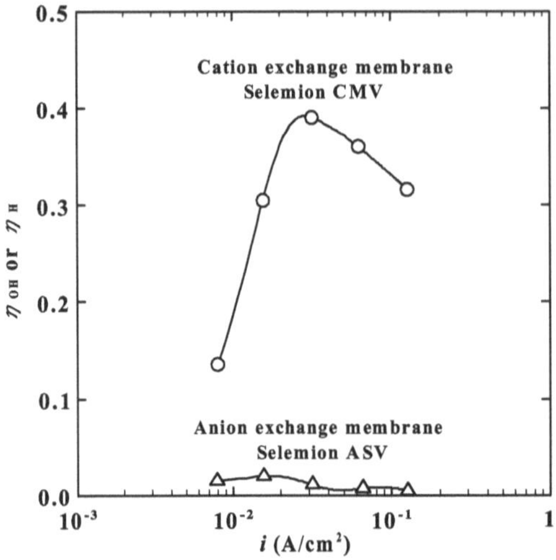

FIGURE 7.12 Precipitation of $Mg(OH)_2$ on a membrane surface and generation of water dissociation. *Tanaka and Seno (1983b), p. 469, Fig. 3.*

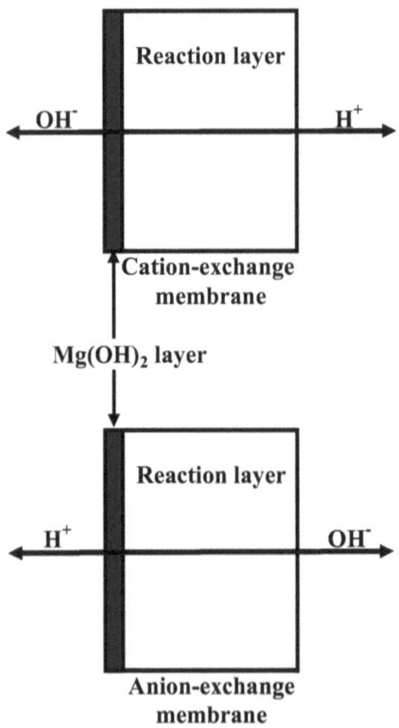

FIGURE 7.13 $Mg(OH)_2$ precipitation on the membrane surface and water dissociation generation (illustration). *Tanaka and Seno (1983b), p. 470, Fig. 5.*

(Aciplex A-102) with $Mg(OH)_2$ supplied into cell D; however, the accelerated water dissociation was not detected.

Next, the cation exchange membrane (Aciplex K-102) was incorporated with D in the apparatus in Fig. 7.8(a). Supplying a 0.02-M NaCl solution suspended 0.1% (W/V) $Fe(OH)_3$ to D and passing 50 mA of an electric current for 5 min, $Fe(OH)_3$ was precipitated on the desalting surface of the cation exchange membrane K*. A 0.02 eq/dm^3 NaCl solution, diluted seawater (Cl^- ion concentration 0.02 eq/dm^3) or a 0.02 eq/dm^3 $MgCl_2$ solution were supplied into D and η_H was measured by applying current density i during 10 min. Plotting of η_H against i (Fig. 7.14) shows that η_H is increased due to the water dissociation caused by the $Mg(OH)_2$ formed by the coupled reaction with OH^- ions generated in the $Fe(OH)_3$ layer and Mg^{2+} ions dissolved in the feeding solution.

From Figs 7.12–7.14, the mechanism of the violent water dissociation in seawater electrodialysis is illustrated in Fig. 7.15. Here, the water dissociation occurs due to the acceleration of concentration polarization and $Fe(OH)_3$ precipitation on the cation exchange membrane. OH^- ions generated by the violent water dissociation combine with Mg^{2+} ions dissolving in seawater to precipitate $Mg(OH)_2$ on the membrane surface. Successively, $Mg(OH)_2$ induces extremely violent water dissociation reaction which is accelerated further by the precipitated $Mg(OH)_2$ itself.

Finally, the Selemion CMV, Aciplex K-102, or Neosepta CH-45T cation exchange membrane was integrated with D in the apparatus in Fig. 7.8(a). Supplying a 0.02-M NaCl solution suspended 0.1% (W/V) $Fe(OH)_3$ and passing an electric current I for 5 min, $Fe(OH)_3$ was precipitated on K*. Then, feeding a 0.02-M NaCl solution and passing a 0.1 A/cm^2 of electric current, η_H was measured. Furthermore, disassembling

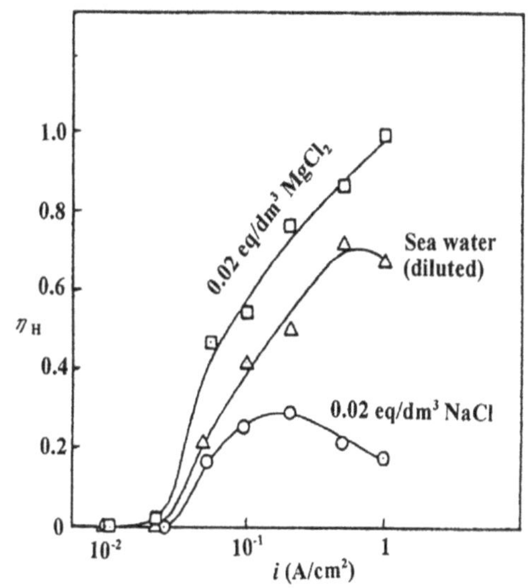

FIGURE 7.14 Effect of Mg^{2+} ions on water dissociation generated on a cation exchange membrane on which $Fe(OH)_3$ is precipitated. Aciplex K-102. *Tanaka (2007a), p. 239, Fig. 9.*

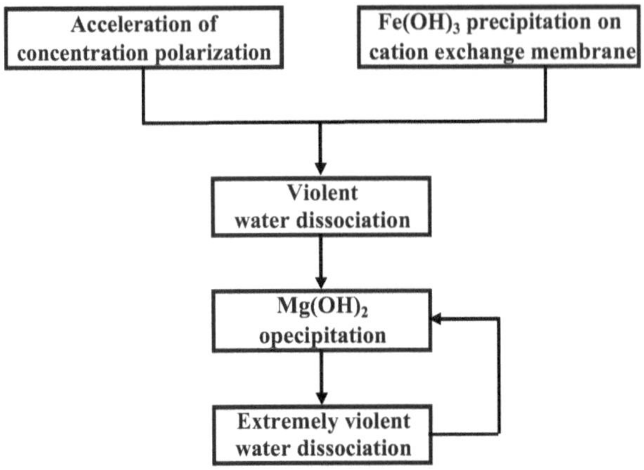

FIGURE 7.15 Mechanism of violent water dissociation generation on the cation exchange membrane. *Tanaka (2007a), p. 239, Fig. 10.*

the apparatus, the K* membrane was washed in a 1-M HCl solution with ultrasonic waves (35 kHz, 15 min). Fe components dissolved into the solution were analyzed with atomic absorption spectrometry. Changing the electric current I, the experiment was repeated (Tanaka, 2007a). The relationship between η_H and the quantity of $Fe(OH)_3$ attached to the membrane surface was measured as indicated in Fig. 7.16, which shows that η_H increases with $Fe(OH)_3$ quantity, i.e., $Fe(OH)_3$ volume. Figure 7.16 demonstrates the catalytic water dissociation generated in the $Fe(OH)_3$ layer (catalyzer).

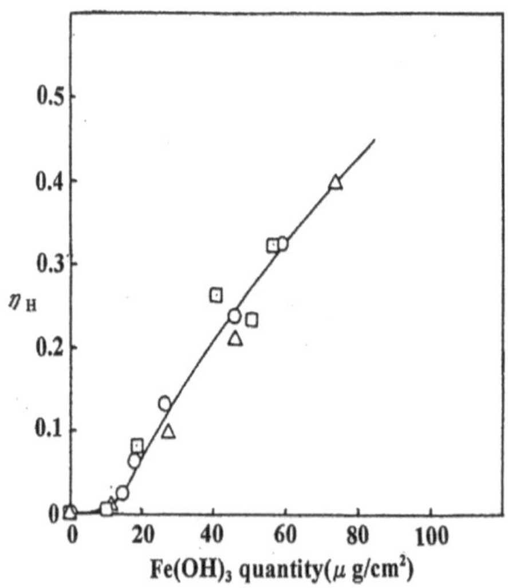

FIGURE 7.16 Precipitation of $Fe(OH)_3$ on the surface of a cation exchange membrane and generation of the water dissociation reaction. 0.02-M NaCl 0.1 A/dm^2. ○, Selemion CMV; △, Aciplex K-102; □, Neosepta CH-45T. *Tanaka (2007a), p. 238, Fig. 8.*

7.7.5 Electric Potential in an Ion Exchange Membrane

Ion exchange groups in an ion exchange membrane immersed in an electrolyte solution are usually electrically neutralized by counter-ions to form nonionic exchange groups, and the electric neutrality is satisfied in the membrane. With an elevated electric current, however, electrolyte concentration of the solution contacting a desalting surface of the membrane is decreased. This phenomenon induces the dissolution of counter-ions combined with the ion exchange groups into the solution phase to form a charged layer on the membrane surface and to form potential difference ψ_m at the membrane/solution interface. The interface potential difference relates to the Donnan potential (Donnan, 1924), repulsion zone (Patel et al., 1977), and space charge (Rubinstein and Shtilman, 1979; Rubinstein, 1981). The electrical potential in the charged layer reflects the electrochemical free energy barrier, i.e., activation energy (Kolyubin et al., 1966). The water dissociation reaction is generated in the charged layer. Thus, the charged layer is also defined as the reaction layer.

When counter-ions are depleted in the charged layer, the electric neutrality is not satisfied. The water dissociation reaction is generated and accelerated due to potential gradient $d\psi/dx$ applied in the charged layer (Section 7.8.3). Furthermore, in order to understand the difference of the water dissociation reaction intensities between a cation exchange membrane and an anion exchange membrane, one has to notice that the reaction is restricted under an electric potential ψ_m due to repulsive force between ionic groups in the charged layer and H^+ or OH^- ions generated in the reaction layer (Section 7.8.4, Fig. 7.25). This suggestion was confirmed by the following experiment (Tanaka, 2010; Tanaka and Seno, 1983a):

The ionic types of the following strong acid and strong base ion exchange membranes were adjusted suitably by immersing the membrane into a 1 eq/dm^3 NaCl solution:

Aciplex K-102 (Sulfonic acid groups, Asahi Chemical Co.)
Aciplex A-102 (quaternary pyridinium groups)
Selemion CMV (Sulfonic acid groups, Asahi Glass Co.)
Selemion AST (quaternary ammonium groups)

The membrane was incorporated with a setup illustrated in Fig. 7.17 with the AgCl electrode E_1 fixed on the membrane surface. The potential of the electrode E_1 is equivalent to that at p on the membrane surface in the figure. Another electrode E_2 (AgCl or Pt) was placed in the cell, keeping the distance from the membrane at 1 mm. Supplying water or a 0.001-M NaCl solution into both cells, the chronopotentiometric potential at p (E_1) was measured on the basis of E_2 while mixing the solution intermittently (Fig. 7.18). The potentials of E_1 and E_2 are equivalent to, respectively, the electric potential in the charged layer in the membrane ψ_m and the potential in the solution ψ_{sol}. The relationship between ψ_m, ψ_{sol}, and the electric potential difference at the membrane/solution interface ψ_{inter} is given by the following equation:

$$\psi_m = \psi_{sol} + \psi_{inter} \tag{7.19}$$

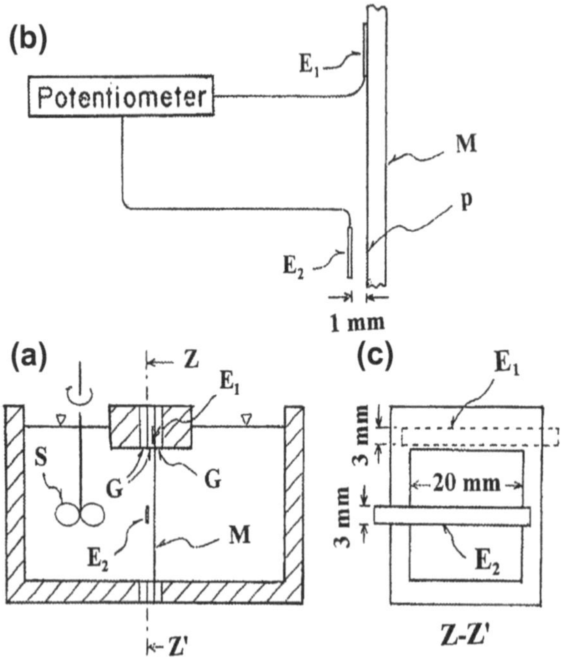

FIGURE 7.17 Apparatus for measuring electric potential in an ion exchange membrane. *Tanaka and Seno (1983a), p. 268, Fig. 1.*

ψ_m was evaluated from potential changes V of the interface potential ψ_{inter} detected on the basis of E_2, i.e., putting $\psi_{sol} = 0$ in Eqn (7.19). Then the setup was reassembled removing the membrane, and V_0 was measured without mixing. ψ_m was determined from $\psi_m = V - V_0$ as listed in Table 7.6, showing that:

1. $|\psi_m|$ is more increased when the membrane is placed in water than in electrolyte solutions.
2. $|\psi_m|$ in the cation exchange membrane is larger than that in the anion exchange membrane.

7.8 Mechanism of the Water Dissociation Reaction

7.8.1 Current Efficiency for the Water Dissociation Reaction (Tanaka, 2007b)

With an unapplied electric potential field, water molecules dissociate to H^+ and OH^- ions as presented by the following equilibrium formula (Eigen, 1954):

$$H_2O \underset{k_b}{\overset{k_a}{\Leftrightarrow}} H^+ + OH^- \tag{7.20}$$

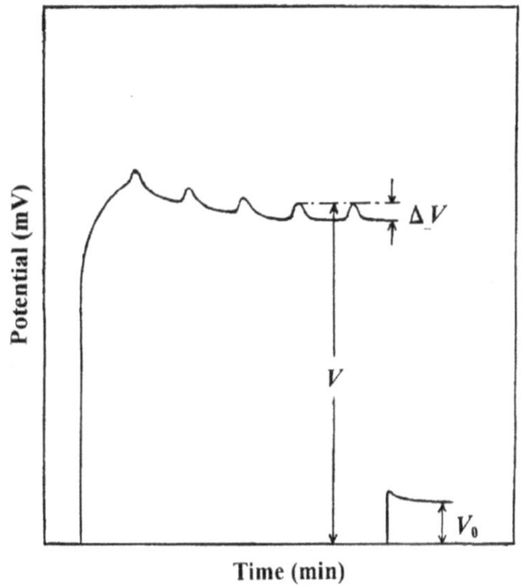

FIGURE 7.18 Chlonopotentiometric change of membrane surface potential. *Tanaka and Seno (1983a), p. 268, Fig. 2.*

Table 7.6 Electric Potential Generated in Ion Exchange Membranes (mV)

	Selemion CMV		Aciplex K-102	
Concentration (meq/dm^3)	0	1	0	1
Counter ion				
H^+	−241.0	−52.7	−244.8	−44.8
Na^+	−227.3	−69.8	−241.0	−88.8
K^+	−219.6	−11.6	−216.7	−51.9
NH_4^+	−207.1	−65.2	−220.2	−46.4
Mg^{2+}	−109.4	−12.2	−134.2	−8.9
Ca^{2+}	−113.0	−29.7	−114.2	−4.7
Ba^{2+}	−100.6	−26.8	−109.6	−4.0
Ni^{2+}	−103.5	−15.7	−138.0	−4.9
Co^{2+}	−119.2	−17.7	−131.8	−10.1

	Selemion CMV		Aciplex K-102	
Concentration (meq/dm^3)	0	1	0	1
Counter ion				
OH^-	+48.6	+8.8	+100.4	+0.3
Cl^-	+11.7	−0.6	+10.5	+0.1
Br^-	+6.2	+0.6	+22.1	+1.5
I^-	+9.6	+5.1	+16.5	+1.1
NO_3^-	+89.9	+3.7	+149.3	+34.8
HCO_3^-	+55.4	+6.0	+102.0	0
SO_4^{2-}	+22.2	0	+47.1	+16.3
SO_3^{2-}	+32.7	+1.7	+81.4	+12.3
$S_2O_3^{2-}$	+30.3	+2.4	+77.0	+63.2

Tanaka and Seno (1983a), p. 269, Table 1.

$$k_a = 2 \times 10^{-5}\ \mathrm{s}^{-1},\ \ k_b = 1.5 \times 10^{14}\ \mathrm{cm^3 mol^{-1} s^{-1}}$$

k_a and k_b are, respectively, the forward and reverse reaction rate constant.

As described in Section 7.7.1, the water dissociation reaction is generated in the membrane. Thus, Eqn (7.20) is generated under an applied potential field within the reaction layer formed in the membrane. The reaction layer is equivalent to the charged layer as described in Section 7.7.5. With an applied potential field, however, the equilibrium in is essentially perturbed. k_a in the strong base anion exchange membrane is considerably increased due to the applied potential field in the reaction layer, while k_b remains being unchanged.

One assumes here that H^+ and OH^- ions are generated by the water dissociation reaction (Eqn (7.20)) at overlimiting current density, and the reaction layer is formed at the inside of the membrane. H^+ ions and OH^- ions are transported in the reaction layer as shown in Fig. 7.19. The generation rate of H^+ ions σ_H and OH^- ions σ_{OH} is:

In the reaction layer formed in the anion exchange membrane:

$$\begin{aligned}\sigma_H &= \int_x^l (k_a C_{H_2O} - k_b C_H C_{OH})dx = k_a C_{H_2O}(l-x) - k_b \int_x^l C_H C_{OH} dx \\ \sigma_{OH} &= \int_0^x (k_a C_{H_2O} - k_b C_H C_{OH})dx = k_a C_{H_2O} x - k_b \int_0^x C_H C_{OH} dx\end{aligned} \tag{7.21}$$

In the reaction layer formed in the cation exchange membrane:

$$\begin{aligned}\sigma_H &= \int_0^x (k_a C_{H_2O} - k_b C_H C_{OH})dx = k_a C_{H_2O} x - k_b \int_0^x C_H C_{OH} dx \\ \sigma_{OH} &= \int_x^l (k_a C_{H_2O} - k_b C_H C_{OH})dx = k_a C_{H_2O}(l-x) - k_b \int_l^x C_H C_{OH} dx\end{aligned} \tag{7.22}$$

where x is an axis drawn in the reaction layer, C_H, C_{OH}, and C_{H_2O} are the concentration of H^+ ions, OH^- ions, and H_2O on x. l is the thickness of the reaction layer and it is equivalent to the thickness of the membrane. C_{H_2O} is constant and $C_H C_{OH}$ is equivalent to the ion product K_W as follows:

$$C_H C_{OH} = K_W \tag{7.23}$$

The transport rate (flux) of H^+ ions J_H and that of OH^- ions J_{OH} on x is expressed by the Nernst–Planck equation in the reaction layer, assuming that the solution convection is negligible in the layer as follows:

$$\begin{aligned}J_H &= -D_H \frac{dC_H}{dx} - \frac{F D_H C_H}{RT}\frac{d\psi}{dx} \\ J_{OH} &= -D_{OH} \frac{dC_{OH}}{dx} + \frac{F D_{OH} C_{OH}}{RT}\frac{d\psi}{dx}\end{aligned} \tag{7.24}$$

Mass balance in Fig. 7.19 is given by the following formulae defined between the generation and transport of H^+ and OH^- ions.

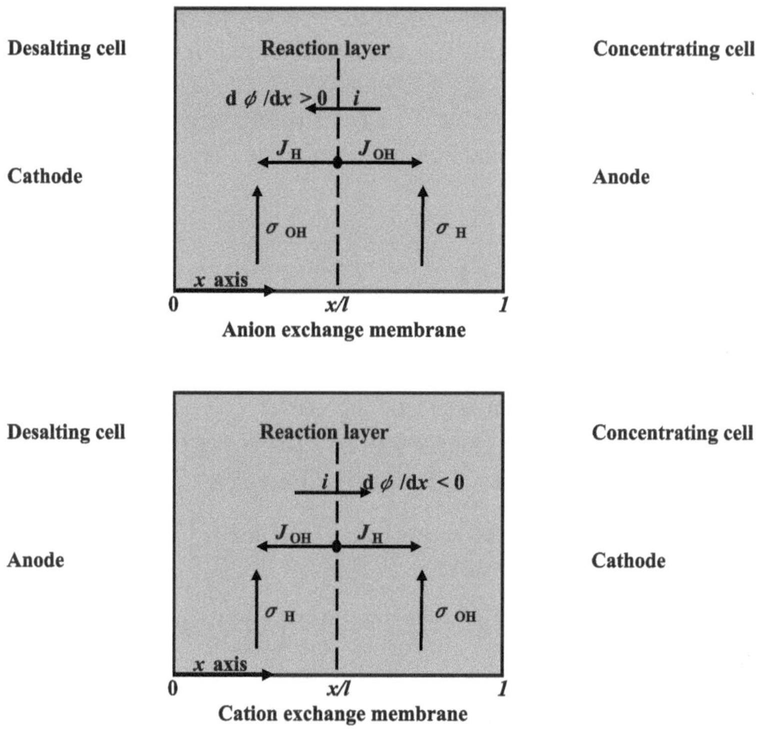

FIGURE 7.19 Generation and transport of H^+ and OH^- ions in the water dissociation reaction layer. *Tanaka (2010), p. 349, Fig. 1.*

In the reaction layer formed in the anion exchange membrane:

$$\sigma_H = -J_H = -\left(\frac{i}{F}\right)\eta_H \quad \sigma_{OH} = J_{OH} = -\left(\frac{i}{F}\right)\eta_{OH} \quad i < 0 \tag{7.25}$$

In the reaction layer formed in the cation exchange membrane:

$$\sigma_H = J_H = \left(\frac{i}{F}\right)\eta_H \quad \sigma_{OH} = -J_{OH} = \left(\frac{i}{F}\right)\eta_{OH} \quad i > 0 \tag{7.26}$$

in which η_H and η_{OH} are current efficiency for H^+ and OH^- ions, respectively; i is current density. When an electrolyte solution dissolves salt cation A and salt anion B, the total current efficiency for the water dissociation reaction η is:

$$\begin{aligned} \eta &= \eta_H + \eta_{OH} \\ \eta + \eta_A + \eta_B &= 1 \end{aligned} \tag{7.27}$$

H^+ ions and OH^- ions generated in the reaction layer are transported by migration and diffusion toward electrodes. Both ions carry an electric current and support on current efficiency $\eta = \eta_H + \eta_{OH}$ perfectly at every point in the reaction layer. If H^+ ions and OH^- ions collide and create water molecules in this circumstance, η must be decreased and the material balance in an electric current must be disturbed.

For maintaining the material balance of ionic species in the electric current and for realizing Eqn (7.27), new water-splitting reactions are generated at every point in the reaction layer.

With Eqns (7.21), (7.22), (7.24)–(7.27), η is introduced as Eqn (7.28) for the anion exchange membrane and cation exchange membrane. The sign of i is plus for both membranes.

$$\left(\frac{i}{F}\right)\eta = \sigma_H + \sigma_{OH} = k_a C_{H_2O} l - k_b \int_0^1 C_H C_{OH} dx = (k_a C_{H_2O} - k_b k_w) l \tag{7.28}$$

7.8.2 Concentration Distribution of H^+ and OH^- Ions in the Reaction Layer, Area Electric Resistance of the Reaction Layer and Forward Reaction Rate Constant (Tanaka, 2007b, 2010)

Current efficiency is also introduced from Eqn (7.24) as:

For the anion exchange membrane:

$$-\frac{i}{F}\eta = -J_H + J_{OH} = -\left\{\left(-\frac{dX_H}{dx} + KX_H\right) + \left(\frac{dX_{OH}}{dx} + KX_{OH}\right)\right\} \quad i < 0 \tag{7.29}$$

For the cation exchange membrane:

$$\left(\frac{i}{F}\right)\eta = J_H - J_{OH} = \left(-\frac{dX_H}{dx} + KX_X\right) + \left(\frac{dX_{OH}}{dx} + KX_{OH}\right) \quad i > 0 \tag{7.30}$$

in which:

$$X_H = D_H C_H \quad X_{OH} = D_{OH} C_{OH} \tag{7.31}$$

$$K = -\frac{F}{RT}\frac{d\psi}{dx} = -\frac{F}{RTl}\frac{d\psi}{d\xi} \tag{7.32}$$

$$\xi = \frac{x}{l} \tag{7.33}$$

K is the potential gradient function, $K < 0$ for the anion exchange membrane, and $K > 0$ for the cation exchange membrane.

Starting from Eqns (7.29)–(7.33), the distribution of C_H and C_{OH} in the reaction layer is introduced as follows:

$$C_H = \sqrt{\frac{u_{OH}}{u_H} K_W \exp(Z)} \tag{7.34}$$

$$C_{OH} = \sqrt{\frac{u_H}{u_{OH}} K_W \exp(-Z)} \tag{7.35}$$

$$Z = -\frac{F}{2RT}(2\xi - 1)\frac{d\psi}{d\xi} \tag{7.36}$$

in which u_H and u_{OH} are, respectively, the mobilities of H^+ ions and OH^- ions. K_W is the ion product of H^+ ions and OH^- ions (Eqn (7.23)).

Assuming a linear change of ψ in the reaction layer:

$$\frac{d\psi}{d\xi} = \Delta\psi = -\nu li \frac{\Delta\psi}{l} = -\nu i \tag{7.37}$$

In electrodialysis experiments, the intensity of the water dissociation reaction is measured from the pH changes in the concentrating cell, which is equivalent to pH values at $x/l = \xi = 1$ in Fig. 7.19.

In the anion exchange membrane, the concentration of OH^- ions $C_{OH,\xi=1}$ is given by putting $\xi = 1$ in Eqns (7.35) and (7.36):

$$C_{OH,\xi=1}(\mathrm{mol/cm^3}) = \sqrt{\frac{u_H}{u_{OH}} K_W \exp\left\{-\left(\frac{F}{2RT}\right) i\nu l\right\}} \quad i < 0 \tag{7.38}$$

$$\frac{u_H}{u_{OH}} = \frac{1.6488 + 4.710 \times 10^{-2} T^{0.9}}{5.271 + 0.2786 T^{0.9}} \tag{7.39}$$

$$-\log K_W = 20.932 - 4.204 \times 10^{-2} T + 1.504 \times 10^{-4} T^2 \tag{7.40}$$

The pH value at $\xi = 1$ is:

$$\mathrm{pH} = -\log K_W + \log C_{OH,\xi=1} - 3 \tag{7.41}$$

The area electric resistance of the reaction layer is introduced from Eqns (7.38) and (7.41):

$$\nu l = \frac{2RT}{Fi \log e}\left\{\mathrm{pH} + 3 - \frac{1}{2}\log\left(\frac{u_H}{u_{OH}}\right)\left(\frac{1}{K_W}\right)\right\} \quad i < 0 \tag{7.42}$$

In the cation exchange membrane, $C_{H,\xi=1}$ is given by putting $\xi = 1$ in Eqns (7.34) and (7.36):

$$C_{H,\xi=1} = \sqrt{\frac{u_{OH}}{u_H} K_W \exp\left\{\left(\frac{F}{2RT}\right) i\nu l\right\}} \quad i < 0 \tag{7.43}$$

$$\frac{u_{OH}}{u_H} = \frac{0.29280 + 1.5478 \times 10^{-2} T^{0.9}}{15.666 + 0.4476 T^{0.9}} \tag{7.44}$$

The pH value at $\xi = 1$ is:

$$\mathrm{pH} = -\log C_{OH,\xi=1} - 3 \tag{7.45}$$

The area electric resistance of the reaction layer is introduced from Eqns (7.43) and (7.45):

$$\nu l = -\frac{2RT}{Fi \log e}\left\{\mathrm{pH} + 3 + \frac{1}{2}\log\left(\frac{u_{OH}}{u_H}\right) K_W\right\} \quad i > 0 \tag{7.46}$$

A 25 °C 0.001-, 0.01-, or 0.1-M NaCl solution was supplied to the setup in Fig. 7.8(a) and (b) incorporated with Selemion ASR and CMR (Asahi Glass Co.). The effective area of A* and K* was reduced to 0.383 cm^2 by a gasket. An electric current was passed for 10 min between Ag and AgCl electrodes. Changing current density i incrementally, current efficiencies for OH^- ions η_{OH} and for H^+ ions η_H were calculated from the pH changes in the solution. The experimental results are described below, disregarding minus signs of current density for anion exchange membranes.

The concentration distribution of H^+ ions C_H and OH^- ions C_{OH} is calculated using Eqns (7.34)–(7.37) and shown in Fig. 7.20. In the cation (anion) exchange membrane, C_H is decreased (increased) and C_{OH} is increased (decreased) at $\xi = 0$, and C_H is increased (decreased) and C_{OH} is decreased (increased) at $\xi = 1$. The concentration changes in the anion exchange membrane is larger than that in the cation exchange membrane because the intensity of the water dissociation reaction in larger in the anion exchange membrane than in the cation exchange membrane.

vl is calculated using Eqns (7.42) and (7.46), and the relationship between vl and i is shown in Fig. 7.21. vl decreases with the increase of i toward a definite value and increases with the decrease of i toward the infinite.

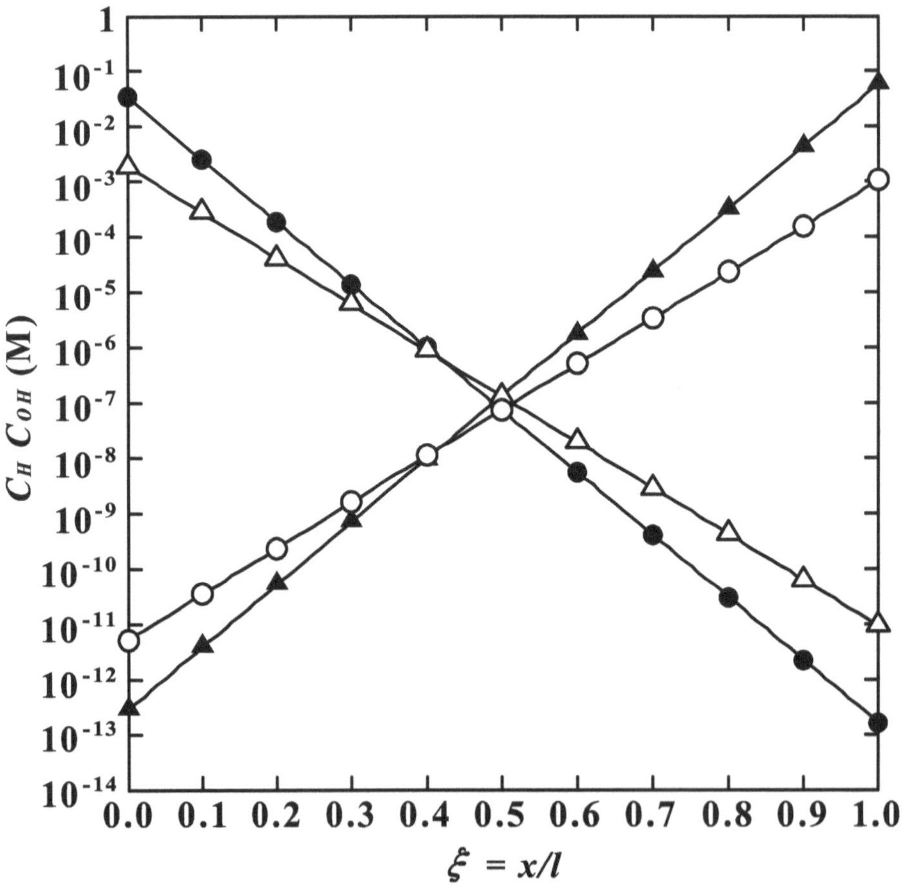

FIGURE 7.20 H^+ ion and OH^- ion concentration distribution in the reaction layer. ○●, C_H; △▲, C_{OH}; Open, Selemion CMR; Filled, Selemion ASR. $C' = 0.01$-M NaCl, $i = 0.20$ A/cm^2, $T = 25$ °C.

Area electric resistance of an ion exchange membrane R is calculated from the parallel electrical circuit consist of the area electric resistance due to the water dissociation reaction νl and area ohmic resistance of the membrane R_Ω as follows:

$$R = \frac{\nu l R_\Omega}{\nu l + R_\Omega} \tag{7.47}$$

The potential gradient in the membrane is:

$$\frac{d\psi}{dx} = Ri = \frac{\nu l R_\Omega i}{\nu l + R_\Omega} \tag{7.48}$$

In Eqn (7.47), extrapolating $i \to 0$ corresponds to $\nu l \to \infty$ (Fig. 7.21):

$$\lim_{i \to 0} R = \lim_{\nu l \to \infty} \frac{\nu l R_\Omega}{\nu l + R_\Omega} = R_\Omega \tag{7.49}$$

Thus, $d\psi/dx$ is extrapolated to $R_\Omega i$ at $i \to 0$.

The specific electric resistance of the reaction layer is calculated by inputting $\nu = \nu l/l$ in Fig. 7.21. l is equivalent to the thickness of the membrane, and 0.011 cm for Selemion ASR and CMR integrated into the experimental setup.

Forward reaction rate constant k_a is introduced from Eqn (7.28) as follows:

$$k_a = \frac{1}{C_{H_2O}} \left(k_b K_W + \frac{i\eta}{Fl} \right) \tag{7.50}$$

k_a is calculated putting into Eqn (7.50) the following parameters and η observed in the experiment, and it is plotted against i (Fig. 7.22). ka is extremely increased compared to $2 \times 10^{-5} s^{-1}$ in Equation (7.20).

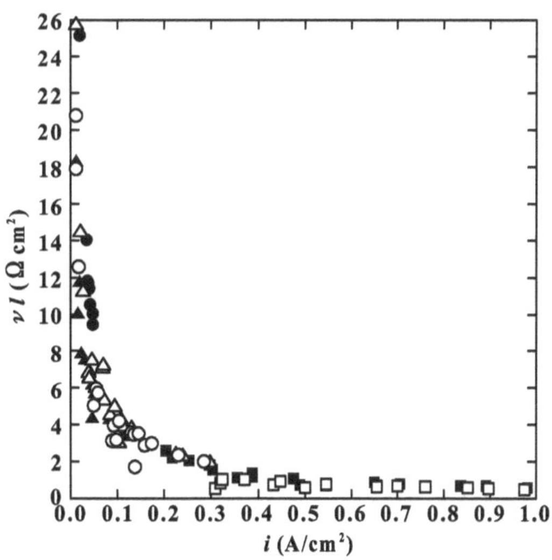

FIGURE 7.21 Area electric resistance of the reaction layer. 0.001-M (○ ●), 0.01-M (△ ▲), 0.1-M (□ ■) NaCl solution. Open, Selemion CMR; Filled, Selemion ASR, 25 °C.

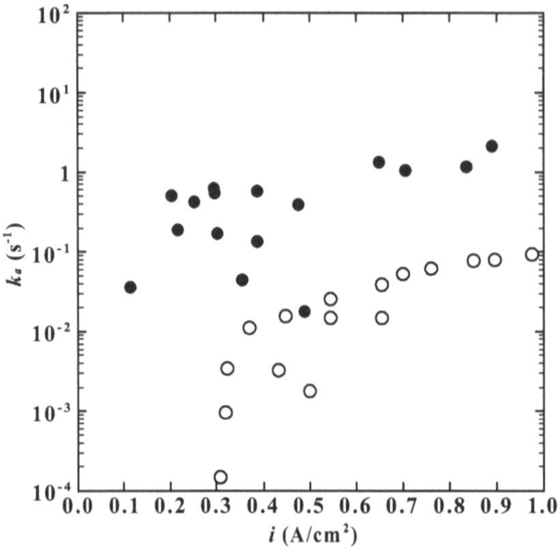

FIGURE 7.22 Forward reaction rate constant. 0.1-M NaCl solution, 25 °C. Open, Selemion CMR; Filled, Selemion ASR.

k_a for the anion exchange membrane is larger than that for the cation exchange membrane.
$C_{H_2O} = 1/18$ mol/cm^3
$k_b = 1.5 \times 10^{14}$ cm^3/mol s
$K_W = 1 \times 10^{-20}$ mol^2/cm^6 (at 25 °C)
$F = 96{,}485$ As/mol
$l = 0.011$ cm (for Selemion CMR and ASR)

7.8.3 Forward Reaction Rate Constant

The increase of the forward reaction rate constant k_a in Fig. 7.22 (compared to $2\times10^{-5}\text{s}^{-1}$) is fundamentally understandable from the catalytic protonation and deprotonation reaction; Eqns (7.11), (7.12), (7.17) and (7.18) suggested by Simons (1984, 1985).

Timashev and Kirganova (1981) and Timashev (1985) presented the chemical reaction model for the protonation and deprotonation reaction, assuming that the water dissociation reaction is catalyzed by a catalytic proton transfer between the charged groups and water at the membrane surface. They introduced the following equation expressing the forward rate constant k_a:

$$k_a(d\psi/dx) = k_a(0)\exp\left\{\frac{F}{RT}l\left(\frac{d\psi}{dx}\right)\right\} \tag{7.51}$$

in which $k_a(d\psi/dx)$ and $k_a(0)$ are k_a with and without an external applied electric field $d\psi/dx$, respectively. F is the Faraday constant and l is the thickness of the reaction layer (membrane). Equation (7.51) shows that the catalytic water dissociation reaction is accelerated by external applied electric field $d\psi/dx$ in the reaction layer (membrane) and it is an advanced expression of Eqns (7.11), (7.12), (7.17) and (7.18). Mafe et al. (1998) discussed proton transfer and water dissociation at the junction in a bipolar membrane and proposed a statistical thermodynamics model to justify the use of Eqn (7.51). Umnov et al. (1999) derived the complete equation developed from Eqn (7.51) for the space charge region in the bipolar membrane, taking into account the catalytic nature of the water dissociation reaction and the recombining reaction of H^+ and OH^- ions.

In order to make clear the mechanism of the water dissociation reaction, it is necessary to discuss at the same time the influence of the electrical potential (activation energy) generated in the reaction layer (charged layer, membrane). This suggestion is introduced from the following Arrhenius equation:

$$k_a\left(\left|\psi_m\right|\right) = k_a(0)\alpha \exp\left(-\frac{F}{RT}\left|\psi_m\right|\right) \tag{7.52}$$

in which R is the gas constant, T is the absolute temperature, $|\psi_m|$ is the electrical potential (activation energy) generated in the reaction layer, and $k_a(0)$ is k_a at $|\psi_m| = 0$. α is the nondimensional coefficient expressing the influence of $|\psi_m|$ to k_a. Equation (7.52) shows that the water dissociation reaction (k_a) (the probability of the water dissociation reaction generation) is suppressed due to the electrical potential generated in the reaction layer $|\psi_m|$. Ramirez et al. (1992) developed the similar equation to discuss the temperature dependency of k_a.

Here, Eqn (7.51) and Eqn (7.52) are combined, and the forward reaction rate constant of the water dissociation reaction generated in an ion exchange membrane k_a is presented by the following equation (Tanaka, 2012):

$$k_a\left(\frac{d\psi}{dx}, \left|\psi_m\right|\right) = k_a(0,0)\alpha \exp\frac{F}{RT}\left(l\frac{d\psi}{dx} - \left|\psi_m\right|\right) \tag{7.53}$$

Equation (7.53) present the fundamental mechanism of the total water dissociation reaction. The water dissociation is catalytically accelerated due to the external potential field $d\psi/dx$ (V/m) in the reaction layer (membrane). l (m) is the thickness of the reaction layer (membrane). At the same time, the water dissociation reaction is suppressed due to the electrical potential generated in the membrane $|\psi_m|$ (V). Equation (7.53) holds in the potential profile displayed in Fig. 7.23, in which subscripts K and A denote, respectively, the cation exchange membrane and anion exchange membrane.

The intensity of the water dissociation reaction is influenced by the ion exchange groups in the membrane. In order to understand the influence of the ion exchange groups, the second term in Eqn (7.53), i.e., Eqn (7.52), must be presented again as follows (Tanaka, 2010):

$$k_a\left(\left|\psi_m\right|\right) = k_a(0)\alpha \exp\left(-\frac{F}{RT}\left|\psi_m\right|\right) \quad (7.52) \tag{7.54}$$

The relationship between $|\psi_m|$ and $k_a|\psi_m|/k_a(0)\alpha$ is given in Fig. 7.24.

The electrical potentials in the ion exchange membranes observed in water and an NaCl solution are extracted from Table 7.6 and listed in Table 7.7. The cation exchange

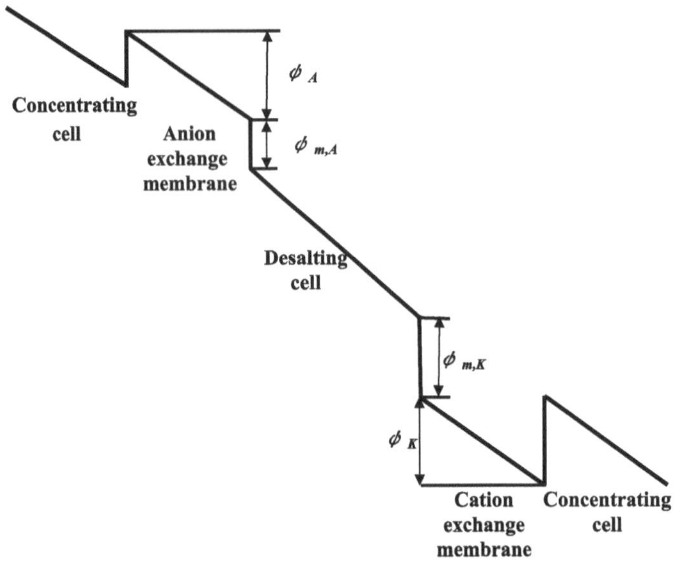

FIGURE 7.23 Potential profile in the reaction layer.

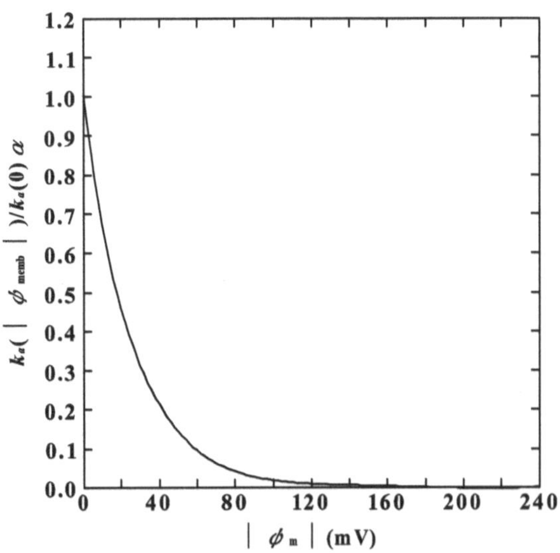

FIGURE 7.24 Influence of electric potential in an ion exchange membrane on the forward rate constant of the water dissociation reaction. 25 °C. *Tanaka (2010), p. 356, Fig. 14.*

Table 7.7 Electric Potential in Ion Exchange Membranes

Solution	Membrane	ψ_m (V)
Water	Cation exchange membrane	
	Aciplex K 102	−0.2410
	Selemion CMV	−0.2273
	Anion exchange membrane	
	Aciplex A 102	0.0105
	Selemion ASV	0.0117
0.001-M NaCl	Cation exchange membrane	
	Aciplex K 102	−0.0888
	Selemion CMV	−0.0698
	Anion exchange membrane	
	Aciplex A 102	0.0001
	Selemion ASV	−0.0006

membrane Aciplex K-102 and Selemion CMV have the sulfonic acid groups. The anion exchange membrane Aciplex A-102 has the quaternary pyridinium groups, and Selemion AST has the quaternary ammonium groups. In Table 7.7, $|\psi_m|$ in the cation exchange membrane is larger than that in the anion exchange membrane. Thus, it is understandable from Eqn (7.54) that the water dissociation reaction (the probability of the water dissociation reaction generation) is more suppressed in the cation exchange membrane as compared with the phenomenon in the anion exchange membrane. In Table 7.7, $|\psi_m|$ is increased with the decrease in the NaCl concentration. Thus, the above phenomenon is promoted with the decrease of the NaCl concentration.

7.8.4 Ion Exchange Groups and the Water Dissociation Reaction

The water dissociation generates H^+ ions and OH^- ions, and both ions become counter-ions and co-ions in the cation exchange membrane, while they become co-ions and counter-ions in the anion exchange membrane. When $|\psi_m|$ is increased, the co-ion generation (the probability of the water dissociation reaction generation) is suppressed due to the generation of repulsive force between ion exchange groups in the reaction layer and co-ions generated by the water splitting reaction. Accordingly, it is estimated that co-ions (OH^- ions) generation in the strong acid cation exchange membrane having sulfonic acid groups is suppressed due to larger $|\psi_m|$ values (ka is lower (Fig. 7.22)), whereas co-ions (H^+ ions) generation in the strong base anion exchange membrane having quaternary ammonium or quaternary pyridium groups is not strongly suppressed (ka is higher (Fig. 7.22)) due to lower $|\psi_m|$ values. Suppression of the catalytic water dissociation reaction in the strong acid cation exchange membrane is displayed in Fig. 7.25.

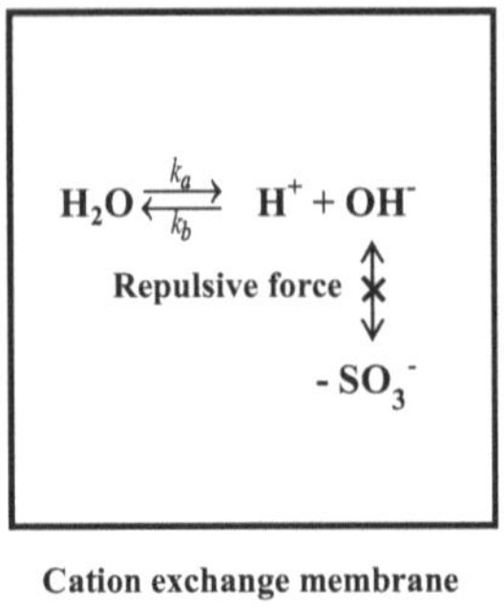

FIGURE 7.25 Suppression of the water dissociation reaction in the strong acid cation exchange membrane.

7.8.5 pK Values of Ion Exchange Groups and the Water Dissociation Reaction (Tanaka, 2010, 2012)

The influence of ion exchange groups to the water dissociation strength is understandable by rearranging the pK values described in ION EXCHANGE (page 86, Table 4.2) edited by Helfferich, (1962) from high to low as follows.

Ion Exchange Group	pK Value
$-N(CH_3)_3$	>13
$-N(C_2H_4OH)(CH_3)_2OH$	>13
$-C_6H_4OH$	9–10
$-NH_2$	7–9
$-NH-$	7–9
$-C_6H_4NH_2$	5–6
$-COOH$	4–6
$-PO_3H_2$	pK_1 2–3
	pK_2 7–8
$-SO_3H$	<<1

The pK value of $-SO_3H$ groups is less than 1 and that of the $-N(CH_3)_3OH$ groups is larger than 13. These groups are strongly ionized, so they must increase $|\psi_m|$ and suppress the probability of the water dissociation reaction generation due to the repulsive force between ion exchange groups in the reaction layer and co-ion generated by the water dissociation reaction itself. Inspecting the intensities of the water dissociation reaction, the current efficiency for the water dissociation reaction η is extremely low (Figs 7.9 and 7.10) and the reaction is generally suppressed. Further, the water dissociation is drastically suppressed in the strong acid cation exchange membrane than in the strong base anion exchange membrane (Fig. 7.6). This phenomenon is presumably due to: (1) the ionization degree of $-SO_3H$ groups is stronger than that of the $-N(CH_3)_3OH$ groups, and (2) $|\psi_m|$ in the strong acid cation exchange membrane is larger than $|\psi_m|$ in the strong base anion exchange membrane (Tables 7.6 and 7.7). From the above discussion, the pK value of $-SO_3H$ groups is assumed to be extremely lower than 1; pK << 1, while that of $-N(CH_3)_3OH$ groups is assumed to be pK > 13.

pK values of weak base groups are less than 13 (pK < 13) and those of acid groups to be larger than 1 (pK > 1). Accordingly, ionization of ion exchange groups and electrical potential $|\psi_m|$ are decreased in the weak base anion exchange membrane and weak acid cation exchange membrane. Thus, the probability of the water dissociation reaction generation is promoted. This phenomenon is due to a decrease of repulsive force between weak base or weak acid ion exchange groups and co-ions generated by the water dissociation reaction. Simons (1984, 1985) recognized that the water dissociation reaction is accelerated by weakly basic groups such as tertiary or secondary amine in an anion exchange membrane or by weakly acidic groups such as carboxylic acid in a cation exchange membrane. The mechanism of these phenomena is understandable from pK values and ionization degrees of ion exchange groups and electrical potential generated in the membrane. Zabolotskii et al. (2014) (Chapter 1, Fig. 1.11) recognized that the water dissociation reaction in the bipolar membrane including $-PO_3H_2$ groups (MB-3) is more accelerated compared to the phenomenon in the membrane including $-SO_3H$ groups (MB-2). The reaction in the membrane including $\equiv N^+$, $=N^+$, $-N^+$ groups (MB-1) is more accelerated compared to the phenomena in the membrane including $-N^+(CH_3)_3$ groups (MB-2).

7.8.6 Catalytic Water Dissociation Reaction

As described in Sections 7.7.2, 7.7.3 and 7.7.4, metallic hydroxides such as $Fe(OH)_3$, $Mg(OH)_2$, $Ni(OH)_2$, $Co(OH)_2$, $Mn(OH)_2$, etc. induce an extremely strong catalytic water dissociation reaction under an applied potential field. The mechanism of the catalytic reaction is explained by the following formula, similar to the protonation and deprotonation reaction.

Kang et al. (2004) suggested the following catalytic reaction caused by $Fe(OH)_3$:

$$Fe(H_2O)_3(OH)_3 + H_2O \rightarrow Fe(H_2O)_2(OH)_4^- + H_3O^+ \tag{7.55}$$

$$Fe(H_2O)_2(OH)_4^- + H_2O \rightarrow Fe(H_2O)_3 + OH^- \tag{7.56}$$

Oda and Yawataya (1968) presented the following catalytic reaction caused by $Mg(OH)_2$:

$$Mg^{2+} + 4H_2O \underset{k_{-1}}{\overset{k_1}{\Leftrightarrow}} Mg(OH)_2 + 2H_2O \tag{7.57}$$

$$Mg(OH)_2 \underset{k_{-2}}{\overset{k_2}{\Leftrightarrow}} Mg^{2+} + 2OH^- \tag{7.58}$$

Ganych et al. (1992) formulate the general water dissociation reaction involving metal ions:

$$R - \left[Me(H_2O)_2\right]^{z+} + H - OH \underset{k_{-1}}{\overset{k_1}{\Leftrightarrow}} R - [Me(H_2O)OH]^{(z-1)+} + H_3O^+ \tag{7.59}$$

$$R - [Me(H_2O)OH]^{(z-1)+} + H - OH \underset{k_{-2}}{\overset{k_2}{\Leftrightarrow}} R - \left[Me(H_2O)_2\right]^{z+} + OH^- \tag{7.60}$$

References

Donnan, F.G., 1924. The theory of membrane equilibria. Chem. Rev. 1, 73–90.

Eigen, M., 1954. Method for investigation of ionic reaction in aqueous solutions with half-times as short as 10^{-9} s, application to neutralization and hydrolysis reaction. Discuss. Faraday Soc. 17, 194–205.

Eigen, M., Maeyer, L.D., 1959. Hydrogen bond structure, proton hydration, and proton transfer in aqueous solutions. In: The Structure of Electrolyte Solutions. Wiley, New York.

Frilette, V.I., 1956. Preparation and characterization of bipolar ion-exchange membranes. J. Phys. Chem. 60, 435–439.

Ganych, V.V., Zabolotskii, V.I., Shel'deshov, N.V., 1992. Electrolytic dissociation of water molecules in systems comprising solutions and MA-40 anion-exchange membranes modified with transition metal ions. Sov. Electrochem. 28, 1138–1143.

de Grotthuss, C.J.T., 1806. Sur la decomposition de l'eau et des corps qu'elle tient en dissociation a l'aide de l' electricite galvanique. Ann. Chim. 58, 54–73 (Paris).

Helfferich, F., 1962. Ion Exchange. McGraw-Hill, New York, p. 86.

Kang, M.S., Choi, Y.J., Moon, S.H., 2004. Effects of inorganic substances on water splitting in ion-exchange membranes, II, 2004, optimal contents of inorganic substances in preparing bipolar membranes. J. Colloid Interface Sci. 273, 533–539.

Kolyubin, A.V., Maksimychev, A.V., Timashev, S.F., 1966. The use of flicker-noise spectroscopy for studying the mechanism of the overlimiting current in a system with cation-exchange membrane. Russ. J. Chem. 32, 206–213.

Kressman, T.R.E., Tye, F.L., 1956. The effect of current density on the transport of ions through ion-selective membranes. Discuss. Faraday Soc. 21, 185–292.

Mafe, S., Ramirez, P., Alcaraz, A., 1998. Electric field associated proton transfer and water dissociation at the junction of a fixed-charge bipolar membrane. Chem. Phys. Lett. 294, 406–412.

Markovitch, O., Agmon, N., 2007. Structure and energetics of the hydronium hydration shells. J. Phys. Chem. A111, 2253–2256.

Oda, Y., Yawataya, T., 1968. Neutrality-disturbance phenomenon of membrane-solution systems. Desalination 5, 129–138.

Onsager, L., 1934. Deviation from Ohm's law in weal electrolyte. J. Chem. Phys. 2, 599–615.

Patel, R.D., Lang, K.C., Miller, I.F., 1977. Polarization in ion-exchange membrane electrodialysis. Ind. Eng. Fundam. 16, 340–348.

Ramirez, P., Aguilella, V.M., Manzanarees, J.A., Mafe, S., 1992. Effects of temperature and ion transport on water splitting in bipolar membranes. J. Membr. Sci. 73, 191–201.

Rosenberg, N.W., Tirrel, C.E., 1957. Limiting currents in membrane cells. Ind. Eng. Chem. 49, 780–784.

Rubinstein, I., 1977. A diffusion model of "water splitting" in electrodialysis. J. Phys. Chem. 81, 1431–1436.

Rubinstein, I., 1981. Mechanism for an electrodiffusion instability in concentration polarization. J. Chem. Soc. Faraday Trans. II 77, 1595–1609.

Rubinstein, I., Shtilman, 1979. Voltage against current curves of cation-exchange membranes. J. Chem. Soc. Faraday Trans. II 75, 231–246.

Simons, R., 1979. Strong electric field effects on proton transfer between membrane-bound amines and water. Nature 280, 824–826.

Simons, R., 1984. Electric field effects on proton transfer between ionizable groups and water in ion exchange membranes. Electrochim. Acta 29, 151–158.

Simons, R., 1985. Water splitting in ion exchange membranes. Electrochim. Acta 30, 275–282.

Tanaka, Y., 1974. Concentration polarization and dissociation of water in the ion exchange membrane electrodialysis. J. Electrochem. Soc. Jpn. 42, 450–456.

Tanaka, Y., 1975. Concentration polarization and dissociation of water in the ion exchange membrane electrodialysis. J. Electrochem. Soc. Jpn. 43, 584–588.

Tanaka, Y., 2007a. Acceleration of water dissociation generated in an ion exchange membrane. J. Membr. Sci. 303, 234–243.

Tanaka, Y., 2007b. Water dissociation in ion-exchange membrane electrodialysis. J. Membr. Sci. 203, 227–244.

Tanaka, Y., 2010. Water dissociation reaction generated in an ion-exchange membrane. J. Membr. Sci. 350, 347–360.

Tanaka, Y., 2012. Mass transport in a boundary layer and in an ion exchange membrane: mechanism of concentration polarization and water dissociation. Russ. J. Electrochem. 48, 739–755.

Tanaka, Y., Matsuda, S., Sato, Y., Seno, M., 1982. Concentration polarization and dissociation of water in ion exchange membrane electrodialysis III, the effects of electrolytes on the dissociation of water. J. Electrochem. Soc. Jpn. 50, 667–672.

Tanaka, Y., Seno, M., 1983a. The concentration polarization and dissociation of water in ion exchange membrane electrodialysis, V. The acceleration of ionic transport on the membrane surface. J. Electrochem. Soc. Jpn. 51, 267–271.

Tanaka, Y., Seno, M., 1983b. Concentration polarization and dissociation of water in ion exchange membrane electrodialysis VI. The effect of insoluble inorganic materials on the dissociation of water under conditions of low concentration and high potential. J. Electrochem. Soc. Jpn. 51, 465–470.

Timashev, S.F., 1985. Role of temperature and entropic factors in the kinetic of membrane processes. DAN USSR 285, 1419–1423.

Timashev, S.F., Kirganova, E.V., 1981. Mechanism of the electrolytic decomposition of water molecules in bipolar ion-exchange membranes. Sov. Electrochem. 17, 366–369.

Umnov, V.V., Shel'deshov, N.V., Zabolotskii, V.I., 1999. Current-voltage curve for the space charge region of a bipolar membrane. Russ. J. Electrochem. 35, 871.

Wien, M., 1928. Uber die abweichungen der electrolyte vom Ohmschen gesetz. Phys. Zeits 29, 751–755.

Wien, M., 1931. Uber leitfahigkeit und dielektrizitatskonstante von elektrolyten bei hochfrequenz. Phys. Zeits 32, 545.

Zabolotskii, V., Sheldeshov, N., Melnikov, S., 2014. Heterogeneous bipolar membranes and their application in electrodialysis. Desalination 342, 183–203.

8

Hydrodynamics

8.1 Introduction

In an ion exchange membrane electrodialysis process, solutions are supplied to a multichannel system consist of desalting cells and concentrating cells, in which spacers are integrated. In an industrial-scale electrodialyzer, solution velocity distribution in the cells causes concentration distribution. Limiting current density of an electrodialyzer is strongly influenced by the distribution of solution flow in desalting cells. Pressure drop in the multichannel system consumes electric energy, which is added to the energy consumption for ion transport in electrodialysis. The above phenomena come into the problems in hydrodynamics and are related to the performance of the electrodialyzer. The spacer has three major roles (Pnueli and Grossman, 1969): (1) it keeps the membranes from touching each other; (2) it takes up any small mechanical force which may arise in the cell and which the membranes cannot support; and (3) it mixes the flow and makes the concentration more homogeneous. In spite of the contribution of the spacers, solution velocities in an electrodialyzer do not become uniform. The static head observed in a practical-scale electrodialyzer is discussed based on the distribution coefficient of a solution flowing into each cell. The total hydraulic head is divided into the static head, velocity head and friction head and discussed based on the Bernoulli theorem.

8.2 Stream Lines Around a Spacer

Hanzawa et al. (1965) observed stream lines in channels and evaluated the solution disturbing effects of a spacer. Figure 8.1 illustrates experimental assembly composed of one sheet of gasket (0.75 mm thick) and two sheets of transparent polyvinyl chloride (PVC) plates (10 mm thick). A diagonal net spacer, a honeycomb net spacer, or a crimp net spacer (Fig. 8.2) was incorporated with the gasket, in which water was supplied. At the same time, coloring materials were injected into the gasket from an injecting set up in Fig. 8.1. The flow lines in the gasket are displayed in Fig. 8.3(a) for the diagonal net spacer and in Fig. 8.3(b) for the honeycomb net spacer or the crimp net spacer. The experiments show evidence that the diagonal net disturbs the solution sufficiently, but the honeycomb net and crimp net do not. The streamlines around the diagonal net are presented in Photo 8.1. The flow pattern for the diagonal net at the bottom of the flow pass in the gasket and at around of an obstacle placed in the gasket are pictured in

Ion Exchange Membranes. http://dx.doi.org/10.1016/B978-0-444-63319-4.00008-0

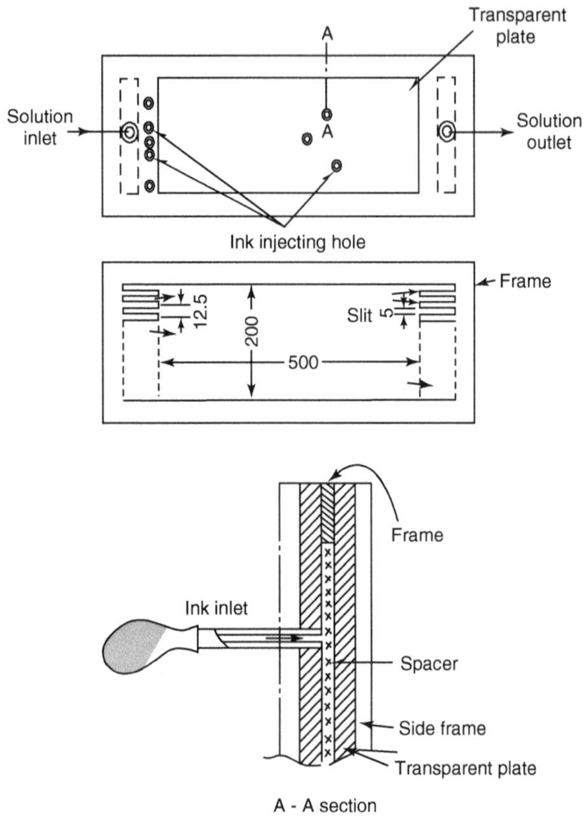

FIGURE 8.1 Assembly for observing stream lines around a spacer. *Hanzawa et al. (1965), p. 17, Fig. 2.*

Fig. 8.4(a) and (b), respectively. The experiment suggests that the structure of the spacer shape is important to make a uniform solution distribution.

8.3 Mass Transport Effect of a Spacer

Kuroda (1993) measured the limiting current density (mass transfer coefficient) using the following oxidation–reduction reaction between ferrocyanide ion $Fe(CN)_6^{4-}$ and ferricyanide ion $Fe(CN)_6^{3-}$ in the simulation electrodialysis.

$$Fe(CN)_6^{3-} + e^- \rightarrow Fe(CN)_6^{4-} \quad \text{(Cathode)} \tag{8.1}$$

$$Fe(CN)_6^{4-} \rightarrow Fe(CN)_6^{3-} + e^- \quad \text{(Anode)} \tag{8.2}$$

The limiting current density in ordinary electrodialysis $i_{lim,ed}$ and in the simulation electrodialysis $i_{lim,sim,ed}$ are expressed by the following equations:

$$i_{lim,ed} = \frac{FDC}{(\bar{t} - t)\delta} \tag{8.3}$$

$$i_{lim,sim,ed} = \frac{DCF}{\delta} = k_L CF \tag{8.4}$$

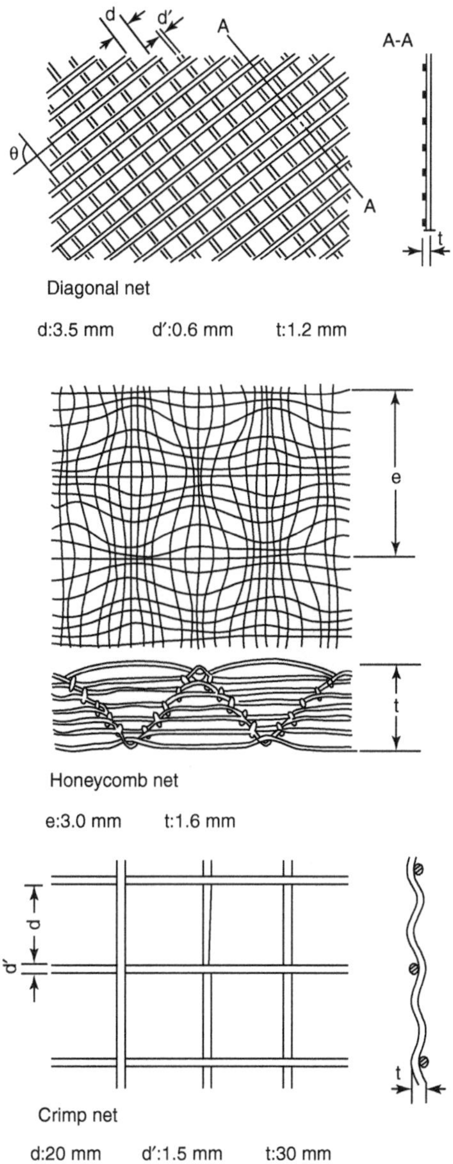

FIGURE 8.2 Structure of spacers. *Hanzawa et al. (1965), p. 16, Fig. 1.*

in which C is the ion concentration of a solution; D is the diffusion constant of electrolytes; F is the Faraday constant; $\bar{t}$ and t are the transport number of counter ions in a membrane and in a solution, respectively; δ is the thickness of a boundary layer; and k_L is the mass transport coefficient. Both equations are equivalent, and $i_{lim,sim,ed}$ is governed by k_L.

In this experiment, spacer A, B, C, or D in Table 8.1 was incorporated with the flow pass (150 mm width, 1540 mm length, 1 mm thickness) and an electrolysis solution (0.01 M $K_3Fe(CN)_6$, 0.01 M $K_4Fe(CN)_6$, 2 M NaOH mixed solution) was supplied. From the current–voltage relationship of the oxidation–reduction reaction, the limiting current

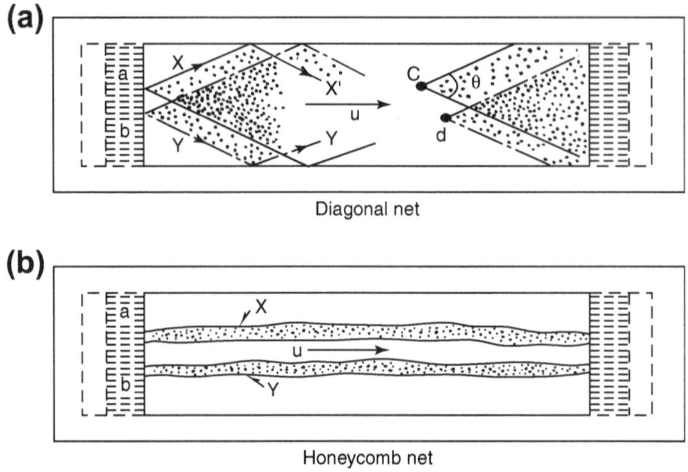

FIGURE 8.3 Stream lines in a flow pass around the diagonal net spacer (a) and the honeycomb net spacer (b). *Hanzawa et al. (1965), p. 19, Fig. 6.*

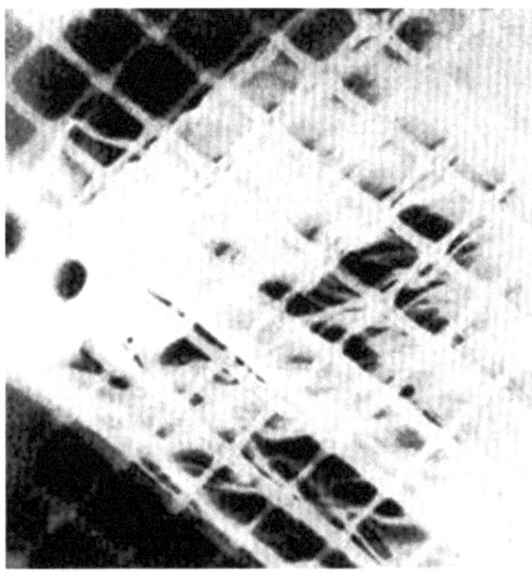

PHOTO 8.1 Stream lines around a diagonal net spacer. *Hanzawa et al. (1965), p. 19, Photo 2.*

density of the simulation electrodialysis $i_{lim,sim,ed}$ was measured. k_L and the following Sherwood number (Sh) were calculated using Eqns (8.4) and (8.5).

$$Sh = k_L De/D \tag{8.5}$$

in which De is the equivalent diameter ($=2dW[1 - \varepsilon_S]/[d + W]$); d and W are the thickness and width of the flow pass, respectively; and ε_S is the volume ratio of the spacer. From the above experiments, the relationship between $ShSc^{-1/3}$ (Schmidt number $Sc = \mu_L/\rho_L D$, μ_L: viscosity, ρ_L: density) that defines the mass transport effect and the

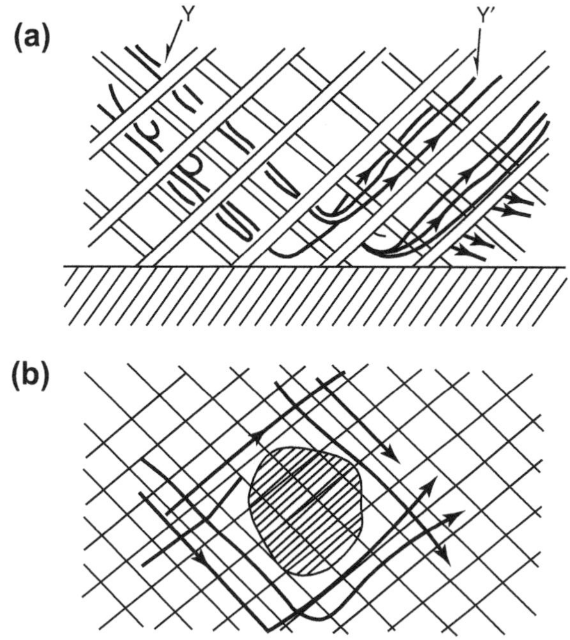

FIGURE 8.4 Stream lines at the bottom of a flow pass (a) and at around of an obstacle (b) with a diagonal net spacer. *Hanzawa et al. (1965), p. 19, Fig. 7.*

Table 8.1 Photograph and Characteristics of Spacers Kuroda (1993), p. 225, Table 3.

	A	B	C	D
Photograph (scale 1:1.25)				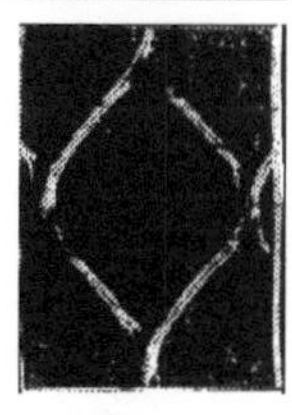
Channel gap $d \times 10^3$ (m)	1.0	2.4	2.4	2.4
Volume fraction in channel $\varepsilon_s(-)$	0.20	0.21	0.22	0.11
Thickness of strings $t_s \times 10^3$ (m)	0.50	1.2	1.4	1.7
Gap between spacer and channel Wall* $g_s \times 10^3$ (m)	0.5	1.2	1.0	0.7
Pitch of mesh** $P_S \times 10^3$ (m)	3.0	6.5	6.9	23.8

*$g_s = d - t_s$.
**$P_S = (ab/2)^{0.5}$.
a, b = Length of two diagonal lines in spacer mesh.

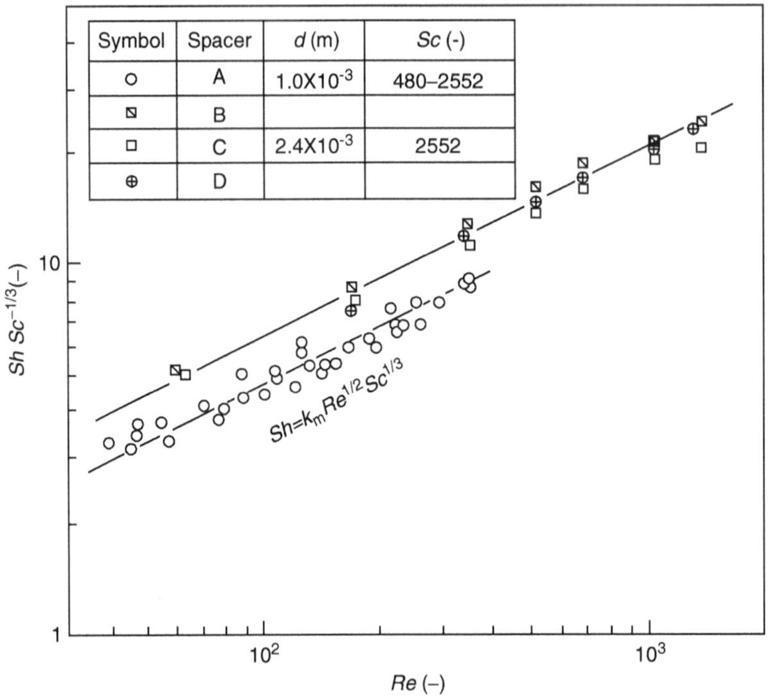

FIGURE 8.5 $Sh/Sc^{1/3}$ versus Re. *Kuroda (1993), p. 256, Fig. 17.*

Reynolds number Re ($=DeU_L\rho_L/\mu_L(1-\varepsilon_S)$, U_L: linear velocity) was introduced as the following equation:

$$Sh\ Sc^{-1/3} = k_m Re^{1/2} \quad (k_m = 0.5) \tag{8.6}$$

The relationships are given in Fig. 8.5, which shows that the intensities of the mass transport effect of each spacer are assessed as B ≒ C ≒ D > A. The solution disturbing effect of the spacer seems to be influenced by the expanding degree of solution flow sideward in the flow pass (cf. Fig. 8.3 and Photo 8.1).

8.4 Dead Space Around a Spacer and Limiting Current Density

Limiting current density of an anion exchange membrane (Aciplex A-172, Asahi Chemical Co.) was measured using the experimental apparatus incorporated with and without a diagonal net spacer in the desalting cell (0.075 cm thick) (Tanaka, 2005). A NaCl solution in the range of 0.002–0.10 M was supplied into the cell at the linear velocity in the range of 1.6–13.6 cm/s. The limiting current densities measured with the spacer are plotted against the values measured without the spacer and presented in Fig. 8.6, which shows that the spacer decreases the limiting current density.

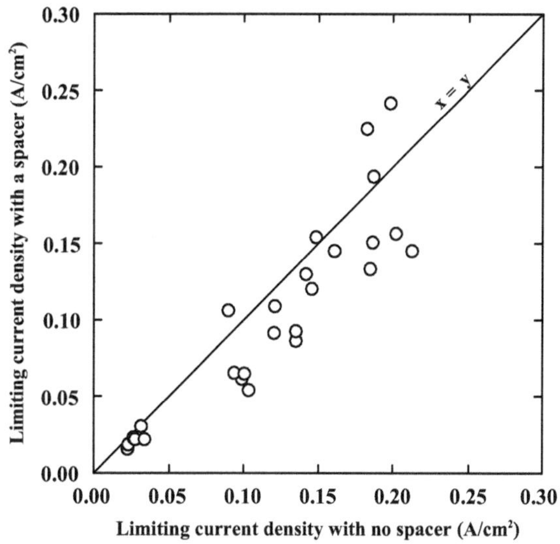

FIGURE 8.6 Effect of a spacer to the limiting current density. *Tanaka (2005), p. 11, Fig. 6.*

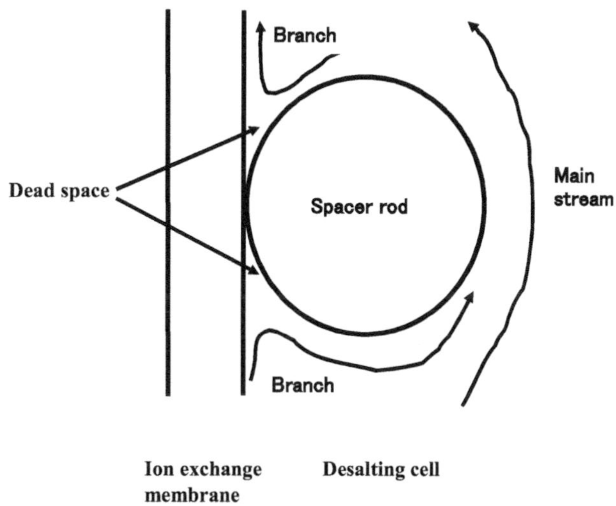

FIGURE 8.7 Solution flow around a spacer string (Illustration). *Tanaka (2005), p. 13, Fig. 10.*

A spacer is usually assumed to function as turbulence promoter. However, Fig. 8.6 suggests that the spacer does not disturb the solution. The spacer rod seems to block the main stream to form branches and dead spaces, as illustrated in Fig. 8.7. In the dead spaces, salt concentration and solution density are estimated to be decreased. The dead spaces decrease the limiting current density, and this phenomenon is supposedly accelerated when substances suspended in a feeding solution are adhered and accumulated in the dead spaces(cf. Fig. 11.12). To increase the limiting current density, it is necessary to increase the Reynolds number to create turbulence flow. It is also necessary to decrease the turbidity of the feeding solution and remove the adhered substances by dissembling the stack and washing the cells.

8.5 Flow Pattern Image in a Flow Channel

Belfort and Gutter (1972) obtained a photographic image of a flow pattern of liquid around the spacer. In this study, an exposed photographic film (Kodak Panatomic X) was placed in an electrodialysis compartment in place of a membrane (Fig. 8.8). The film was exposed to saturation for about 2 min in ordinary fluorescent ceiling light. Photographic developer (22.2 ml/l Rodinal by Agfa) was then pumped through the cell compartment incorporated with the exposed film. Under proper conditions, development of the film occurred at a rate proportional to the rate at which the developer solution flows over and mixes with the surface layer of the photographic emulsion. The photograph of typical mass diffusion contours is shown in Photo 8.2. The lighter areas represent faster flowing

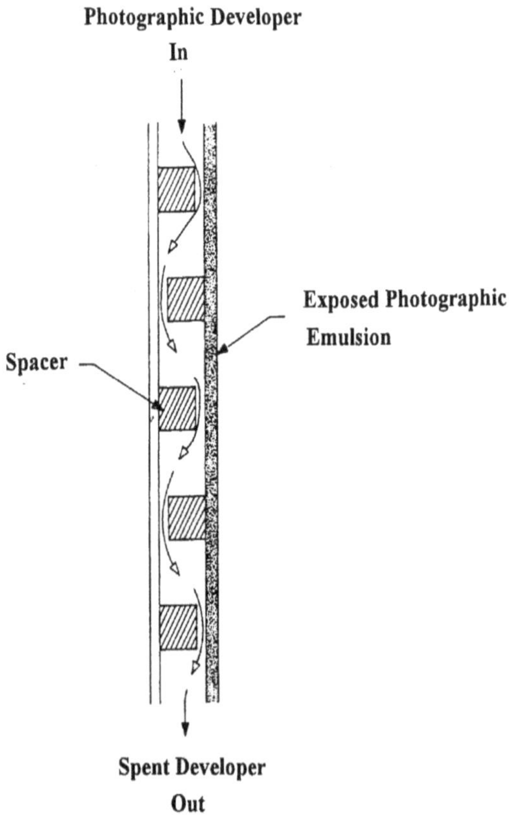

FIGURE 8.8 Surface flow pattern experiment by Astropower photographic technique. *Belfort and Gutter (1972), p. 229, Fig. 5.*

PHOTO 8.2 Sectional photographic (positive) of the film (negative) obtained from Astropower photographic technique.Volume flow rate; 44 ml/min, flow direction; right to left, channel thickness; 0.045 in. *Belfort and Gutter (1972), p. 259, Fig. A-2.*

regions or greater mass diffusion. The two dark strips on top and bottom of the photograph are the channel walls. Dark regular repeated patterns represent the location where the spacer and the surface touch, and irregular dark spots are air bubbles that remained in the stack during the 12 min developer run. All the light streaks can be thought of as vortex screw-like flow and can be followed along the flow path in the flow direction.

To look at the photograph quantitatively, a vertical scan was carried out by the densitometer at a preselected position. Figure 8.9 is an example of this scan for the expanded spacer (Exmet Corp. hexagonal net). The ordinate is a measure of mass diffusion toward the photographic film. The empty channel (black) density–displacement curve is superimposed on the density–displacement curve for the spacer. Any part of the spacer curve that drops below the maximum of the blank curve (optical density = 0.67) can be considered dangerous for scale formation.

8.6 Flow Pattern and Limiting Current Density

Ichiki et al. (1976) assembled an electrodialysis cell incorporated with a cation exchange membrane (Selemion CMV, Asahi Grass Co.) and an anion exchange membrane (Selemion AMV). A hexagonal net spacer, a honeycomb net spacer, and a diamond net spacer were integrated in a desalting cell. A 3000 ppm NaCl solution was supplied to the desalting cell at constant linear velocity u, and a constant electric current I was applied. Changing an electric current incrementally, the limiting current density i_{lim} was measured from the inflection of I–V plots, at the same time the static head drop ΔH in the cell was measured. Changing u in the desalting cell incrementally, Reynolds number Re versus ΔH plots and Re versus i_{lim} plots were obtained.

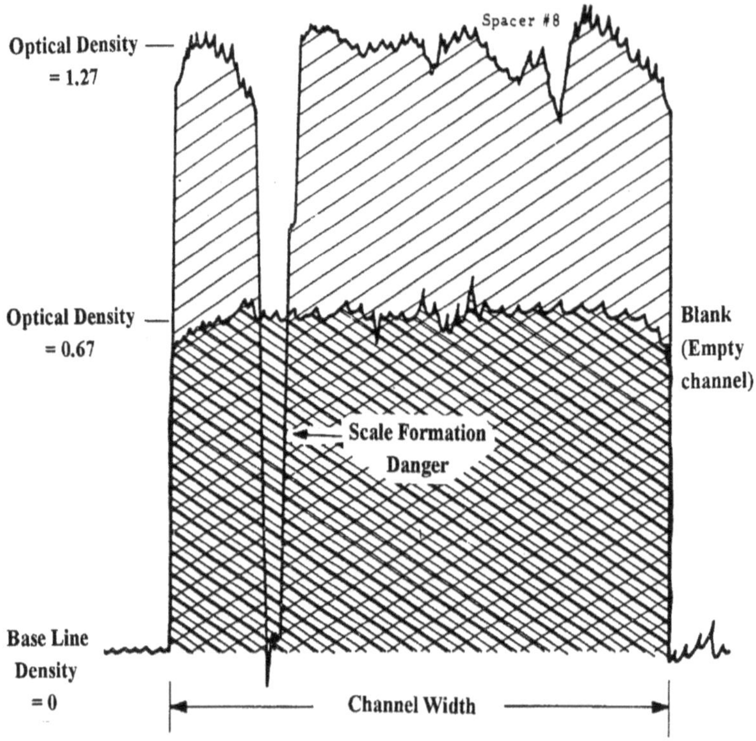

FIGURE 8.9 Plot of density versus displacement across the width of the film for a spacer and for the empty channel.Hexagonal expanded spacer volume flow rate: 44 ml/min superficial velocity: 5.05 cm/s. *Belfort and Gutter (1972), p. 231, Fig. 7.*

The plotting of i_{lim} and ΔH against *Re* in this investigation is illustrated by the model in Fig. 8.10. The model gives the flow pattern of a solution divided at the inflection point B. A–B and B–C correspond to the laminar flow and turbulent flow, respectively. The relationship between *Re* and ΔH is expressed by the following equation, including the constant m and n:

$$\Delta H = mRe^n \tag{8.7}$$

n takes the following values:

$$\begin{aligned} n &= 1.0 - 1.6 \ \text{(laminar flow)} \\ n &= 1.7 - 2.0 \ \text{(turbulent flow)} \end{aligned} \tag{8.8}$$

8.7 Local Flow Distribution in a Flow Channel

Limiting current density of an electrodialyzer depends on a local flow distribution of a solution in a desalting cell. Azechi et al. (1966) observed a local flow distribution using a desalting cell model as follows. A stack of a practical scale unit-cell type electrodialyzer

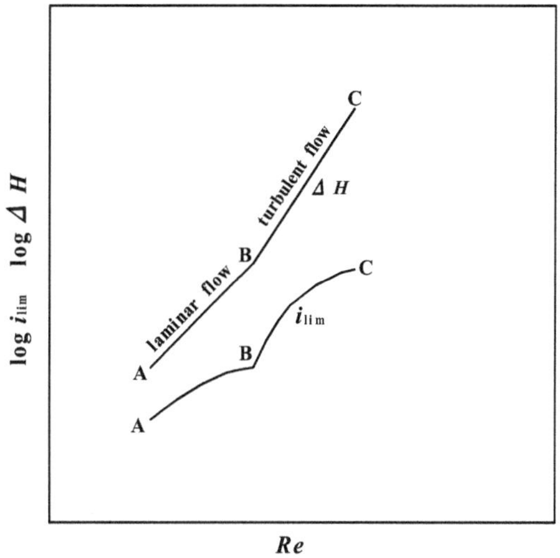

FIGURE 8.10 Re vs ΔH plot (Illustration). *Tanaka (2004), p. 32, Fig. 5.*

being operated for 10 months in a salt-manufacturing plant was supplied for this investigation. The stack had been integrated with cross-piece spacers manufactured by arranging strings (2 mm thick × 2 mm width) at intervals of 25 mm.

In this investigation, first epoxy resin (Three ronge AE and HW, Three Bond Co.) was poured into the desalting cells in the stack and was hardened by a cross-linking reaction. The stack was then disassembled, and the part of the hardened epoxy resin was taken out. The membrane deformation in the stack formed during 10 months of operation was reproduced exactly on the surface of the epoxy resin. A test plate being engraved with uneven membrane deformations on the surface was cut down from the epoxy resin. The desalting cell model was assembled with the resin plate (epoxy resin), a transparent plate (PVC, 10 mm thick), a middle plate (PVC, $d' = 3.4$, 3.6, 4.0 or 4.4 mm thick), and a base plate (PVC, 15 mm thick) (Fig.8.11). The distance between the transparent plate and the resin plate d corresponds to the local flow pass depth, which was measured by adjusting the thickness of the middle plate d'.

Next, water was supplied into the spaces formed between the test plate and transparent plate, and the stream lines were observed by pouring coloring substances in the spaces (Fig. 8.12). The stream lines are shown by arrows originated at a′, b′, c′--k′ positioned on the right-hand side in the figure. Reversed stream lines are also observed by supplying water in a reverse direction. Contour lines due to the membrane deformation are measured by d values presented at the bottom of the figure. Finally, local velocities u and the local flow pass depth d at x,y are divided by their average $\overline{u}$ and $\overline{d}$, and the relationship between $d/\overline{d}$ and $u/\overline{u}$ is computed as indicated in Fig. 8.13 and presented by the following equation:

$$u/\overline{u} = \left(d/\overline{d}\right)^2 \qquad (8.9)$$

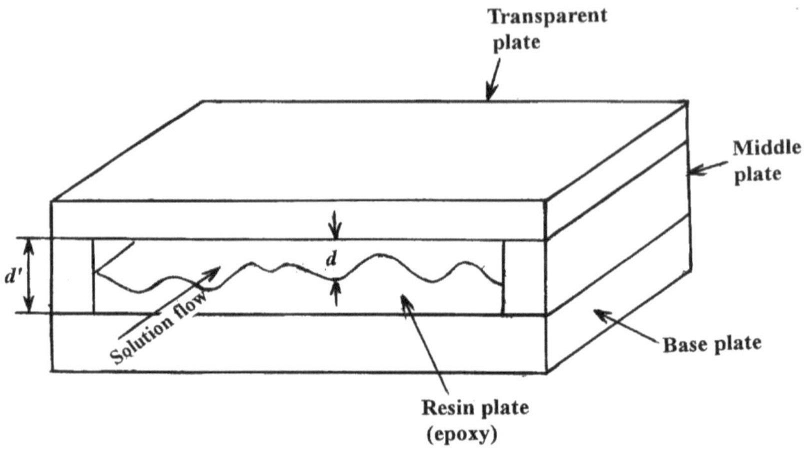

FIGURE 8.11 Assembly for measuring solution velocity distribution in a channel. *Tanaka (2007), p. 225, Fig. 10.19.*

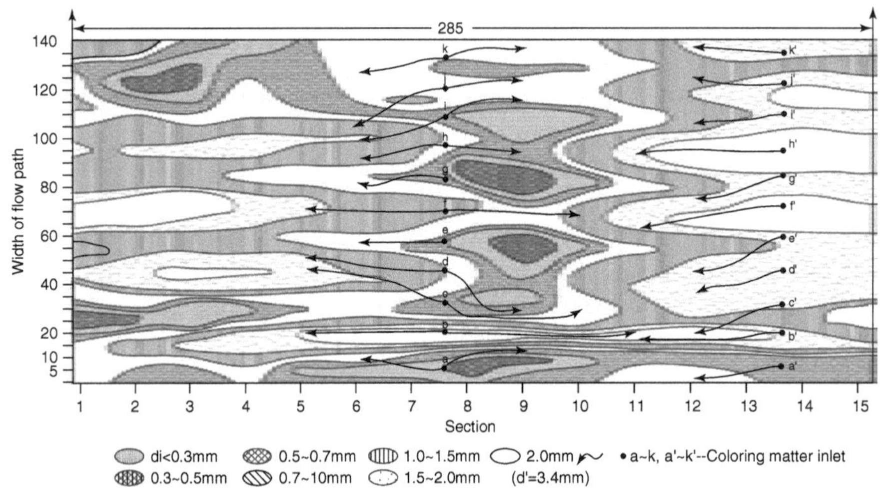

FIGURE 8.12 Stream lines in a channel. *Azechi et al. (1966), p. 21, Fig. 4.*

This study shows that the local solution velocity u depends on the local distance d between a cation and an anion exchange membrane and between a membrane and a spacer.

8.8 Solution Velocity Distribution Between Desalting Cells

Seawater was supplied to desalting cells in a unit-cell type electrodialyzer (Table 8.2) incorporated with Selemion CMV/AST membranes and cross-piece spacers and electrodialyzed at the linear velocity of 3 cm/s and the current density of 2 A/dm^2. The solution velocity distribution in stacks was measured by measuring the solution volume Q^*

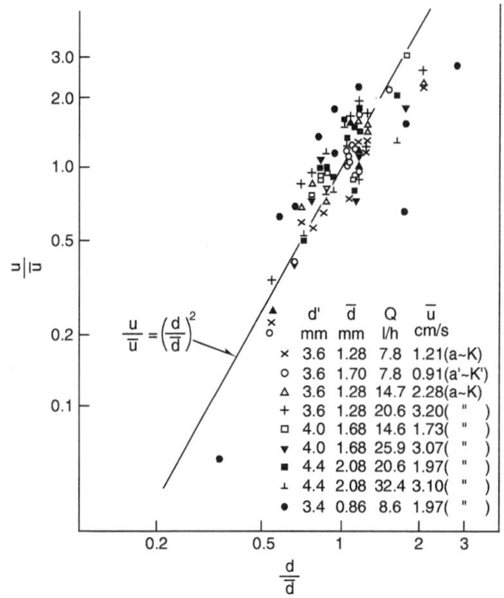

FIGURE 8.13 Relationship between local flow pass depth and local solution velocity in a channel. *Azechi et al. (1966), p. 22, Fig. 7.*

Table 8.2 Specifications of the Electrodialyzer

Membrane	Selemion CMV/AMT
Numbers of stacks in the electrodialyzer	44
Numbers of desalting cells in a stack	12
Membrane area	0.96 m^2 (98 × 98 cm) Cross niece
Distance between membranes in a desalting cell	2 mm
Spacer	Cross pniece

Tanaka (2005), p. 11, Table 2.

flowing out from each stack. After that, the stacks were disassembled and washed, and then the solution velocity distribution was measured again in the same manner (Tanaka, 2005). Frequencies of stacks N_{stack} are plotted against solution volume ratio $\xi = (Q^* - \overline{Q})/\overline{Q}$, ($\overline{Q}$: the average solution volume flowing out from the stacks) and shown in Fig. 8.14, which is equated with the normal distribution. The standard deviation of the normal distribution σ is calculated as 0.079 and 0.037 for respectively before and after disassembling/washing.

Next, the solution velocity distribution in desalting cells in one of the stacks before assembly was observed by injecting coloring liquid into the inlets of each desalting cell and measuring elapsed time until the coloring liquid appears in the outlets of the desalting cells. The average linear velocity $\overline{u}$ was adjusted to 3 cm/s. The frequencies of desalting cells N_{cell} are plotted against the solution velocity ratio $\xi = (u^* - \overline{u})/\overline{u}$, (u*;

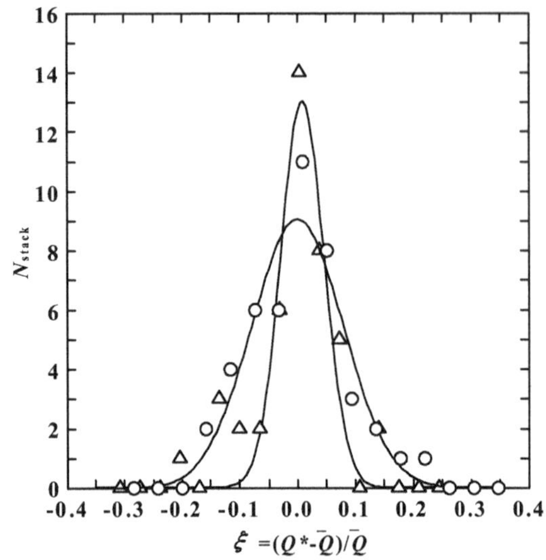

FIGURE 8.14 Velocity distribution of desalted solutions between stacks. ○ before disassembly $\overline{Q} = 23.7$ l/min $\sigma = 0.079$ △ after washing and assembly $\overline{Q} = 28.9$ l/min $\sigma = 0.037$. *Tanaka (2005), p. 12, Fig. 7.*

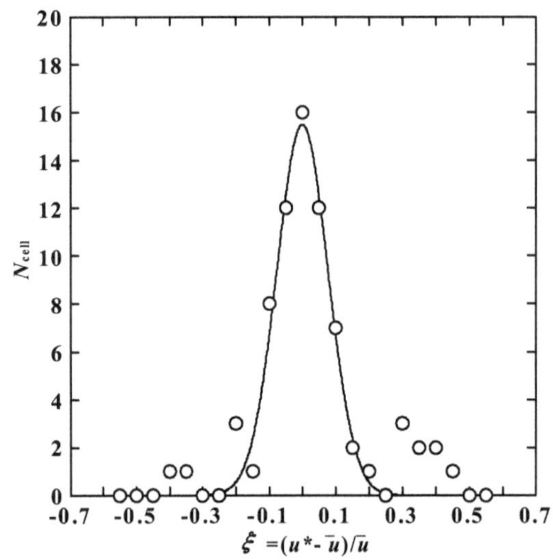

FIGURE 8.15 Velocity distribution of desalted solutions between cells. $\overline{u} = 3$ cm/s $\sigma = 0.078$. *Tanaka (2005), p. 12, Fig. 8.*

solution velocity in each cell) and shown in Fig. 8.15, which is also equated with the normal distribution, and σ is calculated as 0.078.

Seawater was supplied to desalting cells in a filter-press type electrodialyzer (Table 8.3) and the current density of 3 A/dm^2 was passed. The electrodialysis was repeated changing the linear velocity of solutions in desalting cells incrementally. The

Table 8.3 Specifications of the Electrodialyzer

Membrane	Aciplex K–172/A–172
Numbers of stacks in the electrodialyzer	1
Numbers of desalting cells in a stack	36
Membrane area	0.15 m^2 (96.6 x 15 cm)
Distance between membranes in a desalting cell	0.75 mm
spacer	Diagonal net

Tanaka (2005), p. 12, Table 3.

solution velocity in each desalting cell u^* was calculated from the salt concentration difference between at the inlets C_{in} and at the outlets C_{out} of cells. N_{cell} versus $\xi = (u^* - \overline{u})/\overline{u}$ was equated with the normal distribution. Standard deviation σ of solution velocity ratio ξ is determined as presented in Table 8.4 σ is distributed in the range of 0.017–0.222.

In the above experiment, the solution velocity ratio ξ in desalting cells integrated in a stack in an electrodialyzer is defined by Eqn (8.10):

$$\xi = \frac{u^* - \overline{u}}{\overline{u}} \tag{8.10}$$

where u^* is the linear velocity in each desalting cell, and $\overline{u}$ is the average linear velocity in a stack. The frequency distribution of ξ is expressed by the normal distribution. Thus the minimum of ξ and u^* is equated with -3σ and u, respectively, where σ is the standard deviation of the normal distribution and u is the minimum value of linear velocities within all desalting cells in a stack. Putting $\xi = -3\sigma$ and $u^* = u$ in Eqn (8.10) yields Eqn (8.11):

$$u = \overline{u}(1 - 3\sigma) \tag{8.11}$$

Table 8.4 Standard Deviation of Solution Velocity Ratio σ in Desalting Cells

u cm/s	C'_{in} Eq/dm^3	C'_{out} Eq/dm^3	σ
1.12	0.590	0.276	0.117
1.58	0.557	0.276	0.222
1.73	0.606	0.392	0.141
2.07	0.603	0.430	0.017
2.24	0.606	0.426	0.134
3.12	0.603	0.488	0.020
3.54	0.606	0.491	0.122
4.93	0.603	0.535	0.102

Tanaka (2005), p. 12, Table 4.

8.9 Air Bubble Cleaning of a Flow Channel

Electrodialyzers are subject to limiting current density reduction due to membrane fouling. To overcome such a problem, Hitachi Ltd. and Babcock-Hitachi K. K. developed an air bubble cleaning type electrodialyzer (Takahashi et al., 1979). The structure of an electrodialyzer is illustrated in Fig. 8.16. Air bubbles are introduced into the rectangular compartments from dispensers through conduits (holes) provided along the upper and lower edges of the frames and membranes. The bubbles clean the membrane surface

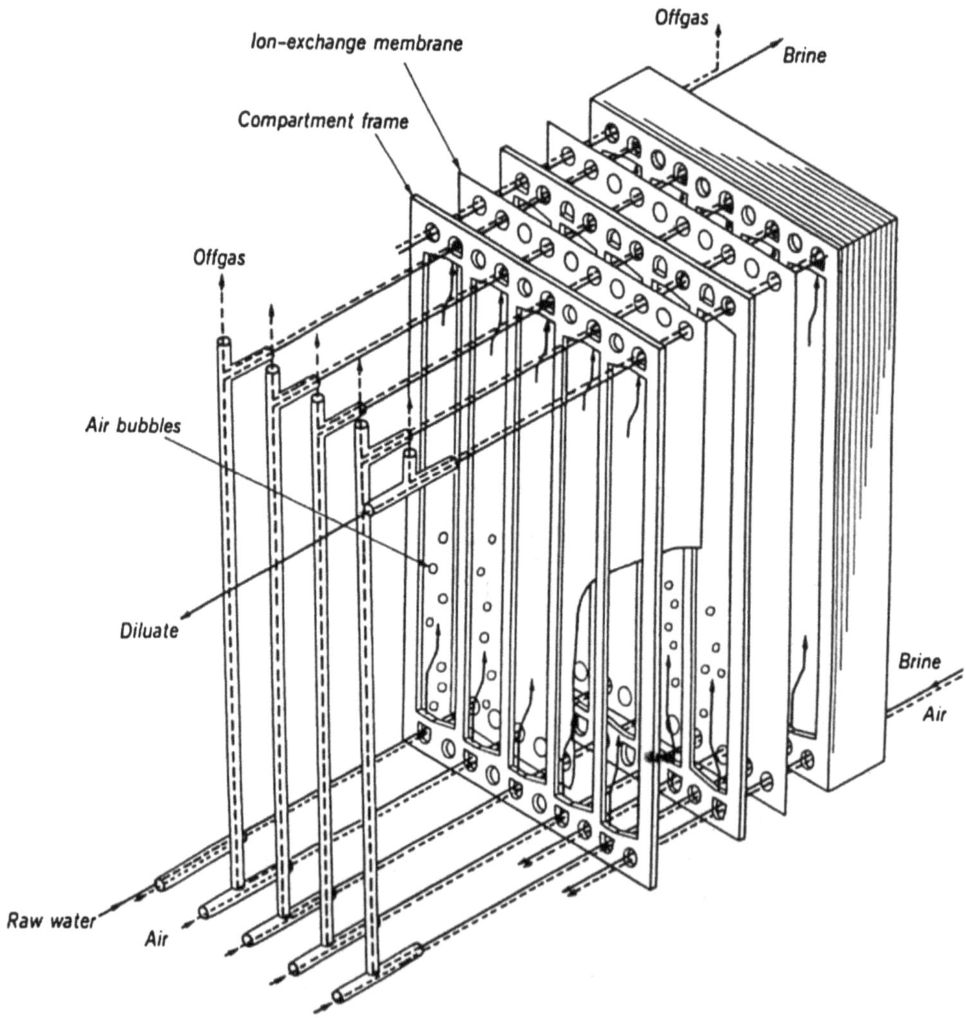

FIGURE 8.16 Structure of an air bubble cleaning type electrodialyzer. *Takahashi et al. (1979), p. 318, Fig. 3.*

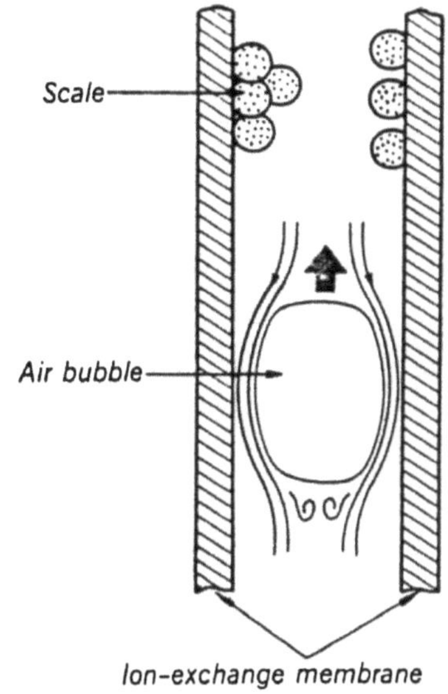

FIGURE 8.17 Principle of air bubble cleaning. *Takahashi et al. (1979), p. 319, Fig. 4.*

and promote turbulence of the solution as illustrated in Fig. 8.17, so that suspended solids readily pass through the electrodialyzer.

Raw water supplied from one end of the compartments is desalinated and discharged from the other end. Figure 8.18 shows limiting current densities in a bubble agitation system (flow rate: 2 cm/s), compared with those in a spacer agitation system observed in NaCl solution electrodialysis. The limiting current density in the bubble agitation system is constant and high even at low flow rates. The current shadowing effect of bubbles is lower than that of spacers and reduces the electric resistance of the electrodialyzer.

8.10 Solution Flow and Mass Transport in a Channel without a Spacer

Sonin and Probstein (1968) developed the hydrodynamic theory of desalination electrodialysis for a multichannel system with steady laminar flow between plane parallel membranes. For sufficiently long channels, it was shown that there are two distinct regions: a "developing region" where the concentration diffusion layers are growing, and a "developed region" where the diffusion layers fill the channel as shown in Fig. 8.19. Parabolic and uniform velocity profiles were considered, and the equations were developed on the basis of the following assumptions:

1. The fluid consists of a unionized solvent and an infinitely dilute, fully ionized salt.
2. The fluid is incompressible and at constant temperature.

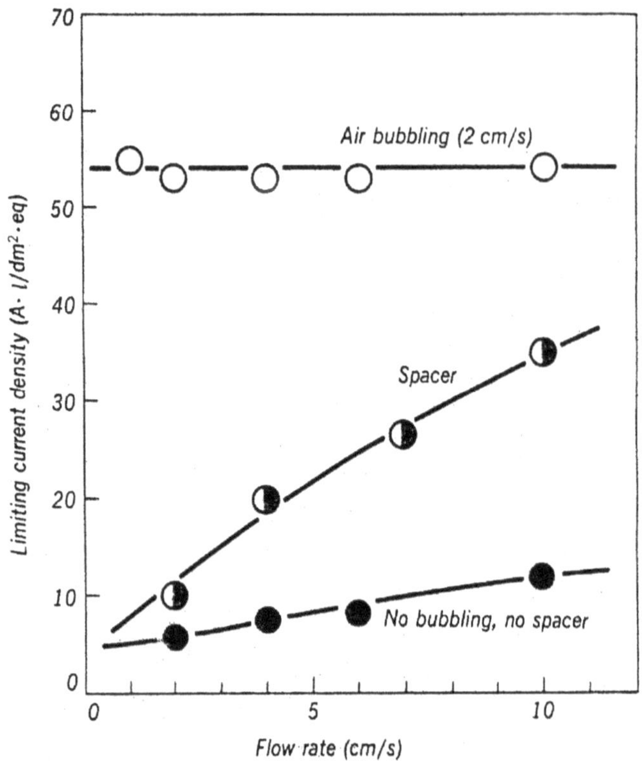

FIGURE 8.18 Effect of air bubbling on limiting current density. *Takahashi et al. (1979), p. 320, Fig. 7.*

3. The membranes are perfectly selective, have constant electrical conductivity with respect to the permeable ion, and are impermeable to water molecules.

It was found that under a wide range of operating conditions, the solution for the total current is represented by the empirical formula:

$$I = \{1 - \exp(-\psi^3)\}^{1/3} \tag{8.12}$$

where I and ψ are, respectively, a dimensionless current and potential. Comparison was made of the calculated limiting total current with experiment.

In the same geometry described above, Probstein et al. (1970) developed a self-consistent theory for fully developed turbulent flow. A set of turbulent transport equations is derived for mean salt concentration and electric field. Closure of the set is brought about by introducing a turbulent diffusion coefficient, the behavior of which is known from empirical mass transfer consideration. An integral method is used to obtain the distribution of salt concentration and current density. Comparison of the theory is made with the experimental results of Cowan and Brown on average limiting current density, and the agreement is found to be satisfactory.

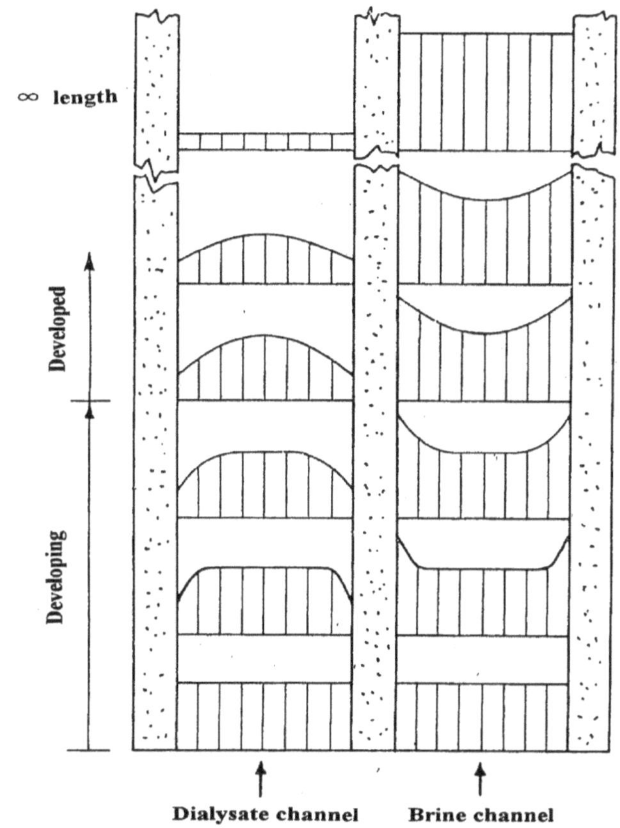

FIGURE 8.19 Illustration of salt concentration in channels. *Sonin and Probstein (1968), p. 303, Fig. 2.*

Mas et al. (1970) discussed the concentration distribution in a boundary layer based on the mass transport equation and the potential difference equation across the boundary layer. They show the concentration polarization and *I–V* curves as shown in Figs. 8.20 and 8.21. Curve 1 represents the concentration profile and *I–V* relationship at the entrance of the desalting cell. Curves 2 and 3 represent the phenomena at increasing distance from the cell entrance. At curve 4, the concentration at the membrane surface is nearly zero. The solution velocity distribution in the desalting cell is assumed to be expressed by a parabolic line.

8.11 Velocity Profile in a Flow Channel with a Spacer

The spacer may have many forms, ranging from extremely fine porous media to coarse nets with holes of the size of the cell width. The flow field and the concentration field depend on these forms. Pnueli and Grossman (1969) investigated the flow between two

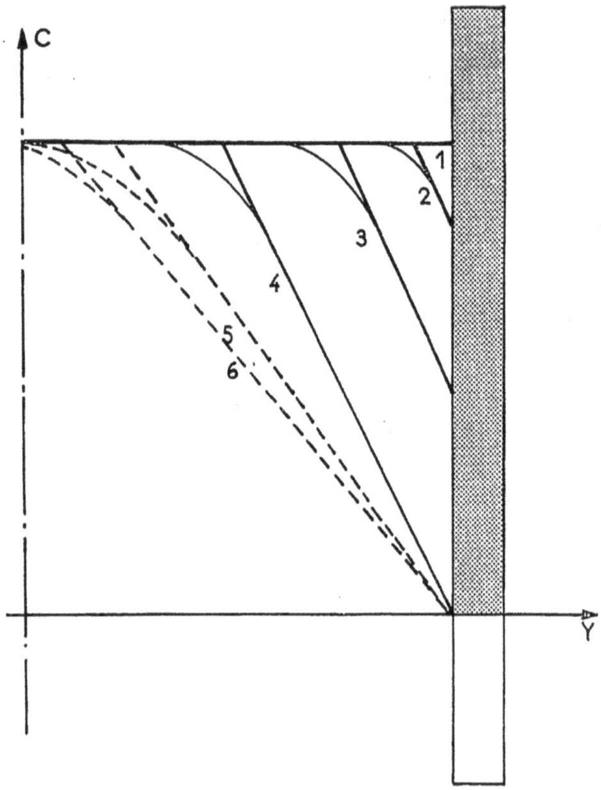

FIGURE 8.20 Concentration profiles for the different distances from the compartment inlet. *Mas et al. (1970), p. 292, Fig. 4(b).*

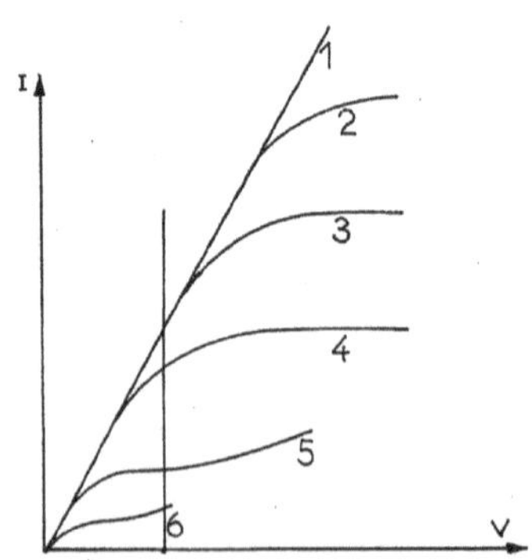

FIGURE 8.21 Current voltage characteristics for different distances from compartment inlet. *Mas et al. (1970), p. 292, Fig. 4(a).*

parallel membranes in the presence of a spacer. A mathematical model for the spacer was introduced, which results in the momentum equation, an extension of the Darcy's law, and was solved as follows to yield velocity profiles.

The force Ku is assumed to act on the control volume through the pores formed between two parallel membranes in the presence of a spacer (Fig. 8.22), and the following balance of forces are assumed to be applied:

$$Pdy + \mu\left(\frac{\partial u}{\partial y} + \frac{\partial^2 u}{\partial y^2}dy\right)dx - \left(P + \frac{\partial P}{\partial x}dx\right)dy - \mu\frac{\partial u}{\partial y}dx - Kudxdy = 0 \tag{8.13}$$

or

$$\frac{\partial P}{\partial x} = -Ku + \mu\,\frac{\partial^2 u}{\partial y^2} \tag{8.14}$$

and superposition of three components:

$$\nabla P = -Kq + \mu\,\nabla^2 q \tag{8.15}$$

where P is the pressure, u is the velocity on an x-axis in flow direction, μ is the viscosity, x and y are two Cartesian components of the coordinates, $\boldsymbol{q}$ is the velocity vector, and K is the constant in the extended Darcy's law. In Figure 8.22, the distance between two parallel membranes is assumed to be 2d and it extends between y = +d and −d. The fully developed flow considered here is completely described by Eqn (8.14) and by the boundary conditions:

$$\begin{aligned} &u = 0 \quad \text{at } y = d \quad \text{and} \quad \text{at } y = -d \\ &\text{or by} \\ &u = 0 \quad \text{at } y = d, \quad \text{and} \quad \partial u/\partial y = 0 \quad \text{at } y = 0 \end{aligned} \tag{8.16}$$

in which d is the half-cell width. On the scale of the model considered here, the flow is fully developed and continuity requires

$$v = w = \frac{\partial P}{\partial y} = \frac{\partial P}{\partial z} = 0 \tag{8.17}$$

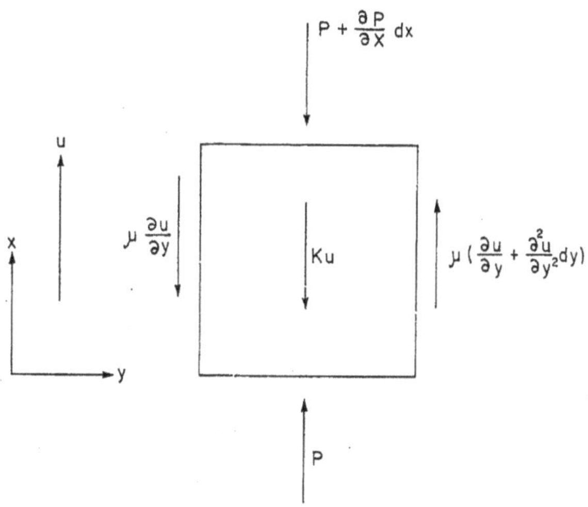

FIGURE 8.22 Differential control volume containing some pores. *Pnueli and Grossman (1969), p. 306, Fig. 2.*

in which v and w are the velocities on a y-axis and a z-axis respectively. Solving Eqn (8.14) with the boundary conditions, Eqn (8.16), the solution velocity u and the volume flux Q are introduced as follows:

$$u = \frac{u_m \left\{ 1 - \frac{\cosh(\beta y/d)}{\cosh \beta} \right\}}{1 - \frac{\tanh \beta}{\beta}} \tag{8.18}$$

$$Q = \int_{-d}^{d} u dy = \left(\frac{2d}{K} \right) \frac{dP}{dx} \left(\frac{\tanh \beta}{\beta} - 1 \right) \tag{8.19}$$

in which $u_m = Q/2d$ is the average solution velocity. β is the nondimensional number, and it characterizes the spacer by the following equation:

$$\beta^2 = Kd^2/\mu \tag{8.20}$$

Some velocity profiles are shown in Fig. 8.23. By definition (see Eqn (8.14) and Eqn (8.20)), β^2 is a measure of the ratio of the "spacer exerted forces" to the "profile shear exerted forces." The denser the spacer, the larger β^2 becomes with a flatter velocity profile. For $\beta = 0$, there is no spacer at all, and the well-known parabolic profile

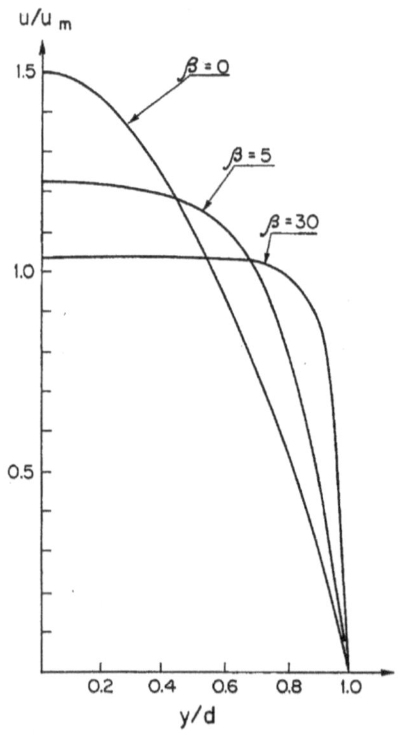

FIGURE 8.23 Velocity profiles in a flow channel. *Pnueli and Grossman (1969), p. 307, Fig. 3.*

is recovered from Eqn (8.18). For $\beta \rightarrow \infty$, the completely flat profile is obtained from Eqn (8.18). Existing systems operate with $3 \leqq \beta \leqq 30$. Under these conditions, Eqn (8.18) can be approximated by the following simpler expression:

$$u = u_m\left\{\frac{\beta}{\beta - 1}\right\}\left[1 - \exp\left\{-\beta\left(1 - \frac{y}{d}\right)\right\}\right] \tag{8.21}$$

8.12 Spacer Mesh Step Model and Mass Transport in a Boundary Layer

Mass transport in electrodialysis is governed by the flow regime, which is determined by the presence of a spacer net ("turbulizer"). The flow is characterized by the shedding of vortices at the discrete threads of the spacer net, and these vortices produce convective mixing, which has a dominant influence on the process. Solan et al. (1971) proposed a "mesh step model" representing the mass transfer process in a laminar concentration boundary layer, the development of which is periodically interrupted.

In this theory, the cell length is divided into mesh steps of length Δx, corresponding to the mesh of the spacer. The basic physical features in the mesh step model are: (a) the spacer produces partial mixing of the flow, and (b) it is characterized by a finite mesh size. At the beginning of the j-th step, the concentration in bulk and the mixing cup average concentration are respectively denoted by $C_{b,j1}$ and $C_{a,j1}$ (subscripts "b" and "a" refer to bulk and average, respectively, and 1 means beginning), and the electric current density and the boundary layer thickness are denoted by i_{j1} and δ_{j1}, respectively. At the end of the j-th step, the variables are denoted by $C_{b,j2}$, $C_{a,j2}$, i_{j2} and δ_{j2} (subscript 2 means end). In this situation, the bulk concentration at the beginning $C_{b,j1}$ is assumed to be equivalent to the value at the exit $C_{b,j2}$:

$$C_{b,j1} = C_{b,j2} \tag{8.22}$$

Further the concentration defect at the membrane surface at the end of the j-th step is expressed by:

$$\Delta C_{j2} = C_{b,j2} - C_{m,j2} \tag{8.23}$$

in which the subscript m refers to membrane. Here the relationship between the concentration defect at the beginning of j+1-th step and that at the end of the j-th step is expressed by the following equation:

$$\Delta C_{(j+1)1} = (1 - k)\Delta C_{j2} \tag{8.24}$$

Here k is the mixing efficiency of the spacer. If $k = 1$, the concentration profile undergoes perfect mixing at the end of each discrete step Δx, while $k = 0$ denotes the complete absence of mixing. Clearly, the mixing process leaves the average concentration unchanged, so that:

$$C_{a,(j+1)1} = C_{a,j2} \tag{8.25}$$

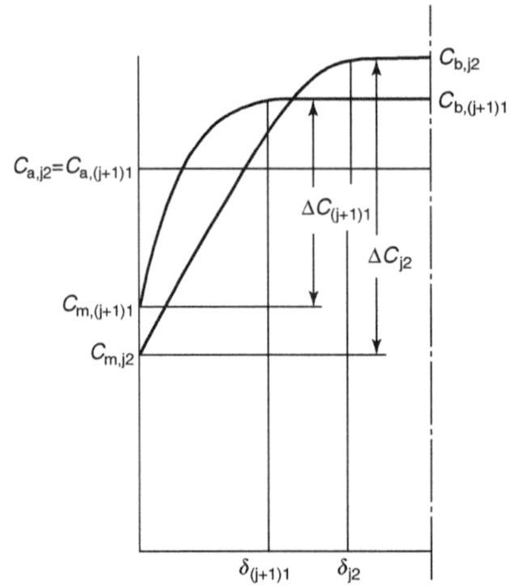

FIGURE 8.24 Concentration profiles before and after mixing. *Solan et al. (1971), p. 93, Fig. 1.*

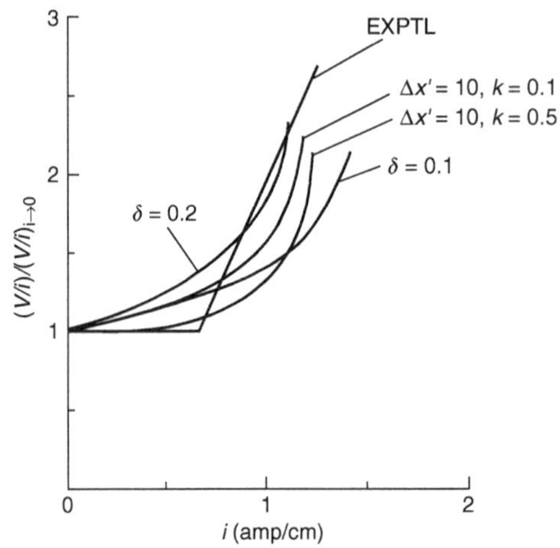

FIGURE 8.25 *V/I* vs *I* curves. *Solan et al. (1971), p. 94, Fig. 3.*

The phenomena mentioned above are schematically illustrated in Fig. 8.24 and analyzed by solving the basic laminar flow equation for nondimensional variables. Figure 8.25 shows representative V/I versus I curves obtained in this manner with experimental, indicating V/I to be decreased by the increase of k and by the decrease of δ. Note here that Δx and δ are the dimensionless values divided by the flow channel thickness.

8.13 Spacer Geometry and Pressure Drop in a Flow Channel

Spacers play an important role in membrane systems referring to problems of mass transfer, homogenizing, and mixing behavior. The specifications of the spacer are defined from a number of geometric parameters, *e.g.* thickness of the grid rods, the distance between the rods, etc. Zimmerer and Kottke (1996) expressed the hydraulic diameter of a spacer-filled flat channel based on the investigation in gas flows in a wind tunnel. Based on the above investigation, the hydraulic diameter of a desalting or a concentrating cell, $d_{H,cell}$, and that of a desalting or a concentrating slot, $d_{H,slot}$, incorporated with a diagonal net spacer (Fig. 8.26) are expressed by the following equation (Tanaka, 2012):

$$d_{H,cell} = \frac{8 - \pi\frac{a}{\chi}}{4\left(\frac{1}{b}+\frac{1}{a}\right) + 2\pi\left(1 - \frac{a}{4b}\right)\frac{1}{\chi}} \tag{8.26}$$

$$d_{H,slot} = \frac{8 - \pi\frac{a}{\chi}}{4\left(\frac{1}{w}+\frac{1}{a}\right) + 2\pi\left(1 - \frac{a}{4w}\right)\frac{1}{\chi}} \tag{8.27}$$

in which, a and b are respectively the flow-pass thickness and flow-pass width in the cell and slot, a and w are respectively the flow-pass thickness and flow-pass width in the slot. χ is the distance between the spacer rods.

The pressure difference between the inlet and the outlet in the cell ΔP_{cell} and that in the slot ΔP_{slot} is expressed by the following equation (Tsiakis and Papageorgiou, 2005):

$$\Delta P_{cell}(Pa) = \frac{3.2\mu\, lu_{cell}}{(d_{H,cell})^2} \tag{8.28}$$

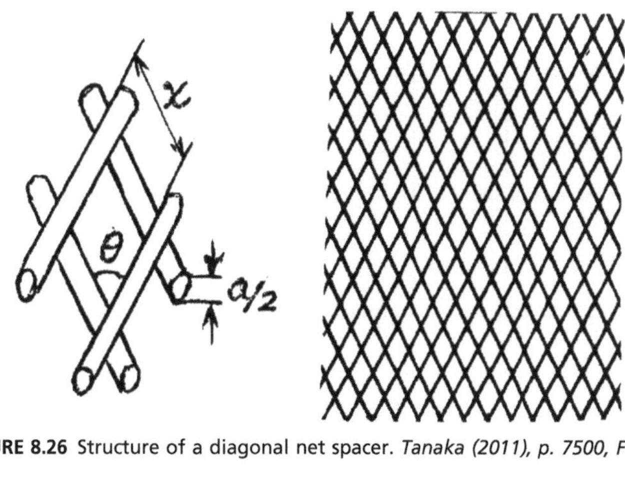

FIGURE 8.26 Structure of a diagonal net spacer. *Tanaka (2011), p. 7500, Fig. 14.*

$$\Delta P_{slot}(Pa) = \frac{3.2\mu\ hu_{slot}}{(d_{H,slot})^2} \tag{8.29}$$

in which l and h are the flow-pass length in the cell and slot, respectively; u_{cell} and u_{slot} are the linear velocity in the cell and slot, respectively; and μ is the viscosity coefficient in a solution.

8.14 Friction Factor of a Solution and Solution Velocity Distribution in Each Desalting Cell (Tanaka, 2004)

Fig. 8.27 illustrates the flow system (System I and System II) of diluate in a stack of an electrodialyzer. In this system, a salt solution is supplied from the outside of the stack to an entrance duct and flows through an entrance slot, a current-passing section, and an

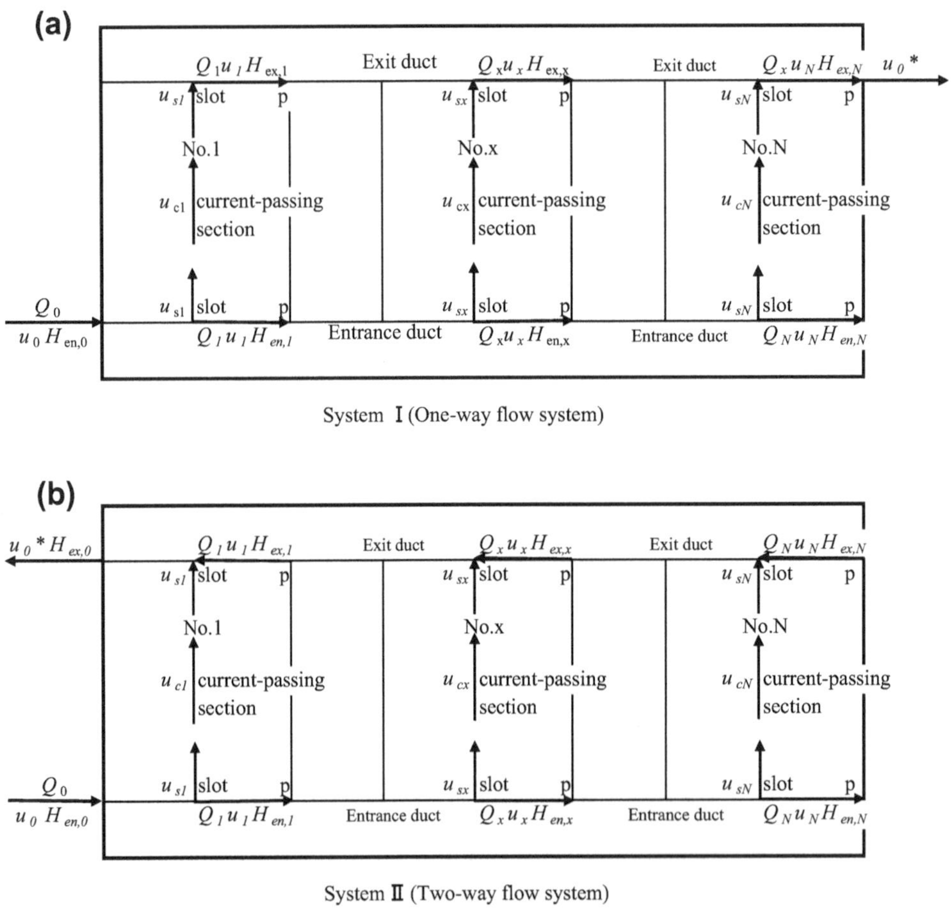

FIGURE 8.27 Diluate flow system in a stack. *Tanaka (2004), p. 24, Fig. 1.*

exit slot, and is discharged from an exit duct to the outside of the stack. Figure 8.28 illustrates the static head of a solution in a desalting cell and a concentrating cell. A salt solution is supplied to a desalting cell by a single-pass flow system. Concentrate is collected by an overflow collecting system from the exit duct at point q, corresponding to the static head of h_{ex}.

The static head difference ΔH between the entrance duct and the exit duct of a desalting cell in Fig. 8.28 is given by Eqn (8.30):

$$\Delta H = \Delta H_c + 2\Delta H_s \tag{8.30}$$

where ΔH_C is the static head difference between the entrance and the exit of a current-passing section, and ΔH_S is the static head difference between the entrance and the exit of the slot. In Eqn (8.30), head loss caused by expansion, reduction, and bending of flow are neglected. Subscript C and S refer to the current-passing section and the slot,

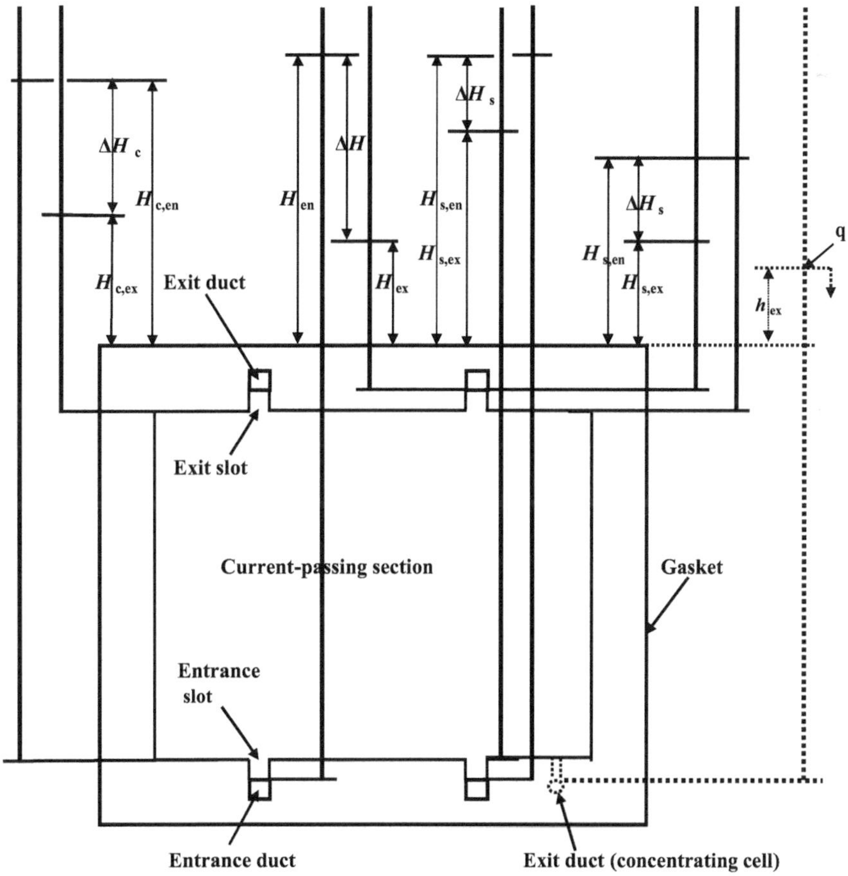

FIGURE 8.28 Static head of a solution in a desalting cell and a concentrating cell. Solid lines and dotted lines represent static head in a desalting cell and a concentrating cell respectively. *Tanaka (2004), p. 25, Fig. 2.*

respectively. Using the Fanning equation, ΔH_c and ΔH_s are expressed by the function of friction head as:

$$\Delta H_c = H_{c,en} - H_{c,ex} = \frac{f_{sol} l_c (a_c + b_c) u_c^2}{g a_c b_c} \tag{8.31}$$

$$\Delta H_s = H_{s,en} - H_{s,ex} = \frac{f_{sol} l_s (a_s + b_s) u_s^2}{g a_s b_s} \tag{8.32}$$

where f_{sol} is the friction factor of the solution, u is the linear velocity of the solution, g is the gravitational acceleration, a is the flow-pass depth, b is the flow-pass width, and l is the flow-pass length.

f_{sol} relates to the Reynolds number Re in Eqn (8.33), because the solution flowing in the current-passing section and in the slot is approximated by laminar flow.

$$f_{sol} = \frac{16}{Re} \tag{8.33}$$

$$Re = \frac{d_e u}{\mu} = \left(\frac{2}{\mu}\right)\left(\frac{ab}{a+b}\right) u \tag{8.34}$$

where d_e is the equivalent diameter, and μ is the kinetic viscosity of the solution. Substituting Eqn (8.34) into Eqns (8.31) and (8.32):

$$\Delta H_c = \left(\frac{8\mu}{g}\right) f_{sol} \left(\frac{a_c + b_c}{a_c b_c}\right)^2 l_c u_c \tag{8.35}$$

$$\Delta H_s = \left(\frac{8\mu}{g}\right) f_{Sol} \left(\frac{a_s + b_s}{a_s b_s}\right)^2 l_s u_s \tag{8.36}$$

Substituting Eqns (8.35) and (8.36) into Eqn (8.30):

$$\begin{aligned} \Delta H &= \left(\frac{8\mu}{g}\right) f_{sol} \left\{ \left(\frac{a_c + b_c}{a_c b_c}\right)^2 l_c u_c + 2\left(\frac{a_s + b_s}{a_s b_s}\right)^2 l_s u_s \right\} \\ &= \left(\frac{8\mu}{g}\right)\left(\frac{a_c + b_c}{a_c b_c}\right)^2 l_c (1 + 2K) f_{sol} u_c \end{aligned} \tag{8.37}$$

$$\begin{aligned} K &= \left\{ \left(\frac{a_c b_c}{a_c + b_c}\right)\left(\frac{a_s + b_s}{a_s b_s}\right) \right\}^2 \left(\frac{l_s}{l_c}\right)\left(\frac{u_s}{u_c}\right) \\ &= \left\{ \left(\frac{a_c b_c}{a_c + b_c}\right)\left(\frac{a_s + b_s}{a_s b_s}\right) \right\}^2 \left(\frac{l_s}{l_c}\right)\left(\frac{b_c}{b_s}\right)\left(\frac{1}{n}\right) \end{aligned} \tag{8.38}$$

where K is the constant that relates to the structure of the current-passing section, and n is the number of the slots in the desalting cell at the entrance or the exit.

Substituting ΔH distribution measured by an electrodialysis experiment into Eqns (8.37) and (8.38), the distribution of $f_{sol}u_c$ and the average of $f_{sol}u_c$; $f_{sol}u_c$(average) are calculated. Further, f_{sol} is calculated from $f_{sol}u_c$ (average) and the average of u_c; u_c (average) in the stack using the following equation:

$$f_{sol} = \frac{f_{sol}u_c\ (average)}{u_c\ (average)} \tag{8.39}$$

In the flow system in Fig. 8.27, the amount of a solution Q_0 supplied to the inlet of an entrance duct corresponds to the total amount of the solution distributed to current-passing sections in cell pair no. 1 to no. N as expressed in Eqn (8.40). Here, N is the number of all desalting cell in the stack:

$$\sum_{x=1}^{N} a_c b_c u_{c,x} = Q_0 = u_0 sn \tag{8.40}$$

where $u_{c,x}$ is the linear velocity in the current-passing section in cell pair no. x, u_0 is the linear velocity at the inlet of an entrance duct, and s is the sectional area of the duct. The amount of a solution Q_x flowing in a duct in cell pair, no. x, is expressed by:

$$Q_x = Q_0 - \sum_{x=1}^{x} a_c b_c u_{c,x} \tag{8.41}$$

The theoretical ratio of Q_x to Q_0 is defined by:

$$\left(\frac{Q_x}{Q_0}\right)_{theo} = 1 - \frac{\sum_{x=1}^{x} a_c b_c u_{c,x}}{Q_0} \tag{8.42}$$

The approximate ratio of Q_x to Q_0 is defined by:

$$\left(\frac{Q_x}{Q_0}\right)_{appro} = \left(1 - \frac{x}{N}\right)^{\theta} \tag{8.43}$$

where θ is the distribution coefficient of a solution flowing into each desalting cell. When $\theta = 1$ holds, the linear velocity in each desalting cell becomes uniform. If the linear velocity is not uniform, $\theta > 1$ or $\theta < 1$ holds. Here, the ratio of $(Q_x/Q_0)_{appro}$ to $(Q_x/Q_0)_{theo}$ is defined by Eqn (10.44). θ is computed to find the situation in which the standard deviation of ε distributed in a stack becomes minimum:

$$\varepsilon = \frac{(Q_x/Q_0)_{appro}}{(Q_x/Q_0)_{theo}} \tag{8.44}$$

Azechi and Fujimoto (1970) measured the performance of an electrodialyzer incorporated with caion exchange membranes Aciplex CK-2 (Asahi Chemical Co.) and anion exchange membranes Aciplex CA-3. The specifications of the electrodialyzer are listed in Table 8.5. The flow system of the desalting solution in this investigation was classified as System I and System II (Fig. 8.27). Seawater was supplied by a single-pass flow system keeping the average linear velocity in desalting cell u_{av} (in a current-passing section) to

Table 8.5 Specifications of an Electrodialyzer. Tanaka (2004), p. 31, Table 1.

Membrane Area S	113.42 dm^2 (106.5 × 106.5 cm)
Ion-exchange membrane	Aciplex CK-2, CA-3, Asahi Chemical Co.
Thickness of a cation-exchange membrane CK-2	0.023 cm
Thickness of an anion-exchange membrane CA-3	0.021 cm
Spacer	Diagonal net
Numbers of cell pairs in a stack N	301
Flow pass length of a current-passing cection l_c	106.5 cm
Flow pass depth of a current-passing cection a_c	0.075 cm
Flow pass width of a current-passing section b_c	106.5 cm
Flow pass length per unit cell pair of a duct t	0.194 cm (=0.023 + 0.021 + 2 × 0.075)
Current density I/S	3×10^{-2} A/cm^2
Desalting Cell	
Numbers of desalting cells in a stack	302
Flow pass length of a slot l_s	4.0 cm
Flow pass depth of a slot a_s	0.075 cm
Flow pass width of a slot b_s	4.5 cm
Dimension of a vertical of a duct α	3 cm
Dimension of a horizontal of a duct β	4 cm
Numbers of entrance or exit ducts of an electrodialyzer n	10
Flow system in an electrodialyzer	Single-pass flow
Concentrating Cell	
Numbers of concentrating cells in a stack	301
Flow pass length of a slot l_s	4.0 cm
Flow pass depth of a slot a_s	0.075 cm
Flow pass width of a slot b_s	1.8 cm
Diameter of a duct	1.8 cm
Numbers of entrance or exit ducts of an electrodialyzer n	4
Flow system in an electrodialyzer	Over flow collection

2.88 cm/s and electrodialyzed by applying constant current density 3.0 A/dm^2. The structure of the desalting cell and the concentrating cell are illustrated in Fig. 8.28. The static head in cell pair no. x in an entrance desalting duct $H_{\text{en,x}}$ and that in an exit duct $H_{\text{ex,x}}$ and their difference ΔH_{x} were measured with manometers, and they are shown in Fig. 8.29 (System I, one-way flow system) and Fig. 8.30 (System II, two-way flow system). In this experiment, Experiment 1 was achieved at first, and then Experiment 2 was achieved after an electrodialytic operation during 300 h. Both experiments were achieved two times changing the flow direction in the ducts such as left → right and right → left.

The static head in cell pair no. x in an entrance desalting cell $H_{\text{en,x}}$, that in an exit duct $H_{\text{ex,x}}$, and their difference ΔH_{x} are plotted against x and are given in Fig. 8.29 (System I, Experiment 1) and Fig. 8.30 (System II, Experiment 1). An arrow in the figure shows the

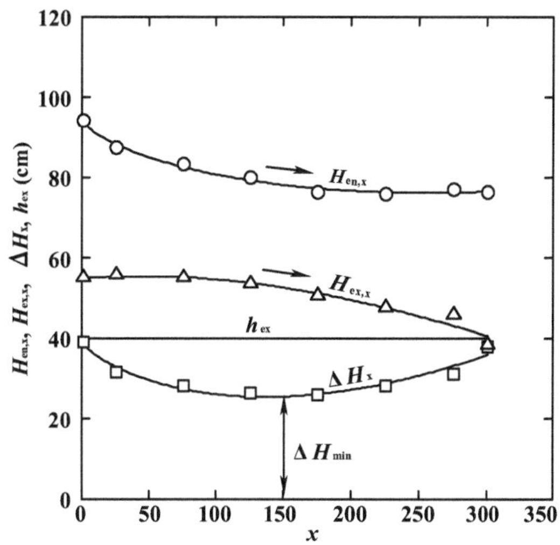

FIGURE 8.29 Distribution of static head of a solution in a desalting duct in a stack (System I). *Tanaka (2004), p. 33, Fig. 7.*

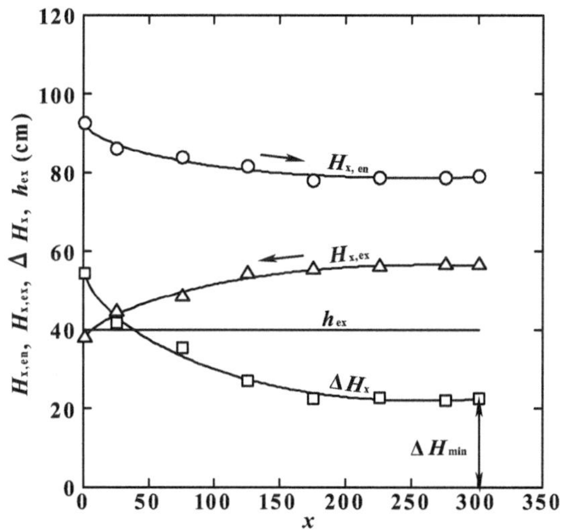

FIGURE 8.30 Distribution of static head of a solution in a desalting duct in a stack (System II). *Tanaka (2004), p. 33, Fig. 8.*

flow direction of a solution in a duct. H_x becomes the lowest point ΔH_{min} at $x = N/2 = 150$, in Fig. 8.29 (System I) and becomes ΔH_{min} at $x = N = 300$ in Fig. 10.30 (System II). Solution velocities on each desalting cell u are proportional to ΔH_x, so that, to achieve a uniform flow rate distribution, it is preferable to operate System I rather

Table 8.6 Solution Friction Factor and Solution Distribution Coefficient in a Current-passing Section in Desalting Cells

Experiment	System		f_{sol}	θ
1	I	①	6.00	0.930
		②	6.16	0.889
		Average	6.08	0.910
	II	①	5.95	1.056
		②	6.01	1.044
		Average	5.98	1.050
	Average		6.03	0.980
2	I	①	7.91	0.952
		②	7.50	0.921
		Average	7.71	0.937
	II	①	7.65	1.015
		②	7.76	1.033
		Average	7.71	1.024
	Average		7.71	0.980

Tanaka (2004), p. 34, Table 2.

than System II or to reduce the number of cell N in a stack. h_{ex} in the figures shows the static head at the exit of a solution in a concentrating duct.

Substituting ΔH_x shown in Figures. 8.29 and 8.30 into Eqns (8.37) and (8.38), $f_{sol}u_c$ is calculated. Substituting $f_{sol}u_c$ calculated above into Eqn (8.39), f_{sol} is calculated in Table 8.6. Further, the solution distribution coefficient θ is calculated by setting the standard deviation of ε (Eqn (8.44)) at the minimum (Table 8.6). In Table 8.6, f_{sol} in System I is the same as that in System II, and $\theta < 1$ holds in System I, and $\theta > 1$ holds in System II. If $\theta = 1$ holds, the solution velocity in desalting cells in a stack becomes uniform. f_{sol} in Experiment 2 is larger than that in Experiment 1. This is probably because the substances suspended in the seawater feed accumulate on the surface of ion exchange membranes or spacers during the operation of the electrodialyzer for 300 h. θ in Experiment 1 is the same as that in Experiment 2. ① and ② in Table 8.6 show two times observations by changing the flow direction in the ducts.

8.15 Pressure Distribution in a Duct in an Electrodialyzer (Tanaka, 2004)

In Figures. 8.27 and 8.28, a solution is supplied to the inlet of an entrance duct at the flow amount of Q_0 and at the linear velocity of u_0. In cell no. x, the solution flows in an entrance duct at the linear velocity of u_x, flows through a slot ($u_{s,x}$), flows in a current-passing section ($u_{c,x}$), and flows out to an exit duct. At the outlet of the exit duct, the solution flows out at the velocity of u^*_0. In this flow system, the static head at point p in cell pair no. x is $H_{en,x}$ in an entrance duct and $H_{ex,x}$ in an exit duct, respectively. The changes of the static head, the velocity head and the friction head in an entrance duct and an exit duct are given using the Bernoulli theorem as follows:

1. In an entrance duct in flow Systems I and II

u_x at a point p in a cell pair no. x in an entrance duct is expressed by the following equation:

$$u_x = u_0\left(1-\frac{x}{N}\right)^{\theta} \tag{8.45}$$

The sum of the static head $H_{en,0}$ and the velocity head $u^2{}_0/2g$ at the inlet of an entrance duct is equal to the sum of every head in the duct described below:

a. The static head at a point p in cell pair number x.
b. The velocity head at a point p in cell pair number x.
c. The total amount of friction head between the inlet and point p in cell pair number x.
d. The total amount of velocity heads of solutions flowing through slots between the inlet and point p in cell pair number x.

The phenomena mentioned above are expressed by Eqn (8.46):

$$H_{en,0}+\frac{u_0^2}{2g} = H_{en,x}+\left(\frac{u_0^2}{2g}\right)\left(1-\frac{x}{N}\right)^{2\theta}+\left(\frac{u_0^2}{2g}\right)\sum_{x=1}^{x}\kappa_{en,x}\left(1-\frac{x}{N}\right)^{2\theta}+\sum_{x=1}^{x}\frac{u_{s,x}^2}{2g} \tag{8.46}$$

where $\kappa_{en,x}$ is the friction coefficient of a solution in the entrance duct.

2. In an exit duct in a flow System I

u_x is indicated by the following equation:

$$u_x = u_0^*\left(\frac{x}{N}\right)^{\theta} \tag{8.47}$$

The relationship among the static head, the velocity head, and the friction head in an exit duct are given by the same manner mentioned in 1, as follows:

$$H_{ex,N}+\frac{(u_0^*)^2}{2g} = H_{ex,x}+\left(\frac{(u_0^*)^2}{2g}\right)\left(\frac{x}{N}\right)^{2\theta}+\left(\frac{(u_0^*)^2}{2g}\right)\sum_{x=1}^{x}\kappa_{ex,x}\left(\frac{x}{N}\right)^{2\theta}+\sum_{x=1}^{x}\frac{u_{s,x}^2}{2g} \tag{8.48}$$

where $\kappa_{ex,x}$ is the friction coefficient of a solution in the exit duct.

3. In an exit duct in a flow System II

Similarly the following equations are introduced:

$$u_x = u_0^*\left(1-\frac{x}{N}\right)^{\theta} \tag{8.49}$$

$$H_{ex,0}+\frac{(u_0^*)^2}{2g} = H_{ex,x}+\left(\frac{(u_0^*)^2}{2g}\right)\left(1-\frac{x}{N}\right)^{2\theta}+\left(\frac{(u_0^*)^2}{2g}\right)\sum_{x=1}^{x}\kappa_{ex,x}\left(1-\frac{x}{N}\right)^{2\theta}-\sum_{x=1}^{x}\frac{u_{s,x}^2}{2g} \tag{8.50}$$

$\kappa_{en,x}$ and $\kappa_{ex,x}$ in Eqns (8.46), (8.48), and (8.50) are related to the friction factors in a duct, $f_{en,x}$, and $f_{ex,x}$ by the Fanning equation as follows:

$$\kappa_{en,x} = 4f_{en,x}\frac{t}{D_{eq}} \tag{8.51}$$

$$\kappa_{ex,x} = 4f_{ex,x}\frac{t}{D_{eq}} \tag{8.52}$$

where t is flow-pass length in a duct in a cell pair, and D_{eq} is the equivalent diameter of a duct section. The shape of a duct section is rectangular, so that D_{eq} is expressed as follows, using the dimension of the wetted perimeter, vertical and horizontal as L, α, and β:

$$D_{eq} = 4\frac{S}{L} = 2\frac{\alpha\beta}{\alpha+\beta} \tag{8.53}$$

Substituting Eqn (8.53) into Eqns (8.51) and (8.52):

$$\kappa_{en,x} = 2f_{en,x}t\,\frac{\alpha+\beta}{\alpha\beta} \tag{8.54}$$

$$\kappa_{ex,x} = 2f_{ex,x}t\,\frac{\alpha+\beta}{\alpha\beta} \tag{8.55}$$

The flow pattern of a solution in a duct is laminar flow, so that $f_{en,x}$ and $f_{ex,x}$ are expressed as follows:

$$f_{en,x} = \frac{16}{Re_{en,x}} \tag{8.56}$$

$$f_{ex,x} = \frac{16}{Re_{ex,x}} \tag{8.57}$$

where $Re_{\text{en,x}}$ and $Re_{\text{ex,x}}$ are the Reynolds numbers in an entrance and an exit duct, respectively, and they are shown taking account of Eqn (8.53) as follows:

$$Re_{en,x} = \left(\frac{D_{eq}}{\chi_{en}}\right)u_0\left(1-\frac{x}{N}\right)^{\theta} = 2\left(\frac{\alpha\beta}{\alpha+\beta}\right)\left(\frac{1}{\chi_{en}}\right)u_0\left(1-\frac{x}{N}\right)^{\theta} \quad \text{(I and II)} \tag{8.58}$$

$$Re_{ex,x} = \left(\frac{D_{eq}}{\chi_{ex}}\right)u_0^*\left(\frac{x}{N}\right)^{\theta} = 2\left(\frac{\alpha\beta}{\alpha+\beta}\right)\left(\frac{1}{\chi_{ex}}\right)u_0^*\left(\frac{x}{N}\right)^{\theta} \quad \text{(I)} \tag{8.59}$$

$$Re_{ex,x} = \left(\frac{D_{eq}}{\chi_{ex}}\right)u_0^*\left(1-\frac{x}{N}\right)^{\theta} = 2\left(\frac{\alpha\beta}{\alpha+\beta}\right)\left(\frac{1}{\chi_{ex}}\right)u_0^*\left(1-\frac{x}{N}\right)^{\theta} \quad \text{(II)} \tag{8.60}$$

where χ_{en} and χ_{ex} are the kinetic viscosities of a solution flowing in an entrance and an exit duct, respectively.

Taking account of Eqn (8.51)–(8.60) in Eqns (8.46), (8.48) and (8.50), the following equations are introduced:

1. In the entrance duct in flow systems I and II

$$H_{en,0} + \left(\frac{u_0^2}{2g}\right) = H_{en,x} + \left(\frac{u_0^2}{2g}\right)\left(1-\frac{x}{N}\right)^{2\theta} + 16\left(\frac{\alpha+\beta}{\alpha\beta}\right)^2\chi_{en}t\left(\frac{u_0}{2g}\right)\sum_{x=1}^{x}\left(1-\frac{x}{N}\right)^{\theta} + \sum_{x=1}^{x}\frac{u_{s,x}^2}{2g} \quad \text{(I and II)} \tag{8.61}$$

Equations (8.61) is simplified as:

$$H_{st,0}+H_{ve,0}=H_{st,x}+H_{ve,x}+H_{fr,x}+H_{ve,s,x} \tag{8.62}$$

where *st, ve,* and *fr* in subscripts refer to the static head, the velocity head, and the friction head in a duct, respectively. *ve, s* refers to the velocity head in a slot.

2. In the exit duct in a flow system I

$$\begin{aligned} H_{ex,N}+\left(\frac{(u_0^*)^2}{2g}\right) &= H_{ex,x}+\left(\frac{(u_0^*)^2}{2g}\right)\left(\frac{x}{N}\right)^{2\theta} \\ &+16\left(\frac{\alpha+\beta}{\alpha\beta}\right)^2\chi_{ex}t\left(\frac{u_0^*}{2g}\right)\sum_{x=1}^{x}\left(\frac{x}{N}\right)^{\theta} \\ &+\sum_{x=1}^{x}\frac{u_{s,x}^2}{2g} \quad \text{(I)} \end{aligned} \tag{8.63}$$

$$H_{st,N}+H_{ve,N}=H_{st,x}+H_{ve,x}+H_{fr,x}+H_{ve,s,x} \tag{8.64}$$

3. In the exit duct in a flow system II

$$H_{ex,0}+\left(\frac{(u_0^*)^2}{2g}\right)=H_{ex,x}+\left(\frac{(u_0^*)^2}{2g}\right)\left(1-\frac{x}{N}\right)^{2\theta}+16\left(\frac{\alpha+\beta}{\alpha\beta}\right)^2\chi_{ex}t\left(\frac{u_0^*}{2g}\right)\sum_{x=1}^{x}\left(1-\frac{x}{N}\right)^{\theta}-\sum_{x=1}^{x}\frac{u_{s,x}^2}{2g} \quad \text{(II)} \tag{8.65}$$

$$H_{st,0}+H_{ve,0}=H_{st,x}+H_{ve,x}+H_{fr,x}-H_{ve,s,x} \tag{8.66}$$

The static head difference ΔH_x between point p in cell pair no. x in an entrance duct and that in an exit duct is shown by:

$$\Delta H_x=H_{en,x}-H_{ex,x} \tag{8.67}$$

The static head distribution in an entrance duct is computed using Eqn (8.61) and that in an exit duct is computed using Eqns (8.63) and (8.65). H_{st}, $H_{st}+H_{ve}$ and $H_{st}+H_{ve}+H_{fr}$ in an entrance duct and an exit duct are plotted against x and are shown in Fig. 8.31 (System I) and Fig. 8.32 (System II). H_{st}, $H_{st,0}+H_{ve,0}$ and $H_{st,N}+H_{ve,N}$ are observed values measured in the experiment described in Section 8.14 (Figs. 8.29 and 8.30).

The velocity head in the slot $H_{ve,s}$ is computed using the following equations introduced from Eqns (8.62), (8.64) and (8.66). $H_{ve,s}$ is computed using the following equations introduced from Eqns (8.62), (8.64) and (8.66):

In the entrance duct in flow system I and II:

$$(H_{ve,s})_{en}=(H_{st,0}+H_{ve,0})_{en}-\left(H_{st}+H_{ve}+H_{fr}\right)_{en} \tag{8.68}$$

In the exit duct in flow system I:

$$(H_{ve,s})_{ex}=(H_{st,N}+H_{ve,N})_{ex}-\left(H_{st}+H_{ve}+H_{fr}\right)_{ex} \tag{8.69}$$

In the exit duct in flow system II:

$$(H_{ve,s})_{ex}=\left(H_{st}+H_{ve}+H_{fr}\right)_{ex}-(H_{st,0}+H_{ve,0})_{ex} \tag{8.70}$$

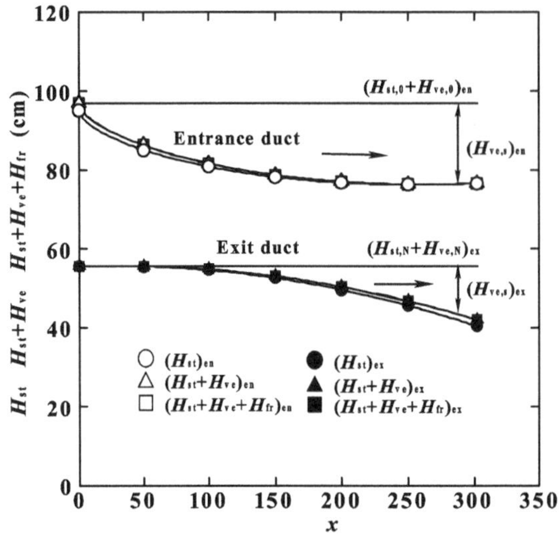

FIGURE 8.31 Pressure distribution in a desalting duct in a stack (System I). *Tanaka (2004), p. 34, Fig. 9.*

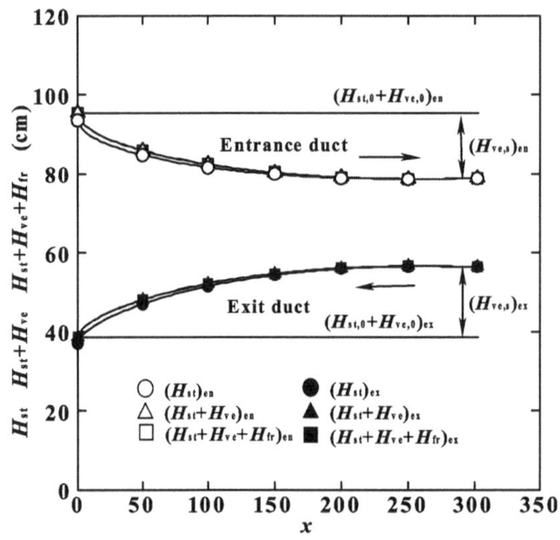

FIGURE 8.32 Pressure distribution in a desalting duct in a stack (System II). *Tanaka (2004), p. 34, Fig. 10.*

Figures 8.31 and 8.32 show the following phenomena.

1. The total hydraulic head in the entrance and exit ducts is governed by the static head; $(H_{st})_{en}$ and $(H_{st})_{ex}$. Velocity head H_{ve} and H_{fr} are negligible.
2. The static head in the entrance duct $(H_{st})_{en}$ is converted to the velocity head in the entrance slots $(H_{ve,s})_{en}$.
3. The velocity head in the exit slots $(H_{ve,s})_{ex}$ is recovered to the static head in the exit ducts $(H_{st})_{ex}$.

4. In order to operate an electrodialyzer stably, it is desirable to achieve uniform static head difference between in the entrance ducts and exit ducts $\Delta H = (H_{st})_{en} - (H_{st})_{ex}$. For this purpose, it is effective to reduce $(H_{ve,s})_{en} - (H_{ve,s})_{ex}$ in Figures 31 (System I) and to keep $\theta = 1$.

References

Azechi, S., Fujimoto, Y., Yuyama, T., 1966. Studies on electrodialytic equipment with ion exchange membrane (XIII). In: Flow Velocity Distribution of the Fluid Passing through the Uneven Flat Flow Path, vol. 11. Scientific Papers of the Odawara Salt Experiment Station, Japan Monopoly Corporation, Odawara, Japan, 20–23.

Azechi, S., Fujimoto, Y., 1970. Flow characteristics of the practical-scale electrodialytic apparatus of filter-press type. Bull. Soc. Sea Water Sci. Jpn. 23, 134–147.

Belfort, G., Gutter, G.A., 1972. An experimental study of electrodialysis hydrodynamics. Desalination 10, 221–262.

Hanzawa, N., Azechi, S., Fujimoto, Y., Nagatsuka, S., 1965. Studies on the electrodialytic equipment with ion exchange membranes (X). In: Comparison of Spacer Used for Electrodialytic Equipment. Scientific Papers of the Odawara Salt Experiment Station, Japan Monopoly Corporation, Odawara, Japan. vol. 10. 16–25.

Ichiki, T., Asawa, T., Hani, H., 1976. An experimental study of turbulent promoters in electrodialysis. Fifth Int. Symp. Fresh Water Sea 3, 89–96.

Kuroda, O., 1993. Study for improvement of efficiency in electrodialysis. Bull. Sea Water Sci. Jpn. 47, 248–258.

Mas, L.J., Pierrard, P.M., Prax, P.A., 1970. Behavior of an electrodialysis unit cell. Desalination 7, 285–296.

Pnueli, D., Grossman, 1969. A mathematical model for the flow in an electrodialysis cell. Desalination 6, 303–308.

Probstein, R.F., Sonin, A.,A., Gur-Arie, E., 1970. A turbulent flow theory of electrodialysis. Desalination 11, 165–187.

Solan, A., Winograd, Y., Katz, U., 1971. An analytical model for mass transfer in an electrodialysis cell with spacer of finite mesh. Desalination 9, 89–95.

Sonin, A.A., Probstein, R.F., 1968. A hydrodynamic theory of desalination by electrodialysis. Desalination 5, 293–329.

Takahashi, S., Arikawa, Y., Teraoka, Y., 1979. Air bubble cleaning type electrodialyzer and its high temperature process. Hitachi Rev. 28, 317–322.

Tanaka, Y., 2004. Pressure distribution, hydrodynamics, mass transport and solution leakage in an ion-exchange membrane electrodialyzer. J. Membr. Sci. 234, 23–39.

Tanaka, Y., 2005. Limiting current density of an ion-exchange membrane and of an electrodialyzer. J. Membr. Sci. 266, 6–17.

Tanaka, Y., 2007. Ion Exchange Membrane; Fundamentals and Applications, Membrane Science and Technology Series, vol. 12. Elsevier, Amsterdam.

Tanaka, Y., 2011. Ion-exchange membrane electrodialysis for saline water desalination and its application to seawater concentration. Ind. Eng. Chem. Res. 50, 7494–7503.

Tanaka, Y., 2012. Ion-exchange membrane electrodialysis program and its application to multi-stage continuous saline water desalination. Desalination 301, 10–25.

Tsiakis, P., Papageorgiou, L.G., 2005. Optimal design of an electrodialysis brackish water desalination plant. Desalination 173, 173–186.

Zimmerer, C.C., Kottke, V., 1996. Effects of spacer geometry on pressure drop, mass transfer, mixing behavior, and residence time distribution. Desalination 104, 129–134.

9 Limiting Current Density

9.1 Introduction

When an electric current passes through an ion exchange membrane, salt concentration on the desalting surface of the membrane decreases because of concentration polarization and reduces to zero at the limiting current density. In this circumstance, there are no more salt ions available to carry the electric current; as a result, the voltage drop across the boundary layer increases drastically and causes higher energy consumption and the generation of water dissociation. The limiting current density is closely related to concentration polarization and water dissociation; these phenomena were discussed in Chapters 6 and 7, respectively. This chapter examines the mechanism of the limiting current density and its application to electrodialysis.

9.2 Nernst Diffusion Model

For ions i, the transport of the ions, J_i, in the boundary layer, is generally described by the extended Nernst–Planck equation:

$$J_i = -D_i\frac{dC_i}{dx} - \frac{FD_iz_iC_i}{RT}\frac{d\psi}{dx} + C_iv \tag{9.1}$$

where C_i is the concentration of i, ψ is the electric potential, D_i is the diffusion constant of i, z_i is the ionic charge number of i, v is the solution velocity caused by convection, F is the Faraday constant, R is the gas constant and T is the absolute temperature.

The solution velocity v in Eqn (9.1) is caused by gravitational convection or electroconvection. Under the forced flowing circumstance in a desalting cell, convection is suppressed, and v is caused only by electroosmosis and volume osmosis passing through the membrane. In this circumstance, the convection term C_iv in Eqn (9.1) substantially disappears, and the concentration distribution profile in the boundary layer is approximated by a straight line as illustrated in Figure 9.1. The boundary layer profile in Figure 9.1 is widely known as the Nernst diffusion layer model, named after and introduced by Nernst (1904). Despite the drastic simplification, as a rule the Nernst model accurately describes diffusion phenomena in the boundary layer as shown by both experimental evidence and theoretical approaches (Vielstich, 1953; Amatore et al., 2001). The solution in bulk is well mixed, so concentration is constant with respect to distance. The diffusion layer (Nernst diffusion layer) is a stationary and stagnant thin layer of a

Ion Exchange Membranes. http://dx.doi.org/10.1016/B978-0-444-63319-4.00009-2

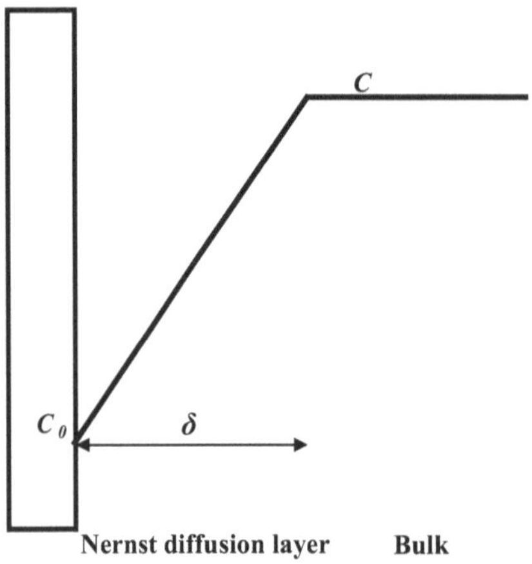

FIGURE. 9.1 Nernst diffusion layer.

solution in contact with the membrane. The thickness of the Nernst diffusion layer δ depends upon the nature of the solution and is estimated from hydrodynamic or kinetic measurements, usually on the order of 10^{-3} to 10^{-2} cm (Helfferich, 1962).

Referring to Figure 9.1, the material balance of counter ions between the membrane phase and the Nernst diffusion layer is represented by the following equations:

$$\frac{i}{F}\bar{t} = \frac{i}{F}t + D\left(\frac{dC}{dx}\right) \tag{9.2}$$

$$\left(\frac{dC}{dx}\right) = \frac{C - C_0}{\delta} \tag{9.3}$$

where $\bar{t}$ and t are transport numbers for counter ions in the membrane and solution, respectively; C and C_0 are the ionic concentrations in bulk (desalting cell) and at $x = 0$ (solution/membrane interface), respectively; and D is the diffusion constant of an electrolyte dissolving in this system.

The limiting current density of the membrane i_{lim} is expressed by Eqn (9.4), introduced by putting $C_0' = 0$ in Eqns (9.2) and (9.3):

$$i_{\text{lim}} = \frac{FD}{\bar{t} - t}\left(\frac{dC}{dx}\right) = \frac{FDC}{(\bar{t} - t)\delta} \tag{9.4}$$

Rosenberg and Tirrel (1957) applied the following Chilton–Coburn transfer factor j_D for analyzing the boundary layer thickness δ in Eqn (9.4):

$$j_D = \left(\frac{k}{u}\right)\left(\frac{\mu}{D}\right)^{2/3} \tag{9.5}$$

where k is the mass transfer coefficient, u is linear velocity in the desalting cell, μ is viscosity, and D is the diffusion coefficient of the electrolyte. The boundary layer thickness δ is thus expressed as follows:

$$\delta = D/k \tag{9.6}$$

The equation for j_D in a flat duct is as follows, for stream line flow (Reynolds number Re $<$ 2100):

$$j_D = 1.85\text{Re}^{-2/3}\left(\frac{l}{d_e}\right)^{-1/3} \tag{9.7}$$

where l is the path length over which the flow is developed, and d_e is the equivalent diameter of the path.

Using Eqns (9.5)–(9.7) and accounting for Re $= d_e u\rho/\mu(\rho = 1)$, boundary layer thickness δ becomes:

$$\delta = \frac{1}{1.85}\left(\frac{Dd_e l}{u}\right)^{1/3} \tag{9.8}$$

Substituting Eqn (9.8) into Eqn (9.4), i_{lim}/C is obtained as:

$$\frac{i_{\text{lim}}}{C} = 1.85\left(\frac{FD}{\bar{t}-t}\right)\left(\frac{u}{Dd_e l}\right)^{1/3} \tag{9.9}$$

For turbulent flow (Re $>$ 2100):

$$j_D = 0.023\text{Re}^{-0.2} \tag{9.10}$$

Values for δ and i_{lim}/C are computed in a manner similar to those used for streamline flow:

$$\delta = \left(\frac{1}{0.023}\right)\left(\frac{D^{1/3}\mu^{2/3}}{u}\right)\text{Re}^{0.2} \tag{9.11}$$

$$\frac{i_{\text{lim}}}{C} = 0.023\left(\frac{Fu}{\bar{t}-t}\right)\left(\frac{D}{\mu}\right)^{2/3}\text{Re}^{-0.2} \tag{9.12}$$

Cowan and Brown (1959) employed δ introduced from the Frank-Kamenetskii equation:

$$\delta = \alpha\left(\frac{\mu\text{Re}^{1/8}}{0.395u}\right) \tag{9.13}$$

where α is the dimensionless constant. Extending further, i_{lim}/C is obtained by substituting Eqn (9.13) into Eqn (9.4):

$$\frac{i_{\text{lim}}}{C} = 0.395\left(\frac{FD}{\bar{t}-t}\right)\left(\frac{\text{Re}^{7/8}}{\alpha\, d_e}\right) \tag{9.14}$$

Yamabe et al. (1967) measured the limiting current density i_{lim} from an inflection of an electric current–electric potential curve under a flowing circumstance in a desalting cell. The diffusion layer thickness δ was calculated by substituting i_{lim} into Eqn (9.4). Results are presented in Table 9.1.

Table 9.1 Diffusion Layer Thickness

NaCl Concentration	Solution Velocity	Diffusion Layer Thickness Selemion CMG-10	Diffusion Layer Thickness Selemion AMG-10
M	cm/s	10^{-2} cm	
0.1	0.026	2.60	2.46
	0.226	2.15	2.29
	0.857	1.05	1.27
0.3	0.025	1.98	2.50
	0.260	1.64	2.65
	0.720	1.25	1.85
0.5	0.024	1.86	1.84
	0.250	1.44	1.58
0.75	0.027	2.01	
	0.226	1.68	
	0.785	0.98	

Yamabe et al. (1967), p.580, Table 1.

9.3 Limiting Current Density Equation Introduced from the Nernst–Planck Equation

Spiegler (1971) introduced the limiting current density equation in an electrolyte solution dissolving monovalent cations A and monovalent anions X from the Nernst–Planck equation as follows:

$$J_+ = -D_+\left\{\left(\frac{dC}{dx}\right) + \left(\frac{FC}{RT}\right)\left(\frac{d\psi}{dx}\right)\right\} \tag{9.15}$$

$$J_- = -D_-\left\{\left(\frac{dC}{dx}\right) - \left(\frac{FC}{RT}\right)\left(\frac{d\psi}{dx}\right)\right\} \tag{9.16}$$

In the membrane, the numerical value of the flux ratio J_+/J_- is equal to the transport number ratio $\bar{t}_+/\bar{t}$:

$$\frac{J_+}{J_-} = \frac{\bar{t}_+}{\bar{t}_-} = \frac{\bar{t}_+}{1-\bar{t}_+} \tag{9.17}$$

From Eqns (9.15)–(9.17):

$$J_+ = \frac{-2D_+}{1-\frac{D_+\bar{t}_-}{D_-\bar{t}_+}}\frac{dC}{dx} \tag{9.18}$$

$$J_- = \frac{-D_-}{1-\frac{D_-\bar{t}_+}{D_+\bar{t}_-}}\frac{dC}{dx} \tag{9.19}$$

It is often more convenient to express the fluxes in terms of the diffusion coefficient of the electrolyte D rather than the ionic diffusion coefficients D_+ and D_-. Note that in a free solution:

$$\frac{D_+}{D_-} = \frac{t_+}{t_-} = \frac{t_+}{1 - t_+} \tag{9.20}$$

Using the Nernst expression for the diffusion coefficient of 1:1 electrolyte in a dilute solution results in the following:

$$D = \frac{2D_+D_-}{D_+ + D_-} = 2D_+(1 - t_+) \tag{9.21}$$

Taking into account Eqns (9.20) and (9.21), Eqns (9.18) and (9.19) can be rewritten as follows:

$$J_+ = -\frac{D\bar{t}_+}{(\bar{t}_+ - t_+)} \frac{dC}{dx} \tag{9.22}$$

$$J_- = -\frac{D\bar{t}_-}{(\bar{t}_- - t_-)} \frac{dC}{dx} \tag{9.23}$$

The electric current density i is calculated from the Faraday law as follows:

$$i = F(J_+ - J_-) \tag{9.24}$$

Substituting Eqns (9.22) and (9.23) into Eqn (9.4) for a cation exchange membrane results in:

$$i = -\frac{FD}{(\bar{t}_+ - t_+)} \frac{dC}{dx} \tag{9.25}$$

Making the same substitution for an anion exchange membrane results in:

$$i = \frac{FD}{(\bar{t}_- - t_-)} \frac{dC}{dx} \tag{9.26}$$

Since i, D, and the transport numbers are constants, the concentration gradient dC/dx in Eqns (9.25) and (9.26) is linear. The limiting current density i_{lim} presented by Eqn (9.4) is introduced by substituting $dC/dx = C/\delta$ into Eqns (9.25) and (9.26).

9.4 Limiting Current Density Equation Introduced by Means of Chemical Engineering Techniques

9.4.1 Analysis Based on the Stanton Number, Pecret Number and Potential Difference Number

Kitamoto and Takashima (1967) expressed the mass transport and energy consumption in an electrodialysis system by the Stanton number St, Pecret number Pe, and potential difference number Ψ:

$$St = \frac{i/F}{Cu} \tag{9.27}$$

$$Pe = \frac{au}{D} \tag{9.28}$$

$$\Psi = \frac{F\Delta\psi}{RT} \tag{9.29}$$

Here, *St* stands for the dimensionless ratio of the material flux transporting across the membrane i/F against that in the solution flowing into the desalting cell *Cu*. Continuing, *Pe* is the dimensionless solution velocity in the desalting cell u, while a is the thickness of the desalting cell, and D is the diffusion constant. Finally, Ψ is the dimensionless potential difference between the potential at the cation exchange membrane surface and at the anion exchange membrane surface in the concentrating cell, and $\Delta\psi$ is approximated by the difference between the potential difference applied to a membrane pair, $\Delta\psi_{applied}$, and the membrane potential, as follows:

$$\Delta\Psi = \Delta\Psi_{applied} - \frac{2RT}{F}\ln\frac{C_{con}}{C_{de}} = \left(\frac{a_{de}}{\kappa_{de}} + \frac{a_{con}}{\kappa_{con}} + r_K + r_A\right)i \tag{9.30}$$

where C_{de} and C_{con} are the electrolyte concentrations in a desalting and a concentrating cell, respectively; a_{de} and a_{con} are the thicknesses of a desalting and a concentrating cell, respectively; κ_{de} and κ_{con} are the specific conductivities of an electrolyte solution flowing in a desalting and a concentrating cell, respectively; and r_K and r_A are the electric resistance of a cation and an anion exchange membrane, respectively.

Plotting *St* against Ψ/Pe was confirmed to be expressed by the straight line as shown in Figure 9.2 by means of the electrodialysis of a NaCl solution using the apparatus (effective membrane area: 5 × 50 or 8 × 24 cm; thickness of a desalting cell: 0.71–5.6 cm) integrated with Selemion CMV-10/AMT-10 (Asahi Glass Co.) or Aciplex CK-1/CA-1,CA-2 (Asahi Chemical Co.) membranes with spacers and feeding a 10^{-3}–0.6 M NaCl solution. Figure 9.2 is represented by the following equation:

$$St = 1.04\left(\frac{\Psi}{Pe}\right) = 1.04\left(\frac{FD}{au}\right)\left(\frac{\Delta\psi}{RT}\right) \tag{9.31}$$

Further, Kitamoto and Takashima (1968) measured the mass transport in a desalting cell applying the limiting current density (from *I–pH* or *I–V* curves), and obtained Figure 9.3 and the following equation:

$$\begin{aligned} \Psi_{\lim} &= 0.09Pe^{0.65} \\ \left(\frac{\Psi}{Pe}\right)_{\lim} &= 0.09Pe^{-0.35} \end{aligned} \tag{9.32}$$

Substituting Eqn (9.31) into Eqn (9.32), the limiting current density in a NaCl solution system is obtained as follows:

$$St_{\lim} = 0.094Pe^{-0.35} \tag{9.33}$$

or

$$i_{\lim} = 1.88 \times 10^2 Cu^{0.65} D^{-0.35} \tag{9.34}$$

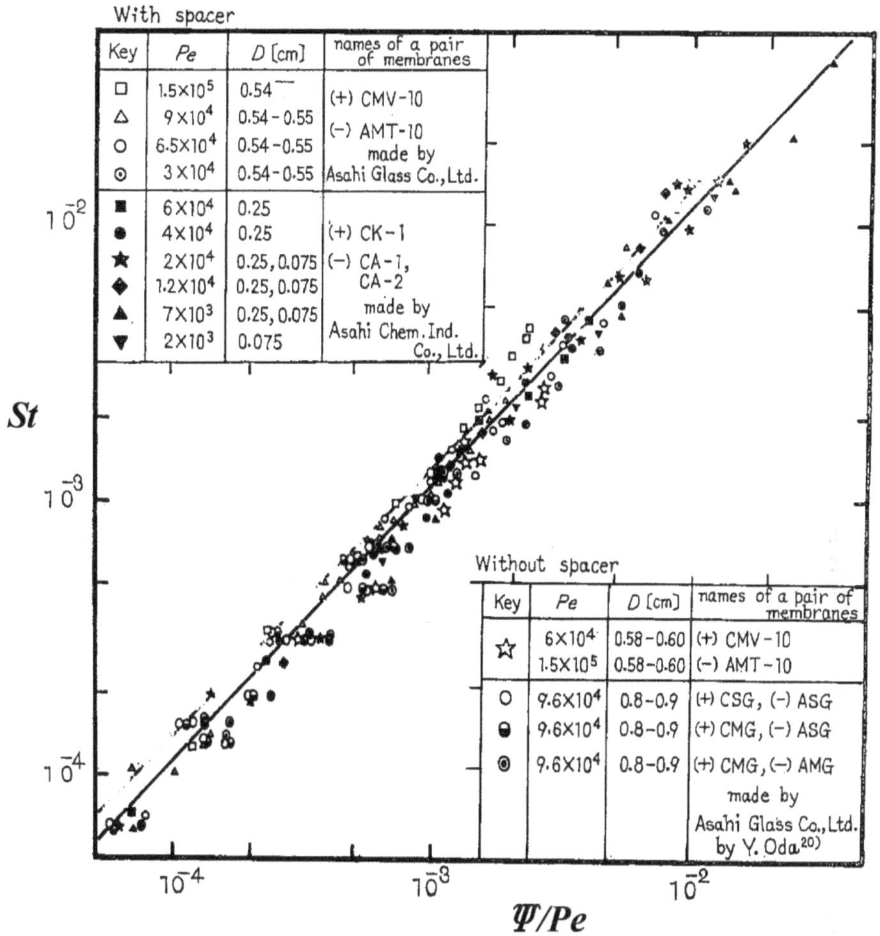

FIGURE. 9.2 Relationship between *St* and Ψ/*Pe*. *Kitamoto and Takashima (1967), p. 1205, Fig. 3.*

9.4.2 Analysis Based on the Sherwood Number

Miyoshi et al. (1968) expressed the ionic electrolyte flux J_i, on a y-axis perpendicular to the membrane surface in a desalting cell incorporated with spacers, with the following equation:

$$J_i = \frac{t}{F} i + D_e \frac{\partial C}{\partial y} \tag{9.35}$$

where t is the transport number of the ionic electrolytes, and D_e is the eddy diffusivity of the electrolytes in the solution. Based on Eqn (9.35) and the equation of continuity, the limiting current density i_{lim} was introduced in the form of the Sherwood number Sh,

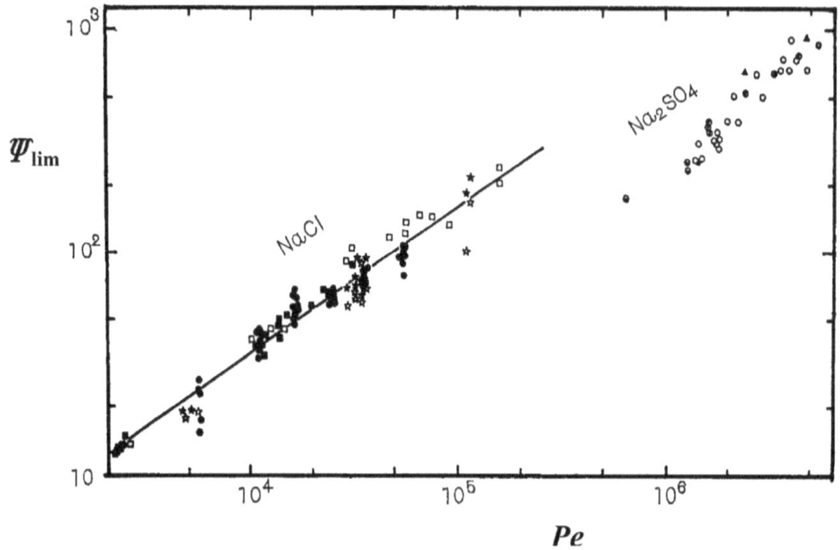

FIGURE. 9.3 Relationship between $\Psi_{\rm lim}$ and *Pe*. *Kitamoto and Takashima (1968), p. 76, Fig. 3.*

exhibiting the dimensionless ratio of limiting current density $i_{\rm lim}$ against electrolyte concentration in a desalting cell *C*:

$$Sh = \left(\frac{a}{DF}\right)\left(\frac{i_{\rm lim}}{C}\right) = \frac{\alpha}{\bar{t}-t}\left(\frac{a}{l}\right)^{1/3} M^{1/3}{\rm Re}^{(1+2\beta)/3}Sc^{1/3} \tag{9.36}$$

in which *Sc* is the Schmidt number equal to ν/D, $\bar{t}$ and t are the transport number of counter ions in the membrane and in the solution respectively. *M* is a variable, and α and β are constants.

Equation (9.36) was experimentally analyzed using the electrodialyzer (membrane area: 20 cm length × 4.0 cm width; desalting cell thickness: 0.08–0.40 cm), in which cation-exchange membranes (Neosepta CL-25T, Tokuyama Co.) and anion exchange membranes (Neosepta AV-4T) were incorporated with honeycomb-net or pointed-twill-net spacers. Supplying a 0.005–0.10 eq/dm^3 NaCl solution or 0.05 eq/dm^3 KCl, $MgCl_2$, $CaCl_2$, $NaHCO_3$, or Na_2SO_4 solution into the electrodialyzer, $i_{\rm lim}$ was measured from the inflection of *V*/*I* versus 1/*I* plots. Based on the experiment mentioned above and Eqn (9.36), *Sh* was introduced in the following equation and is presented in Figure 9.4:

$$Sh = \frac{0.095}{(\bar{t}-t)\{n(d_s-d_f)\}^{1/2}}\ \frac{(a/l)^{1/3}M^{1/3}}{\left\{(1-\varepsilon)^2/\varepsilon^3\right\}^{1/5}} \times {\rm Re}^{1/2}Sc^{1/3} \tag{9.37}$$

in which d_s is the thickness of a spacer, *n* and d_f are the number and fiber diameter of the spacer, ε is the void fraction of the spacer, and *M* is a spacer parameter represented by using the eddy viscosity parameter *m* originated by the spacer, as follows:

$$M = \frac{m^3}{(m+1)^2\ln(m+1) - 1.5m^2 - m} \tag{9.38}$$

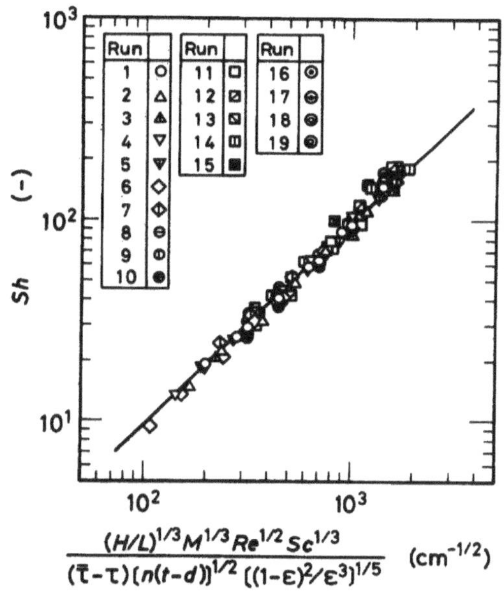

FIGURE. 9.4 Relationship between Re, *Sc*, and *Sh*. *Miyoshi et al. (1968), p. 598, Fig. 6.*

$$m = 2.1 \times 10^5 \{n(d_s - d_f)\}^{2.4} \left\{ \frac{(1-\varepsilon)^2}{\varepsilon^3} \right\} \tag{9.39}$$

9.4.3 Analysis Based on the Reynolds Number, Schmidt Number, and Shape Factor

Huang and Yu (1988) studied the effects of the Reynolds number, Schmidt number, and shape factor of the electrodialysis cell on the mass transfer rate in electrodialysis for various systems at limiting current density. In this study, the electrolyte concentration C in the electrodialysis cell was expressed by:

$$\frac{\partial C}{\partial t} + v\nabla C = \nabla(D\nabla C) \tag{9.40}$$

where v is the flow velocity of an electrolyte solution, and D is the molecular diffusion coefficient of the electrolytes in a solution. Considering the flow of an electrolyte solution through two flat parallel ion exchange membranes at steady state, for flow velocity in the x-direction v_x, and constant molecular diffusion coefficient D, Eqn (9.40) is written on the coordinate along the membrane surface from the upstream end x, and normal to the membrane surface y, as:

$$v_x = \frac{\partial C}{\partial x} = D\frac{\partial^2 C}{\partial y^2} \tag{9.41}$$

with the boundary conditions under limiting current density:

$$\begin{aligned} &C = 0, \quad x > 0, \ y = B \\ &C = C^0, \quad x = 0, \ -B < y < B \\ &C = C^0, \quad x > 0, \ y = -\infty \end{aligned} \tag{9.42}$$

Here, B is the half-thickness of the channel, and C^0 is the electrolyte concentration in bulk. If the Schmidt number is large, the thickness of the diffusion boundary layer will be smaller than that of the viscous boundary layer, so the velocity component v_x can be approximated by:

$$v_x = \langle 3v \rangle \left(1 - \frac{y}{B}\right) \tag{9.43}$$

Where $\langle v \rangle$ is the average value of v.

Next, if the diffusivity in the solution of the cation is smaller than that of the anion, concentration polarization develops first on the surface of the cation exchange membrane. In this situation, the dimensionless mass transfer rate, the Nusselt number Nu, showing the dimensionless ionic flux across the membrane N against the electrolyte concentration in the desalting cell C on the cation exchange membrane, is defined as:

$$Nu = \frac{N_+|_{y=B} d_e}{C_+^0 D_+} \tag{9.44}$$

$$N_+ = \frac{i_{\text{lim}}}{z_+ F} = -v_+ \left(1 - \frac{z_+}{z_-}\right) D_+ \frac{\partial C}{\partial y}\bigg|_{y=B} \tag{9.45}$$

$$N_- = 0 \tag{9.46}$$

Here, N_+ and N_- are molar fluxes of cations and anions, respectively, d_e is the equivalent diameter of the channel, v_+ is the number of cations produced by the dissociation of one molecule of electrolytes in the solution, z_+ and z_- are the valence of cations and anions, respectively, C_+^0 is the concentration of cations in bulk, and D_+ is the diffusion constant of cations.

Solving Eqns (9.41)–(9.46), Nu on the cation exchange membrane is introduced as follows:

$$Nu = \left(1 - \frac{z_+}{z_-}\right) \frac{(4/3)^{1/3}}{\Gamma(4/3)} \left(\frac{\text{Re}Scd_e}{x}\right)^{1/3} \tag{9.47}$$

The average value of the Nu over the length L, $(Nu)_{ave}$ of the cation exchange membrane is introduced as:

$$(Nu)_{ave} = \frac{1}{L}\int_0^L Nudx = 1.849\left(1 - \frac{z_+}{z_-}\right)\left(\frac{\text{Re}Scd_e}{L}\right)^{1/3} \tag{9.48}$$

which is expressed by the function of the Reynolds number Re, Schmidt number Sc, and dimension factor d_e/L.

On the other hand, if the diffusivity in the solution of the anion is smaller than that of the cation, concentration polarization develops first on the surface of the anion

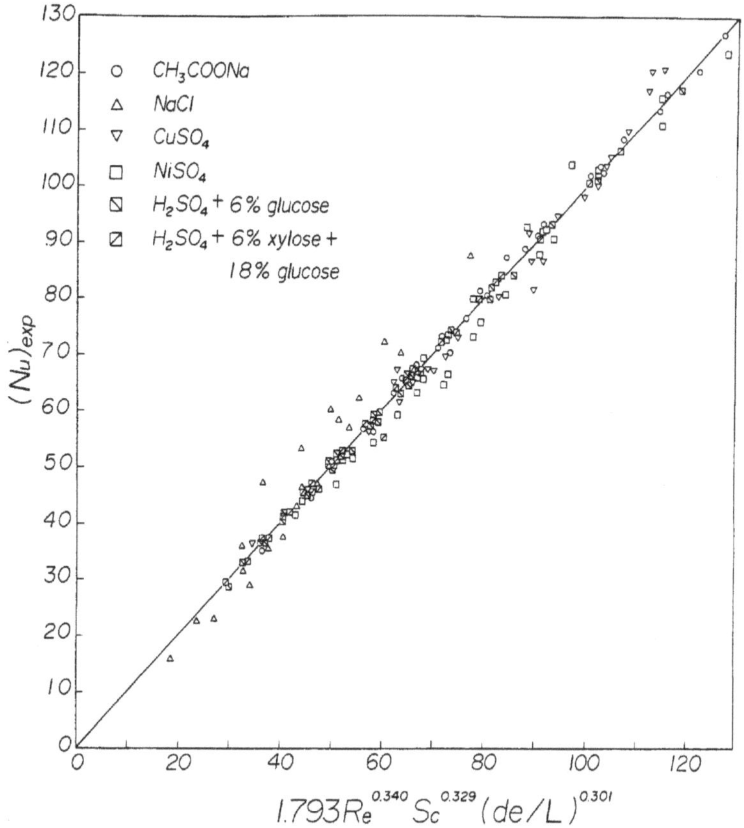

FIGURE. 9.5 Relationship between *Nu*, *Sc*, and d_e/L. *Huang and Yu (1988), p. 203, Fig. 7.*

exchange membrane. In this case, $(Nu)_{ave}$ on the anion exchange membrane is derived in a similar manner as Eqn (9.48); however, Nu, N_+, and N_- are defined as:

$$Nu = \frac{-N_-|_{y=-B} d_e}{C_-^0 D_-} \tag{9.49}$$

$$N_+ = 0 \tag{9.50}$$

$$N_- = \frac{i_{\lim}}{z_- F} = -\nu_-\left(1 - \frac{z_-}{z_+}\right) D_- \frac{\partial C}{\partial y}\bigg|_{y=B} \tag{9.51}$$

The theory presented above was examined through measurement of the limiting current density using the electrodialysis apparatus (effective membrane area: 20 cm^2) incorporated with a desalting channel and an ion exchange membrane pair (Selemion CMV/AMV or Aciplex K-102/A-102). Supplying a CH_3COONa, NaCl, $CuSO_4$ or $NiSO_4$ solution, or a mixed solution of H_2SO_4 + glucose or xylose, into the channel, and passing an electric current, the limiting current density was determined from the first inflection point of the current density versus the voltage curve. The relationship between the experimental Nusselt number $(Nu)_{\exp}$ and Re, Sc, and d_e/L is shown in Figure 9.5, and represented by the following equation, which is similar to the theoretical Eqn (9.48).

$$\frac{(Nu)_{\exp}}{\left(1-\frac{z_+}{z_-}\right)} \text{ (for cation exchange membrane)} = \frac{(Nu)_{\exp}}{\left(1-\frac{z_-}{z_+}\right)} \text{ (for anion exchange membrane)} = 1.793\mathrm{Re}^{0.341}Sc^{0.329}\left(\frac{d_e}{L}\right)^{0.301} \quad (9.52)$$

9.5 Dependence of Limiting Current Density on Salt Concentration, Solution Velocity, and Temperature

A drawback of Eqn (9.4) is that definite diffusion layer thickness δ is not determined beforehand. In order to determine the limiting current density of an ion exchange membrane, it is practically reasonable to measure the limiting current density directly and evaluate the effects of electrodialysis conditions on the limiting current density. An example of such an experiment (Tanaka, 2005) is introduced below.

The experimental apparatus shown in Figure 9.6(a) was assembled and incorporated with cation exchange membranes (Aciplex K-172, Asahi Chemical Co.) and anion exchange membranes (Aciplex A-172). The thickness of cell D was adjusted to 0.075 cm; width and length of the flow pass in cell D were adjusted to 1 and 2 cm, respectively (Figure 9.6(b)); a diagonal net spacer was put into cell D; and a 25 °C NaCl solution was supplied to cell D. Passing an electric current and changing the current

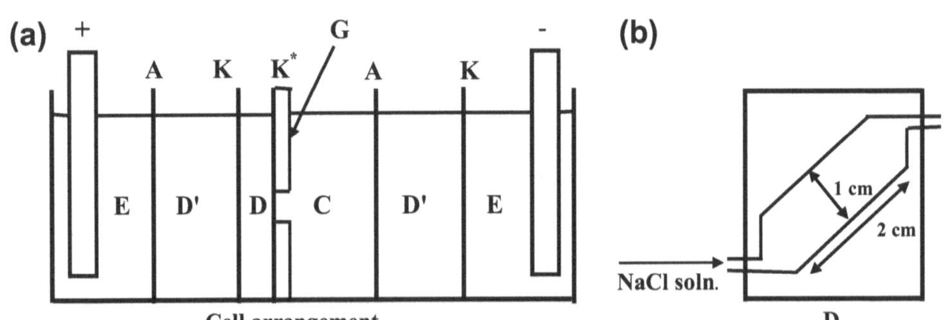

FIGURE. 9.6 Apparatus for measuring limiting current density of a cation exchange membrane. K: Cation exchange membrane. K*: Cation exchange membrane for measuring limiting current density. A: Anion exchange membrane. D: Cell for measuring limiting current density. D′: Desalting cell. C: Concentrating cell. G: Gasket. E: Electrode cell. *Tanaka (2005), p. 9, Fig. 1.*

density incrementally, the limiting current density i_{lim} of the cation exchange membrane (K* in the figure) was measured from the inflection of V/I versus $1/I$ plot. The experiment was repeated, by changing NaCl concentration C and solution velocity u in D, step by step; in the same manner, the i_{lim} of the anion exchange membrane was measured. The i_{lim} is plotted against C with u as a parameter, and shown in Figure 9.7 (cation exchange membrane, $i_{\text{lim},K}$) and Figure 9.8 (anion exchange membrane, $i_{\text{lim},A}$). The plots except for $u = 0$ are expressed by Eqn (9.53):

$$i_{\text{lim}} = \text{m}C^{\text{n}} \tag{9.53}$$

where m and n are expressed by the following functions of u (Figure 9.9), for cation exchange membrane Aciplex K-172:

$$\begin{aligned} \text{m} &= 83.50 + 24.00u \\ \text{n} &= 0.7846 + 8.612 \times 10^{-3}u \end{aligned} \tag{9.54}$$

and for anion exchange membrane Aciplex A-172:

$$\begin{aligned} \text{m} &= 66.36 + 14.72u \\ \text{n} &= 0.7404 + 3.585 \times 10^{-3}u \end{aligned} \tag{9.55}$$

The electrodialyzer was incorporated with Neosepta CL-25T/ACH-45T membranes (Tokuyama Co.) and a diagonal net spacer. A NaCl solution was supplied to the desalting cell changing temperature T, linear velocity u, and NaCl concentration C as follows.

$$\begin{aligned} T &= 10, 15, 20, 25, 30, 35, 40\ ^\circ C \\ u &= 6, 9, 12\ \text{cm/s} \\ C &= 0.01 - 0.05\ \text{eq/dm}^3 \end{aligned}$$

Passing an electric current I and changing the current density incrementally, the limiting current density of CL-25T cation exchange membrane i_{lim} was measured from the

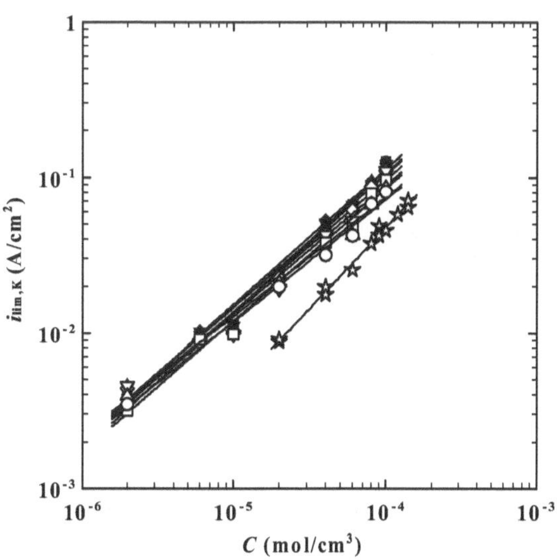

FIGURE. 9.7 C versus $i_{\text{lim},K}$ plot. Aciplex K-172, 25 °C, u: (○) 1.6 cm/s, (△) 2.2 cm/s, (□) 3.2 cm/s, (▽) 4.9 cm/s, (◇) 6.3 cm/s, (●) 7.8 cm/s, (▲) 9.1 cm/s, (■) 10.6 cm/s, (▼) 12.2 cm/s, (◆) 13.6 cm/s, (☆) 0 cm/s. *Tanaka (2005), p. 10, Fig. 3.*

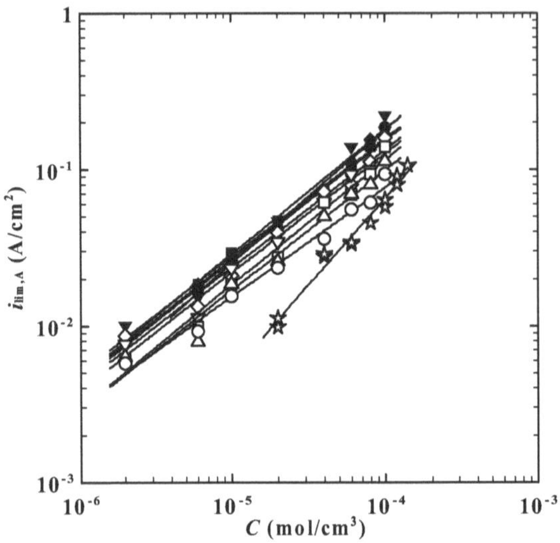

FIGURE. 9.8 C versus $i_{lim,A}$ plot. Aciplex A-172, 25 °C, u: (○) 1.6 cm/s, (△) 2.2 cm/s, (□) 3.2 cm/s, (▽) 4.9 cm/s, (◇) 6.3 cm/s, (●) 7.8 cm/s, (▲) 9.1 cm/s, (■) 10.6 cm/s, (◆) 13.6 cm/s, (☆) 0 cm/s. *Tanaka (2005), p. 10, Fig. 4.*

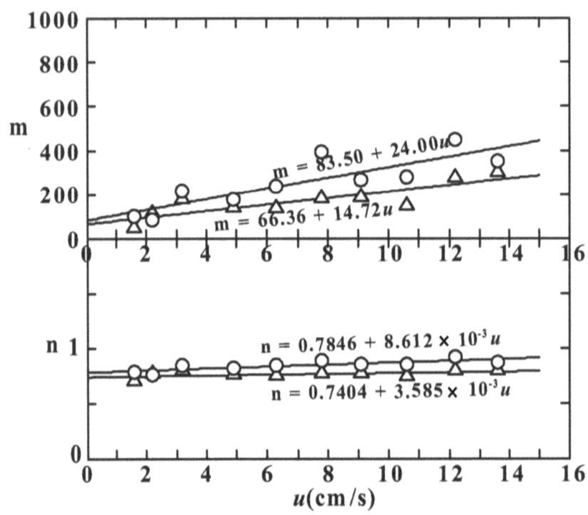

FIGURE. 9.9 Coefficients m and n in the limiting current density equation. ○: Aciplex K-172, △: Aciplex A-172. *Tanaka (2005), p. 10, Fig. 5.*

inflection of V/I versus $1/I$ plot. The value $k = (i_{lim}/C)_T/(i_{lim}/C)_{25}$ is plotted against $T/25$ and shown in Figure 9.10 and the following equation (Tanaka, 2013):

$$k = l_1 + l_2\left(T/25\right) + l_3(T/25)^2$$
$$l_1 = 0.5950, \quad l_2 = 0.2731, \quad l_3 = 0.1310 \tag{9.56}$$

In electrodialysis of inorganic salt solution, limiting current density of a cation exchange membrane $i_{lim,K}$ is less than that of an anion exchange membrane $i_{lim,A}$. In this situation,

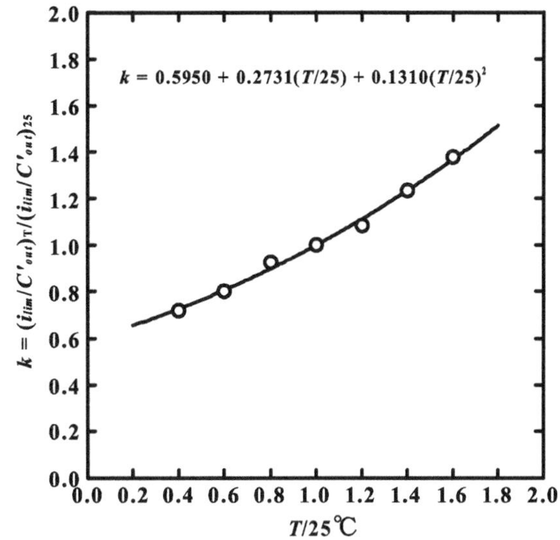

FIGURE. 9.10 Temperature dependence of limiting current density. Neosepta CL-25T. *Tanaka (2013), p. 132, Fig. A1.*

the limiting current density equation at T (°C) is expressed by the limiting current density of the cation exchange membrane, and it is introduced from Eqns (9.53)–(9.56) as follows:

$$i_{\lim}(A/\mathrm{cm}^2) = \left\{ l_1 + l_2\left(\frac{T}{25}\right) + l_3\left(\frac{T}{25}\right)^2 \right\}(m_1 + m_2 u)C^{n_1+n_2 u} \tag{9.57}$$

$$l_1 = 0.5950,\ \ l_2 = 0.2731,\ \ l_3 = 0.1310,\ \ m_1 = 83.50,\ \ m_2 = 24.00,\ \ n_1 = 0.7846,\ \ n_2 = 8.612 \times 10^{-3}$$

9.6 Limiting Current Density of an Electrodialyzer

When current density reaches the limit of an ion exchange membrane $i_{\lim}$ at the outlet of a desalting cell at its lowest linear velocity and electrolyte concentration, the average current density applied to an electrodialyzer is defined as its limiting current density $(I/S)_{\lim}$, in which I is the total electric current and S is the membrane area. The limiting current density $(I/S)_{\lim}$ is strongly related to the circumstances distributing in an apparatus (Tanaka, 2005); namely, $(I/S)_{\lim}$ is influenced by the distribution of solution velocity (cf. Section 8.8). It is also influenced by the distribution of electrolyte concentration in desalting cells and the electric current in a stack. Further, it is naturally influenced by the limiting current density of an ion exchange membrane $i_{\lim}$.

For an electrodialyzer, $i_{\lim}$ is introduced by replacing C and u in Eqn (9.57) with electrolyte concentration C'_{out} and linear velocity u_{out} at the outlet of the desalting cell with the lowest linear velocity. Then, $(I/S)_{\lim}$ is introduced as follows:

$$\left(\frac{I}{S}\right)_{\lim} = \frac{i_{\lim}}{\zeta_{out}} = \left\{ l_1 + l_2\left(\frac{T}{25}\right) + l_3\left(\frac{T}{25}\right)^2 \right\} \frac{(m_1 + m_2 u'^{\#}_{out})}{\zeta_{out}} \left(C'^{\#}_{out}\right)^{n_1+n_2 u'^{\#}_{out}} \tag{9.58}$$

in which the superscript # denotes the least value in desalting cells in a stack, and ζ_{out} is the outlet current density ununiformity coefficient defined by:

$$\zeta_{out} = \frac{i_{out}}{I/S} \tag{9.59}$$

where i_{out} is current density at the outlet of desalting cells integrated in an electrodialyzer.

References

Amatore, C., Szunerits, S., Thouin, L., Warkocz, J.S., 2001. The real meaning of Nernst's steady diffusion layer concept under non-forced hydrodynamic conditions. A simple model based on Levich's seminal view of convection. J. Electroanal. Chem. 500, 62–70.

Cowan, D.A., Brown, J.H., 1959. Effect of turbulence on limiting current in electrodialysis cells. Ind. Eng. Chem. 51, 1445–1448.

Helfferich, F., 1962. Ion Exchange, p. 253. McGraw-Hill, NewYork.

Huang, T.C., Yu, I.Y., 1988. Correlation of ionic transfer rate in electrodialysis under limiting current density conditions. J. Membr. Sci. 35, 193–206.

Kitamoto, A., Takashima, Y., 1967. Mass transfer by electrodialysis using ion-exchange membranes—studies on electrodialytic concentration of electrodialysis. J. Chem. Eng. Jpn. 31, 1201–1207.

Kitamoto, A., Takashima, Y., 1968. Studies on electrodialysis, maximum attainable concentration, limiting current density and on energy efficiency in electrodialysis using ion-exchange membranes. J. Chem. Eng. Jpn. 32, 74–82.

Miyoshi, H., Fukumoto, T., Kataoka, T., 1968. A method for estimating the limiting current density in electrodialysis. Sep. Sci. Technol. 23, 585–600.

Nernst, W.Z., 1904. Theorie der reaktions geschwindigkeit in heterogenen systemen. Z. Phys. Chem. 47, 52–55.

Rosenberg, N.W., Tirrel, C.E., 1957. Limiting currents in membrane cells. Ind. Eng. Chem. 49, 780–784.

Spiegler, K.S., 1971. Polarization at ion exchange membrane-solution interfaces. Desalination 9, 367–385.

Tanaka, Y., 2005. Limiting current density of an ion-exchange membrane and of an electrodialyzer. J. Membr. Sci. 266, 6–17.

Tanaka, Y., 2013. Development of a computer simulation program of batch ion-exchange membrane electrodialysis for saline water desalination. Desalination 320, 118–133.

Vielstich, W., 1953. Der Zusammenhang zwischen Nernstscher diffusionsschicht und Prandlscher Stromungsgrenzschicht. Z. Elektrochem. 57, 646.

Yamabe, T., Yamana, S., Yamagata, K., Takai, S., Seno, M., 1967. On the concentration polarization effect in the electrodialysis using ion-exchange membrane (IV) Concentration polarization effect multi-compartment electrodialysis cell. J. Electrochem. Soc. Jpn. 35, 578–582.

10

Leakage

10.1 Introduction

In an electrodialyzer, a part of an electric current does not pass through ion exchange membranes, and it flows between electrodes through slots and ducts. Electric current leakage is an ineffective and inevitable phenomenon that increases energy consumption in an electrodialysis process. Section 10.2 describes equivalent circuit developed by Wilson (1960). The circuit is applied to calculate the electric current leakage in a practical-scale electrodialyzer operated in the computer simulation program.

Dimensions of all parts of an electrodialyzer are not always consistent with the values in the specifications. Small pinholes can open in an electrodialyzer because the strength of ion exchange membranes is relatively low. Gaps may occur between the materials composing the electrodialyzer (desalting cells, concentrating cells, ion exchange membranes, and spacers) in the assembly works of an electrodialyzer. A pressure difference between the desalting cells and concentrating cells causes solutions leak through the pinholes in the membranes during the operation of the electrodialyzer. The size of the pinholes is extremely larger comparing to that of micro-pinholes in the membranes. Section 10.3 describes the principle of the solution leakage and its observation.

10.2 Electric Current Leakage

10.2.1 Electric Current Leakage Equation

An electrodialyzer is assumed to be assembled by arranging desalting cells and concentrating cells alternately with ion exchange membranes and electrode cells placed on both ends of the electrodialyzer (Figure 10.1). Diluate and concentrate are supplied to each desalting and concentrating cell through ducts and slots incorporated at the bottom (inlet) of the electrodialyzer. The solutions flow upward in the desalting and concentrating cells, pass through the slots and ducts incorporated at the head (outlet) of the electrodialyzer and are discharged to the outside of the electrodialyzer. In this flow system, each desalting and concentrating cell are connected through the solutions passing across slots and ducts. Accordingly, when an electric current is supplied from the electrodes, an unavailable electric current (leakage), which does not pass through the effective area of membranes, is generated as shown by the arrows in Figures 10.1 and 10.2 (Equivalent circuit). R is the electric resistance (Ω) of an effective portion, including

Ion Exchange Membranes. http://dx.doi.org/10.1016/B978-0-444-63319-4.00010-9

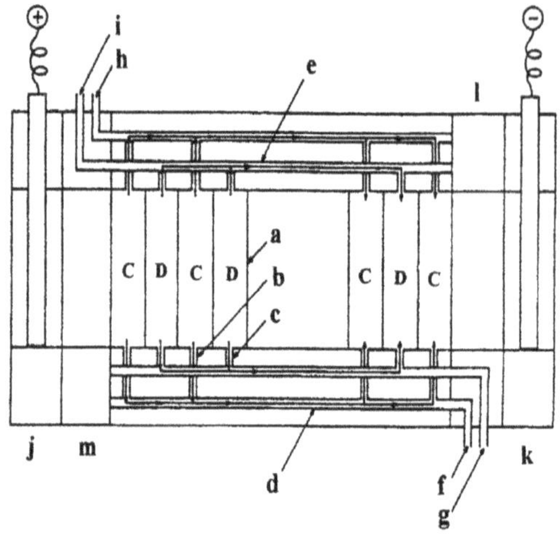

FIGURE 10.1 Leakage current in an electrodialyzer. a: Ion exchange membrane, b: Slot of a concentrating cell, c: Slot of a desalting cell, d: Duct for concentrating cells, e: Duct for desalting cells, f: Inlet for concentrating cells, g: Inlet for desalting cells, h: Outlet for concentrating cells, i: Outlet for desalting cells, j: Anode, k: Cathode, l: Feeding cell, m: Discharging cell, C: Concentrating cell, D: Desalting cell. *Tanaka, 2007, p. 272, Fig. 12.2.*

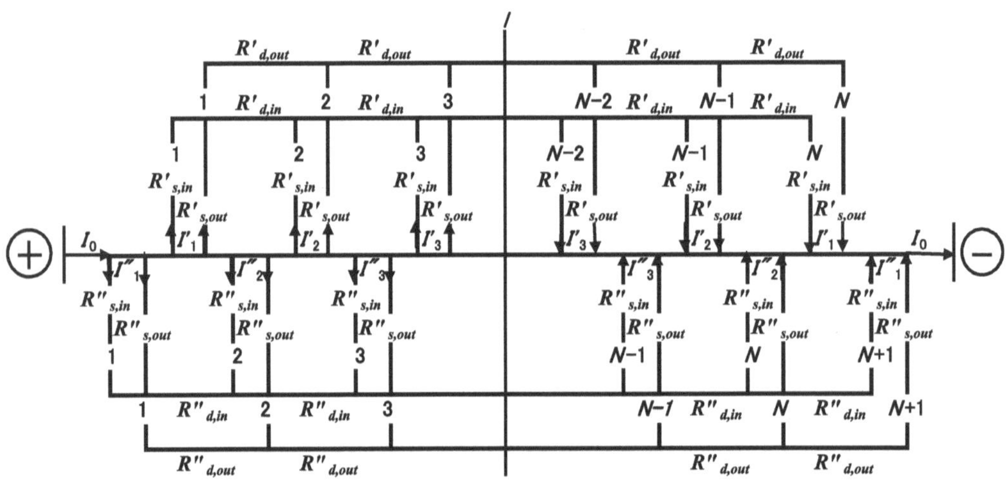

FIGURE 10.2 Equivalent circuit of a leakage current in an electrodialyzer. *Wilson, 1960, p. 269, Fig. 6.24.*

a desalting cell r', a concentrating cell r'', a cation exchange membrane r_K, and an anion exchange membrane r_A as follows:

$$R = r' + r'' + r_K + r_A + \frac{e_M}{I} \tag{10.1}$$

in which e_M is the sum of membrane potentials generated at the cation and anion exchange membrane, I is an electric current.

In the equivalent circuit Figure 10.2, R'_s and R''_s represent the electric resistance of slots. R'_d and R''_d represent those of ducts. Subscripts "in" and "out" refer to inlet and outlet, respectively. Superscripts, single quotation marks, and double quotation marks refer to desalting cell and concentrating cell. I_0 is the electric current supplied from electrodes. The number of desalting cells and concentrating cells is, respectively, N and $N + 1$. Line l is drawn at the center in the equivalent circuit. The leakage currents I'_n and I''_n passing through desalting slot number n and concentrating slot number n from the anode are equivalent to the values at slot number n from the cathode. Thus, one is allowed to calculate the leakage occurring only in the left half section of the circuit.

Here, the overall electric resistance of the slots R_s and of the ducts R_d is defined for simplification as follows:

$$\frac{1}{R_s} = \left(\frac{1}{R'_{s,in}} + \frac{1}{R'_{s,out}}\right) + \left(\frac{1}{R''_{s,in}} + \frac{1}{R''_{s,out}}\right) \tag{10.2}$$

$$\frac{1}{R_d} = \left(\frac{1}{R'_{d,in}} + \frac{1}{R'_{d,out}}\right) + \left(\frac{1}{R''_{d,in}} + \frac{1}{R''_{d,out}}\right) \tag{10.3}$$

Furthermore, one defines the overall slot electric resistance ratio R_S^* and overall duct electric resistance ratio R_d^* taking electric resistance of a cell pair R as the standard as follows:

$$R_S^* = \frac{R_s}{R} \tag{10.4}$$

$$R_d^* = \frac{R_d}{R} \tag{10.5}$$

The electric current in the left circuit in Figure 10.2 is simplified by applying Eqns (10.2)–(10.5) and assuming N to be extremely large as shown in Figure 10.3, in which the number of both desalting and concentrating cells becomes $N/2$. I_n is the electric current passing across the membrane at desalting cell number n and concentrating cell number n. $I_n^\#$ is the sum of the leakage current passing through the slot in desalting

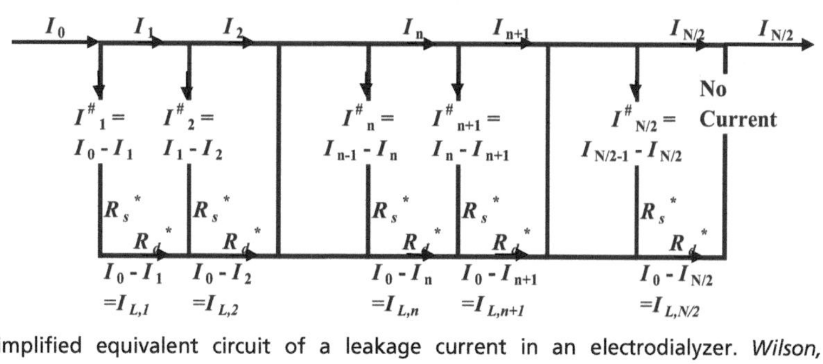

FIGURE 10.3 Simplified equivalent circuit of a leakage current in an electrodialyzer. *Wilson, 1960, p. 269, Fig. 6.25.*

cell number n and that in concentrating cell number n. Applying the Kirchhoff equation to the equivalent circuit in Figure 10.3, the electric current leakage is computed as follows:

At first, the electric current in circuit number n is expressed by applying the Kirchhoff equation as follows, within the range of $1 < n < (N/2) - 1$:

$$I_{n-1}R_S^* = I_n(2R_S^* + 1 + R_d^*) - I_{n+1}R_S^* - I_0R_d^* \tag{10.6}$$

For $n = N/2$, substituting $I_n = I_{N/2} = I_{n+1} = I_{N/2+1}$ due to no current in Eqn (10.6):

$$I_{N/2-1}\,R_S^* = I_{N/2}(R_S^* + 1 + R_d^*) - I_0\,R_d^* \tag{10.7}$$

For $n = N/2–1$, Eqn (10.6) becomes:

$$I_{N/2-2}R_S^* = I_{N/2-1}(2R_S^* + 1 + R_d^*) - I_{N/2} - I_0R_d^* \tag{10.8}$$

Substituting Eqn (10.7) into Eqn (10.8):

$$I_{N/2-2}R_S^* = I_{N/2}\frac{R_S^{*2} + 3(1 + R_d^*)R_S^* + (1 + R_d^*)^2}{R_S^*} + I_0R_d^*\frac{3R_S^* + 1 + R_d^*}{R_S^*} \tag{10.9}$$

Comparison of the electric resistance of a slot with those of a membrane cell and a duct is expressed as:

$$R_S^* >> 1 \tag{10.10}$$

$$R_S^* >> R_d^*$$

Taking account of Eqn (10.10) in Eqn (10.9):

$$R_S^{*2} + 3(1 + R_d^*)R_S^* >> (1 + R_d^*)^2$$
$$3R_S^* >> 1 + R_d^*$$

So, Eqn (10.9) becomes:

$$I_{N/2-2}R_S^* = I_{N/2}\{R_S^* + 3(1 + R_d^*)\} - 3I_0R_d^* \tag{10.11}$$

Next, for $N/2–2$, Eqn (10.6) becomes:

$$I_{N/2-3}R_S^* = I_{N/2-2}(2R_S^* + 1 + R_d^*) - I_{N/2-1}R_S^* - I_0R_d^* \tag{10.12}$$

Substituting Eqns (10.7) and (10.11) into Eqn (10.12):

$$I_{N/2-3}R_S^* = I_{N/2}\frac{R_S^{*2} + 6(1 + R_d^*)R_S^* + 3(1 + R_d^*)^2}{R_S^*} - 3I_0R_d^*\frac{2R_S^* + 1 + R_d^*}{R_S^*} \tag{10.13}$$

Taking account of Eqn (10.10) in Eqn (10.13):

$$R_S^{*2} + 6(1 + R_d^*)R_S^* >> 3(1 + R_d^*)^2$$
$$2R_S^* >> 1 + R_d^*$$

So Eqn (10.13) becomes:

$$I_{N/2-3}R_S^* = I_{N/2}\{R_S^* + 6(1 + R_d^*)\} - 6I_0R_d^* \tag{10.14}$$

Taking into account that the coefficients of $1 + R_d^*$ and $I_0\, R_d^*$ in Eqns (10.11) and (10.14) are $3 = 1 + 2$ and $6 = 1 + 2 + 3$, respectively, one arrives at the following general equation:

$$I_{n-1}R_S^* = I_{N/2}\left\{R_S^* + (1+R_d^*)\sum_{n=n}^{N/2}\left(\frac{N}{2}-n+1\right)\right\} - I_0 R_d^* \sum_{n=n}^{N/2}\left(\frac{N}{2}-n+1\right) \tag{10.15}$$

Putting $n = 1$ in Eqn (10.15):

$$I_0\left(R_S^* + R_d^*\sum_{n=1}^{N/2} n\right) = I_{N/2}\left\{R_S^* + (1+R_d^*)\sum_{n=1}^{N/2} n\right\} \tag{10.16}$$

Now in Figure 10.3, the leakage current passing through slot number n in the circuit is expressed as follows:

$$I_n^{\#} = I_{n-1} - I_n \tag{10.17}$$

Substituting Eqn (10.15) into Eqn (10.17):

$$I_n^{\#} = \left\{I_{N/2}(1+R_d^*) - I_0 R_d^*\right\}\frac{(N/2)-n+1}{R_S^*} \tag{10.18}$$

Substituting for $I_{N/2}$ in Eqn (10.18) from Eqn (10.16) gives:

$$I_n^{\#} = \frac{(N/2-n+1)I_0}{R_S^* + (1+R_d^*)\sum_{n=1}^{N/2} n} = \frac{(N/2-n+1)I_0}{R_S^* + (1/8)(1+R_d^*)N(N+2)} \tag{10.19}$$

Next, in the circuit in Figure 10.3, the leakage current $I_n^{\#}$ in slot number n passes through from duct numbers n to $N/2$. Thus, the contribution of $I_n^{\#}$ to the total leakage current in the electrodialyzer $I_{L,n}$ is presented by:

$$I_{L,n} = I_n^{\#}\frac{N/2-n+1}{N/2} \tag{10.20}$$

The total leakage current I_L in the electrodialyzer is the electric current passing the duct number $N/2$ in Figure 10.3 and it is introduced by summing up from $I_{L,1}$ in slot number 1 to $i_{L,N/2}$ in slot number $N/2$ as follows:

$$I_L = \sum_{n=1}^{N/2} I_{L,n} = \sum_{n=1}^{N/2} I_n^{\#}\frac{N/2-n+1}{N/2} \tag{10.21}$$

Wilson (1960) introduced the leakage current ratio I_L/I_0 by substituting Eqns (10.19) and (10.20) into Eqn (10.21) as follows:

$$\frac{I_L}{I_0} = \frac{\sum_{n=1}^{N/2} n^2}{(N/2)\left\{R_S^* + (1+R_d^*)\sum_{n=1}^{N/2} n\right\}} = \frac{2(N+1)(N+2)}{24R_S^* + 3(1+R_d^*)N(N+2)} \tag{10.22}$$

10.2.2 Computation of Electric Current Leakage

A salt solution is supplied to an electrodialyzer and electrodialyzed with a single-pass continuous mode passing a constant electric current. Structure of the desalting cell and concentrating cell is illustrated in Figure 10.4.

Specifications of the electrodialyzer are listed below:

Flow-pass thickness in a desalting and a concentrating cell; $a = 0.05$ cm
Flow-pass width in a desalting and a concentrating cell; $b = 100$ cm
Flow-pass length in a desalting and a concentrating cell; $l = 100$ cm
Spacer rod volume ratio (Diagonal net spacer); $\varepsilon = 0.07557$
Number of desalting cells; $N = 300$
Number of concentrating cell; $N + 1 = 301$
Slot width and duct diameter in a desalting cell; $w' = 4.0$ cm
Slot length in a desalting cell; $h' = 4.0$ cm
Slot and duct number in a desalting cell; $n' = 5 \times 2$ (inlet and outlet) $= 10$

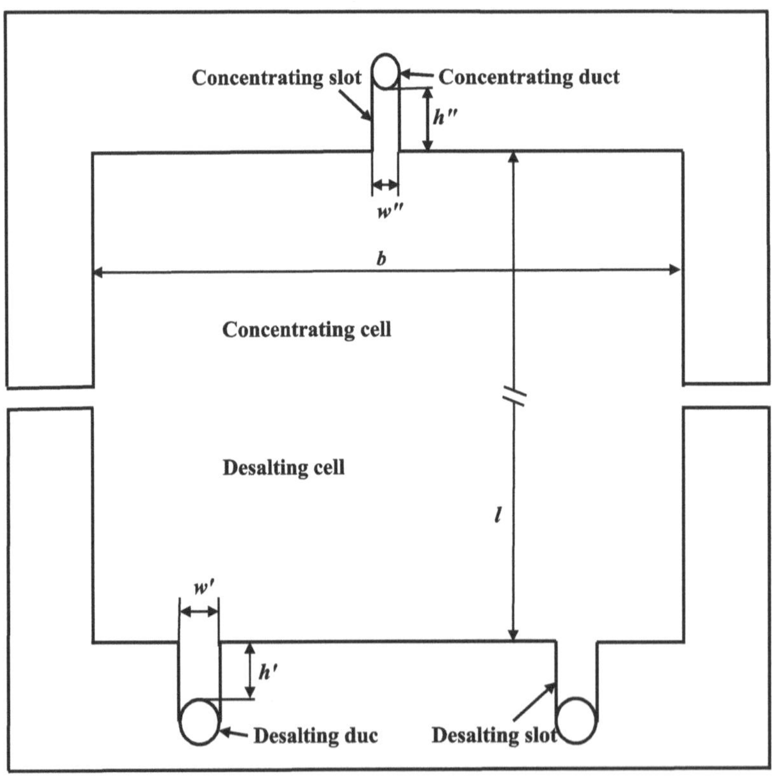

FIGURE 10.4 Structure of a desalting cell and a concentrating cell. *Tanaka, 2012, p. 11, Fig. 1.*

Slot width and duct diameter in a concentrating cell; $w'' = 2.0$ cm
Slot length in a concentrating cell; $h'' = 4.0$ cm
Slot and duct number in a concentrating cell; $n'' = 2 \times 2$ (inlet and outlet) $= 4$
Duct length in a cell pair; $d = 0.05$ (De) $+ 0.05$ (Con) $+ 0.013$ (K) $+ 0.013$ (K) $= 0.126$ cm

Operating conditions of the electrodialyzer are:

Salt concentration at the inlets of desalting cells; $C'_{in} = 2000$ mg/dm^3
Salt concentration at the inlets of concentrating cells; $C''_{in} = 2000$ mg/dm^3
Linear velocity at the inlets of desalting cells; $u'_{in} = 10$ cm/s
Linear velocity at the inlets of concentrating cells; $u''_{in} = 1$ cm/s
Temperature; $T = 25$ °C
Current density; $I/S = 0.5$ A/dm^2
Electric current; $I_0 = 0.5$ A/dm$^2 \times 100$ dm$^2 = 50$ A

The following electric resistance of each part in the electrodialyzer is computed by inputting the above data into the electrodialysis program (Applications, Chapter 13 and Chapter 24, Companion site 1).

Specific resistance of diluate
- in the inlet slots; $\kappa'_{s,in} = 300.00$ Ωcm
- in the inlet ducts; $\kappa'_{d,in} = 277.33$ Ωcm
- in the outlet slots; $\kappa'_{s,out} = 414.08$ Ωcm
- in the outlet ducts; $\kappa'_{d,out} = 382.79$ Ωcm

Specific resistance of concentrate
- in the inlet slots; $\kappa''_{s,in} = 300.00$ Ωcm
- in the inlet ducts; $\kappa''_{d,in} = 277.33$ Ωcm
- in the outlet slots; $\kappa''_{s,out} = 82.84$ Ωcm
- in the outlet ducts; $\kappa''_{d,out} = 76.58$ Ωcm

Ohmic resistance of diluate in desalting cells; $r' = 1.792 \times 10^{-3}$ Ω
Ohmic resistance of concentrate in concentrating cells; $r'' = 6.037 \times 10^{-4}$ Ω
Ohmic resistance of ion exchange membrane pair; $r_K + r_A = 2.063 \times 10^{-3}$ Ω
Ohmic resistance equivalent to membrane potential; $e_M/I = 7.302 \times 10^{-4}$ Ω
Electric resistance of an effective portion $R = r' + r'' + r_K + r_A + e_M/I = 5.189 \times 10^{-3}$ Ω

The overall electric resistance and electric current leakage are calculated from the above data as follows:

$$R'_{s,in} = \frac{\kappa'_{s,in} h'}{a w' n'} = 600.00\Omega$$

$$R'_{s,out} = \frac{\kappa'_{s,out} h'}{a w' n'} = 828.16\Omega$$

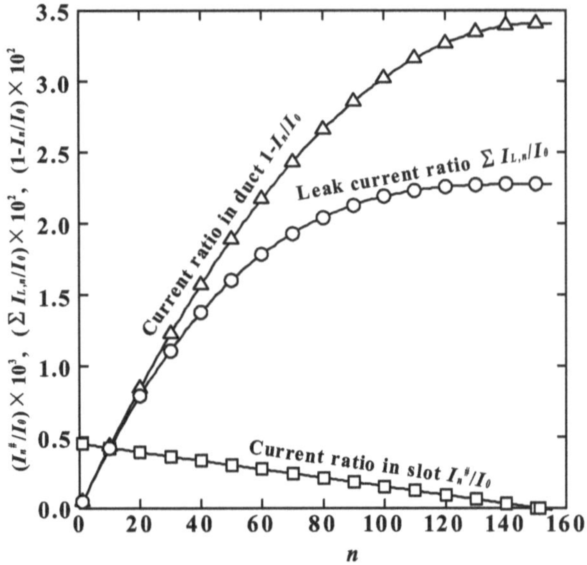

FIGURE 10.5 Electric current leakage in an electrodialyzer.

$$R''_{s,in} = \frac{\kappa''_{s,in} h''}{a w'' n''} = 3000.00\Omega$$

$$R''_{s,out} = \frac{\kappa''_{s,out} h''}{a w'' n''} = 828.40\Omega$$

$$R_s = \frac{1}{\left(\frac{1}{R'_{s,in}} + \frac{1}{R'_{s,out}}\right) + \left(\frac{1}{R''_{s,in}} + \frac{1}{R''_{s,out}}\right)} = 226.52\Omega$$

$$R_S^* = R_S / R = 43653$$

$$R'_{d,in} = \frac{\kappa'_{d,in} d}{\frac{1}{4}\pi w'^2 n'} = 0.27807\Omega$$

$$R'_{d,out} = \frac{\kappa'_{d,out} d}{\frac{1}{4}\pi w'^2 n'} = 0.38381\Omega$$

$$R''_{d,in} = \frac{\kappa''_{d,in} d}{\frac{1}{4}\pi w''^2 n''} = 2.78071\Omega$$

$$R''_{d,out} = \frac{\kappa''_{d,out} d}{\frac{1}{4}\pi w''^2 n''} = 0.76785\Omega$$

$$R_d = \frac{1}{\left(\frac{1}{R'_{d,in}} + \frac{1}{R'_{d,out}}\right) + \left(\frac{1}{R''_{d,in}} + \frac{1}{R''_{d,out}}\right)} = 0.12717\Omega$$

$$R_d^* = R_d / R = 24.507$$

Substituting I_0, N, R_s^*, and R_d^* into Eqn (10.22):

$$I_{L,N/2} = 1.139 \text{ A}$$
$$I_{L,N/2}/I_0 = 2.278 \times 10^{-2}$$

$I_n^{\#}$ and $I_{L,n}$ are calculated inputting cell pair number n into Eqs, (10.19) and (10.20). Electric current ratio in slots $I_n^{\#}$, electric current ratio in ducts; $1 - I_n/I_0$ and leak electric current ratio $\Sigma I_{L,n}/I_0$ are plotted against n and shown in Figure 10.5. At $n = 150 = N/2$, $\Sigma I_{L,n}/I_0$ is confirmed to become $I_{L,N/2}/I_0 = 2.278 \times 10^{-2}$.

10.3 Solution Leakage (Tanaka, 2004)

10.3.1 Overall Mass Transport Equation and Solution Leakage

In an electrodialyzer, ions and solutions are transported across the membranes. When solutions leak, the flux of ions J_S and a solution J_V across a membrane pair are expressed by the following equation, which is introduced from the overall mass transport equation (Chapter 5) by adding the terms of leakage:

$$J_S = \lambda(I/S) - \mu(C'' - C') + C'q'_L - C''q''_L \tag{10.23}$$
$$J_V = \phi(I/S) + \rho(C'' - C') + q'_L - q''_L \tag{10.24}$$
$$C'' = J_S/J_V \tag{10.25}$$

Here, I is electric current, S is membrane area, I/S is average current density, and C' and C'' are electrolyte concentration in a desalting cell and a concentrating cell, respectively. λ is the overall transport number, μ is the overall solute permeability, ϕ is the overall electroosmotic permeability, and ρ is the overall volume osmotic permeability. q'_L is the solution leakage moving from a desalting cell to a concentrating cell; q''_L is that moving from a concentrating cell to a desalting cell. Terms $\lambda(I/S)$, $\mu(C'' - C')$ and $C'q'_L - C''q''_L$ in Eqn (10.23) stand for the migration, diffusion and leakage of ions, respectively. Terms $\phi(I/S)$, $\rho(C'' - C')$ and $q'_L - q''_L$ in Eqn (10.24) correspond to the electro-osmosis, volume-osmosis, and leakage of solutions, respectively. Using Eqn (10.23)–(10.25), q'_L and q''_L are introduced as follows:

$$q'_L = \left(\frac{\lambda - \phi C''}{C'' - C'}\right)\left(\frac{I}{S}\right) - (\mu + \rho C'') \tag{10.26}$$

$$q''_L = \left(\frac{\lambda - \phi C'}{C'' - C'}\right)\left(\frac{I}{S}\right) - (\mu + \rho C' + J_V) \tag{10.27}$$

q'_L and q''_L are known by measuring the parameters included in Eqns (10.26) and (10.27).

10.3.2 Measurement of Solution Leakage

In order to measure q'_L and q''_L in a practical scale electrodialyzer using Eqns (10.26) and (10.27), the membrane pair characteristics λ, μ, ϕ, and ρ must be measured before hand in an experimental scale electrodialyzer in which the solution leakage is

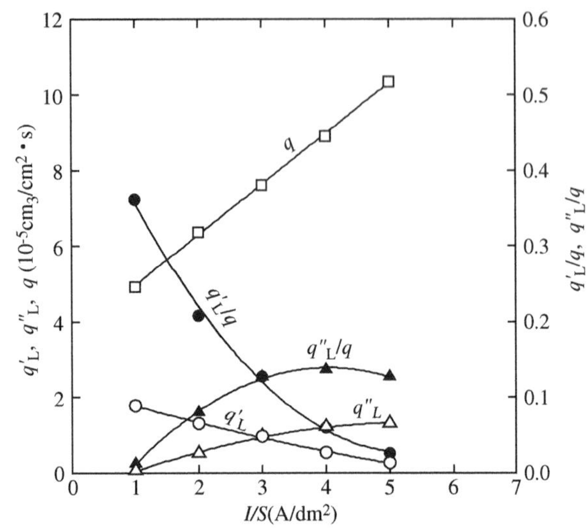

FIGURE 10.6 Relationship between current density and solution leakage in an electrodialyzer. $S = 113$ dm^2, $N = 301$ pairs, $u = 4.7$ cm/s, 25 °C. *Tanaka, 2004, p.20, Fig. 7.*

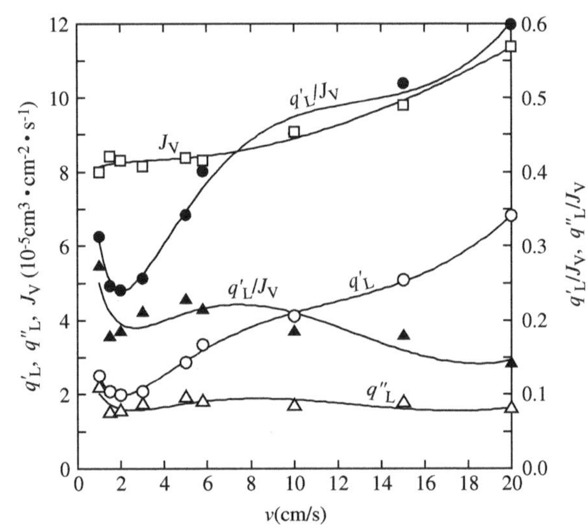

FIGURE 10.7 Relationship between linear velocity in desalting cells and solution leakage in an electrodialyzer. $S = 14.5$ dm^2, $I/S = 3$ A/dm^2, 27 °C Aciplex K-172/A-172. *Tanaka, 2004, p. 21, Fig. 8.*

confirmed to be zero, under the same operating conditions of the practical scale electrodialyzer. On the premise of the experimentation described above, the following results are obtained:

Figure 10.6 shows the effect of current density I/S on q'_L and q''_L in a practical scale electrodialyzer. The specifications of the electrodialyzer are: ion exchange membrane; Aciplex CK-2/CA-3 (Asahi Chemical Co.), thickness of a desalting and a concentrating cell; 0.075 cm, effective membrane area; 113 dm^2, number of membrane pairs; 300. q'_L and q'_L/J_V are recognized to be decreased with the increase of i, but

q''_L and q''_L/J_V are increased with the increase of i. These phenomena are estimated to occur because the increase of current density gives rise to an increase in solution density and static head in concentrating cells, and a resultant decrease in q'_L and increase in q''_L.

Figure 10.7 shows the influence of linear velocity u in desalting cells on q'_L and q''_L in a middle scale electrodialyzer. The specifications of the electrodialyzer are, ion exchange membrane; Aciplex K-172/A-172, thickness of a desalting cell and a concentrating cell; 0.075 cm, effective membrane area; 14.5 dm^2, number of membrane pairs; 36. q'_L and q'_L/J_V are recognized to be increased with the increase of u. However, q''_L and q''_L/J_V are decreased to some extent with the increase of u. These changes are understandable because the increase of linear velocity in desalting cells brings about a static head increase in desalting cells.

Presently, 1.0 million tons of edible salt per year is produced in Japan using ion exchange membrane electrodialyzers. The effective membrane area is 1–2 m^2, and the numbers of membrane pairs incorporated are 200–300 in a stack and 2000–3000 between electrodes. The temperature of feed seawater changes with time in a year. Current density is kept at a relatively higher level in a summer season and a lower level in a winter season. The flux of a solution across a membrane pair and electrolyte concentration in a concentrated solution are changed under the influence of current density i and seawater temperature T. q'_L and q''_L were calculated by substituting the membrane pair characteristics λ, μ, ϕ, and ρ measured in experimental electrodialysis and the performance (i, T, C', C'', and J_V) of electrodialyzers for concentrating seawater operating in salt manufacturing plants (Japan Tobacco & Salt Public Corporation, 1993) into Eqns (10.26) and (10.27). q'_L, q''_L, i, T, C'' and J_V are plotted against operating time t. The plots for Plant E and Plant F are shown in Figures 10.8 and 10.9 for

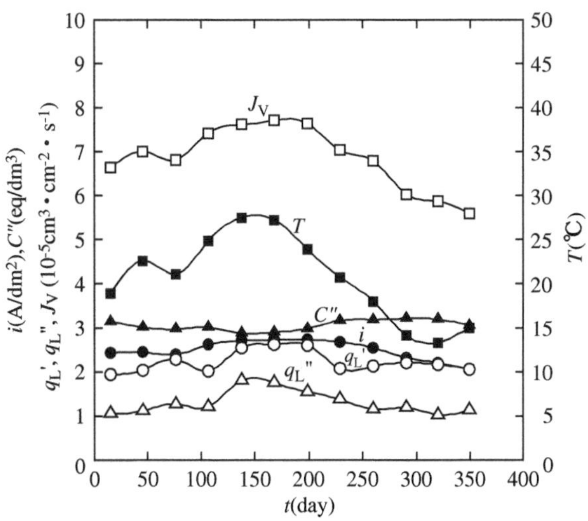

FIGURE 10.8 Solution leakage in electrodialyzers operating in a salt manufacturing plant (Plant E). *Tanaka, 2004, p. 21, Fig. 9.*

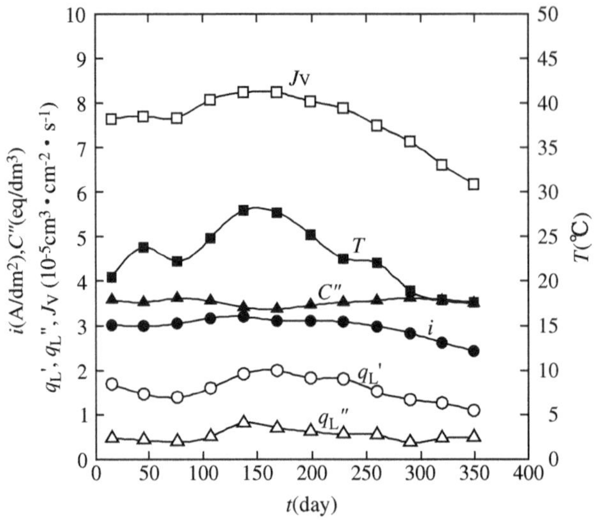

FIGURE 10.9 Solution leakage in electrodialyzers operating in a salt manufacturing plant (Plant F). *Tanaka, 2004, p. 21, Fig. 10.*

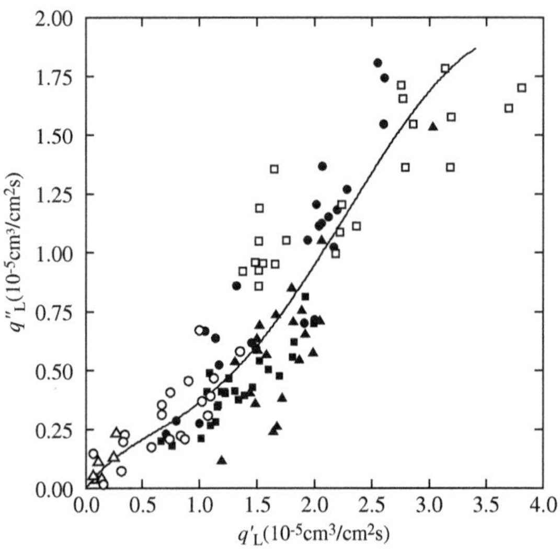

FIGURE 10.10 Solution leakage in electrodialyzers operating in salt manufacturing plants. ○: Plant A, △: Plant B, □: Plant C, ●: Plant D ▲: Plant E, ■: Plant F. I/S = 2.18–3.34 A/dm^2, T = 13.6–33.8 °C, C' = 0.428–0.554 eq/dm^3, C'' = 2.873–3.799 eq/dm^3. *Tanaka, 2004, p. 22, Fig. 13.*

examples. q_L''s in electrodialyzers in every salt-manufacturing plant are plotted against q_L's and are indicated in Figure 10.10. q_L'' is recognized to increase along with the increase of q_L', and both values in each plant are confirmed to change within the large scope.

References

Japan Tobacco & Salt Public Corporation, 1993. Salt Production Technical Report.

Tanaka, Y., 2004. Overall mass transport and solution leakage in an ion-exchange membrane electrodialyzer. J. Membr. Sci. 235, 15–24.

Tanaka, Y., 2007. Ion Exchange Membranes: Fundamentals and Applications, Membrane Science and Technology Series 12. Elsevier, Amsterdam, p. 272.

Tanaka, Y., 2012. Ion-exchange membrane electrodialysis program and its application to multi-stage continuous saline water desalination. J. Membr. Sci. 301, 11.

Wilson, J.R., 1960. Demineralization by Electrodialysis. Buther-worths Scientific Publication, London, pp. 265–274.

11 Membrane Deterioration

11.1 Introduction

The greater part of practically available ion exchange membranes consists of organic materials with internal ion exchange groups. Their physical and chemical properties are not tough, so the membranes may deteriorate under severe circumstances. Small particles suspended in a feed solution invade an electrodialyzer and are deposited onto the membrane surface (surface fouling). Suspended organic colloidal substances precipitate onto anion exchange membranes (organic fouling). Membrane deterioration, surface fouling, and organic fouling are inevitable phenomena that shorten the membrane lifetime and lower plant economic viability. To operate an industrial electrodialyzer stably, it is necessary to understand the mechanism of these phenomena.

11.2 Membrane Characteristic Stability Against Various Agents

Kneifel and Hattenbach (1980) measured property changes with elapsed time under several chemical environments for the following commercially available membranes.

1. Tokuyama Soda Co. Ltd. (Japan)

   Anion exchange membrane AF-4T/AFS-4T/AV-4T/AC-158, Cation exchange membrane C66-5T/CH-45T/CH-2T/CL-25T.

2. Asahi Glass Co. Ltd. (Japan)

   Anion exchange membrane AMV/ASV, Cation exchange membrane CMV.

3. Asahi Chemical Ind. Co. Ltd. (Japan)

   Anion exchange membrane A-111, Cation exchange membrane K-101.

4. Ionac Chemical Company (USA)

   Anion exchange membrane MA 3148/MA 3475 R/IM 12, Cation exchange membrane MC 3470.

Ion Exchange Membranes. http://dx.doi.org/10.1016/B978-0-444-63319-4.00011-0

5. Ionics Inc. (USA)

Anion exchange membrane 111BZL183/103PZL183/103QZL183, Cation exchange membrane 61AZL183/61CZL183.

6. Forschungsinstitut Berghof GmbH (Israeli)

Anion exchange membrane PEN 4/NPES A3, Cation exchange membrane NPES C2.

7. American Machine and Foundry (USA)

Cation exchange membrane AMF-C100/AMF-C311.

8. Rhone-Poulenc Chemie GmbH (Germany)

Anion exchange membrane ARP, Cation exchange membrane CRP.

The membrane properties in the literature and those in reference sheets of the membrane producers are not comparable because measurements are often conducted with different methods and under different conditions. To obtain comparable values, the values of the exchange capacity, water content, membrane resistance, thickness, and permselectivity were determined and made to comply under uniform conditions in Table 11.1 for cation exchange membranes and in Table 11.2 for anion exchange membranes. Here, the permselectivity P is given by $P = (\bar{t} - t)/(1 - t)$, in which $\bar{t}$ and t are the transport number of counter-ions in the membrane and in the solution, respectively, and $\bar{t}$ was measured from membrane potentials. The results indicated in the tables are average values from tests of membranes from various membrane deliveries, and hence possibly of different production charges.

Table 11.1 Properties of Cation Exchange Membranes

Membrane	Exchange Capacity (eq/kg)	Water Content (%)	Electric Resistance 1 N NaCl 25 °C (Ω cm²)	Thickness (mm)	Permselectivity 1.0/0.5 N KCl	Permselectivity 0.1/0.5 N KCl	Transport Number 1.0/0.5 N KCl	Transport Number 0.1/0.5 N KCl
MC 3470	1.5	35	6–10	0.6	0.68	0.94	0.83	0.97
MC 3142	1.3	25	20–30	0.22	0.65	0.93	0.82	0.97
CMV	2.4	25	2.9	0.15	0.95	0.98	0.98	0.99
K-101	1.4	24	2.1	0.24	0.91	0.98	0.95	0.99
CL-25T	2.0	31	2.9	0.18	0.81	0.97	0.90	0.98
C66-5T	3.3	34	1.5	0.16	0.89	0.97	0.94	0.98
CH-45T	2.3	27	2.1	0.17	0.92		0.96	
CH-2T	2.4	32	2.2	0.17	0.86	0.97	0.93	0.98
61AZL 183	2.0	40	4.5	0.6	0.62	0.93	0.81	0.97
61 CZL 183	2.6	34	7.2	0.7	0.81	0.96	0.95	0.98
AMF-C311	0.6	16		0.3	0.84	0.98	0.92	0.99
AMF-C100	1.1	22	4.8	0.22	0.77	0.98	0.88	0.99
NPES C/2	1.9	15–20	3–7	0.4	0.91		0.96	
CRP	2.6	40	6.3	0.6	0.65		0.82	

Kneifel and Hattenbach (1980), p. 78, Table 1.

Table 11.2 Properties of Anion Exchange Membranes

Membrane	Exchange Capacity (eq/kg)	Water Resistance (%)	Electric Resistance 1 N NaCl 25 °C (Ω cm²)	Thickness (mm)	Permselectivity		Transport Number	
					1.0/0.5 N KCl	0.1/0.05 N KCl	1.0/0.5 N KCl	0.1/0.05 N KCl
MA 3148	0.8	18	12–70	0.20	0.85	0.98	0.92	0.99
MA 3475 R	1.4	31	5–13	0.6	0.70	0.94	0.85	0.97
IM 12	0.5	22	8	0.18	0.4	0.70	0.71	0.85
AMV	1.9	19	2–45	0.14	0.92	0.99	0.96	0.99
ASV	2.1	24	2.1	0.15	0.91	0.99	0.95	0.99
A-111	1.2	3.1	2–3	0.21	0.45	0.91	0.73	0.95
AF-4T	2.0	27	1.7	0.16	0.90	0.98	0.96	0.99
AFS-4T	1.8	25	3.0	0.19	0.95	1.00	0.97	1.00
AV-4T	1.4	24	2.4	0.15	0.90	0.98	0.95	0.99
AVS-4T	1.5	20	5.1	0.17	0.93	0.99	0.97	1.00
AC-158			2.5		0.82	0.98	0.91	0.99
111 BZL 183				0.5	0.52	0.97	0.77	0.97
103 PZL 183	1.2	38	4.9	0.6	0.43	0.90	0.72	0.95
103 QZL 219	1.5	30	8.0	0.7	0.70	0.85	0.95	0.98
PEN 4	0.1–0.3	31	3–12	0.3	0.5–0.7		0.77	
NPES A/3	0.2–0.7	27	3–12	0.4	0.4–0.8		0.6–0.9	
ARP	1.8	34	6.9	0.5	0.79		0.90	

Kneifel and Hattenbach (1980), p. 78, Table 1.

In this experiment, the membrane samples were subjected to the various agents at room and elevated temperature for up to 5 years in the laboratory, and membrane characteristics such as electric resistance, permselectivity, and ion exchange capacity were measured during the test. Figure 11.1 shows the electric resistance for six membrane types after being subjected to different agents at room temperature. The strongest variations for most membranes are caused by the effect of an NaOH solution, especially applied to the heterogeneous membranes MA 3148 and MC 3142 (strong decrease in resistance values) and the anion exchange membranes AF-4T (strong increase in resistance values) and AMV (crumbled). Good stability is exhibited by the membrane C 66-5T. The effect of the oxidizing solution on the electric resistance at room temperature is shown in Fig. 11.2 and indicates a particularly strong decrease in the resistance value of the membrane type CMV. All membranes with the exception of MC 3470 and MA 3475 R were destroyed at 85 °C.

The dependence upon duration test in 1 N NaCl at 85 °C of resistance is shown in Fig. 11.3, of permselectivity in Fig. 11.4, and of exchange capacity in Fig. 11.5. The strong increase in resistance of the membrane types AMV, AV-4T, CMV, AF-4T, and CH-45T is striking. The membrane types 103 QZL 219 and 61 CZL 183 showed no significant changes. Variations in the permselectivities in a trial of up to 100 days were slight with the exception of membrane type MA 3148. However, a longer duration showed a marked decrease in values in the case of membrane types AFS-4T, C 66-5T,

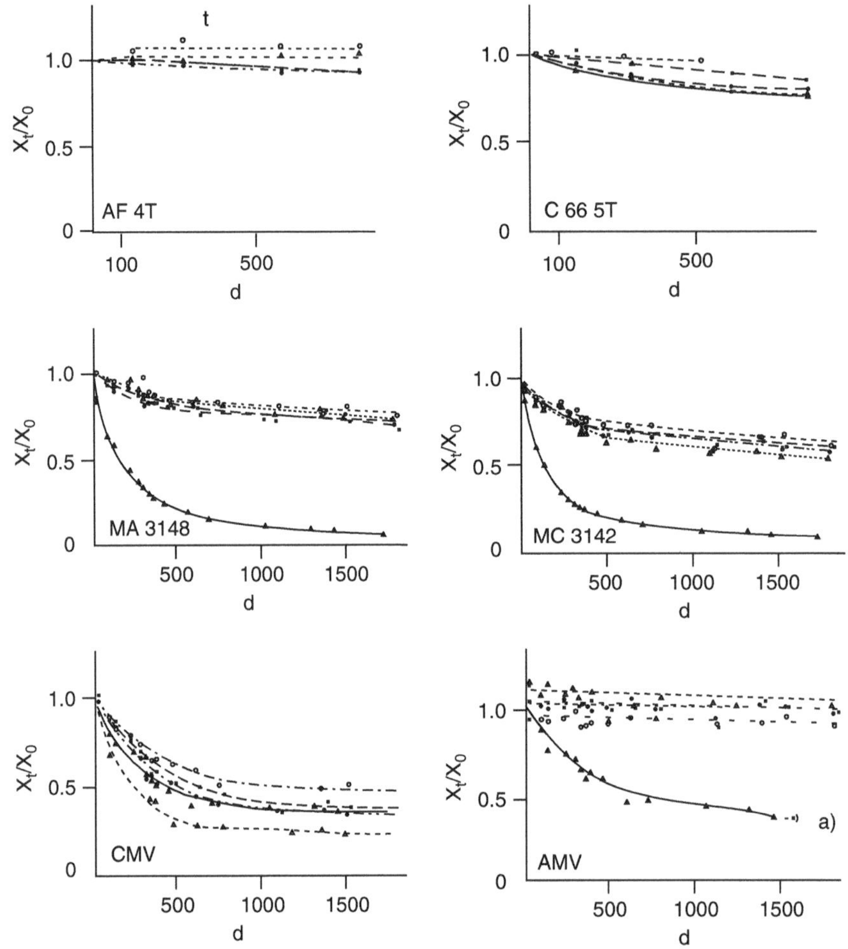

FIGURE 11.1 Membrane resistance after exposure to various agents at room temperature. ● Distilled water, ■ 0.2 N NaCl, ○ 0.1 N NaCl, △ 0.1 N HNO_3, ▲ 0.1 N NaOH. a) Indicates membrane destruction. *Kneifel and Hattenbach (1980), p. 85, Fig. 2.*

AF-4T, MA 3148, CRP, and ARP. The samples of types AF-4T and MA 3148 crumbled and type CMV shrank strongly. The values of exchange capacity, with one exception (IM 12), changed slightly for trial times up to 100 days. In longer trials, the exchange capacities—with the exception of K-101—were reduced to less than 60% of the initial values.

Most experimental data at room temperature and at 85 °C were obtained for membrane types AMV, CMV, AF-4T, C66-5T, and K-101. Results for these membranes in long-term exposure tests in salt solution at room temperature indicated good stability. At 85 °C the membranes AMV, CMV, and AF-4T showed low stability, type C66-5T fair stability, and K-101 good stability.

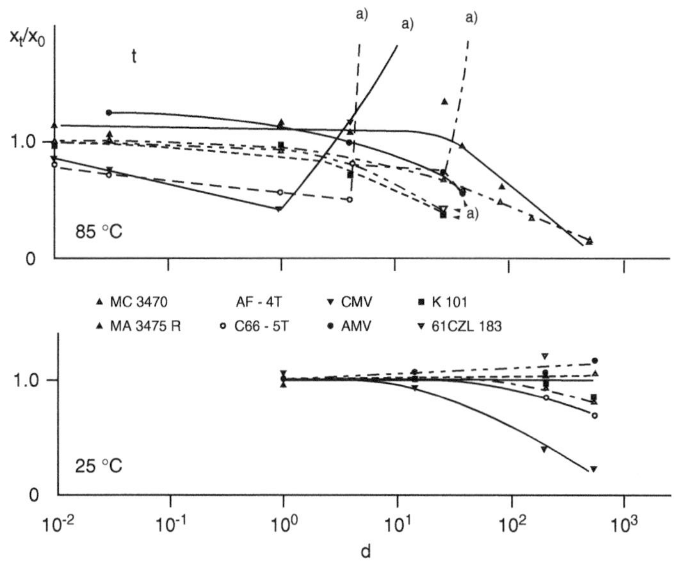

FIGURE 11.2 Membrane resistance after exposure to an oxidizing agent. 0.1 N K_2CrO_4 + 1 N HCl + 1 N NaCl. a) Indicates membrane destruction. *Kneifel and Hattenbach (1980), p. 86, Fig. 3.*

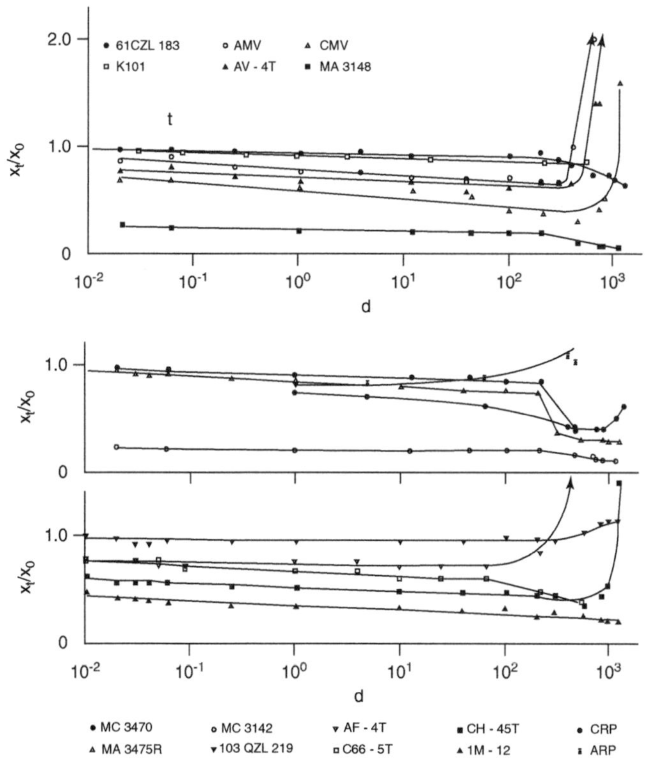

FIGURE 11.3 Membrane resistance after exposure to 1 N NaCl at 85 °C. *Kneifel and Hattenbach (1980), p. 87, Fig. 4.*

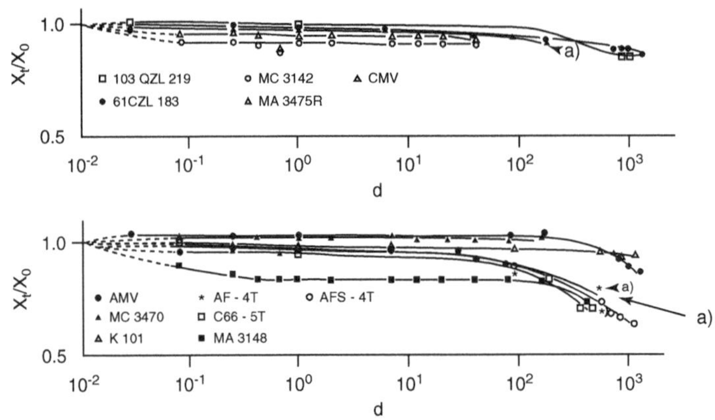

FIGURE 11.4 Permselectivity after exposure to 1 N NaCl at 85 °C. a) Indicates membrane destruction. *Kneifel and Hattenbach (1980), p. 88, Fig. 5.*

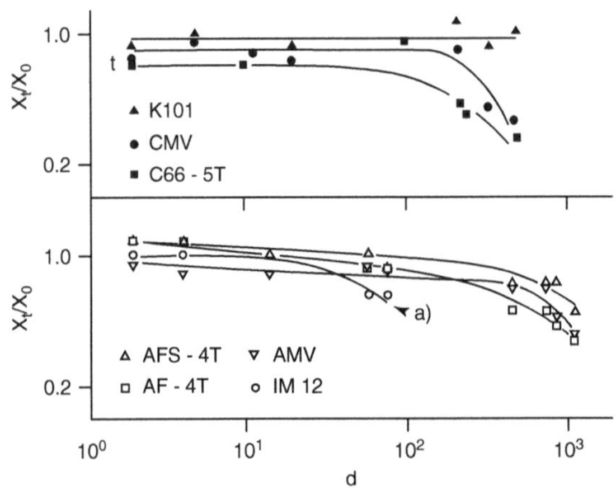

FIGURE 11.5 Exchange capacity after exposure to 1 N NaCl at 85 °C. *Kneifel and Hattenbach (1980), p. 88, Fig. 6.*

11.3 Performance Changes of Ion Exchange Membranes in Long-term Seawater Electrodialysis

Long-term seawater electrodialysis was carried out using a small-scale electrodialyzer (effective membrane area, 2 dm^2) integrated with commercially available homogeneous ion exchange membranes and heterogeneous membranes produced in the laboratory (Hanzawa et al., 1966). The types of membranes (cation exchange membrane/anion exchange membrane) applied in this experiment were:

1. Tokuyama Soda Co.

Neosepta CL-25T (St·DVB·PVC)/AVS-4T (St·DVB·VP·PVC), including PVC fiber reinforcement.

2. Asahi Chemical Ind. Co.

   Aciplex CK-2 (St·DVB)/CA-3 (St·DVB), including no reinforcement.

3. Asahi glass Co.

   Selemion CMV (St·BD)/AST (St·BD), including glass fiber reinforcement.

4. Japan Monopoly Corp.

   Amberlite XE-69·PVC/Amberlite XE-119·PVC; Heterogeneous membranes produced from strong acidic cation exchange resin Amberlite XE-69 and strong basic anion exchange resin XE-119, with PVC powder to form sheets including no reinforcement.

The membranes described above are produced from the following materials: St, styrene; DVB, divinyl benzene; PVC, polyvinyl chloride; VP, vinylpyridine; and BD, butadiene.

Sand-filtered seawater was supplied to the electrodialyzer and was electrodialyzed at an current density of 1 A/dm^2 for up to 4 years. The electrolyte concentration in a concentrating cell C'' (Baume's solution density Be'), volume flux across a membrane pair q, current efficiency η, and feeding seawater temperature T are plotted against running time t.

The plots are shown in Fig. 11.6 (Neosepta CL-25T/AVS-4T), Fig. 11.7 (Aciplex CK-2/CA-3), Fig. 11.8 (Selemion CMV/AST), and Fig. 11.9 (Amberlite XE-69·PVC/XE-119·PVC), demonstrating periodic changes of C'' and q caused by the seasonal change of seawater temperature T. After $t = 25{,}000$ h, chlorine gas was unfortunately contaminated into the feeding seawater. Because of this, η is known to be decreased during this period. Except for data affected by this problem, the performance of the homogeneous membranes

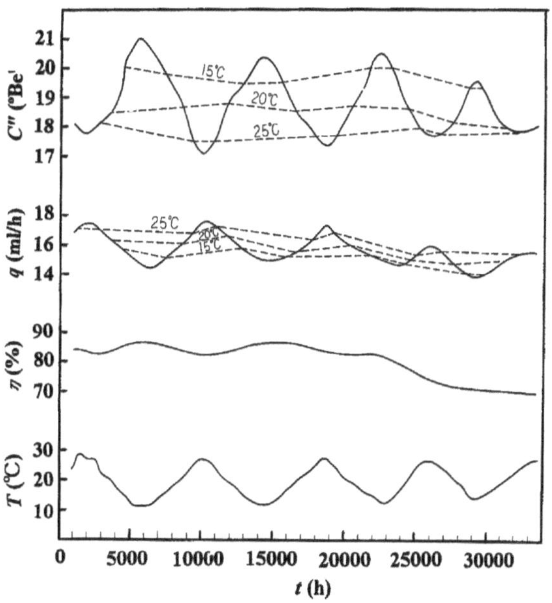

FIGURE 11.6 Seawater electrodialysis test for a long time. Neosepta CL-25T/AVS-4T. *Hanzawa et al. (1966), p. 4, Fig. 2.*

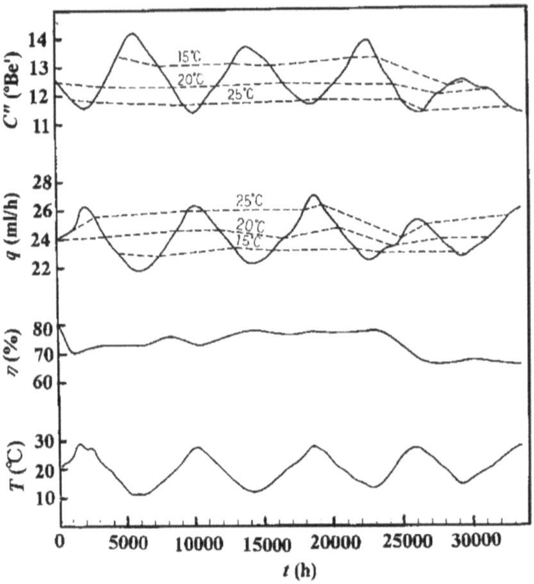

FIGURE 11.7 Seawater electrodialysis test for a long time. Aciplex CK-2/CA-3. *Hanzawa et al. (1966), p. 5, Fig. 4.*

(Figs 11.6–11.8) was stable. However, C'', q, and η of heterogeneous membrane (Fig. 11.9) decreased with time. In parallel with this experiment, long-term seawater electrodialysis was performed setting the current density at 2–4 A/dm^2. However, the performance and durability were not evaluated because the membranes were destroyed by the precipitation of $CaCO_3$ and $CaSO_4$ onto the membrane surfaces.

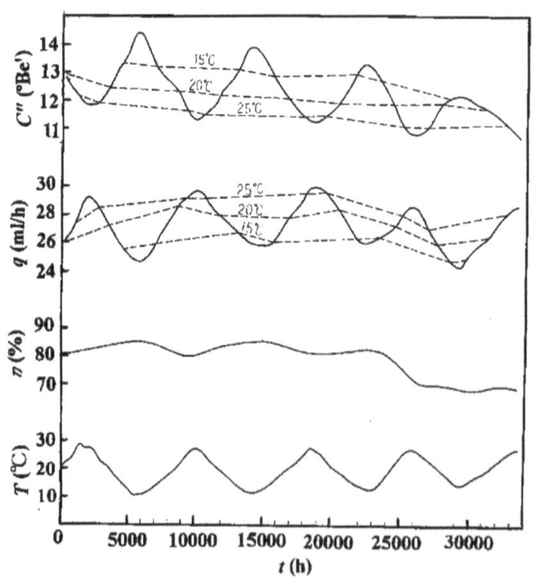

FIGURE 11.8 Seawater electrodialysis test for a long time. Selemion CMV/AST. *Hanzawa et al. (1966), p. 6, Fig. 6.*

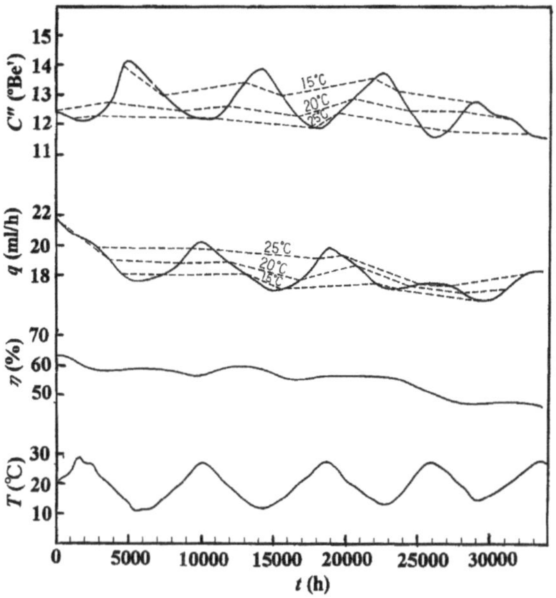

FIGURE 11.9 Seawater electrodialysis test for a long time. Amberlite XE-69·PVC/XE-119·PVC. *Hanzawa et al. (1966), p. 7, Fig. 9.*

11.4 Surface Fouling

11.4.1 Mechanism of Surface Fouling

Surface fouling is caused by deposits of small particles suspended in a feed solution on membranes. Grossman and Sonin (1973) discussed the mechanism of surface fouling as follows, assuming that the membranes are covered with thin surface films of the deposited materials.

Figure 11.10 shows the salt concentration distribution in a desalting and a concentrating cell (flow-pass thickness: *a*). The linear concentration distribution in the diffusion layers is an approximation that results from the Nernst-diffusion model (Section 9.2) for the convective-diffusion process. In the Nernst-diffusion model, the salt ions are assumed to arrive at the membrane surface by diffusion across hypothetical diffusion layers of thickness δ. Using Eqns (9.2) and (9.3), one obtains for salt concentration C_0 at the desalting surface of the membrane:

$$C_0 = C' - \frac{\bar{t} - t}{FD} i\delta \tag{11.1}$$

in which i is current density, C' is the salt concentration in the desalting cell outside the diffusion layer, D is the salt diffusion coefficient in the solution, $\bar{t}$ and t are the transport number of counter-ions in the membrane and in the solution, respectively, and F is the Faraday constant.

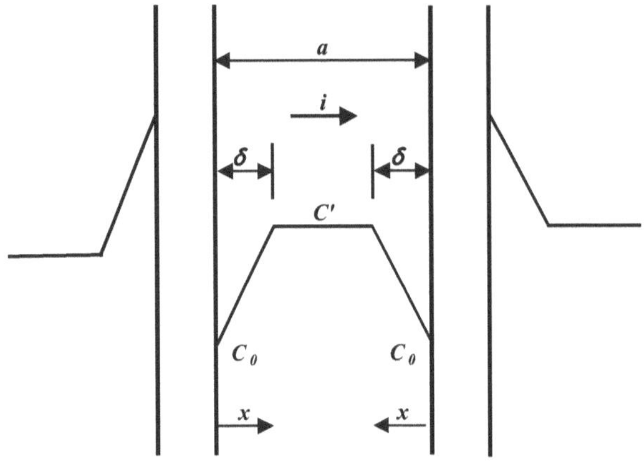

FIGURE 11.10 Salt concentration distribution in electrodialysis channels according to the Nernst-diffusion model. *Grossman and Sonin (1973), p. 110, Fig. 1.*

When the surface film is formed on the membrane, as shown in Fig. 11.11, the material balance in the diffusion layer is expressed based on the Nernst-diffusion model introduced from Eqn (11.1) as:

$$C' - C^* = \frac{\bar{t} - t}{FD} i\delta \tag{11.2}$$

The material balance in the film is expressed by:

$$C^* - C_0 = \frac{\bar{t} - t}{FD_f} i\Delta \tag{11.3}$$

C^* is the salt concentration at the interface between the diffusion layer and the film, Δ is the film thickness, D_f is the salt diffusion coefficient in the film, and t is the transport number in the film. In Eqn (11.2), materials composing the surface film are assumed to be uncharged (neutral), so t in the film (in Eqn (11.3)) is approximated by t in the solution (in Eqn (11.2)). At the limiting current density, substituting $i = i_{\lim}$ and $C_0 = 0$ into Eqns (11.2) and (11.3) leads to:

$$C' - C^* = \frac{\bar{t} - t}{FD} i_{\lim}\delta \tag{11.4}$$

$$C^* = \frac{\bar{t} - t}{FD} i_{\lim}\Delta \tag{11.5}$$

Canceling C^* from Eqns (11.4) and (11.5), the limiting current density of the membrane with the film is introduced as follows:

$$i_{\lim} = \frac{FDC'}{(\bar{t} - t)\delta}\left(\frac{1}{1 + fp}\right) \tag{11.6}$$

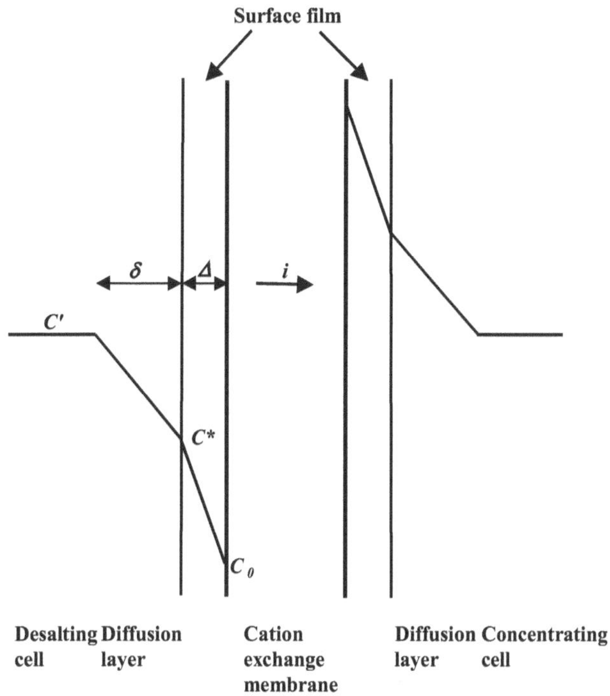

FIGURE 11.11 Salt concentration distribution with surface films on a cation exchange membrane. *Grossman and Sonin (1973), p. 112, Fig. 2.*

Here, *fp* is the fouling parameter defined by Eqn (11.7), which is dimensionless and indicates the effect of the film on the limiting current density.

$$fp = \frac{D\Delta}{D_f\delta} \tag{11.7}$$

It is clear from Eqns (11.6) and (11.7) that a fouling film depresses the limiting current density only if quantity *fp* is comparable with unity or larger; that is, when the film thickness is comparable with the Nernst diffusion layer thickness and/or the film diffusion coefficient is small compared with the diffusion coefficient in the solution. *fp* represents the ratio of the Ohmic resistance of the film to the Ohmic resistance of the fluid within the Nernst diffusion layer. Because in many practical cases the diffusion layer is thin (Table 9.1) compared with the thickness of the desalting cell (*a* in Fig. 11.10), *fp* can be an order of unity even if the Ohmic resistance of the fouling layer is small compared with that of the desalting cell. In other words, the film may bring about a significant reduction in the limiting current density and yet contribute negligibly to the Ohmic resistance of the system as a whole. However, if Δ is comparable to the thickness of the desalting cell, or if D_f is small, a film capable of depressing the limiting current would also increase the resistance of the system.

11.4.2 Formation of Films on the Membrane Surface

Surface fouling discussed above is caused by depositing suspended matters in a feeding solution onto the membrane surfaces to form films. Constituents of the substances attached to the membrane surface in seawater electrodialysis consist of inorganic acid-soluble and acid-insoluble substances with ignition losses as shown in Table 11.3 (Azechi et al., 1964). Small particles suspended in a feeding solution are usually removed by means of sand filtration, membrane filtration, coagulation sedimentation filtration, etc. However, extremely small particles pass through the filter, enter into an electrodialyzer, and deposit onto the membrane surface. Microorganisms passed through the filter enter into an electrodialyzer and breed on the membrane surface to form the primary film, into which inorganic particles and plant fragments are trapped and adherent complexes are formed. Ohwada et al. (1981) measured viable bacteria count, chlorophyll, and organic components in substances attached to membranes integrated into the electrodialyzer operating in salt manufacturing plants (Table 11.4). They observed the membrane surface after disassembling the electrodialyzer and reported that the films are extended as a mosaic pattern.

From these investigations, the suspended particles are assumed to be partially excluded by spacers and trapped at the interface between the spacer rods and the membranes, as illustrated in Fig. 11.12. The solution velocity in the boundary layer is usually accelerated by the convection owing to the forced flow or gravitational flow, and the mass transport in the boundary layer is expressed by the extended Nernst–Planck equation. However, in the attached substance film, the convection is limited because it is caused merely by electro-osmotic flow and volume-osmotic flow passed through the

Table 11.3 Constituents of Substances Attached to an Ion Exchange Membrane

Components	Cation Exchange Membrane (g/100 g)	Anion Exchange Membrane (g/100 g)
H_2O	84.85	85.62
Acid Soluble		
$Fe(OH)_3$	4.41	3.85
$Cu(OH)_2$	0.03	0.03
Etc.	0.82	0.80
Total	5.26	4.68
Acid Insoluble		
SiO_2	2.51	2.32
Fe_2O_3	1.42	1.14
CuO	0.02	0.03
Al_2O_3	0.12	0.34
Etc.	0.50	0.54
Total	4.57	4.37
Ignition loss	3.20	3.49

Azechi et al. (1964), p. 27, Table 7.

Table 11.4 Deposition of Substances on Ion Exchange Membranes and Microorganisms in Deposit

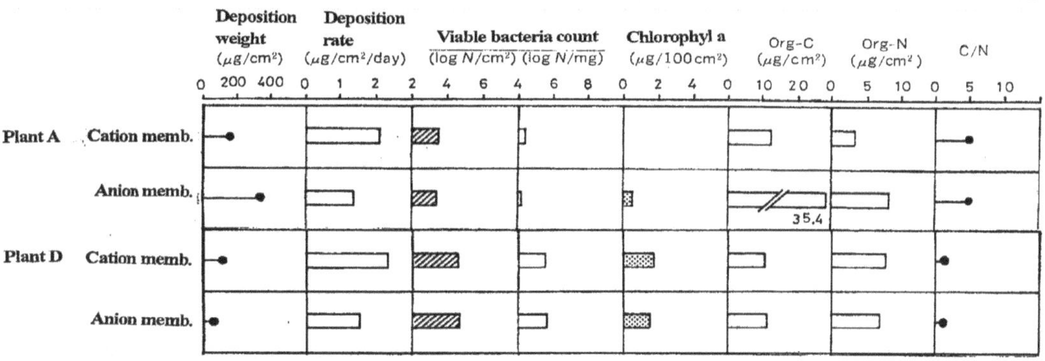

Ohwada et al. (1981), p. 369, Figs 2 and 3.

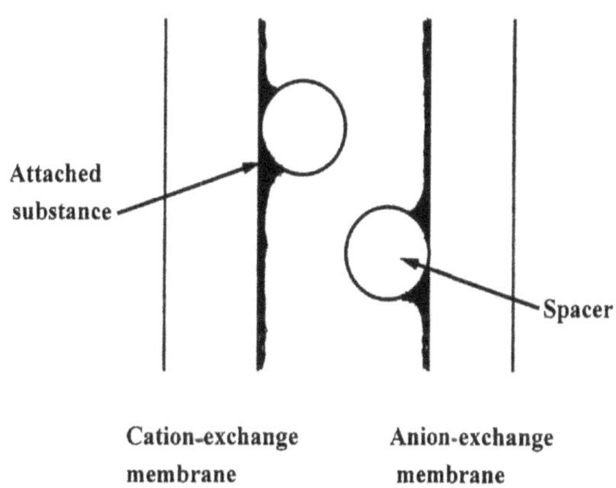

FIGURE 11.12 Attached substances on the membrane surface. *Tanaka (1997), p. 233, Fig. 5.*

membrane. The mass transport in the film is expressed by the Nernst–Planck equation. In the film, the concentration polarization is accelerated and the limiting current density is decreased because of limited convection (cf. Fig. 8.7) (Tanaka, 1997).

11.4.3 Removal of Films on the Membrane Surface

The deposit is removed by stack disassembling (Applications Section 12.3.4), periodic current reversal (Applications Section 16.6) and washing with a chemical reagent, which is introduced in this section.

Urabe and Doi (1987) washed the membranes with a mixed solution of alkalis, mineral salts, and organic solvents. In this research, cation exchange membrane Neosepta CL-25T and anion exchange membrane ACH-45T were integrated into an electrodialyzer (effective membrane area, 0.25 m^2; number of membrane pairs, 500). Seawater was supplied and desalinated passing an electric current. In the course of the electrodialysis operation, the cell voltage was increased from 0.7 to 1.2 V owing to the precipitation of fine substances to the membrane surfaces. A 25–30 °C washing solution mixed with tap water (500 l) and methanol (500 l) dissolving NaCl (90 kg) and NaOH (40 kg) was circulated through the desalting and concentrating cells for 3 h. Then, the cell voltage was restored to the original 0.7 V without a decrease in the current efficiency, permselectivity, and strength of the membranes. In the washing operation, NaCl, NaOH, and methanol are estimated to dissolve substances precipitated onto the membrane into the external solution.

Yamashita (1976) washed membranes with ammonia water dissolving citric acid and ethylenediamine tetra acetic acid (EDTA). In this study, filtrated industrial waste water was supplied to an electrodialyzer (Selemion CMV/AMV; effective membrane area, 5 dm^2; number of membranes, 20 pairs) maintaining the solution velocity and static head loss at 40 l/min and 154 mm Hg, respectively. After 166 m^3 of the solution was electrodialyzed, it became difficult to operate the apparatus because the static head increased to 452 mm Hg owing to the adhesion of suspended substances in the feeding solution to flow paths in the electrodialyzer. Therefore, a 30 °C aqueous solution dissolving 2% citric acid and 1% EDTA was supplied to the electrodialyzer at a flow velocity of 40 l/min, keeping the pH at 4.5 by adding ammonia water. The static head was decreased to 162 mm Hg after 45 min; then, after operating for 15 min, the electrodialyzer was disassembled and it was found that the attached substances disappeared. Finally, the electrodialyzer was reassembled and the operation proceeded supplying the solution at the static head of 162 mm Hg. In this experiment, complex forming agents such as citric acid and EDTA dissolved the substances attached to the membrane at a reasonable pH value.

Ueno et al. (1980) supplied seawater to an electrodialyzer incorporated with Selemion CMV and ASV membranes. The static pressure at the inlet of the apparatus was 0.6 kg/cm^2 G at first. In the course of operation, the pressure gradually increased as a result of the attachment of slime to the membrane surfaces in desalting cells; eventually, it became impossible to operate because the pressure attained a limiting value of 1.2 kg/cm^2 G. To avoid the problem, 900 ml of an aqueous solution dissolving 0.5% Neopelex Paste (Kao Atlas Co., anion-surfactant, effective component: sodium dodecyl-benzene sulfate) and 0.5% hydrazine was supplied into 1 m^3 of water in a washing line and was circulated through the circulating tank and the electrodialyzer for 30 min. Next, 25.7 kg of 35% hydrogen peroxide was added to the circulating tank and the washing solution was circulated through the electrodialyzer for 3 h, keeping the temperature at 25–30 °C. Consequently, the pressure was decreased to 0.6 kg/cm^2 G and it became possible to operate the electrodialyzer again. After the experiment, the apparatus was disassembled and it was confirmed that the slime had been perfectly removed. The electric resistance of the ion exchange membranes was not altered by washing.

11.5 Organic Fouling

11.5.1 Organic Fouling Phenomena

Organic fouling of ion exchange membranes is a major problem in electrodialysis. It is caused by the precipitation of organic colloidal substances onto the membranes, and because most colloidal substances present in natural water are negatively charged, they almost always precipitate onto anion exchange membranes. Korngold et al. (1970) experimentally examined the process of organic fouling using multi-compartment electrodialysis cells and humate as the fouling agents, as follows.

Fouling was defined as the rise in voltage across anion exchange membranes as a result of fouling agents. The apparatus used for the experiments was a multi-compartment system with forced circulation. Each compartment was a circular hole in a 1-cm-thick Perspex plate with a diameter of 2 cm and opposite tangential inflow and outflow tubes 3 mm in diameter. To measure voltage across the membrane, two electrodes were inserted in each compartment so that they were near the membrane surfaces adjoining the compartments. The membranes were clamped tightly between two compartments.

Figure 11.13 shows the fouling curves of an AMF anion exchange membrane (American Machine Foundry (USA)) in 0.1 N KCl solutions containing 1.0–0.1% humate,

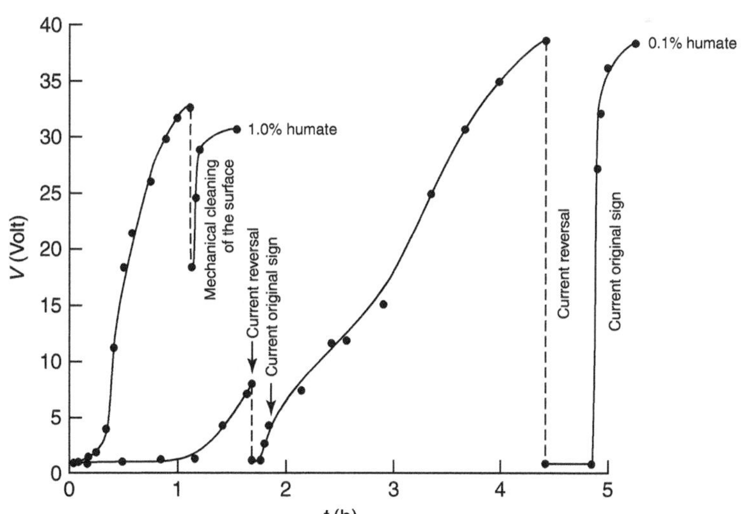

FIGURE 11.13 Fouling of AMF A-63 membranes in a mechanically stirred apparatus by Na-humate containing 0.1 N KCl solutions. $i = 19$ mA/cm^2, $E =$ voltage across the membrane. *Korngold et al. (1970), p. 199, Fig. 1.*

respectively. Reversal of the current immediately reduced the voltage on the membrane to its original values, but its reversal to its original direction raised it again almost instantaneously as high as it had been before reversal. Mechanical cleaning of the membrane surface also helped for only a short time. There was, of course, much more humic acid precipitated in the 1% solution. Figure 11.14 shows the influence of current density on the organic fouling. The abscissa is not the time but the amount of electricity passed across an AMF A-63 membrane in the multi-compartment apparatus. Fouling is given on the ordinate as the voltage drop across the membrane. Figure 11.15 shows the result of experiments in which the fouling was performed in solutions of increasing KCl concentration, keeping all other parameters unchanged. It is clear that fouling increased quickly with a decrease in KCl concentration.

The following experiment showed that fouling acidifies the solution on the side of the anion exchange membrane on which humic acid precipitates and makes the solution on its other side alkaline. In this experiment using a six-compartment cell (Fig. 11.16), an AMF anionic membrane was put between cells 3 and 4 with voltage measuring electrodes near its surface. Both cells were separated by cation AMF membranes from their adjoining compartments that contained 0.1 N NaCl to insulate them from the cells with the current electrodes. Then, HCl was added to the cathode cell (pH ~2) and NaOH to the anode cell (pH ~11). The electrolyte in measuring cells 3 (desalting cell) and 4 (concentrating cell) was 0.1 N in each cell in the control experiment, and cell 3 received an amount of 0.06% sodium humate in the main experiment. A current of 60 mA ($i = 19\ mA/cm^2$) was passed through the system in both experiments and the voltage across the membrane and the pH in both measuring compartments were measured as a function of time. Then, NaCl was added every 10 min to cell 3 to compensate for what was lost by electrodialysis. The pH in cells 2 and 5 was determined every 10 min to

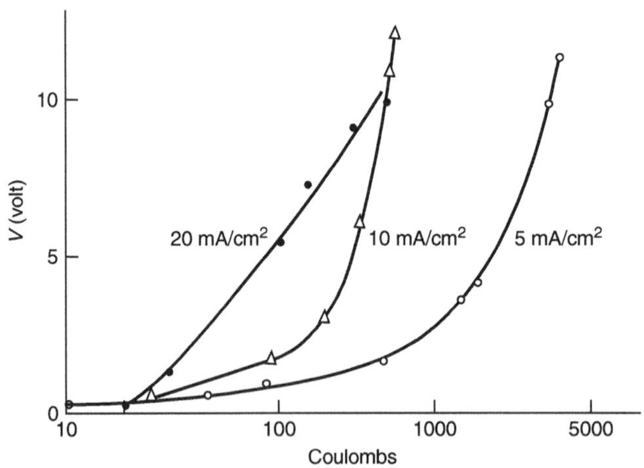

FIGURE 11.14 Influence of current density on fouling. AMF A-63 membranes. 0.04 N KCl containing 1000 ppm Na-humate. Flow velocity, 1.8 cm/s, pH 8.4. *Korngold et al. (1970), p. 199, Fig. 2.*

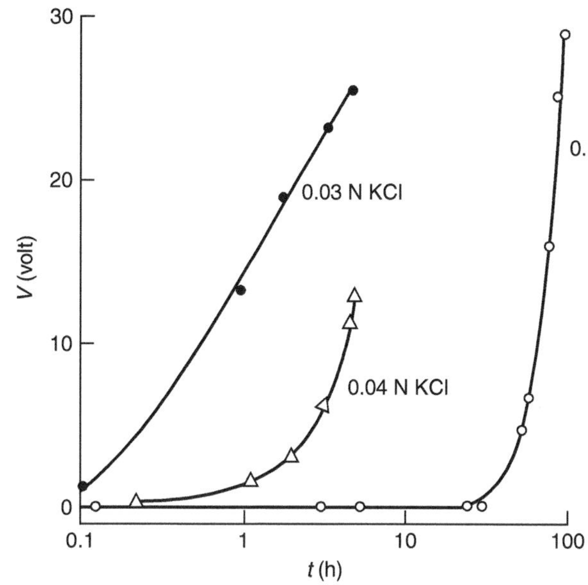

FIGURE 11.15 Influence of KCl concentration on fouling AMF A-63 membranes. 1000 ppm Na-humate. 10 mA/cm^2. Flow velocity 1.8 cm/s, pH 8.4. *Korngold et al. (1970), p. 200, Fig. 3.*

ensure that they did not change and they isolated the experimental cells from the electrode compartments. Cells 3 and 4 were stirred mechanically at constant speed. Table 11.5 shows the results. It is clear that fouling caused changes of pH on both sides of the membrane. These results indicate that fouling by anionic colloids of insoluble acids is a catalytic process triggered by polarization on the anion exchange membrane and

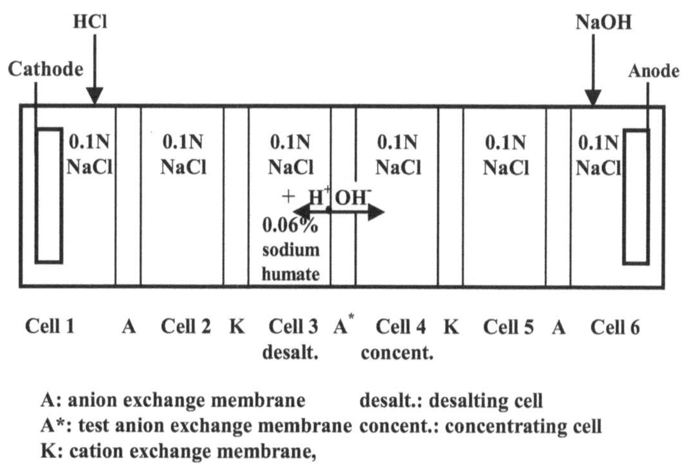

FIGURE 11.16 Six-compartment cell for measuring pH changes caused by organic fouling.

Table 11.5 Acid Generated on Cathode Side of Anion Exchange Membrane and Alkali Generated on Anode Side

	Control, No Humate			With Humate		
Time (min)	V	pH in Cell 3 (Desalt)	pH in Cell 4 (Concent)	V	pH in Cell 3 (Desalt)	pH in Cell 4 (Concent)
0	2.2	5.5	5.5	2.5	6.6	5.5
5	2.2	5.4		4.5	5.7	
17	2.3	5.2		7.5	5.5	6.0
30	2.3	5.1		8.2	5.5	6.0
40	2.3	5.0	5.6	8.7	5.5	6.0
63	2.3	4.8		9.0	4.2	6.5
80	2.4	4.5		10.5	3.4	7.0
95	2.5	4.5		10.7	3.2	7.0
110	2.5	4.5	6.3	11.8	3.0	10.9
130	2.5	4.6		12.5	2.9	
150	2.5	4.6	6.3	12.2	2.8	11.1
165	2.5	4.5		14.0	2.7	

Desalt, desalting cell; concent., concentrating cell. 600 ppm humate in 0.1 N NaCl. 19 mA/cm^2. Mechanical stirring.
Korngold et al. (1970), p. 202, Table 1.

increasing the polarization on it so as to increase fouling more rapidly after an incubation period. The underlying chemical process must be the precipitation of humic acid on the membrane. The layer of this cation-active colloid on the anion-active membrane creates a composite sandwich membrane that becomes depleted of salt ions from both sides because it receives the current in this closing direction. Thus, an increasing part of current conduction is taken over by the ions of water owing to water dissociation, making the humate side (desalting side) acid and the other side (concentrating side) alkaline (Fig. 11.16). The H^+ ions entering the humate solution catalytically increase the precipitation of humic acid so that fouling increases rapidly once it has started.

11.5.2 Water Dissociation Reaction Generated Due to Organic Fouling

Korngold et al. observed pH changes resulting from the organic fouling of an anion exchange membrane, as described in Section 11.5.1. They suggested the formation of a sandwich membrane structure. Salt ions are depleted from both sides of the sandwich structure and the water dissociation is suggested to be generated at the interface between the humic acid layer and the anion exchange membrane.

Based on the concept on the mechanism of the water dissociation reaction discussed in Chapter 7, water dissociation is estimated to be generated in the fouled anion exchange membrane, as illustrated in Fig. 11.17(a), in which + denotes quaternary ammonium groups in the water dissociation reaction layer formed in the anion exchange membrane and − denotes anionic groups in the fouling substance (fumic acid

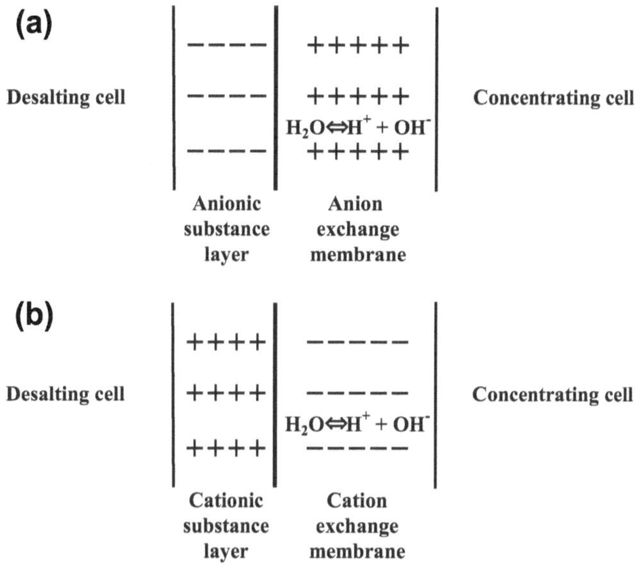

FIGURE 11.17 Water dissociation reaction generated in a fouled ion exchange membrane. (a) Fouled anion exchange membrane. (b) Fouled cation exchange membrane.

in Section 11.5.1) layer. The water dissociation generates H^+ ions and OH^- ions which become co-ions and counter-ions, respectively, in the reaction layer. Co-ion (H^+ ion) generation (water dissociation reaction) is suppressed in principle in the electric potential generated in the reaction layer owing to repulsive force between co-ions and ion exchange groups (Tanaka, 2010). However, in Fig. 11.17(a), co-ion (H^+ ion) generation is not strongly suppressed because of lower repulsive forces between co-ions (H^+ ions) and positively charged ion exchange groups (quaternary ammonium groups) in the reaction layer.

Here, the water dissociation reaction in the fouled cation exchange membrane is discussed based on the concept described in Chapter 7. In the fouled cation exchange membrane, water dissociation is estimated to be strongly suppressed at a higher electric potential owing to elevated repulsive forces between co-ions (OH^- ions) and negatively charged ion exchange groups (sulfonic acid groups) in the reaction layer, as illustrated in Fig. 11.17(b), in which − denotes sulfonic acid groups in the reaction layer and + denotes positive charged groups in the fouling substance layer.

The structure in Fig. 11.17(b) is conceptually equivalent to the polycation layer formed on the monovalent cation selective ion exchange membrane (Fig. 2.10). In the electrodialysis experiment in Section 2.5.3, monovalent cation selectivity is given by forming a di-cyandiamide formaldehyde condensate (polycation) layer, which includes many quaternary ammonium positive groups in the molecule (Fig. 2.13). The experiment shows that cell voltage and energy consumption were extremely increased when the monovalent ion ratio in a concentrated solution ε increased beyond 0.99 (Fig. 2.16), whereas the water dissociation did not occur. The experiment demonstrates that the water dissociation reaction is suppressed in the cation exchange membrane.

If one assumes in Fig. 2.10 that the polycation layer is fouled by the cation exchange membrane, Fig. 2.10 becomes equivalent to Fig. 11.11(a) and the water dissociation reaction is assumed to be generated in the polycation layer. Such a phenomenon might be generated but the intensity of the reaction is estimated to be low because the thickness of the polycation layer is extremely thin and the intensity of the water dissociation reaction is proportional to the thickness of the reaction layer (Eqn (7.28)).

11.5.3 Anti-organic Fouling Membranes

1. Macroreticular anion exchange membrane

In order to prevent organic fouling, Kusumoto et al. (1976) developed the following macroreticular (macroporous) anion exchange membrane. In this investigation, a syrupy monomer solution was prepared by mixing styrene-divinylbenzene with polybutadiene rubber (viscosity increasing agent) and *t*-amylalcohol (macroporous structure forming agent). The macroreticular membrane was prepared by coating the mixed solution with a polypropyrene reinforcement and polymerizing it under pressure. The membranes were synthesized by changing the ratio of *t*-amylalcohol, as shown in Table 11.6.

Figure 11.18 shows the influence of the amylalcohol ratio to the electric resistance, transport number, and diffusion constant of the membranes synthesized above, and that the macroporous structure develops with an increase in *t*-amylalcohol content. The anti-organic fouling performance was measured with a two-cell electrodialysis unit filled with a 0.05 N NaCl + 100 ppm dodecyl-benzene sulfonate solution in the cathode chamber (desalting side) and a 0.05 N NaCl solution in the anode chamber (concentrating side). Passing an electric current of 2 mA/cm^2, electric resistance change owing to the organic fouling of the membrane was detected, as shown in Fig. 11.19, by measuring the potential difference V_m across the membrane. The result shows that the electric resistance increase caused by organic fouling is suppressed with the development of a macroporous structure.

Table 11.6 Composition of Monomer Mixed Solution[a]

No.	St[b]	DVB[c]	PB[d]	*t*-AmOH[e]
1	0.7	0.3	0.1	0.4
2	0.7	0.3	0.1	0.5
3	0.7	0.3	0.1	0.8
4	0.7	0.3	0.1	1.0

[a]Weight.[b]Styrene.[c]Divinylbenzene.[d]Polybutadene.[e]*t*-Amylalcohol.
Kusumoto et al. (1976), p. 3207, Table 1.

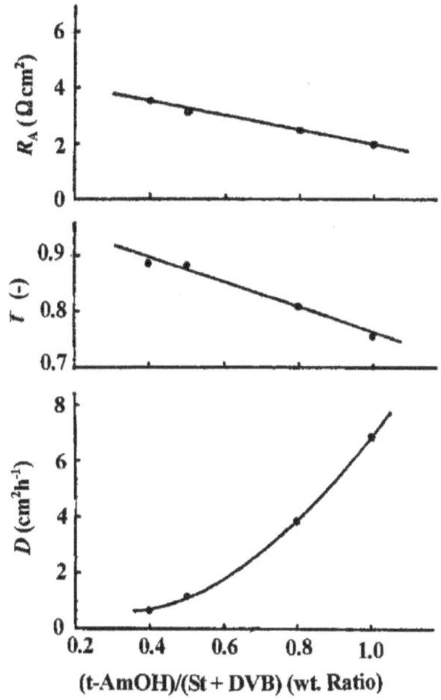

FIGURE 11.18 Electrochemical properties of a macroreticular anion exchange membrane. R_A: electric resistance (25 °C, 0.5 N NaCl). $\bar{t}$: transport number (25 °C, 0.5/2.5 N NaCl). D: NaCl diffusion constant (25 °C). *Kusumoto et al. (1976), p. 3210, Fig. 3.*

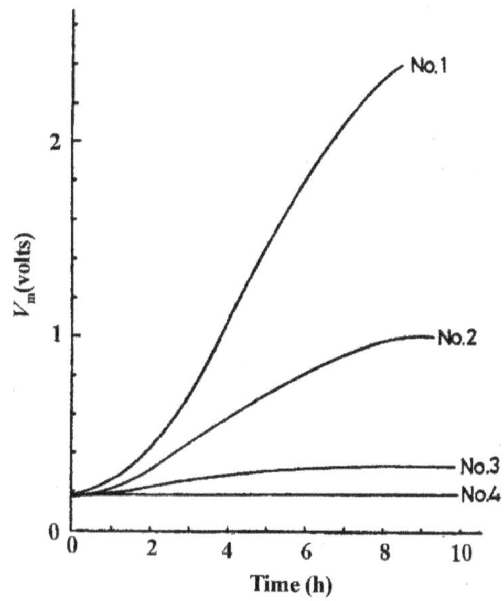

FIGURE 11.19 Anti-organic fouling of a macroreticular anion exchange membrane. *Kusumoto et al. (1976), p. 3211, Fig. 4.*

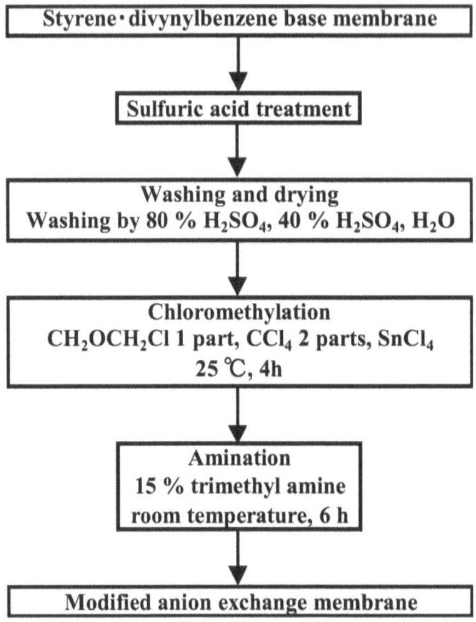

FIGURE 11.20 Synthetic process of a modified anion exchange membrane.

Hodgdon and Sudbury (1971) applied for a patent of "Ion exchange membranes having a macroporous surface area," in which the membrane was fabricated with a polymeric macroporous surface that lessened the tendency to foul when employed in the electrodialysis of solutions containing a fouling constituent.

2. Formation of thin cation exchange layer on anion exchange membrane

Another anti-organic fouling membrane was developed by sulfonating the surface of an anion exchange membrane and forming a thin cation exchange layer on the anion exchange membrane (Kusumoto and Mizutani, 1975). The synthetic process starts from sulfuric acid treatment of a styrene-divinylbenzene base membrane and the modified membrane is prepared via chloromethylation and amination (Fig. 11.20). Table 11.7 shows that the electric resistance R_A and transport number $\bar{t}$ was not influenced by

Table 11.7 Sulfuric Acid Treatment and Membrane Characteristics

No.	t^a (h)	A·S^b (meq/g dry memb.)	$R_A{}^c$ (Ω cm^2)	$\bar{t}^d$
1	0	–	2.2	0.91
2	0.5	Not detected	2.3	0.91
3	1.25	Trace	2.3	0.91
4	3.0	0.01	2.3	0.91
5	5.0	0.08	2.4	0.91

[a]Treatment time in 98% sulfuric acid.[b]Sulfonic acid concentration.[c]Electric resistance: 0.5 N NaCl, 25 °C.[d]Transport number: 0.5/2.5 N NaCl, 25 °C.

Kusumoto and Mizutani (1975), p. 115, Table 1.

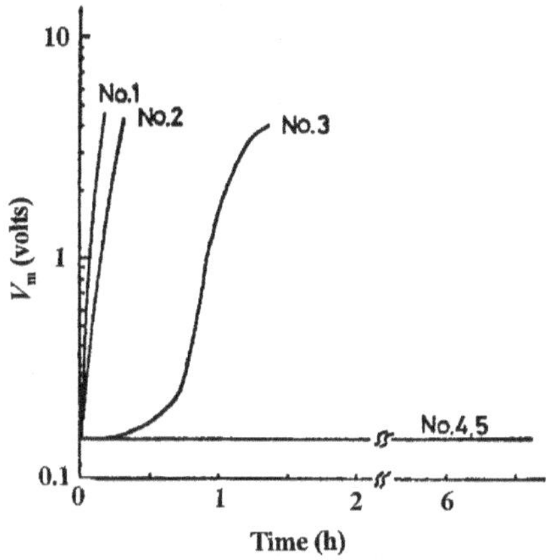

FIGURE 11.21 Anti-organic fouling of a modified anion exchange membrane. *Kusumoto and Mizutani (1975), p. 115, Fig. 2.*

sulfuric acid treatment time *t*. Figure 11.21 gives the influence of *t* to the potential difference V_m in the electrodialysis experiment caused by the organic fouling owing to dodecyl-sulfonic acid. The anti-organic fouling performance of membranes 4 ($t = 3.0$ h) and 5 ($t = 5.0$ h) is excellent.

References

Azechi, S., Akiyama, M., Suwa, K., 1964. Studies on the Concentration Operation by Ion Exchange Membrane Method (VI). In: Effects of Suspended Matter in Sea Water on Electrodialytic Unit and Concentration Results (Part 1), Scientific Papers of the Odawara Salt Experiment Station, vol. 9. Japan Monopoly Corporation, pp. 18–33.

Grossman, G., Sonin, A.A., 1973. Membrane fouling in electrodialysis: a model and experiments. Desalination 12, 107–125.

Hanzawa, N., Yuyama, T., Suzuki, K., Nakayama, M., 1966. Studies on Durability of Ion Exchange Membrane (IV) Long Term Electrodialytic Concentration Test. In: Scientific Papers of the Odawara Salt Experiment Station, vol. 11. Japan Monopoly Corporation, pp. 1–13.

Hodgdon, R.B., Sudbury, Jr., 1971. Ion Exchange Membranes Having a Macroporous Surface Area. US 3,749,655.

Kneifel, K., Hattenbach, K., 1980. Properties and long-term behavior of ion exchange membranes. Desalination 34, 77–95.

Korngold, E., Korosy, F.D.E., Rahav, R., Taboch, M.F., 1970. Fouling of anionselective membranes in electrodialysis. Desalination 8, 195–220.

Kusumoto, K., Mizutani, Y., 1975. New anion-exchange membrane resistant to organic fouling. Desalination 17, 111–120.

Kusumoto, K., Ihara, H., Mizutani, Y., 1976. Preparation of macroreticular anion exchange membrane and its behavior relating to organic fouling. J. Appl. Polym. Sci. 20, 3207–3213.

Ohwada, K., Shimizu, U., Taga, N., 1981. Microorganism and organic matter deposited on the ion exchange membrane. Bull. Sea Water Sci. Jpn. 34, 367–372.

Tanaka, Y., 1997. Concentration polarization in the substance layer attached to an ion exchange membrane. Bull. Sea Water Sci. Jpn. 51, 228–236.

Tanaka, Y., 2010. Water dissociation reaction generated in an ion exchange membrane. J. Membr. Sci. 350, 347–360.

Ueno, K., Ozawa, T., Ooki, H., Ishida, T., Nakajima, K., Sudo, T., 1980. Washing Method of Ion-Exchange Membranes. JP S55-33662.

Urabe, S., Doi, K., 1987. Washing Method of Ion-Exchange Membranes. JP S62-52624.

Yamashita, I., 1976. Removing Method of Fouling Substances in an Electrodialyzer. JP S51-131477.

PART 2

Applications

12

Electrodialysis

12.1 Overview of Technology

Industrial application of ion exchange membranes first started in the field of electrodialysis (ED) and induced the development of the fundamental theory. This is manifested from many phenomena explained in Fundamentals, which are described taking the ED into account. This chapter describes the structure of an electrodialyzer and maintenance technology, and introduces several practical examples of ED technology. The operation of an electrodialyzer is discussed in Chapters 13–15, Electrodialysis program.

12.2 Electrodialyzer

12.2.1 Structure of an Electrodialyzer

The basic structure of the vertical sheet-flow type module consists of stacks in which cation exchange membranes, anion exchange membranes, and gaskets (desalting cells and concentrating cells) are arranged alternately (Fig. 12.1). Fastening frames are put on both outsides of the stack, which is fastened together through cross-bars set in the frames. Deformation of the membranes is prevented by regulating hydrostatic pressure in the fastening frames. Inlet duct/slots and outlet duct/slots are prepared at the bottoms and heads of the gaskets, respectively. Spacers are incorporated with the gaskets to prevent the contact of cation exchange membranes with anion exchange membranes. Many stacks are arranged through the fastening frames. Electrode cells are put on both ends of the electrodialyzer, which are fastened by pressing on the outsides of electrode cells (Fig. 12.2).

A salt solution to be desalinated is supplied from the solution feeding frames to the entrance ducts, flows through the entrance slots, current-passing portions, and exit slots, and discharged from the exit ducts to the outside of the stack (Figs 12.1 and 12.2). The salt solution is supplied to the concentrating cells and discharged to the outside of the stack.

The effective membrane area is in the range of less than 0.5 m^2 to a maximum of about 2 m^2. To reduce energy consumption, it is desirable to decrease the electric resistance of the membrane and gasket thickness. The gasket material is selected from synthesized rubber, polyethylene, polypropylene, polyvinyl chloride and ethylene-vinyl acetate copolymer, etc. The spacer is incorporated with the gasket and a solution flows dispersing along the spacer net.

Ion Exchange Membranes. http://dx.doi.org/10.1016/B978-0-444-63319-4.00012-2

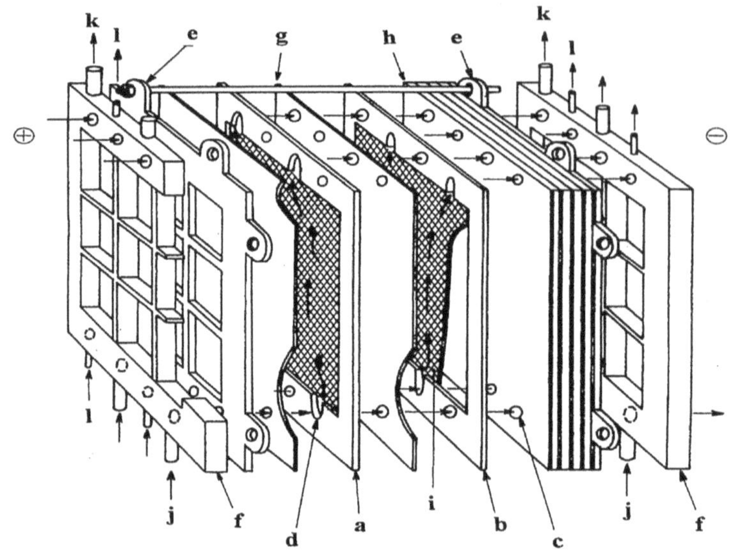

FIGURE 12.1 Structure of a stack (sheet-flow type). a, Desalting cell; b, concentrating cell; c, duct; d, slot; e, fastening frame; f, feeding frame; g, cation exchange membrane; h, anion exchange membrane; i, spacer; j, feeding solution; k, desalted solution; l, concentrated solution. *Azechi (1980), p. 78, Fig. 17.*

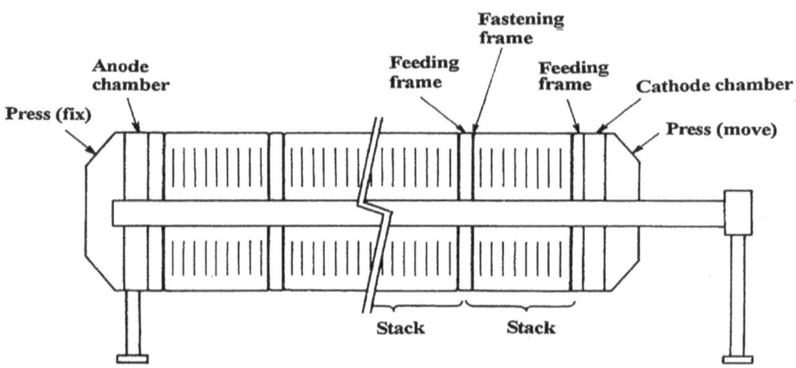

FIGURE 12.2 Sheet-flow type electrodialyzer. *Azechi (1980), p. 78, Fig. 16.*

12.2.2 Parts of an Electrodialyzer

The electrodialyzer is composed as follows (Urabe and Doi, 1978).

12.2.2.1 Fastening Frame

A maximum of 2000 pairs of membranes are arranged between electrodes in an electrodialyzer, so disassembling and assembling works are easy. The membrane array is divided further into several stacks including 50–400 pairs. Fastening frames are fixed by bolts on both ends of the stack. The fastening frame serves as a solution feeding frame so that diluate and concentrate are supplied to each gasket cell from the feeding frame incorporated in every stack. The material of the fastening frame is selected from polyvinyl chloride, polypropylene and rubber-lining iron, etc.

12.2.2.2 Solution Feeding Frame

A solution feeding frame is integrated for feeding solutions to each desalting and concentrating cell. Duct holes are prepared at corresponding positions of the holes in the gasket. Solutions are supplied through the ducts to each stack.

12.2.2.3 Gasket

The shape of the gasket is illustrated in Fig. 12.3. A solution is supplied from the inlet duct prepared at the bottom, flows through the slot, and is fed into the current passing portion. Then the solution is discharged through the outlet slot to the duct prepared at the head. The gasket has the following functions: (1) prevent solution leakage from the inside to the outside of the electrodialyzer, (2) adjust the distance between a cation exchange membrane and an anion exchange membrane, and (3) prevent solution leakage between a desalting cell and a concentrating cell occurring at slot sections. To prevent solution leakage, it is best to integrate a soft material for the gasket; it is desirable to integrate a hard and stable material to avoid dimension changes during long-term operations. The material of the gasket is selected from rubber, ethylene vinyl acetate copolymer, polyvinyl chloride, polyethylene, etc. The thickness of the gasket is in the range of 0.5–2.0 mm.

12.2.2.4 Slot

It is important to reduce inside solution leakage (Fundamentals Section 10.3), which takes place through pinholes and cracks in the membranes or through gaps as a result of

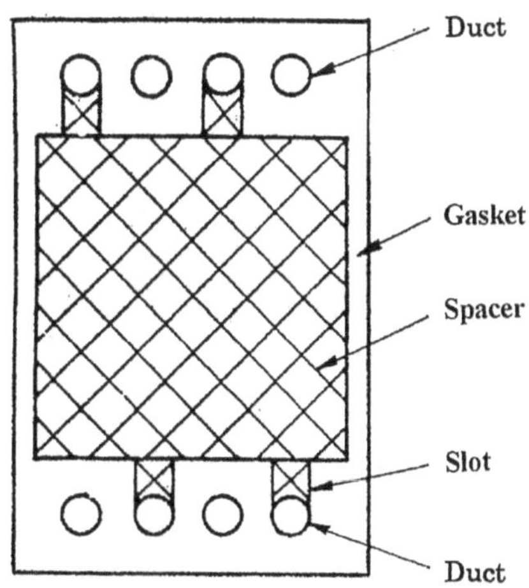

FIGURE 12.3 Gasket. *Urabe and Doi (1978), p. 25, Fig. 3.*

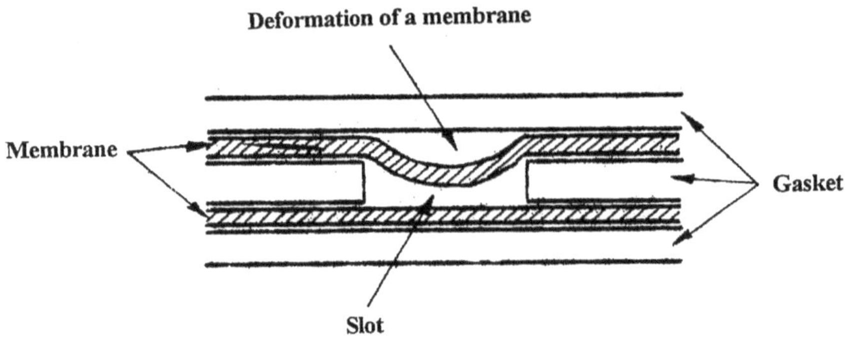

FIGURE 12.4 Deformation of an ion exchange membrane. *Urabe and Doi (1978), p. 25, Fig. 4.*

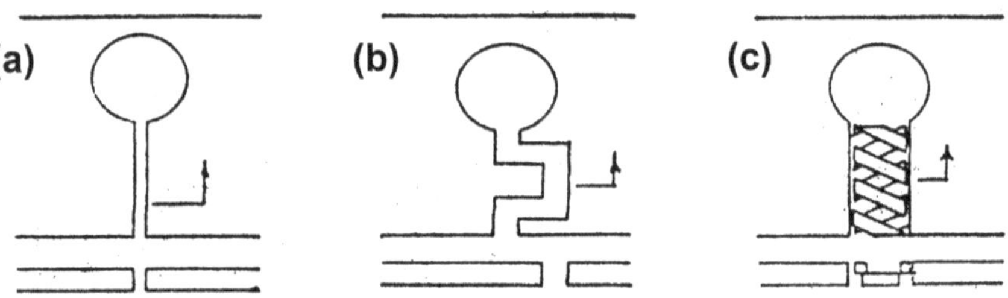

FIGURE 12.5 Structure of slots. *Urabe and Doi (1978), p. 25, Fig. 5.*

membrane deformation at the slot, as shown in Fig. 12.4. To prevent these problems, a lot of devices are proposed, as exemplified in Fig. 12.5 that (a) decrease the width of the slot, (b) bend the slot, or (c) insert the support into the slot.

12.2.2.5 Spacer

The function of a spacer is to keep a distance between membranes. In addition, the spacer increases the limiting current density resulting from solution disturbance (Fundamentals Sections 8.2–8.5, 8.11–8.13). The spacer is selected taking account of requirements such as (1) low friction head loss, (2) low electric current screening effect, (3) easy air discharge, and (4) less blocking of flow pass caused by the precipitation of fine particles suspended in a feeding solution. The structures of a spacer are listed in Fig. 12.6 as (a) expanded polyvinyl chloride, (b) wave porous plate, (c) diagonal net, (d) Mikoshiro texture, and (e) honeycomb net.

12.2.2.6 Electrode and Electrode Chamber

Platinum plated titanium, graphite, or magnetite is employed for anode material, and stainless steel or iron is employed for cathode material. The shape of electrodes is classified into net, bar, and flat. A partition is inserted between an electrode chamber and a stack to prevent mixing of solutions. In an anode chamber, oxidizing substances

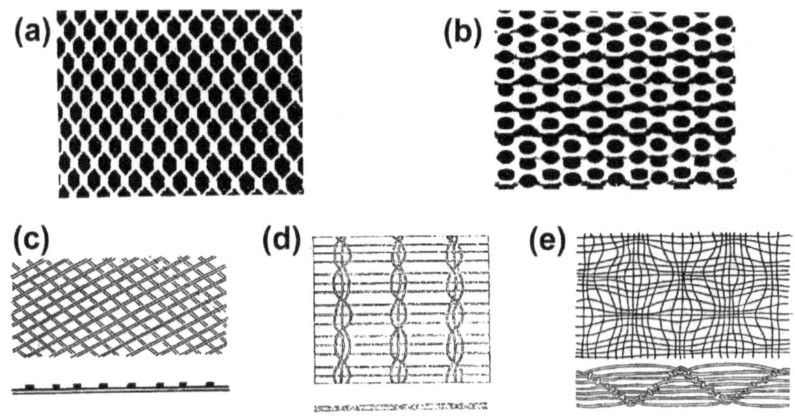

FIGURE 12.6 Structure of spacers. (a) Expanded polyvinyl chloride, (b) wave porous plate, (c) diagonal net, (d) Mikoshiro texture, and (e) honeycomb net. *Urabe and Doi (1978), p. 26, Fig. 7.*

such as chlorine gas evolve. An ion exchange membrane is easily deteriorated by contact with oxidizing substances, so it is necessary to use two sheets of partitions and put a buffer chamber between the two partitions. The material of the partition is an ion exchange membrane, an asbestos sheet, or a battery partition.

An acid solution is added into a cathode solution and the electrodialyzer is operated with the pH control of the cathode solution to prevent magnesium hydroxide precipitation in the cathode chamber. A feeding solution or a concentrated solution is supplied into the electrode chamber. The concentration of oxidizing substances in the anode solution is reduced by adding sodium sulfite or sodium thiosulfate into the solution. Sometimes a sodium sulfate solution is supplied to an anode and a cathode chamber to achieve neutralization by mixing the effluent from both chambers.

12.2.2.7 Press

An oil pressure press is usually used, adjusting the pressure to 5–10 kg/cm^2.

12.2.3 Requirements for Improving the Performance of an Electrodialyzer

This section describes definite problems of lowering the circumstances in an electrodialyzer and requirements for improving the performance of an electrodialyzer (Urabe and Doi, 1978).

1. Solution velocity distribution between desalting cells

In an electrodialyzer, ion exchange membranes and desalting and concentrating cells are arranged alternately and a solution is supplied into the desalting cells. In this flow system, the solution velocity distribution in desalting cells does not become uniform.

This phenomenon causes concentration distribution and current density distribution in the electrodialyzer and gives rise to a decrease in the limiting current density of the electrodialyzer (Fundamentals Sections 8.8, 9.6). To operate the electrodialyzer stably, it is necessary to make the solution velocities between desalting cells uniform.

2. Solution leakage in an electrodialyzer

The dimensions of all parts of an electrodialyzer are not always consistent with the values in the specifications. Small pinholes can open in an electrodialyzer because the strength of ion exchange membranes is relatively low. Gaps may occur between the materials composing the electrodialyzer in the assembly works of an electrodialyzer (Section 12.3.4). In these circumstances, a pressure difference between the desalting cells and concentrating cells causes solution leakage through the pinholes in the membranes during the operation of the electrodialyzer (Fundamentals Section 10.3). To avoid these problems, pinholes and gaps in the electrodialyzer are removed and the pressure difference between desalting cells and concentrating cells is controlled.

3. Distance between membranes

Electrical resistance and energy consumption are decreased with distance between the membranes. While it causes friction loss with an increase in solution flow, suspended material blocking in a feeding solution and pumping motive power increase. Thus, it is necessary to keep an optimal distance between membranes. The optimal distance is takes into account the electric resistance of ion exchange membranes and that of electrolyte solutions in desalting and concentrating cells.

4. Spacer

The main function of a spacer is to create spaces between a cation exchange membrane and an anion exchange membrane. With the decrease in solution velocity and the Reynolds number, the hydrodynamic pattern becomes laminar flow and the disturbing effect of the spacer is reduced. To increase the limiting current density, turbulent flow should be induced by increasing the Reynolds number (Fundamentals Section 8.6).

5. Electric current leakage

An electric current flows through slots and ducts, causing ineffective current leakage. Current leakage is increased with the number of cell pairs integrated into a stack and the sectional area of slots and ducts (Fundamentals Section 10.2). However, these phenomena are related to the solution velocity distribution between cells described in 1.

6. Simplicity of the structure of an electrodialyzer

Disassembling and assembling work is a specific feature in an electrodialyzer operation (Section 12.3.4). The excellent durability of ion exchange membranes results from careful handling in this work. Thus, simplicity of structure is a requirement for this work.

12.3 Maintenance Technology

12.3.1 Pretreatment of a Feeding Solution

Pretreatment steps for feeding solutions are classified as follows (Itoi et al., 1978).

1. Ground water

When well water is supplied:
Well water → Sand filter (or Granular activated carbon or Ion exchange column) → ED

2. River water

When river water is purified to produce drinking water:
River water → Coagulation sedimentation → Oxidation → Sand filter → Activated carbon → Chlorination → De-chlorination → ED
When river water dissolves a lot of hardness components:
River water → Lime soda softening → Sand filter → Activated carbon → ED

3. Lake water

Lake water generally dissolves a lot of organic substances.
Lake water → Contact oxidation → Chlorination → Coagulation sedimentation → Sand filter → De-chlorination → ED

4. Reuse of drainage from industrial water

Drainage → Coagulation sedimentation → Sand filter → Active carbon → ED

5. Seawater

Seawater → Sand filter → ED

12.3.2 Turbidity Measurement

12.3.2.1 Optical Method

Turbidity reduces the transparency owing to suspended particles, colloidal material, or both. The principles of the measurement are: (1) turbidimetry, (2) photometric photometry, and (3) integrated photometric photometry. Process liquid is sampled and the actual turbidity is measured visually by lab personnel or with a lot of bench-top–style turbidity meters. Sampling is influenced by other parameters such as time, settling, and temperature. Currently, online and in-line turbidity meters have become automated. Online or in-line measurement achieves continuous measurement in real time and provides a more representative assessment of actual process conditions.

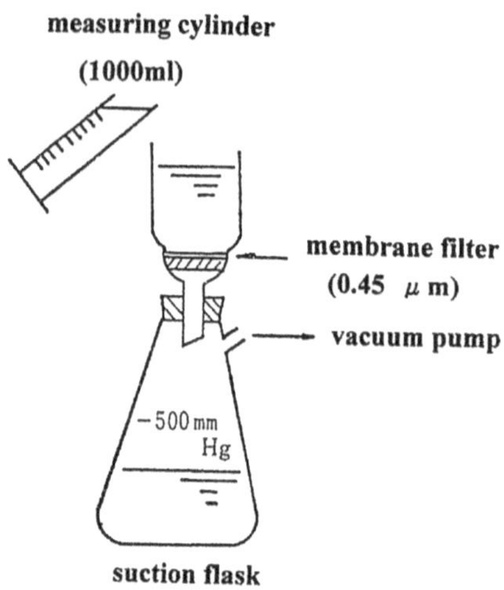

FIGURE 12.7 Membrane filter value measuring method. *Kunisada (1979), p. 185, Fig. 1.*

12.3.2.2 Membrane Filter Method

A sample is filtered through a membrane filter (MF) (for example, HAWP 0.45 μm (47 mmϕ); Millipore Co.). Turbidity is measured from the filtering speed as follows (Kunisada, 1979).

1. MF value

Elapsed time (seconds) is measured to filtrate 1 dm^3 of a sample solution under a decreased pressure (50 cm Hg) with the unit in Fig. 12.7. Data are standardized by viscosity at 25 °C.

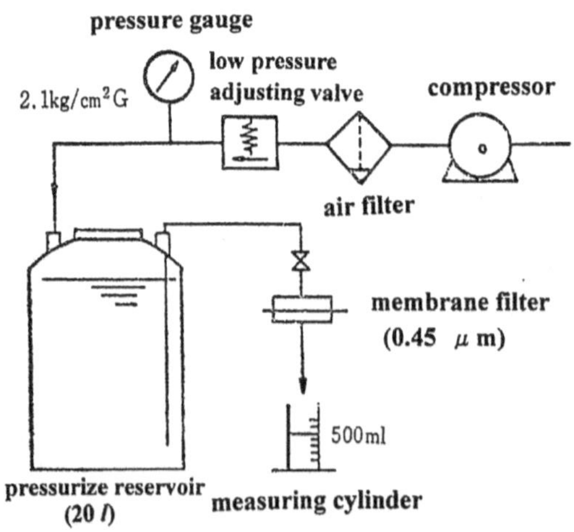

FIGURE 12.8 PI and FI value measuring method. *Kunisada (1979), p. 185, Fig. 2.*

2. PI (Plugging index) and FI (Fouling index) value

A sample is filtered pressurizing to 2.1 kg/cm^{2}G. with the unit in Fig. 12.8, and:

1. Elapsed time t_0 is measured to filtrate the first 500 cm^3 of the solution.
2. After 15 min from the start of the filtration, elapsed time t_{15} is measured to filtrate the next 500 cm^3 of the solution.

PI (Plugging index) and FI (Fouling index) are calculated using the following equations:

$$\mathrm{PI} = (1 - t_0/t_{15}) \times 100 \tag{12.1}$$

$$\mathrm{FI} = \mathrm{PI}/15 \tag{12.2}$$

12.3.3 Scale Trouble Prevention

12.3.3.1 $CaCO_3$ Precipitation, Langelier Saturation Index (LSI), and Acid Dosage

Carbonate equilibrium equations in a solution are given as follows:

$$(\mathrm{CO_2})\mathrm{g} \Leftrightarrow (\mathrm{CO_2})\mathrm{aq} \tag{12.3}$$

$$(\mathrm{CO_2})\mathrm{aq} + \mathrm{H_2O} \Leftrightarrow (\mathrm{H_2CO_3})\mathrm{aq} \tag{12.4}$$

$$(\mathrm{H_2CO_3})\mathrm{aq} \Leftrightarrow (\mathrm{H^+})\mathrm{aq} + (\mathrm{HCO_3^-})\mathrm{aq} \tag{12.5}$$

$$(\mathrm{HCO_3^-})\mathrm{aq} \Leftrightarrow (\mathrm{H^+})\mathrm{aq} + (\mathrm{CO_3^{2-}})\mathrm{aq} \tag{12.6}$$

$$(\mathrm{CO_3^{2-}})\mathrm{aq} + (\mathrm{Ca^{2+}})\mathrm{aq} \Leftrightarrow (\mathrm{CaCO_3})\mathrm{s} \tag{12.7}$$

The equilibrium constant of Eqn (12.6) K_a is:

$$K_a = \frac{\gamma_{\mathrm{H^+}}[\mathrm{H^+}]\gamma_{\mathrm{CO_3^{2-}}}[\mathrm{CO_3^{2-}}]}{\gamma_{\mathrm{HCO_3^-}}[\mathrm{HCO_3^-}]} \tag{12.8}$$

The solubility product K_{sp} is introduced from Eqn (12.7):

$$K_{sp} = \gamma_{\mathrm{Ca^{2+}}}\left[\mathrm{Ca^{2+}}\right]\gamma_{\mathrm{CO_3^{2-}}}[\mathrm{CO_3^{2-}}] \tag{12.9}$$

From Eqns (12.8) and (12.9), pH at saturation in calcium carbonate $\mathrm{pH_S}$ is introduced as:

$$\mathrm{pH}_s = -\log\left(\frac{K_a\gamma_{\mathrm{Ca^{2+}}}[\mathrm{Ca^{2+}}]\gamma\mathrm{HCO_3^-}[\mathrm{HCO_3^-}]}{\gamma_{\mathrm{H^+}}K_{sp}}\right) \tag{12.10}$$

The LSI is:

$$\mathrm{LSI} = \mathrm{pH} - \mathrm{pH}_s = \mathrm{pH} + \log\left(\frac{K_a\gamma_{\mathrm{Ca^+}}[\mathrm{Ca^{2+}}]\gamma_{\mathrm{HCO_3^-}}[\mathrm{HCO_3^-}]}{\gamma_{\mathrm{H^+}}K_{sp}}\right) \tag{12.11}$$

In these equations, the activity coefficient γ is calculated by the following equation:

$$\log(\gamma) = -\frac{0.5(z_i)^2\sqrt{I}}{1+\sqrt{I}} \tag{12.12}$$

where z_i is the charges of ion i and I is the following ionic strength.

$$I = 2.5 \times 10^{-5}\ \text{TDS} \tag{12.13}$$

TDS is the total dissolved solids in mg/l or g/m^3.

If LSI < 0, the water dissolves $CaCO_3$; no potential to scale.

If LSI > 0, $CaCO_3$ precipitation may occur; scale can form.

If LSI $= 0$, borderline scale potential.

From the principle of these equilibrium reactions, it is understandable that $CaCO_3$ precipitation is prevented by acid dosage causing a decease in LSI; i.e., $CaCO_3$ precipitation is prevented by HCO_3^- ion decomposition to evolve CO_2 gas by dosing with an aqueous HCl or H_2SO_4 solution into the process liquid.

$$HCO_3^- + H^+ \rightarrow CO_2\uparrow + H_2O \tag{12.14}$$

12.3.3.2 Precipitation Controlling Agent Dosage

Small amounts of precipitation controlling agents such as condensed sodium phosphate $Na_2[Na_4(PO_3)_6]$ are dosed into the solution, resulting in the absorption of crystalline nuclei to the agents and dissolution of $CaSO_4$ or $CaCO_3$ owing to the following chelate reaction:

$$\begin{gathered} Na_2\left[Na_4(PO_3)_6\right] + CaX \rightarrow Na_2\left[Na_2Ca(PO_3)_6\right] + Na_2X \\ CaX\text{: } CaSO_4 \text{ or } CaCO_3 \end{gathered} \tag{12.15}$$

Carboxyl methyl cellulose or poly-acrylic acid are also available instead of condensed sodium phosphate.

12.3.4 Disassembling and Assembling Works

Despite the filtration, a small quantity of fine particles passes through a filter and invades an electrodialyzer. Sometimes, fine organisms pass through the filter and breed in an electrodialyzer. The fine particles are adhered to the surface of membranes and spacers in desalting cells (Fundamentals Section 11.4.2), causing an increase in the flow resistance of a solution in the desalting cell with acceleration of concentration polarization on the membrane surface (Fundamentals Section 11.4.1). To avoid these problems, an electrodialyzer is usually disassembled and washed periodically. The disassembling and washing process is generally as follows (Tanaka, 1987).

1. Electric current interruption
2. Solution feeding interruption
3. Solution discharge
4. Stack extraction
5. Stack disassembling
6. Membrane surface washing
7. Desalting cell, concentrating cell and spacer washing
8. Stack assembling
9. Leak test

10. Integrating stack into an electrodialyzer
11. Solution feeding
12. Electric current passing

12.4 Practice

12.4.1 Potable Water Production from Brackish Water

Residents on an isolated island experience serious water shortages because of their high dependence on rain water. To solve this problem, Asahi Chemical Co. constructed the ED plant (2500 m^3/day) illustrated in Fig. 12.9 in 1990 on Ohshima Island, Tokyo (Fukuhara et al., 1993). Currently, Ohshima Town supplies potable water to residents. The specifications of this plant are shown in Table 12.1.

Raw brackish water pumped up from wells is fed to the suspended solid filter (SSF), and then it is supplied to the first-stage electrodialyzer unit (EDU-A1 and EDU-B1) through the first-stage desalination tank (DST-1). Part of the filtrated solution is supplied to the concentrated solution tank (CST). The desalted solution in DST-1 is further

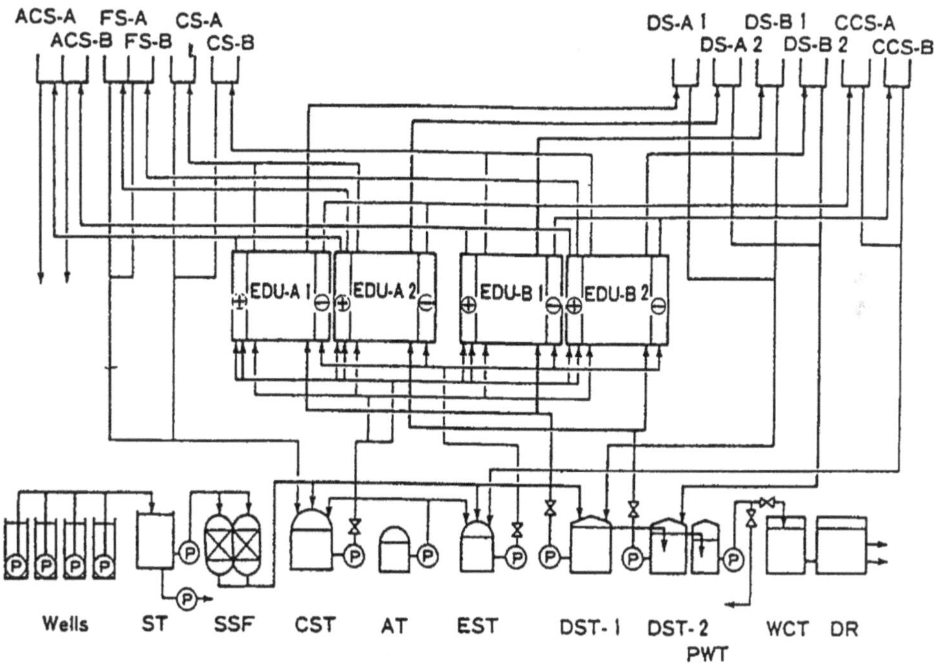

FIGURE 12.9 Saline water desalting process (Ohshima island). ACS, anode compartment solution; AT, acid tank; CCS, cathode compartment solution; CS, concentrated solution; CST, concentrated solution tank; DR, distributing reservoir; DS, desalinated solution; DST, desalination tank; EDU, electrodialyzer unit; EST, electrode solution tank; FS, frame solution; PWT, product water tank; SSF, suspended solid filter; ST, stock tank; WCT, water cleaning tank. *Fukuhara et al. (1993), p. 164, Fig. 7.*

Table 12.1 Specifications of Electrodialysis Plant

Site	Ohshima, Tokyo
Capacity, product water	2500 m^3/day, from 1200 ppm TDS raw water 1000 m^3/day, from 3000 ppm TDS raw water
Raw water source	4 Brackish wells
Product water	TDS: 450 ppm or less Chloride ions: 150 ppm or less
Electrodialyzers	Asahi Chemical SS-0
Conductive area	1780 m^3 (318 pairs/stack, 4 stack)
Thickness	Dilution compartment: 0.5 mm Concentration compartment: 0.5 mm
Mode of operation	Dual-train, two steps per line Automatic, continuous

TDS, total dissolved solids.
Fukuhara et al. (1993), p. 164, Table 1.

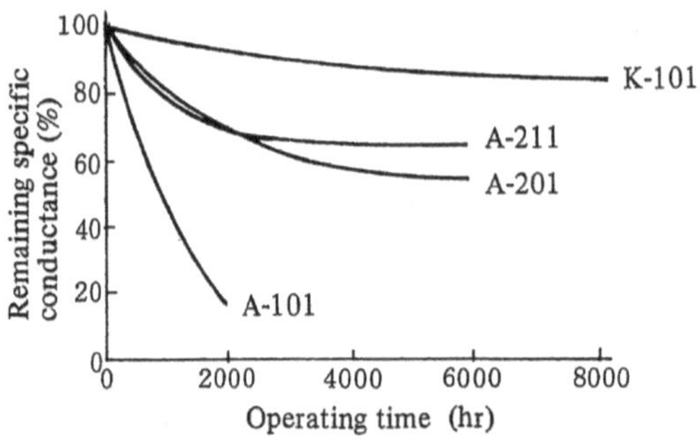

FIGURE 12.10 Membrane conductivity changes with operating time. *Okada et al. (1975), p. 9, Fig. 6.*

Table 12.2 Operational Results

Production rate	2610 m^3/day (from 1200 ppm TDS raw water)
Product water	TDS: 440 ppm Chloride ions: 145 ppm pH: 6.8
Electric consumption	0.8 kWh/m^3

TDS, total dissolved solids.
Fukuhara et al. (1993), p. 166, Table 2.

desalted at the second-stage electrodialyzer unit (EDU-A2 and EDU-B2) at about 400 mg/dm^3, and then is supplied to the water cleaning tank (WCT).

Organic fouling owing to organic acid including in raw brackish water is avoided by integrating anti-fouling anion exchange membrane Aciplex A-201 into the electrodialyzer instead of standard type Aciplex A-101 membrane (cf. Fig. 12.10).

Table 12.3 Analysis of Raw and Product Water

	Raw Water[a]	Product Water
Visual inspection	Colorless, transparent	Colorless, transparent
Turbidity (°)	<1.0	<1.0
Color (°)	<1.0	<1.0
pH	7.2	7.3
Electrical conductivity (μS/cm)	2500	650
Total hardness ($CaCO_3$, ppm)	685	83
Calcium hardness ($CaCO_3$, ppm)	406	33
Evaporation residue (ppm)	1631	418
Silica (SiO_2, ppm)	43	42
Chloride ion (Cl, ppm)	632	131
Sulfuric ion (SO_4, ppm)	262	40
Hydrocarbonate ion (HCO_3, ppm)	210	82
Calcium ion (Ca, ppm)	51	14
Potassium ion (K, ppm)	17	3
Magnesium ion (mg, ppm)	48	12
Sodium ion (Na, ppm)	368	94

[a]From single well.
Fukuhara et al. (1993), p. 166, Table 3.

Table 12.4 Operating Cost

	Units	Projected	Actual
Volume of product water	m^3/day	1950	1853
Raw water TDS	mg/l	1600	1250
Water recovery	%	80.0	86.5
Electric consumption	yen/m	27.82	10.04
Sterling agents	yen/m^3	0.38	3.73
Membrane replacement	yen/m^3	11.80	–
Total	yen/m^3	40.00	13.77

TDS, total dissolved solids.
Fukuhara et al. (1993), p. 167, Table 4.

Operational results are shown in Table 12.2. An analysis of raw and product water is provided in Table 12.3. The values listed in both tables meet the potable water standards established by the Ministry of Welfare, Japan. The actual operating cost is determined based on the plant operation within a year, as indicated in Table 12.4.

12.4.2 ED Desalination System Powered by Photovoltaic Power Generation

Babcock-Hitachi K.K. and Hitachi Ltd. established an ED system combined with sunlight photovoltaic power generation on Oshima Island and at Sakiyama, Nagasaki (Inoue and

Kuroda, 1993). In this system, the electrodialyzer was operated with a direct current power generator using solar batteries, and produced potable water.

Figure 12.11 indicates the system on Oshima Island, showing a two-stage seawater desalting process for producing 10^3/day of potable water. In the first stage, if it is fine, seawater (35,000 mg/l-TDS) is desalted during the daytime to obtain an intermediate desalted solution (5000 mg/l-TDS); at the same time, electric power surplus is stored in batteries. In the second stage, the intermediate desalted solution is further desalted to obtain potable water (400 mg/l-TDS) on a rainy or cloudy day or at night using the stored electric power.

Figure 12.12 shows the system in Sakiyama for producing 200 m^3/day of potable water from brackish water. In this system, electric consumption for pumping up raw brackish water is larger. Thus, on a fine day, raw brackish water (1500 mg/l-TDS) is pumped up and desalted to obtain potable water (400 mg/l-TDS). At the same time, brackish water and electricity are stored on a fine day for operating the electrodialyzer on

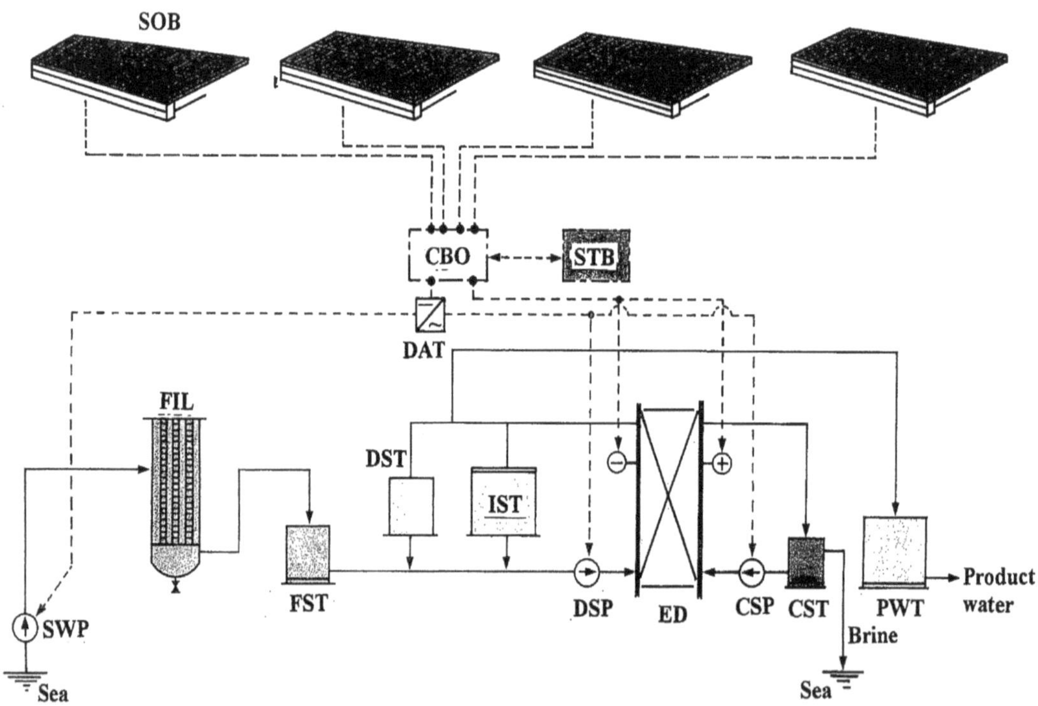

FIGURE 12.11 Seawater desalination process powered by solar generation (Oshima island). CBO, control board; CSP, concentrated solution pump; CST, concentrated solution tank; DAT, direct/altering current transducer; DSP, desalted solution pump; DST, desalted solution tank; ED, electrodialyzer; FIL, filter; FST, filtrated solution tank; IST, intermediately concentrated solution tank; PWT, product water tank; SOB, solar battery; STB, storage battery; SWP, seawater pump. *Inoue and Kuroda (1993), p. 153, Fig. 3.*

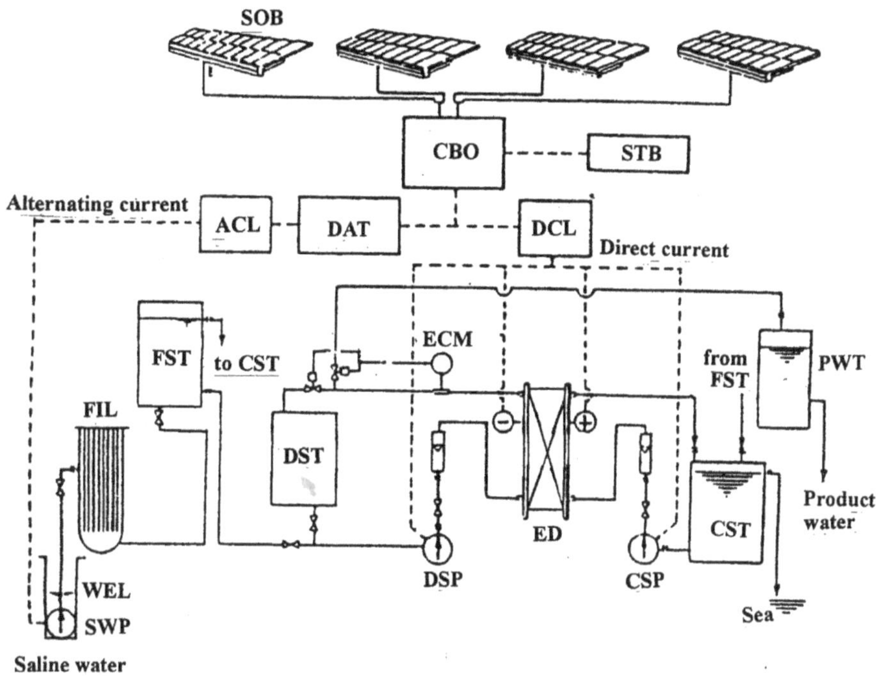

FIGURE 12.12 Saline water desalination process powered by solar generation (Sakiyama). ACL, altering current load; CBO, control board; CSP, concentrated solution pump; CST, concentrated solution tank; DAT, direct/altering current transducer; DCL, direct current load; DSP, desalted solution pump; DST, desalted solution tank; ECM, electro-conductivity meter; ED, electrodialyzer; FIL, filter; FST, filtrated solution tank; PWT, product water tank; SOB, solar battery; STB, storage battery; SWP, seawater pump, WEL, well. *Inoue and Kuroda (1993), p. 153, Fig. 4.*

a rainy or a cloudy day and at night. The specifications of both systems are shown in Table 12.5.

An electric power consumption pattern from the Oshima plant is illustrated in Fig. 12.13. In the batch desalting system, electric power consumption is larger at the beginning and decreases with elapsed time. Thus, when a large amount of electric power is generated on a fine day, seawater is electrodialyzed to obtain an intermediate desalted solution, which is temporarily stored in the tank (high concentration operation). At the same time, electric power surplus is stored in the batteries. On a cloudy or a rainy day and at night, the stored solution is further desalted to obtain potable water (low concentration operation), passing the stored electric power.

Figure 12.14 shows an electric power consumption pattern of the Sakiyama plant, which is automatically controlled by sunlight photovoltaic power generation, a storage battery, and ED. Electric power generated in the solar system is directly consumed in the electrodialyzer and the electric power surplus is charged in the battery.

The operating performance of both plants is indicated in Table 12.6.

Table 12.5 Specifications of the Desalting Processes

	Oshima	Sakiyama
Solar Generator		
Type	Silicon single crystal	Silicon single crystal
Capacity	25 kW	65 kW
Module number	64 W × 390 module	47 W × 1380 module
Storage Battery		
Type	Lead storage battery	Lead storage battery
Capacity	115 kWh	230 kWh
Voltage	96 V	192 V
Electrodialyzer		
Capacity	10 m^3/day	200 m^3/day
Raw water	Seawater (35,000 mg/l-TDS)	Saline water (1500 mg/l-TDS)
Type	Filter-press	Filter-press
Gasket dimension	185 × 2000 mm	370 × 2000 mm
Effective membrane area	0.238 m^2	0.476 m^2
Distance between membranes	0.8 mm	0.8 mm
Number of cell pairs	250	600
System control	Three-mode operation 1. High-concentration operation 2. Low-concentration operation 3. Stand-by operation	Four-mode operation 1. Pump-up + desalting operation 2. Pump-up operation 3. Desalting operation 4. Stand-by operation

TDS, total dissolved solids.
Inoue and Kuroda (1993), p. 153, Table 1.

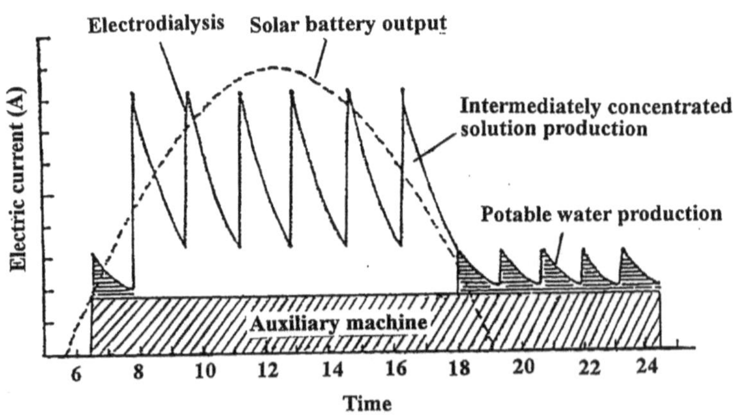

FIGURE 12.13 Electric power consumption pattern of the seawater desalination plant (Oshima island). *Inoue and Kuroda (1993), p. 154, Fig. 5.*

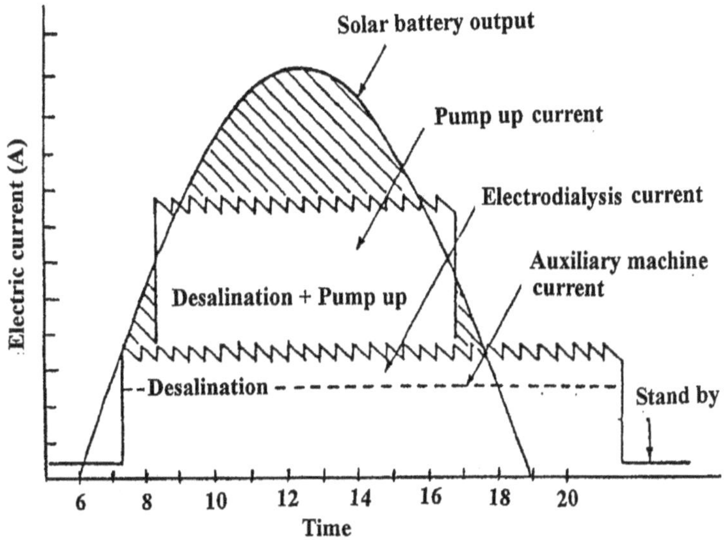

FIGURE 12.14 Electric power consumption pattern of saline water desalination plant (Sakiyama). *Inoue and Kuroda (1993), p. 154, Fig. 6.*

Table 12.6 Operating Performance of Electrodialyzer

	Oshima			Sakiyama	
Year	**1986**	**1987**	**1988**	**1900**	**1901**
Sunlight quantity (kWh/m^2)	3.60	3.73	3.86	4.08	3.60
Generation quantity (kWh/day)	49.3	50.6	57.0	162	145
Water production (m^3/day)	3.45	3.44	4.04	229	194
Electric power consumption (kWh/m^3)	14.3	14.7	14.1	0.71	0.75

Inoue and Kuroda (1993), p. 155, Table 2.

12.4.3 Electrodialytic Recovery of Waste Water from a Metal Surface Treatment Process

ED is applicable to recovering waste water and nickel in the metal surface treatment process developed by Asahi Glass Co., as follows (Itoi et al., 1986).

The nickel plating process includes several rinsing processes. In this process, Ni concentration in the effluent from the first rinsing stage is high, so the effluent is usually returned to the electroplating bath and the effluent from the final rinsing stage is discharged. The ED process is designed to collect Ni ions in the first rinsing stage and return them to the electroplating bath for the purpose of increasing the recovery ratio of Ni and decreasing the Ni content in the waste in the final rinsing stage.

Figure 12.15 is the continuous process operating in a car components factory. The process includes two stages of rinsing baths and one unit of an electrodialyzer that is

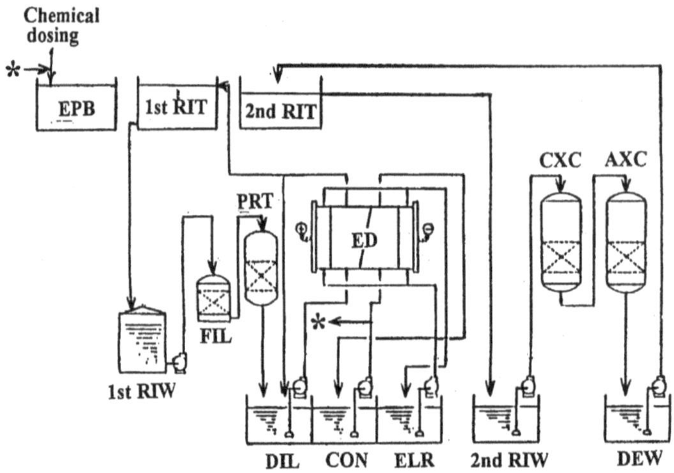

FIGURE 12.15 Electrodialysis Ni^{2+} electroplating waste water recovery process. AXC, anion exchange column; CXC, cation exchange column; CON, concentrate; DEW, demineralized water; DIL, diluate; ED, electrodialyzer; ELR, electrode rinse; EPB, electro plating bath; FIL, filter; PRT, pretreatment; RIT, rinse tank; RIW, rinse waste. *Itoi et al. (1986), p. 207, Fig. 1.*

Table 12.7 Specifications of Electrodialysis Plant

Electrodialyzer	Model DU-111
Ion exchange membrane	Selemion CMV/AMV
Size of membrane	0.49 × 0.98 m
Effective membrane area	0.336 m^2
Number of membrane pairs	40 pairs
Distance between membranes	2.0 mm
Flow velocity	3.0 cm/s
Current density	1.0 A/dm^2
Maximum voltage	50 V

Itoi et al. (1986), p. 208, Table 2.

designed to maintain the Ni concentration in the first-stage rinsing bath to about 5 g/l when the concentration in the Ni plating bath is about 84 g/l. The specifications of the electrodialyzer are presented in Table 12.7. The limiting current density for $NiSO_4$ or $NiCl_2$ is extremely low compared with that for Na_2SO_4 or NaCl. Thus, it is important to control the current density to prevent the precipitation of $Ni(OH)_2$ caused by water dissociation. Furthermore, organic substances in the solution added into the electroplating bath possibly generate the organic fouling of anion exchange membranes (Fundamentals Section 11.5). Therefore, it is necessary to remove the organic substances by means of special pretreatment.

The material balance of Ni^{2+} ions and water in the process is indicated in Fig. 12.16, which shows that the recovery ratio is larger than 90% and a diluted solution is perfectly

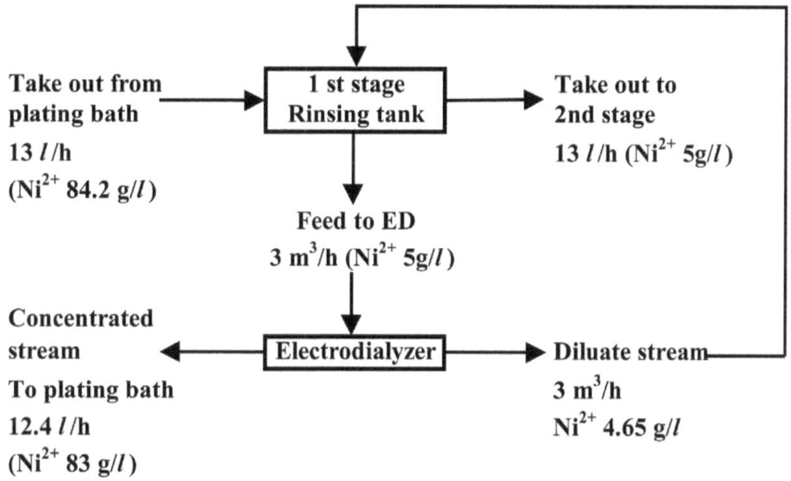

FIGURE 12.16 Ni^{2+} and water balance in the electrodialysis process. *Itoi et al. (1986), p. 209, Fig. 2.*

Table 12.8 Cost Estimation of Electrodialytic Recovery of Rinsing Waste in Nickel Electroplating Process

Nickel recovery rate (as $NiSO_4 \cdot 6H_2O$)	3460 kg/month
Electricity consumption (as $NiSO_4 \cdot 6H_2O$)	0.7 kWh/kg
Equipment installation cost	15 million yen
Electricity unit cost	10 yen/kWh
Purchasing price of nickel salt (as $NiSO_4 \cdot 6H_2O$)	360 yen/kg
Profit of nickel salt recovery	
360 yen/kg × 3460 kg/month =	1,245,600 yen/month
Running cost	
Electricity 10 yen/kWh × 0.7 kWh/kg × 3460 kg/month =	24,220 yen/month
Maintenance and consumable item (annually 3%) 15,000,000 × 0.03/12 month =	37,500 yen/month
Amortization (7 years) 15,000,000 yen/(7 year × 12 month) =	178,570 yen/month
Interest (annually 9%) 15,000,000 yen × 0.09/12 month =	112,500 yen/month
Running cost total	352,790 yen/month
Profit	892,810 yen/month

Itoi et al. (1986), p. 210, Table 5.

recycled to the first-stage rinsing bath. Current efficiency and electric power consumption of ED were, respectively, greater than 90% and 2 kWh/kg Ni ion.

The cost estimation is presented in Table 12.8.

12.4.4 Reuse of Waste Water by Electrodialytic Treatment

Ion exchange membrane ED technology is applicable to waste water treatment to save resources and protect the environment. Tokuyama Co. developed technology to

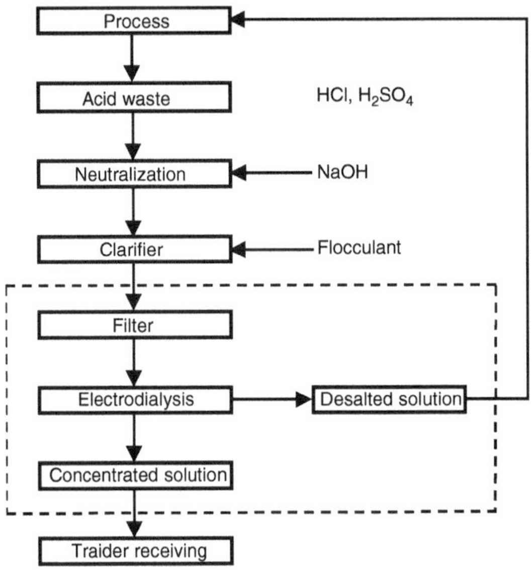

Designed process is surrounded by the dotted line.

FIGURE 12.17 Electrodialytic treatment process of industrial waste. *Matsunaga (1986), p. 192, Fig. 1.*

Solution	Na	Ca	Cl	SO_4
Circulating solution	0.255N	13 ppm	0.168 N	0.087 N
Concentrated solution	3.870	220	3.329	0.541

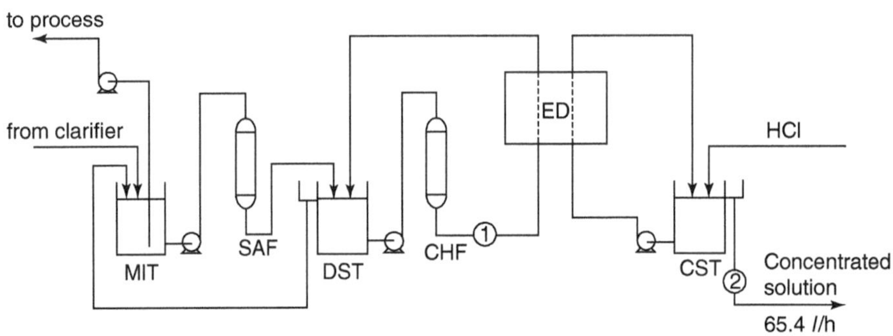

FIGURE 12.18 Electrodialytic treatment process of industrial waste. CHF, check filter; CST, concentrated solution tank; DST, desalted solution tank; ED, electrodialyzer; MIT, middle tank; SAF, sand filter. *Matsunaga (1986), p. 193, Fig. 2.*

establish a closed system by means of the electrodialytic reuse of waste water in the plating process, as shown in Figs 12.17 and 12.18 (Mastunaga, 1986).

1. Typical components of a raw feeding solution in this system:

   pH, 9.5. NaCl, 17.6 g/l. Na_2SO_4, 7.1 g/l. Ca, 48 mg/l. SS, 20 mg/l. Main component: $Fe(OH)_3$. Others: Fe, Cu, Zn, P, etc.

2. Conditions of ED treatment:

Quantity of a desalted solution, 10 ton/month. Degree of desalination, maximum. Degree of concentration, maximum. Temperature, normal. Operating time, 24 h/day, 30 days/month, 12 months/year. Operating system, feed-and-bleed mode (Chapter 14). Ion exchange membrane, Neosepta C5S-8T (monovalent cation selectively permeable cation exchange membrane) and Neosepta ACH-45T (anion exchange membrane).

3. Operating results:

Current density, 3.0 A/dm^2. Cell voltage, 0.5 V/pair (18–20 °C). Electrolyte concentration in a desalted solution, $C_{Na} = 0.25$–0.32 eq/dm^3, $C_{Ca} = 13$–15 mg/l. Electrolyte concentration in a concentrated solution, $C_{Na} = 3.80$–4.02 eq/dm^3, $C_{Ca} = 220$–240 mg/l. Quantity of a concentrated solution, 60–66 l/h (1.44–1.58 m^3/day). Quantity of a desalted solution, 10.9–12.1 ton/month.

4. Running cost:

Electric power, 262 kWh/day × 15 yen/kWh = 3930 yen/day. Labor, 0.2 people/day × 15,000 yen/people = 3000 yen/day. Others (Ion exchange membrane, HCl, electrode, filter), 1330 yen/day. Total, 8260 yen/day.

12.4.5 Electrodialytic Recovery of Acid

Acid is recovered from a waste acid applying an H^+ ion permselective cation exchange membrane placed on an anion exchange layer on a cation exchange membrane, as illustrated in Fig. 12.19. Here, multivalent cations M^{n+} do not pass through the membrane owing to repulsive effects between the cations and the anion exchange layer (cf. Fundamentals Section 2.5.3). This concept is applicable to recover an acid from an aqueous solution dissolving, for instance, $Fe(NO_3)_3$ with HNO_3, as shown in Fig. 12.20. In this system, cation exchange membranes correspond to the membrane illustrated in Fig. 12.19, and they permeate H^+ ions selectively and do not permeate Fe^{3+} ions. On the other hand, anion exchange membranes permeate NO^- ions selectively rather than H^+

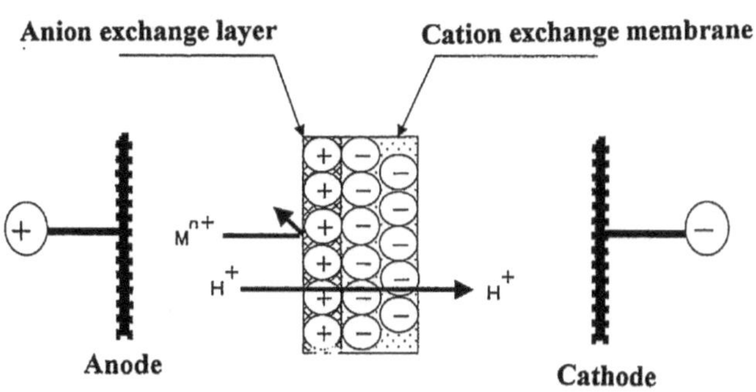

FIGURE 12.19 H^+ ion permselective cation exchange membrane. Mn^+, n valent metallic ion. *Katayama (2004), p. 156, Fig. 6.1.4.*

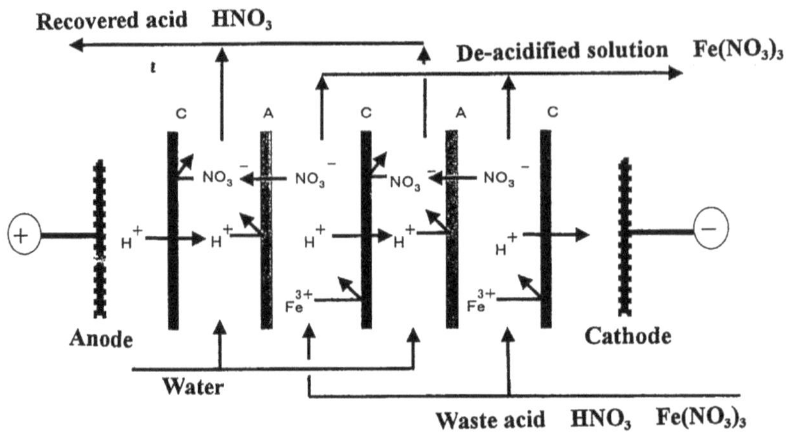

FIGURE 12.20 Acid recovery by means of ion exchange membrane electrodialysis. C, H^+ ion permselective cation exchange membrane; A, H^+ ion low-permselective anion exchange membrane. *Katayama (2004), p. 156, Fig. 6.1.5.*

Table 12.9 Material Balance in Electrodialyzer for Acid Recovery

Solution	Solution Quantity	Composition			
		H^+	Al^{3+}	SO_4^{2-}	Cl^-
	l/h	M	M	M	M
Waste acid (before electrodialysis)	742	0.54	0.30	0.65	0.16
De-acid solution (after electrodialysis)	671	0.07	0.33	0.50	0.08
Water	115	0	0	0	0
Recovered acid	186	1.90	0.01	0.79	0.35

Katayama (2004), p. 158, Table 6.1.2.

ions. Consequently, HNO_3 is recovered in the concentrating chamber and $Fe(NO_3)_3$ remains in the desalting chamber.

Table 12.9 shows the material balance in an electrodialyzer (effective membrane area, 90 m^2) developed by Tokuyama Co., which treated a wasted acid solution (0.54 M acid solution including H_2SO_4 and HCl with 0.30 M of Al^{3+} ions) generated in an etching process of aluminum products (Katayama, 2004). The recovered acid concentration is 1.90 M including 0.01 M of Al^{3+} ions. The acid recovering ratio and leak ratio are, respectively, 88% and 0.6%. Energy consumption and current efficiency are, respectively, 12 kW and 55%.

12.4.6 Seawater Concentration for Salt Production

ED is used to concentrate seawater to produce salt in Japan. The seawater concentrating process is illustrated in Fig. 12.21 (Tomita, 1995). Seawater pumped up from

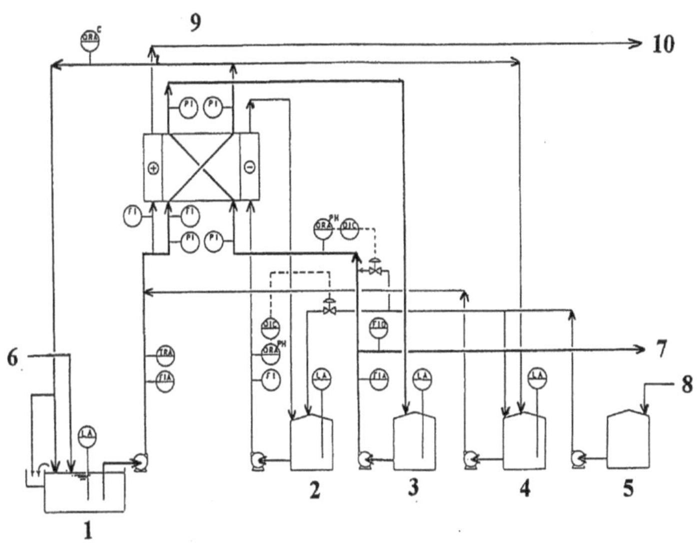

FIGURE 12.21 Electrodialytic seawater concentrating process. 1, Diluted seawater tank; 2, cathode solution tank; 3, concentrated seawater tank; 4, washing solution tank; 5, HCl tank; 6, filtrated seawater; 7, concentrated seawater output; 8, HCl; 9, electrodialyzer; 10, anode solution. *Tomita (1995), p. 93, Fig. 5.4.4.*

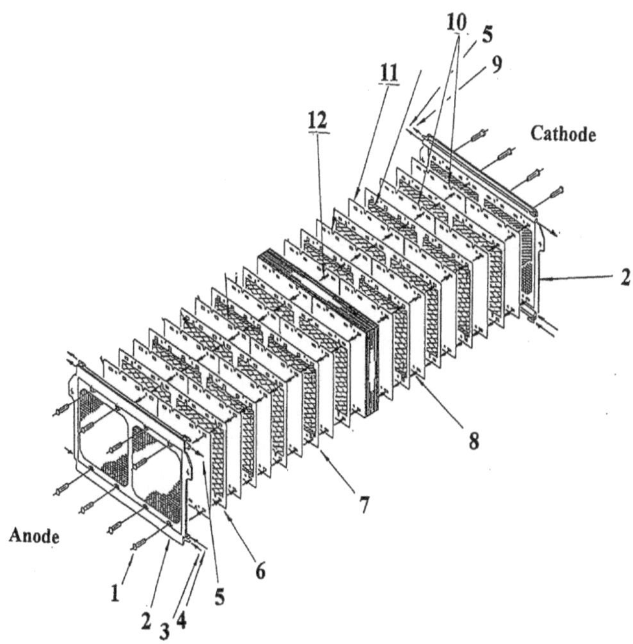

FIGURE 12.22 Composition of an electrodialyzer (Tokuyama). 1, Fastening bolt; 2, fastening and feeding frame; 3, concentrated seawater inlet; 4, diluted seawater inlet; 5, diluted seawater outlet; 6, concentrating cell; 7, desalting cell; 8, desalting cell manifold; 9, concentrated seawater outlet; 10, cation exchange membrane; 11, anion exchange membrane; 12, concentrating cell manifold. *Tomita (1995), p. 89, Fig. 5.4.2(A).*

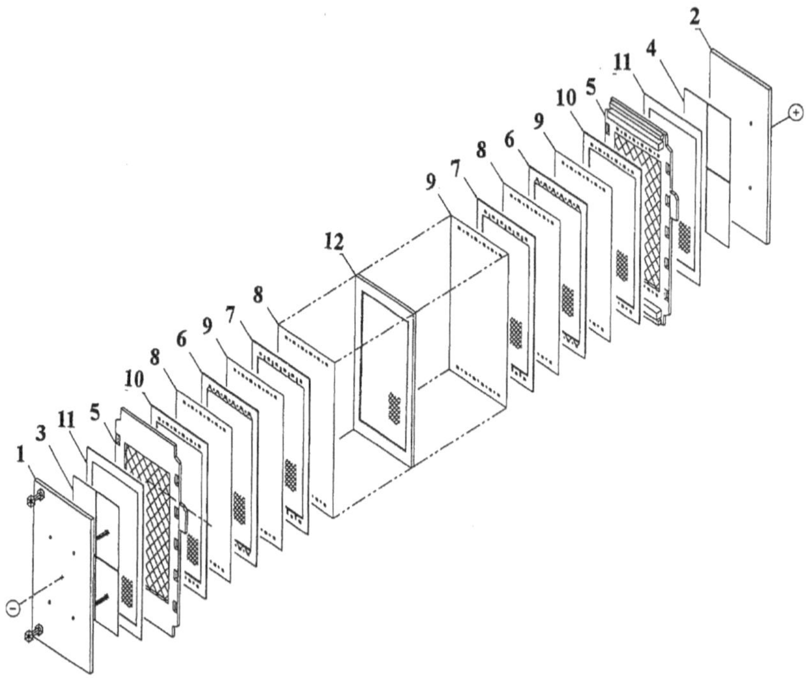

FIGURE 12.23 Composition of an electrodialyzer (Asahi Glass). 1, Cathode chamber; 2, anode chamber; 3, cathode plate; 4, anode plate; 5, intermediate frame; 6, desalting cell; 7, concentrating cell; 8, cation exchange membrane; 9, anion exchange membrane; 10, packing cell frame; 11, intermediate packing; 12, blind cell frame. *Tomita (1995), p. 90, Fig. 5.4.2(B).*

sea is filtered and supplied to the electrodialyzer via the filtered solution tank. Concentrated seawater is circulated between the concentrated seawater tank and the electrodialyzer, and its gain is supplied to an evaporating process to obtain salt crystals. $CaCO_3$ scale precipitation in concentrating cells is prevented by adding hydrochloric acid to concentrated seawater to decompose HCO_3^- and CO_3^{2-} ions (Section 12.3.3). Part of the filtrated seawater is supplied to anode chambers. Titanium is used as the anode material. To avoid membrane damage owing to Cl_2 and HClO generated by an anode reaction, a perfluorinated ion exchange membrane is integrated between the cathode chamber and the adjacent stack. The waste solution from the anode chamber is mixed with filtrated seawater to suppress the growth of microorganisms in seawater (Fundamentals Section 11.4.2). The cathode material is plated with Pt. An HCl solution is supplied to the cathode chamber to neutralize OH^- ions generated by the cathode reaction. A washing system with acid or chemical reagents is provided to dissolve adhered substances in desalting cells (Fundamentals Section 11.4.3).

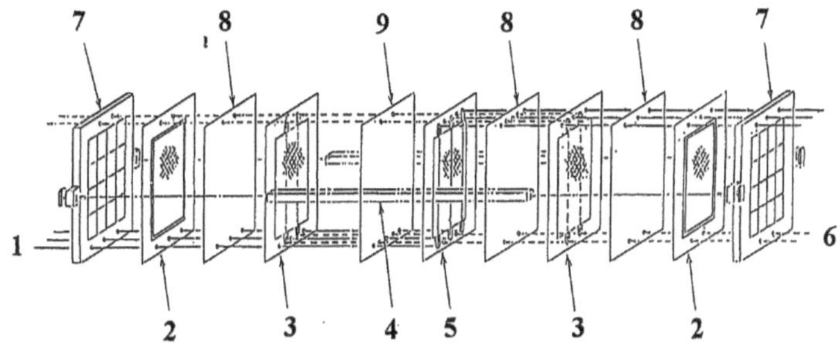

FIGURE 12.24 Composition of an electrodialyzer (Asahi Chemical). 1, Diluted seawater; 2, special gasket; 3, concentrating cell; 4, turnbuckle; 5, desalting cell; 6, concentrated seawater; 7, fastening frame; 8, cation exchange membrane; 9, anion exchange membrane. *Tomita (1995), p. 91, Fig. 5.4.2(C).*

Table 12.10 Performance of Electrodialyzer for Concentrating Seawater (Plant A, 1987)

Current density	A/dm^2	2.66
Temperature	°C	25.3
Cl current efficiency	%	88.6
Na current efficiency	%	80.8
Energy consumption	kWh/t NaCl	178.6
Concentrated solution purity[a]	%	90.90
Constitution of concentrated solution		
NaCl	g/dm^3	190.6
Cl	eq/dm^3	3.568
SO_4	eq/dm^3	0.015
Ca	eq/dm^3	0.064
Mg	eq/dm^3	0.169
K	eq/dm^3	0.095
Na	eq/dm^3	3.256

[a]NaCl (g)/total electrolyte (g).
Tanaka (1991), p. 214, Table 4.10.

When the turbidity of raw seawater is 2 ppm, it is decreased to about 0.05 ppm by filtering through two-stage sand filters. Despite the filtration, fine particles pass through the filter and invade the electrodialyzer and precipitate onto the membrane surfaces. $Fe(OH)_3$ components precipitated on the membrane possibly give rise to water dissociation (Fundamentals Sections 7.7.4, 7.8.6). Sometimes, fine organisms pass through the filter and breed in the electrodialyzer. These problems are avoided by disassembling and washing the electrodialyzer at intervals of 4–6 months (Section 12.3.4). To increase current efficiency and avoid $CaSO_4$ scale precipitation, membrane surfaces are treated to give monovalent ion perm-selectivity (Fundamentals Section 2.5.3).

The compositions of electrodialyzers developed by Tokuyama Co., Asahi Glass Co., and Asahi Chemical Co. are illustrated in Figs 12.22–12.24 (Tomita, 1995). The typical performance of an electrodialyzer is exemplified in Table 12.10 (Tanaka, 1991).

12.4.7 Desalination of Amino Acid and Amino Acidic Seasonings

Amino acid is an amphoteric electrolyte and its behavior is different from that of usual electrolytes. Itoi and Utsunomiya (1965) electrodialyzed the following aqueous solutions of methionine and glycine containing sodium formate: methionine 25 g/l, HCOONa 20 g/l; and glycine 70 g/l, HCOONa 20 g/l. The solution was supplied to the electrodialyzer (membrane pairs, 9; effective membrane area, 209 cm^2) incorporated with Selemion CSG and AST. Changing pH and keeping cell voltage at 15 V (average current density, nearly 1 A/dm^2), amino acid permeation ratio (the ratio of amino acid transported across a membrane pair against that dissolving in a feeding solution), was measured (Fig. 12.25). The amino acid permeation ratio becomes minimal near the isoelectric point of amino acid. Changing current density and keeping pH near the isoelectric point of the amino acid, amino acid permeability (quantity ratio

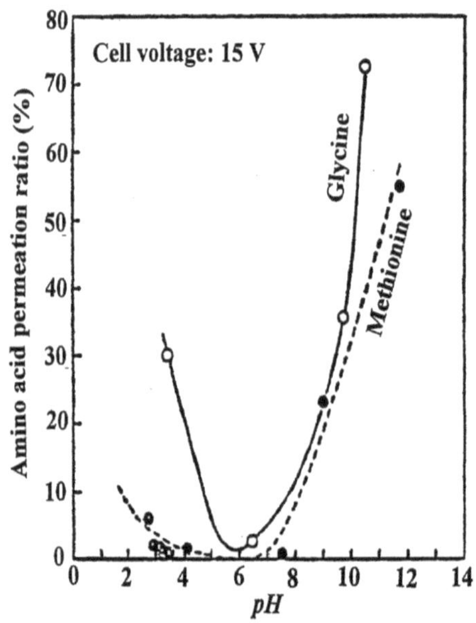

FIGURE 12.25 Relationship between solution pH and amino acid permeation ratio through ion exchange membranes. ○, Glycine 70 g/l, HCOONa 20 g/l; ●, methionine 25 g/l, HCOONa 20 g/l. *Itoi and Utsunomiya (1965), p. 8, Fig. 2.*

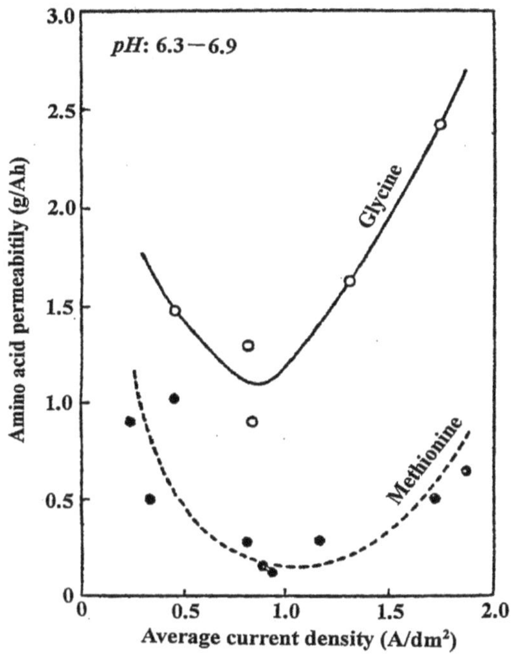

FIGURE 12.26 Relationship between current density and transport rate of amino acid. ○, Glycine 70 g/l, HCOONa 20 g/l; ●, methionine 25 g/l, HCOONa 20 g/l. *Itoi and Utsunomiya (1965), p. 5, Fig. 3.*

of amino acid transported across a membrane pair against electricity), was measured (Fig. 12.26). The amino acid permeability becomes minimal by applying the limiting current density. Concentration changes in NaCl and the essences in the batch system ED of soy sauce are listed in Tables 12.11 and 12.12.

Table 12.11 Electrodialytic Demineralization of Soy Sauce

	Start	End
Solution quantity (l)	11.6	
NaCl concentration (g/l)	191	31
Essence concentration (g/l)	163	191
pH	4.7	4.7
Density	1.181	1.103
Current efficiency (%)		90

Effective membrane area: 209 cm^2; numbers of membranes: 10 pairs; current density: 3.5 A/dm^2; applied voltage: 25–58 V.
Itoi (1983), p. 160, Table 5.4.3.

Table 12.12 Concentration Changes of Components in Electrodialysis (ED) of Soy Sauce

Component	Ratio (Before ED/After ED)
NaCl	0.800
Total amino acid	0.94
Glutamic acid	0.93
Asparatic acid	0.95
Lysine	0.98
Leucine	0.98
Isoleucine	0.98
Alanine	0.96
Phenylalanine	0.87
Valine	0.95
Total nitrogen	0.96
Essence part	1.009

Itoi (1983), p. 161, Table 5.4.4.

12.4.8 Desalination of Natural Essences

In an extraction process of natural essences, several types of salt, acid, and alkali are added, and salt is obtained as a by-product in a final stage. Salt content in the natural essences is expected to be controlled to maintain taste and human health. Tokuyama Co. developed the following desalination processes of natural essences (Ideue, 1986; Yamamoto, 1993).

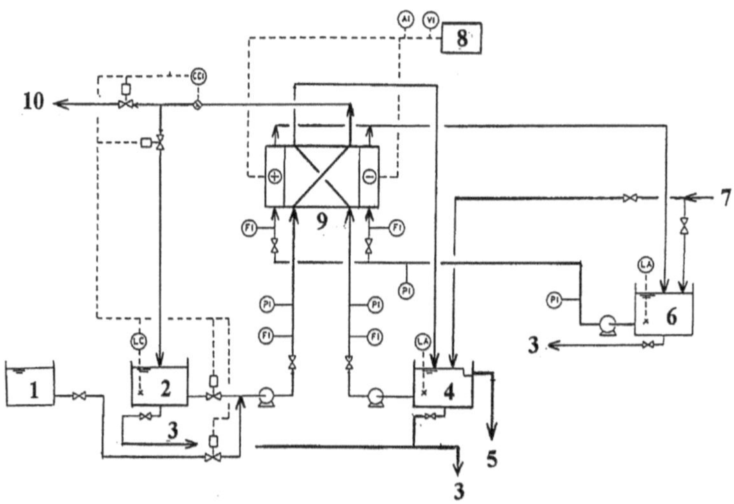

FIGURE 12.27 Electrodialysis process of natural essences. 1, Raw solution tank; 2, desalted solution tank; 3, drain; 4, concentrated solution tank; 5, concentrated waste solution; 6, electrode solution tank; 7, water supply; 8, ammeter; Ai, ampere indicator; CCi, conductivity control indicator; Fi, flow indicator; LA, level alarm; LC, level control; Pi, pressure indicator; Vi, voltage indicator. *Yamamoto (1993), p. 184, Fig. 2.*

Table 12.13 Desalination of Fish Essences

		Raw Essence	Desalted Essence	Desalting Ratio
Conductivity		15.4 mS/cm	2.74 mS/cm	82.2%
Viscosity		4 cp	6.4 cp	
Ash	Na	2975 ppm	560 ppm	81.2%
	K	241	19	92.1
	Ca	38	2.4	93.1
	Mg	45	1.9	95.8
	Cl	3299	583	82.3

Yamamoto (1993), p. 186, Table 6.

An ED system was set up using ion exchange membranes that met food hygiene standards established by the Ministry of Welfare, Japan, and was incorporated with polyvinyl chloride or polypropylene materials suitable for food production. It was further necessary to pay attention to prevent solution stagnation and cultural contamination. Further cleaning in place (CIP) was necessary to operate the apparatus stably. The electrodialyzer was operated in a batch system (Chapter 15) using the ED process indicated in Fig. 12.27, in which a conductivity control indicator was set at an exit of the electrodialyzer to detect the concentration of a desalted solution. The solution feed and solution discharge were automatically controlled by switch valves operating with the conductivity control indicator.

Natural essences include meat essences, seafood essences, flesh essences, and so forth. They were extracted with the aid of an NaCl solution. Salt added in the extraction process was desalinated by means of ED, as mentioned above. Constituent changes in desalination of extracted fish essences are shown in Table 12.13. Specifications of an electrodialyzer for seafood essences are shown in Table 12.14.

Table 12.14 Specifications of Electrodialyzer for Desalinating Seafood Essences

Requirement	
Raw solution	
NaCl	14 g/l
Protein	100 g/l
Specific gravity	1.05
Viscosity	4.0 cp
pH	5.5
Treating amount	0.5 t/h
Specifications	
Effective membrane area	256 m^2
Number of membrane pairs	200 pairs
Ion exchange membrane	Neosepta CM-1, AM-1
Model	TS-25-200
Operating system	Automatic batch system
Running cost	1517 yen/t

Yamamoto (1993), p. 187, Table 8.

12.4.9 ED of Milk and Whey

12.4.9.1 Composition of Milk and Whey (Ideue, 1986; Tomita et al., 1986)

Whey is a by-product of cheese production. The output of whey amounts to nine times of that of cheese, so it is regarded as important in the dairy industry to use whey effectively. The effective use of permeates derived from an ultra-filtration process of milk and whey also needs to be determined. The composition of milk, whey, and ultra-filtration permeates is listed in Table 12.15. Ash content in the dry matter of the whey and permeate is so high that it is expected to be reduced by ED.

Powdered milk for babies is prepared from cow milk. The constituents of breast milk and cow milk are not the same, as shown in Table 12.16. Total ash and casein compositions of cow milk are, respectively, 3.4 and 4.4 times those of breast milk. Thus, powdered milk is prepared by adding whey to cow milk. Before this mixing operation, it is necessary to extract excessive ash from the milk or whey by ED.

12.4.9.2 Limiting Current Density

In the ED of a dairy product solution, protein is condensed onto the desalting surface of an anion exchange membrane at above limiting current density owing to pH changes caused by the water dissociation reaction (Fundamentals Section 11.5). Insoluble salts such as calcium phosphate, etc., precipitate onto the concentrating surface of the anion exchange membrane. Thus, it is necessary to increase the limiting current density to operate a practical electrodialyzer stably. Nagasawa et al. (1973, 1974) measured the limiting current density in ED of a skim milk solution as follows.

The limiting current density i_{lim} was measured from the relationship between current density and voltage (Fundamentals Section 6.2). The relationship between the conductivity of the skim milk solution κ and i_{lim} was expressed as $i_{\text{lim}} = 1.08 \times 10^{-3}\ \kappa$. $(i/\kappa)_{\text{lim}}$ was proportional to the 0.6th power of the solution velocity of skim milk V. $(i/\kappa)_{\text{lim}}/V^{0.6}$

Table 12.15 Typical Composition of Milk and UF-Permeate

	Fat (%)	Protein (%)	Lactose (%)	Ash (%)	Water (%)	Ash/Dry Matter (%)
Cow milk	3.3	2.9	4.5	0.7	88.6	6.1
Cheese whey	0.3	0.7	4.5	0.6	93.7	9.5
Permeate derived from UF of milk	0	0.2	4.4	0.5	94.8	9.6

Tomita et al. (1986), p. 147, Table 1.

Table 12.16 Comparison of Composition between Breast Milk and Cow Milk

	Water (%)	Ash (%)	Whey Protein (%)	Casein (%)	Fat (%)	Lactose (%)
Breast milk	88.0	0.2	0.68	0.42	3.5	7.2
Cow milk	88.6	0.7	0.69	2.21	3.3	4.5

Tomita et al. (1986), p. 148, Table 2.

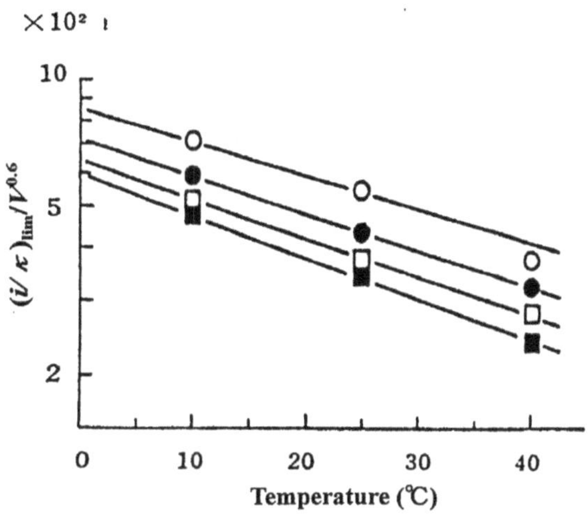

FIGURE 12.28 Effects of temperature and total solid content on $(i/\kappa)_{lim}/V^{0.6}$. S%: ○, 0.3; ●, 8.3; □, 16; ■, 25. *Nagasawa et al. (1974), p. 581, Fig. 4.*

versus temperature T and dry matter content of skim milk S was given in Fig. 12.28. i_{lim} was not influenced by the flow length and distance between the membranes. Thus, the limiting current density equation was introduced as:

$$i_{lim} = Q \times S^{-1/6} \times \left(1.01665S^{0.1}\right)^{-T} \times V^{0.6}\,\kappa \tag{12.16}$$

where Q is a constant. S, T, and V can be voluntarily fixed, so this equation is simplified as:

$$i_{lim} = R\kappa \tag{12.17}$$

in which R is a constant. i_{lim} is known from κ. The electrodialyzer was operated stably at under i_{lim} estimated from Eqn (12.17).

12.4.9.3 Ion Exchange Membranes for the Demineralization of Milk and Whey

In the demineralization of skim milk or whey including rich proteinacious materials or other organic matters, conventional membranes are not applicable because the performance of the membranes is easily deteriorated. In order to demineralize milk or whey, the following membranes were prepared (Okada et al., 1975).

ANTI-ORGANIC FOULING

The specific conductance of various membranes κ was measured to demineralize Gouda cheese whey. The membranes are, Aciplex K-101 (conventional cation exchange membrane, Asahi Chemical Co.), A-101 (conventional anion exchange membrane) and A-201 and A-211 (both are newly developed anion exchange membranes). κ of K-101 is not changed largely with running time as shown in Fig. 12.10. κ of A-101 shows very rapid

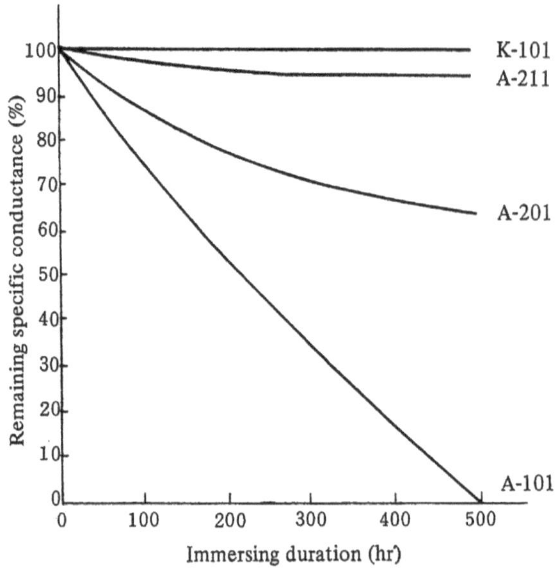

FIGURE 12.29 Membrane conductivity changes with time in alkaline circumstances. 1% NaOH at 37 °C. *Okada et al. (1975), p. 9, Fig. 7.*

decrease. However, κ of A-201 and A-211 membranes does not change significantly for 6000 h. A-201 and A-211 membranes show excellent anti-fouling.

ANTI-ALKALINE CIRCUMSTANCE

Organic fouling owing to deposition of proteinacious materials or fatty substances from the dairy products is cleaned by washing in an alkaline detergent by means of the CIP method. Exposure of the membrane to OH^- ions generated from water dissociation vitiates its performance. Therefore, the lifespan of the membrane is shortened by contacting the membrane with a basic solution. Figure 12.29 shows the durability evaluated by measuring the specific conductance κ of membranes immersed in a 1% NaOH solution. The κ of conventional cation exchange membrane K-101 does not change; however, that of anion exchange membrane A-101 decreases remarkably with the duration of immersion, while the decrease in κ is suppressed for anti-organic fouling anion exchange membranes A-201 and A-211. In particular, the A-211 membrane shows excellent anti-alkaline performance.

PERMEABILITY TO LARGER ORGANIC IONS OF AN ANION EXCHANGE MEMBRANE

An anion exchange membrane that does not permeate larger organic anions causes lowered pH in the demineralization of dairy products. This is because OH^- ions caused by the water dissociation reaction in an anion exchange membrane transfer toward a concentrating cell instead of the larger organic anions. This phenomenon induces an

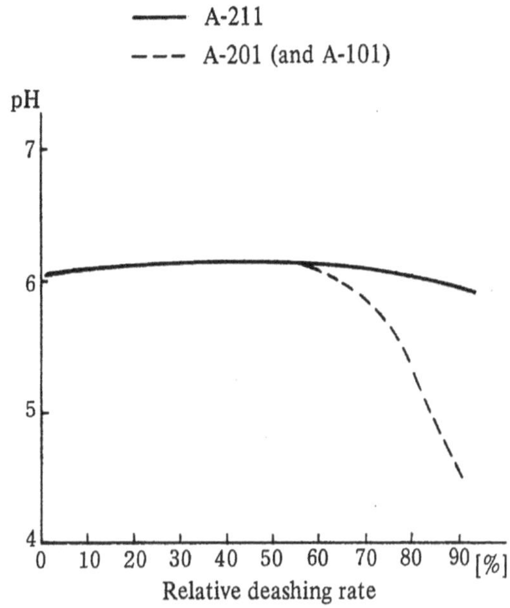

FIGURE 12.30 pH changes in whey in demineralization. *Okada et al. (1975), p. 10, Fig. 8.*

increase in pH in the concentrating cell and the precipitation of inorganic salts such as $Ca_3(PO_4)_2$, $CaCO_3$, and $CaSO_4$. The pH in the desalting cell is lowered because H^+ ions transfer toward the desalting cell. Figure 12.30 shows pH changes upon 90% ash reduction of Gouda cheese whey for three types of anion exchange membranes. The A-211 membrane scarcely causes pH decrease. This is because the pore radius in the A-211 membrane is large enough that organic acids such as citric acid, lactic acid, amino acid, and phosphoric acid easily permeate the membrane and water dissociation does not easily occur.

This experiment shows that A-211 is the most suitable anion exchange membrane. The CIP operations of an electrodialyzer incorporated with A-211 membranes achieve 90% ash reduction and extension of lifespan.

12.4.9.4 ED System for Demineralization of Milk or Whey

Figure 12.31 illustrates the four-stack ED process MED SV1/2 4-4 developed by Morinaga Milk Industry Co. (Okada et al., 1975). The specifications and operating conditions of this system are listed in Tables 12.17 and 12.18, respectively. In Fig. 12.31, the pump P_1 feeds the whey to balance tank WB-1. Next, the whey is supplied to the first stack of the ED system, demineralized to some extent, and overflows from WB-1 to WB-2. The whey in WB-2 circulates through pump P_3 and the second stack, a part of which then overflows into WB-3. Similarly, the whey flows via P_4, third stack, WB-4, fourth stack, WB-5, and finally demineralized whey flows

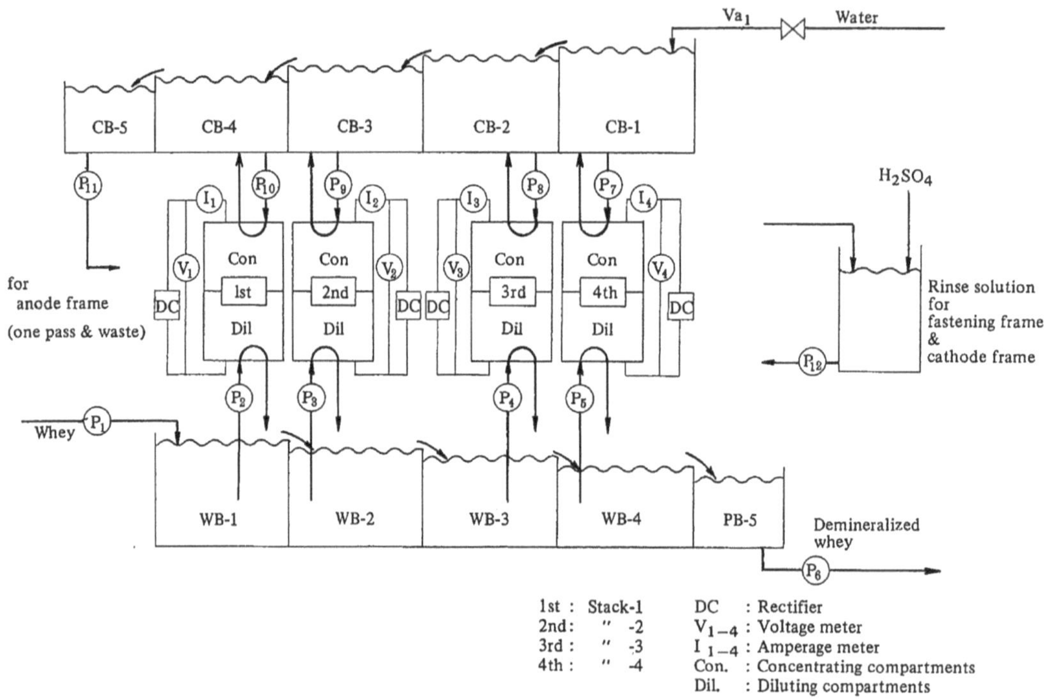

FIGURE 12.31 Flow diagram of Morinaga continuous electrodialysis process. *Okada et al. (1975), p. 14, Fig. 14.*

Table 12.17 Specifications of MED SV1/2 4-4

1. Center-press	1
2. Stack	4
3. Electrode	4 pairs
4. Cell pair	150 cell pairs/stack
5. Distance between membranes	0.75 mm
6. Effective membrane area	50 dm^2/cell
7. Spacer	Sheet-flow type
8. Number of channels	1
9. Width of channel	500 mm
10. Length of channel	1000 mm

Okada et al. (1975), p. 17, Table 3.

out via P_6 and is supplied to the next process such as pasteurization, evaporation, and drying. The conductivity of the whey is monitored in each balance tank to control the desalting rate. The feed rate of the whey to this system is automatically controlled by means of measuring conductivity in the effluent of the demineralizing stream.

Table 12.18 Operating Conditions of Demineralization of Whey (Morinaga ED System)

1. Quality of whey to be treated	
1.1 Total solids	20%
1.2 Ash content	1.60%
1.3 Specific conductance	013 S/cm
1.4 Sediment test (200 ml, 1000 G)	Less than 0.1 ml
2. Operating conditions	
2.1 Level of applied-current density	$3000 \times \kappa^{a}$ (mA/cm^2)
2.2 Linear velocity in the compartment	12 cm/s
2.3 Operating temperature	20 °C

[a]κ, Specific conductance.
Okada et al. (1975), p. 17, Table 4.

12.4.10 Desalination of Sugar Liquor

Sugar is produced as the cane sugar, using a fine sugar manufacturing system; and as a beet sugar manufacturing system. The raw material for the former is sugar cane; that for the latter is beet sugar. In the sugar manufacturing process, the greater part of organic non-sugar components in raw sugar is removed by means of defecation, carbonation, adsorption (bone char, active carbon, ion exchanger, etc.). In these treatments, inorganic components are not removed and are transferred to an evaporation process in which residual organic non-sugar components are separated from sugar crystals and remain in molasses. In the course of repeated evaporation and separation, inorganic and organic non-sugar components are gradually accumulated in the molasses and are finally discharged to the outside of the system as waste molasses. The waste molasses includes considerable amount of sugar components, so it is used as raw material for fermentation or animal food, however, its economical value is extremely low compared with that of sugar itself. Because of this background, technology was expected to be developed to prevent the sugar component transfer to waste molasses and increase the sugar recovering ratio. Further desalting technology employing ion exchange membrane ED has become attractive because the sugar recovering ratio is influenced by residual inorganic components in syrup.

The application of ED in the sugar manufacturing industry has been investigated since the latter half of the 1950s. However, it is difficult to put this program into practice because water dissociation and organic fouling are apt to occur in anion exchange membranes. To avoid these problem, Taito Co. and Asahi Chemical Co. developed technology employing neutral membranes consisting of polyvinyl alcohol instead of anion exchange membranes (Sugiyama et al., 1982; Kokubu et al., 1983). Advantages and disadvantages of this method (Transport depletion method) are as follows.

Advantages:

1. Neutral membranes do not deteriorate owing to organic fouling.
2. Sugar does not decompose because a pH decrease resulting from water dissociation does not occur in the neutral membrane.
3. Current density can be increased because water dissociation does not occur.

Disadvantages:

1. The removing efficiency of anions is low.
2. The current efficiency η is low because of the transport number of anions of a neutral membrane $\bar{t}_A = 0.5$. If one assumes the transport number of a cation exchange membrane $\bar{t}_K = 1.0$, $\eta = \bar{t}_K + \bar{t}_A - 1 = 0.5$.

In the first stage in a sugar manufacturing process, raw molasses (original syrup) is treated to remove organic materials, and evaporated to produce A sugar and A molasses. In the next stage, B sugar and B molasses are produced from A molasses through a similar process. Finally, C sugar and C molasses are produced from B molasses. Table 12.19 shows the ED experiment of B molasses, which is centrifuged two times after

Table 12.19 Effect of Electrodialysis on Pretreatment for Molasses

			I			II		I−II
Desalting ratio (%)			66.18			63.60		2.58
Current efficiency (%)			43.55			34.01		9.54
Current density (A/dm^2)			3.04			3.07		−0.03
		Start	**End**	**Difference**	**Start**	**End**	**Difference**	
Molasses volume (l)		10.00	9.76	△ 0.24	10.00	9.76	△ 0.24	0.00
Molasses concentration (°Bx)		51.35	46.85	△ 4.50	50.45	46.45	△ 4.00	△ 0.50
Molasses purity (%)		51.70	59.01	7.31	52.66	58.45	5.79	1.52
pH		6.35	6.35	0.00	6.40	6.40	0.00	0.00
			I			II		
		Start	**End**	**Desalting Ratio**	**Start**	**End**	**Desalting Ratio**	**I − II**
Ash (% on solid)	CaO	0.19	0.06	68.42	0.22	0.13	40.91	27.51
	MgO	0.74	0.33	55.41	0.72	0.42	41.67	13.74
	K_2O	5.47	1.59	70.93	5.98	2.01	66.39	4.54
	Cl	3.99	0.15	96.24	3.65	0.19	94.79	1.45
	SiO_2	0.41	0.33	19.51	0.44	0.41	6.82	12.69
	SO_3	0.69	0.41	40.58	1.22	1.00	18.03	22.55
	P_2O_5	0.18	0.19	△ 5.56	0.21	0.18	14.29	△ 19.84
	CO_2	0.19	0.14	26.32				
	Sulfate ash	12.42	4.73	61.92	12.78	5.19	59.39	2.53

I, Double-stage centrifuging after $CaCl_2$ added; II, Double-stage centrifuging without $CaCl_2$ added.
Kokubu et al. (1983), p. 29, Table 1.

adding $CaCl_2$ (case I) and without $CaCl_2$ (case II). The experiment indicates that the desalting ratio and current efficiency in case I are increased comparing with those in case II. This is because minerals such as CaO, SiO_2, SO_3, and P_2O_5 are removed by $CaCl_2$ treatment in case I.

Based on the experiment described above, Taito Co. designed an ED treatment process of B molasses as shown in Fig. 12.32. In this process, B molasses is treated first by defecation mixing with $Ca(OH)_2$ and $CaCl_2$. After heating and two stages of centrifuging, the second centrifuging molasses is supplied to the electrodialyzer via the refrigerator to obtain desalted molasses (C molasses). Then, C sugar is crystallized in the evaporation of C molasses. By integrating the ED step in the sugar manufacturing process, it has become possible to obtain D sugar from C molasses.

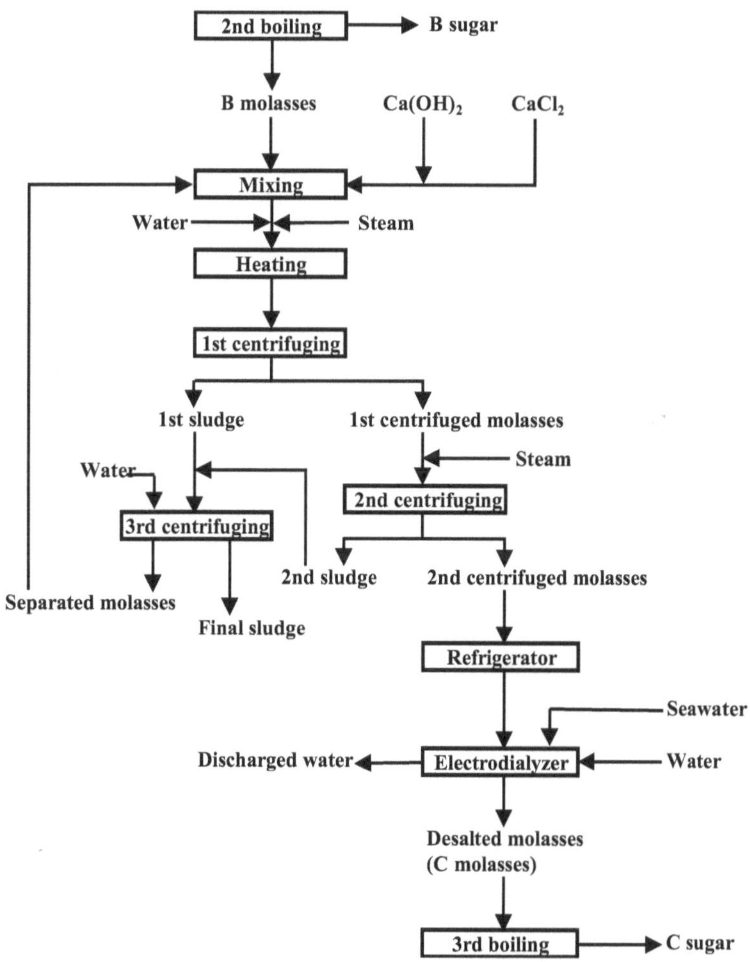

FIGURE 12.32 Desalination process of B molasses. *Kokubu et al. (1983), p. 30, Fig. 3.*

References

Azechi, S., 1980. Electrodialyzer. Bull. Soc. Sea Water Sci. Jpn. 34, 77–83.

Fukuhara, K., Hamada, M., Azuma, I., 1993. Efficient desalination of brackish water by electrodialysis. In: Industrial Application of Ion Exchange Membranes. Research Group of Electrodialysis and Membrane Separation Technology, Soc. Sea Water Sci., Jpn, vol. 2, pp. 159–167.

Ideue, K., 1986. Desalination with ion exchange membranes in food industry. Food Dev. 21, 54–59.

Inoue, S., Kuroda, O., 1993. Electrodialysis desalination system powered by photovoltaic power generation. In: Industrial Application of Ion Exchange Membranes. Research Group of Electrodialysis and Membrane Separation Technology, Soc. Sea Water Sci., Jpn, vol. 2, pp. 151–157.

Itoi, S., 1983. Electrodialytic demineralization of soy sauce and amino acid seasonings. In: Food Industry and Membrane Utilization. Saiwai Shobo Inc., Tokyo.

Itoi, S., Komori, R., Terada, Y., Hazama, Y., 1978. Basis of electrodialyzer design and cost estimation. Ind. Water 239, 29–40.

Itoi, S., Nakamura, K., Kawahara, T., 1986. Electrodialytic recovery of waste water from metal surface treatment process. In: Industrial Application of Ion Exchange Membranes. Research Group of Electrodialysis and Membrane Separation Technology, Soc. Sea Water Sci., Jpn, vol. 1, pp. 206–211.

Itoi, S., Utsunomiya, T., 1965. Electrodialysis of aqueous solution of amino acid containing electrolyte by ion exchange membrane. Asahi Glass Res. Rep. 15, 171–178.

Katayama, S., 2004. Waste water treatment and recovering of useful substances. In: Seno, M., Tanioka, A., Itoi, S., Yamauchi, A., Yoshida, S. (Eds.), Functions and Applications of Ion Exchange Membranes. Industrial Publishing & Consulting Inc., Tokyo, pp. 151–169.

Kokubu, T., Yamauchi, T., Miyagi, S., 1983. Application of electrodialysis in sugar industry. Food Ind. 9, 26–31.

Kunisada, Y., 1979. Measurement technology of turbidity in sea water. Bull. Soc. Sea Water Sci. Jpn. 33, 184–189.

Matsunaga, Y., 1986. Reuse of waste water by electrodialytic treatment. In: Industrial Application of Ion Exchange Membranes. Research Group of Electrodialysis and Membrane Separation Technology, Soc. Sea Water Sci., Jpn, vol. 1, pp. 188–196.

Nagasawa, T., Okonogi, S., Tomota, M., Tamura, Y., Mizota, T., 1973. Demineralization of skim milk by means of electrodialysis with ion permselective membrane, I. Relationship between limiting current density and specific conductivity of demineralizing solution. Jpn. J. Zootech. Sci. 44, 426–431.

Nagasawa, T., Okonogi, S., Tomita, M., Tamura, Y., Mizota, T., 1974. Demineralization of skim milk by means of electrodialysis with ion selective membrane, II. The effects of linear velocity, temperature and total solid content etc. on the limiting current density. Jpn. J. Zootech. Sci. 45, 578–584.

Okada, K., Tomita, M., Tamura, Y., 1975. Electrodialysis in the treatment of dairy products. In: Symposium "Separation Processes by Membranes, Ion-exchange and Freeze-Concentration in Food Industry," Paris, March 13–14.

Sugiyama, M., Takatori, Y., Touyama, R., Nakamura, A., Yamauchi, T., Kawate, H., Yamaguchi, A., 1982. On the desalination process of sugar solutions by electrodialysis using neutral membranes. J. Sugar Refin. Technol. 30, 26–32.

Tanaka, Y., 1987. Concentration of seawater using electrodialysis. In: Kawasaki, J., Kunime, T., Sakai, K., Hakuta, T. (Eds.), Membrane Separation Technology Hand Book. Science Forum, Tokyo, pp. 211–215.

Tanaka, Y., 1991. Membrane separation. In: Seno, M., Abe, M., Suzuki, T. (Eds.), Ion Exchange. Kodansya Scientific Co., Tokyo, pp. 211–227.

Tomita, A., 1995. Electrodialyzer. In: Ogata, N. (Ed.), Engineering in Salt Manufacturing, Electrodialysis, vol. 2. Japan Salt Industry Association, pp. 85–101.

Tomita, M., Tamura, Y., Mizota, T., 1986. Electrodialysis of milk and whey. In: Industrial Application of Ion Exchange Membranes. Research Group of Electrodialysis and Membrane Separation Technology, Soc. Sea Water Sci., Jpn, vol. 1, pp. 147–158.

Urabe, S., Doi, K., 1978. Electrodialyzer. Ind. Water 239, 24–28.

Yamamoto, Y., 1993. Desalination of natural essences by electrodialysis. In: Industrial Application of Ion Exchange Membranes. Research Group of Electrodialysis and Membrane Separation Technology, Soc. Sea Water Sci., Jpn, vol. 2, pp. 181–188.

13

Computer Simulation Program: Single-Pass (Continuous) Program

13.1 Introduction

An electrodialysis (ED) processes is applied widely for saline water desalination. The computer simulation program described in this chapter is developed for discussing the performance of an electrodialyzer for saline water desalination. The ED processes are classified to a single-pass (continuous) process, a feed-and-bleed process, and a batch process. Among the programs of these processes, the single-pass program is the most fundamental one, and the performance of the process is discussed from various points as follows.

Belfort and Daly (1970) constructed optimization routine for a single-pass ED plant. The algorithm was applied to the Office of Saline Water test bed plant at Webster, South Dakota and compared the actual cost and operating conditions. Avriel and Zelighter (1972) developed a mathematical model for preliminary engineering design and economical evaluation of a single-pass electrodialysis plant. Detailed cost computations were performed, resulting in capital investment and annual operating costs. Lee et al. (2002) developed a computer simulation program for describing a single-pass process and estimated investment and operation costs. Furthermore, ED plant was designed and optimized in terms of overall costs and the different parameters. Moon et al. (2004) investigated ionic transport across membranes based on the one- and two-dimensional single-pass ED modeling by using the principles of electrochemistry, transport phenomena, and thermodynamics. Fidaleo and Moresi (2005) simulated mass transfer, mass balance, potential drop, and limiting current density in a single-pass operation based on the Nernst–Planck equation. Sadrzadeh et al. (2007) modeled continuous flow desalination starting from a differential equation of steady-state mass balance and gives salt concentration in dilute compartments or separation percent for various voltages, flow rates, and feed concentrations. Nikonenko et al. (2008) described ED or electrodeionization characteristics (mass transfer coefficient, Sherwood number, degree of desalination, and others). The reasonability of the functions was discussed with experimental measurements. Brauns et al. (2009) developed a simulation model through solver software. Experimental verification of the software was performed using an industrial-type pilot plant. Limiting current density was theoretically evaluated in the model calculations for design purposes and corrected with the experimental results.

Ion Exchange Membranes. http://dx.doi.org/10.1016/B978-0-444-63319-4.00013-4

The single-pass program becomes the basis of the feed-and-bleed program (Chapter 14) and the batch program (Chapter 15). This is because the single-pass program describes the function of the electrodialyzer. This chapter describes the single-pass program established based on the principles of electrochemistry and ED experiments being supplied strong electrolyte solutions (Tanaka, 2012a). Calculation is carried out in the spreadsheet with a use of common software (Excel) and ordinary hardware (computer). The program is integrated in websites. So, readers can operate the program in the websites (Companion Site; Companion sites 1, 2 and 3) by inputting the source code, i.e., optional process specifications and operating conditions. The program aims to function as a pilot plant operation.

13.2 Single-Pass (Continuous) ED Process

The one-stage single-pass ED process is illustrated in Fig. 13.1. The electrodialyzer is incorporated with desalting cells, concentrating cells in the stack marked with gray. The number of cells is N for desalting cells and $N+1$ for concentrating cells. The anode and cathode cells are placed at both outsides of the stack, and an electric current is supplied between the electrodes. A feeding solution of salt concentration C'_{in} is supplied to desalting cells at linear velocity of u'_{in} at the inlets. C'_{in} and u'_{in} are decreased to C'_{out} and u'_{out} at the outlets of desalting cells, respectively. A part of the feeding solution is also supplied to concentrating cells for preventing scale formation in the concentrating cells. The salt concentration and linear velocity in concentrating cells are respectively $C''_{in} = C'_{in}$ and u''_{in} at the inlets and C''_{out} and u''_{out} at the outlets. Partition cells are

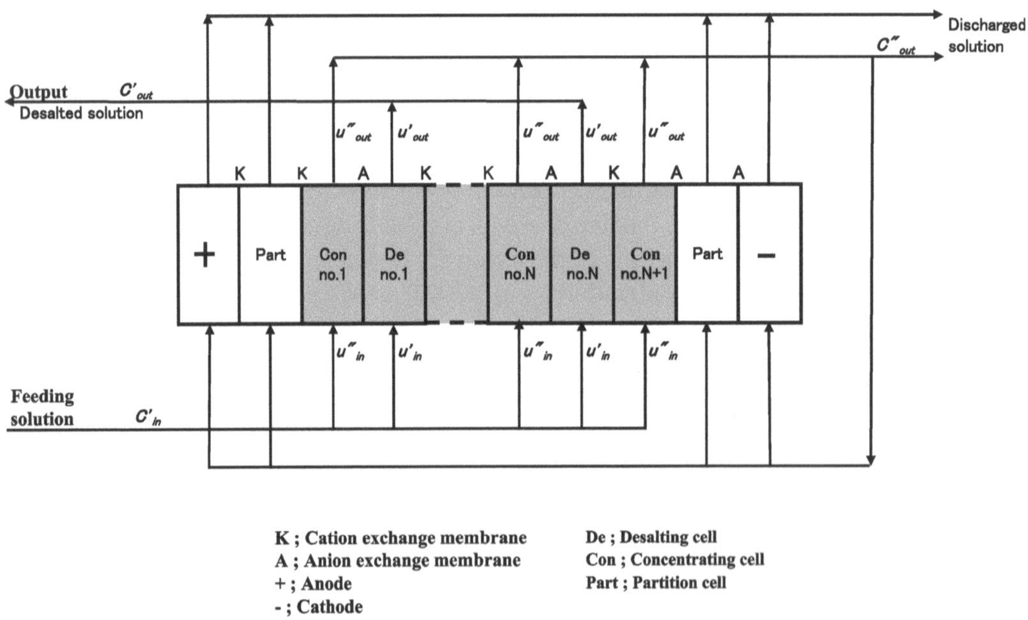

FIGURE 13.1 One-stage single-pass (continuous) electrodialysis process. *Tanaka, 2012a, p. 17, Fig. 13.3.*

incorporated between the stack and electrode cells for preventing the influence of electrode reactions to the performance of the electrodialyzer. A part of a concentrated solution flowing out from the outlets of concentrating cells is supplied to electrode cells and partition cells. A multi-stage ED process is formed by arranging the single-pass process in Fig. 13.1 in series.

13.3 Mass Transport in Single-Pass (Continuous) Process

Figure 13.2 illustrates the mass transport in the single-pass process. A salt solution (raw salt solution, concentration: C'_{in}) is supplied to the inlets of desalting cells (De) at average linear velocity of u'_{in}. For preventing scale formation in concentrating cells, a part of a raw salt solution is supplied also to the inlets of concentrating cells (Con) at the average linear velocity of u''_{in}. By supplying an electric current I, ions and solutions are transferred from desalting cells to concentrating cells across an ion exchange membrane pair and their flux is defined by J_S and J_V, respectively. In desalting (concentrating) cells, salt concentration is decreased (increased) along the flow-pass from $C'_{in}(C''_{in} = C'_{in})$ under applied average current density I/S and reaches average salt concentration $C'_{out}(C''_{out})$ at the outlets of desalting (concentrating) cells. Salt concentration change in desalting cells

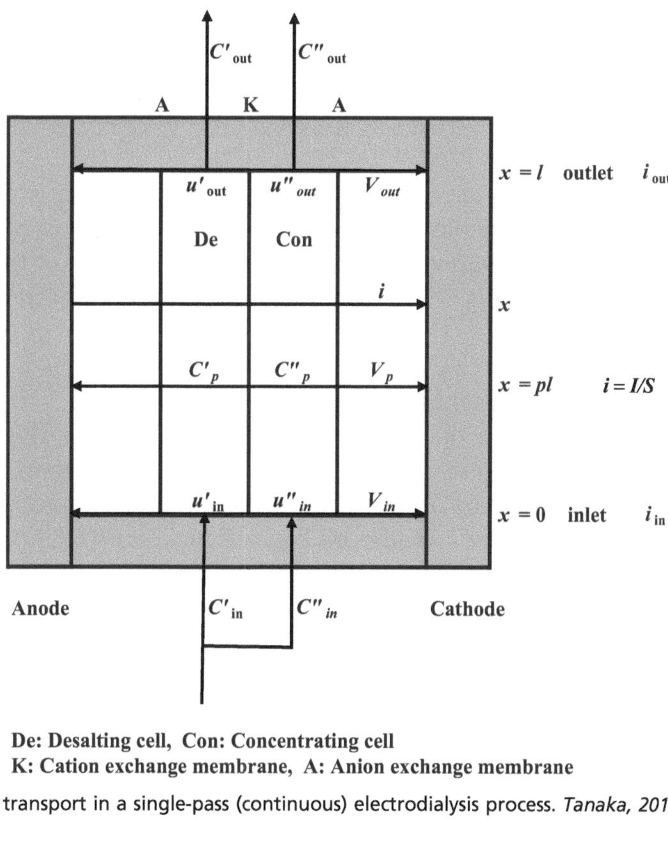

FIGURE 13.2 Mass transport in a single-pass (continuous) electrodialysis process. *Tanaka, 2012a, p. 12, Fig. 13.2.*

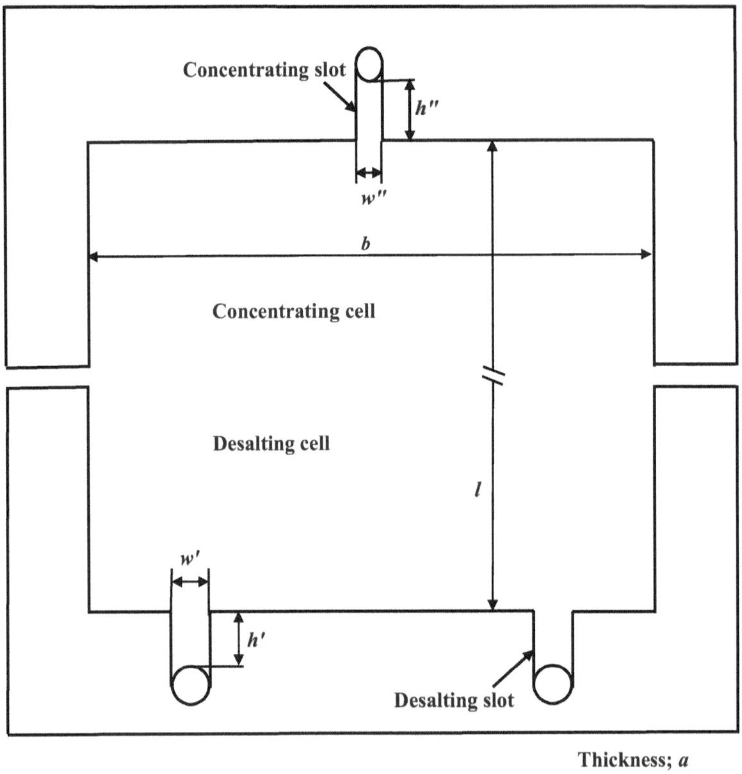

FIGURE 13.3 Structure of a desalting cell and a concentrating cell. *Tanaka, 2012a, p. 11, Fig. 13.1.*

causes current density change along the flow-pass from i_{in} at the inlets to i_{out} at the outlets. The current density becomes i at x distant from the inlets of desalting cells. $I/S, J_S, J_V, C'_p, C''_p, u'_p$, and u''_p are altogether the values at $x = pl$ distant from the inlets of desalting and concentrating cells. V_{in}, V_{out}, and V_p are voltage difference between electrodes, respectively, at the inlets $(x = 0)$, the outlets $(x = l)$ and $x = pl$ of desalting cells $(V_{in} = V_{out} = V_p)$.

Figure 13.3 illustrates the desalting cell and concentrating cell.

Structure of the flow-pass in the cell is classified to an *I*-shape (Figure 13.3) and a *U*-shape (Figure 16.4). Computer simulation program described in this chapter is developed for an *I*-shape cell. In order to apply the program to a *U*-shape cell, it is necessary to create an imaginary *I*-shape cell equivalent to the *U*-shape cell.

13.4 Specifications and Operating Conditions of an Electrodialyzer

1. Flow-pass thickness in a desalting and a concentrating cell; a (cm)
2. Flow-pass width in a desalting and a concentrating cell; b (cm)

3. Flow-pass length in a desalting and a concentrating cell; l (cm)
4. Membrane area; $S = bl$ (cm^2)
5. Number of stacks in an electrodialyzer; 1
6. Number of desalting cells, concentrating cells and membrane pairs integrated in a stack; N, $N+1$, N
7. Probe electrodes are inserted into concentrating cells integrated at the ends of a stack for measuring cell voltage
8. Average current density; I/S (A/cm^2)
9. Salt concentration at the inlets of desalting cells; C'_{in} (eq/cm^3)
10. Salt concentration at the inlets of concentrating cells; $C''_{in} = C'_{in}$ (eq/cm^3)
11. Linear velocity at the inlets of desalting and concentrating cells; u'_{in}, u''_{in} (cm/s)
12. Standard deviation of the normal distribution of solution velocity ratio; σ
13. Flow system in desalting and concentrating cells; single-pass flow
14. Diagonal net spacers are integrated in desalting and concentrating cells and in desalting and concentrating slots. Dimensions of a spacer are:
   Diameter of a spacer rod d_S = half thickness of the cell; $a/2$
   Distance between the rods; χ
   Crossing angle of the rods; θ
15. Number and dimension of the cells and slots in a cell pair are:

	Number	Thickness	Width	Length
Desalting cell	1	a	b	l
Concentrating cell	1	a	b	l
Desalting slot	n'	a	w'	h'
Concentrating slot	n''	a	w''	h''

13.5 Overall Mass Transport Equation and Membrane Characteristics (Sections 5.2 and 5.3)

Fluxes of ions J_S and a solution J_V across an ion exchange membrane pair at $x = pl$ distant from the inlets of desalting and concentrating cells are expressed by the following overall mass transport equation (Tanaka, 2006):

$$J_S = \lambda(I/S) - \mu\left(C''_p - C'_p\right) = (t_K + t_A - 1)(I/S)/F - \mu\left(C''_p - C'_p\right) = \eta(I/S)/F \qquad (13.1)$$

$$J_V = \phi(I/S) + \rho\left(C''_p - C'_p\right) \qquad (13.2)$$

in which λ (eq/C) is the overall transport number, μ (cm/s) is the overall solute permeability, ϕ (cm^3/C) is the overall electro-osmotic permeability, and ρ (cm^4/eq s) is the overall volume osmotic permeability. t is the transport number of counter-ions in the membrane. η is the current efficiency and F is the Faraday constant.

ρ versus λ, μ, and ϕ plots are given by the following empirical equations:

$$\lambda = 9.208 \times 10^{-6} + 1.914 \times 10^{-5}\rho \quad (13.3)$$

$$\mu = 2.005 \times 10^{-4}\rho \quad (13.4)$$

$$\phi = 3.768 \times 10^{-3}\rho^{0.2} - 1.019 \times 10^{-2}\rho \quad (13.5)$$

ρ versus alternating current electric resistance of an ion exchange membrane pair $r_{alter} = r_{alter,K} + r_{alter,A}$ is expressed by the following equation (Tanaka, 2011a):

$$r_{alter} = r_{alter,K} + r_{alter,A} = 1.2323\rho^{-(1/3)} \quad (13.6)$$

The relationship between ρ and solution temperature T (°C) is approximated by the following equation (Tanaka, 2011b):

$$\rho = 3.421 \times 10^{-3} + 3.333 \times 10^{-4}T \quad (13.7)$$

Equations (13.3)–(13.6) mean that ρ is the leading parameter and it represents all of the overall membrane characteristics. λ, μ, ϕ, $r_{alter,K} + r_{alter,A}$, and ρ are determined by setting T in Eqn (13.7). Equations (13.3)–(13.7) are empirical equations developed using electrodialyzers integrated with commercially available homogeneous ion exchange membranes and supplying inorganic strong electrolyte solutions. The equations are practically applicable under the above situations.

13.6 Salt Concentration and Linear Velocity in Desalting Cells

Salt solutions are supplied to desalting cells and pass through the cells by single-pass flow.

This phenomenon is expressed by the following equations:

$$q'_{in}\ (\mathrm{cm^3/s}) = ab \times u'_{in}\ (\mathrm{cm/s}) \quad (13.8)$$

$$q'_{out}\ (\mathrm{cm^3/s}) = q'_{in} - q \quad (13.9)$$

$$q\ (\mathrm{cm^3/s}) = blJ_V \quad (13.10)$$

$$u'_{out}\ (\mathrm{cm/s}) = q'_{out}/ab \quad (13.11)$$

Assuming the linear velocities to be changed linearly in desalting cells, average linear velocity at $x = pl$ distant from the inlets of desalting cells u'_p is:

$$u'_p = u'_{in} - p(u'_{in} - u'_{out}) = u'_{in} - p(q/ab) \quad (13.12)$$

p in Eqn (13.12) is calculated using Eqns (13.68), (13.81) and (13.82) in “Section 13.11” in the course of the trial-and-error calculation.

C'_{out} is introduced as follows:

$$C'_{out}\ (\mathrm{eq/cm^3}) = C'_{in} - \left(\frac{\eta l}{aF}\right)\left(\frac{1}{u'_p}\right)\left(\frac{I}{S}\right) \tag{13.13}$$

Linear velocities in desalting cells are not uniform between the cells. The limiting current density $(I/S)_{\lim}$ is strongly influenced by the solution velocity distribution between desalting cells. Solution velocity ratio in desalting cells is defined as:

$$\xi = \frac{u - \overline{u}}{\overline{u}} \tag{13.14}$$

u is the linear velocity in each desalting cell and $\overline{u}$ is the average linear velocity in desalting cells integrated in a stack. One assumes here that the frequencies Y_j of ξ in group j; ξ_j are expressed by the normal distribution (standard deviation; σ) (Fundamentals Section 8.8) (Tanaka, 2005).

$$Y_j = \frac{N}{\sqrt{2\pi}} \int_{\xi_j-\Delta\xi_j}^{\xi_j+\Delta\xi_j} \left(-\frac{\xi^2}{2}\right) d\xi \tag{13.15}$$

N is the number of desalting cells. ξ_j is the solution velocity ratio in desalting cell group j. $\Delta\xi_j$ is 1/2 of ξ_j value range. ξ_j is divided into $n+1$; ξ_0, ξ_1, ξ_2 ...ξ_n as:

$$\xi_j = -3\sigma\left(1 - \frac{j}{n/2}\right), \quad j = 0, 1, 2\ldots n \tag{13.16}$$

From the above equations, the linear velocity at $x = pl$ distant from the inlet of the jth desalting cell is introduced as:

$$u'_{p,j} = u'_p\left(\xi_j + 1\right) \tag{13.17}$$

Accordingly,

$$C'_{out,j} = C'_{in,j} - \left(\frac{\eta l}{aF}\right) \times \frac{1}{u'_p(\xi_j + 1)}\left(\frac{I}{S}\right) \tag{13.18}$$

in which, $C'_{in,j} = C'_{in}$.

Salt concentration at $x = pl$ distant from the inlets of desalting cells C'_p is computed by summing up C'_{pj} from $j = 0$ to $j = n$ and calculating the average of the summation for N cells.

$$C'_p = \frac{1}{N}\sum_{j=0}^{n} Y_j C'_{p,j} \tag{13.19}$$

in which

$$C'_{p,j} = C'_{in,j} - \left(\frac{\eta l}{aF}\right) \times \frac{1}{u'_{p/2}(\xi_j + 1)}\{a_1 p + (a_2/2)p^2 + (a_3/3)p^3\} \tag{13.20}$$

13.7 Salt Concentration and Linear Velocity in Concentrating Cells

For preventing scale formation by diluting salt concentration in concentrating cells, salt solutions supplied to the desalting cells are also fed to the inlets of concentrating cells at the average linear velocity of u''_{in} The solutions pass through the concentrating cells by single-pass flow. Linear velocities are also not uniform between concentrating cells as in the desalting cells. However, the solution velocity distribution in concentrating cells never influences to the limiting current density, so one assumes the uniform velocity distribution between concentrating cells. The equations are introduced as follows.

The material balance of solutions and ions transferring from desalting cells to concentrating cells at $x = pl$ is given by the following equations:

Solutions

$$abN\left(u''_{out} - u''_{in}\right) = SNJ_V = lbN\left\{\phi\left(\frac{I}{S}\right) + \rho\left(C''_p - C'_p\right)\right\} \tag{13.21}$$

Ions

$$abN\left(C''_{out}u''_{out} - C''_{in}u''_{in}\right) = SNJ_S = lbN\left\{\lambda\left(\frac{I}{S}\right) - \mu\left(C''_p - C'_p\right)\right\} \tag{13.22}$$

Assuming linear salt concentration and solution velocity changes in concentrating cells, C''_{out}, C_p, u''_{out}, and u''_p are introduced from Eqns (13.21) and (13.22) as follows:

$$C''_{out} = \frac{\sqrt{M^2 + 4LN} - M}{2L} \tag{13.23}$$

$$C''_p = C'_{in} + p\left(C''_{out} - C''_{in}\right) \tag{13.24}$$

$$u''_{out} = \frac{1}{2}\left(\sqrt{M^2 + 4LN} + M\right) \tag{13.25}$$

$$u''_p = u''_{in} + p\left(u''_{out} - u''_{in}\right) \tag{13.26}$$

in which

$$L = \frac{l}{a}\rho p \tag{13.27}$$

$$M = \frac{l}{a}\left[\phi\left(\frac{I}{S}\right) - \rho\left\{C'_p - (1-p)C''_{in}\right\}\right] + u''_{in} \tag{13.28}$$

$$N = C'_{in}u''_{in} + \frac{l}{a}\left\{\lambda\left(\frac{I}{S}\right) - \mu\left(C''_p - C'_p\right)\right\} \tag{13.29}$$

Salt concentration of a solution passing through the membrane pair at $x = pl$; C''_p is introduced from Eqns (13.1) and (13.2) as follows:

$$C''_p = \frac{1}{2\rho}\left(\sqrt{A^2 + 4\rho B} - A\right) \tag{13.30}$$

$$A = \phi(I/S) + \mu - \rho C_p' \tag{13.31}$$

$$B = \lambda(I/S) + \mu C_p' \tag{13.32}$$

13.8 Physical Properties of Solutions in Desalting and Concentrating Cells

The unit of salt concentration C (eq/cm^3) described in Sections 13.5 and 13.6 is changed to C_1 (eq/dm^3), C_2 (salt g/solution dm^3) and C_3 (salt g/solution kg) using the following equations:

$$C_1\ (\text{eq/dm}^3) = C\ (\text{eq/cm}^3) \times 10^3 \tag{13.33}$$

$$C_2\ (\text{g salt/dm}^3\ \text{solution}) = 57.87C_1 \tag{13.34}$$

$$C_3\ (\text{g salt/kg solution}) = \frac{\sqrt{Q^2 + 4PC_2} - Q}{2P} \tag{13.35}$$

$$P = \left(7.881 - 1.368 \times 10^{-2}T + 8.978 \times 10^{-5}T^2\right) \times 10^{-4} \tag{13.36}$$

$$Q = 1.001 - 1.101 \times 10^{-4}T - 3.356 \times 10^{-6}T^2 \tag{13.37}$$

The density d (kg/dm^3) (Sato and Matsuo, 1974), specific electric conductance κ (S/cm) (Akiyama and Tanaka, 1992), and electrolyte (NaCl) activation coefficient γ of an electrolyte solution (Harned and Owen, 1998) are given by the following functions of temperature T (°C) and salt concentration C_3 (g salt/kg solution):

$$d = 1.001 - 1.101 \times 10^{-4}T - 3.356 \times 10^{-6}T^2 + \left(7.881 - 1.368 \times 10^{-2}T + 8.978 \times 10^{-5}T^2\right) \times 10^{-4}C_3 \tag{13.38}$$

$$\begin{aligned}\kappa =& \left(0.9383 + 3.463 \times 10^{-2}T\right) \times 10^{-3}C_3 - \left(1.655 + 3.863 \times 10^{-2}T\right) \times 10^{-6}C_3^2 \\ &- \left(1.344 + 3.160 \times 10^{-2}T\right) \times 10^{-9}C_3^3\end{aligned} \tag{13.39}$$

$$\gamma = 0.5927 + 0.4355C_3^{-0.5} - 7.201 \times 10^{-5}C_3 + 3.503 \times 10^{-6}C_3^2 \tag{13.40}$$

13.9 Electric Resistance of an Ion Exchange Membrane Pair and Solutions in Desalting and Concentrating Cells (Section 2.6.3)

The alternating current electric resistance of an ion exchange membrane r_{alter} is measured conventionally at 25 °C by passing a 1000 Hz alternating current.

For measuring the direct current electric resistance of an ion exchange membrane, the membrane is set in a two-cell apparatus and a low-concentration NaCl solution (specific conductivity, κ' (S/cm)) is supplied into both cells. Direct current electric resistance of the membrane r^*_{dire}(Ω cm^2) is measured at 25 °C by passing a direct current. The relationship between κ' and r^*_{dire}/r_{alter} is expressed by the following empirical equation (Tanaka et al., 2003):

$$\log\left(\frac{r^*_{dire}}{r_{alter}}\right) = 0.3380 + 0.6386 \log \kappa' + 0.2961(\log \kappa')^2 \tag{13.41}$$

A low-concentration NaCl solution (specific conductivity, κ') is supplied to the desalting side, and a high-concentration NaCl solution (specific conductivity, κ'') is supplied to the concentrating side of the above apparatus. The direct current electric resistance of the membrane r_{dire} is measured at 25 °C by passing a direct current and subtracting the effect of membrane potential. The empirical relationship between κ''/κ' and r_{dire}/r^*_{dire} is (Tanaka et al., 2003):

$$\frac{r_{dire}}{r^*_{dire}} = 1.000 - 0.1359 \log\left(\frac{\kappa''}{\kappa'}\right) \tag{13.42}$$

Direct electric current of a membrane pair r_{memb} is estimated by multiplying Eqn (13.41) by Eqn (13.42):

$$r_{memb} = r_K + r_A = \left(\frac{r^*_{dire}}{r_{alter}}\right) \times \left(\frac{r_{dire}}{r^*_{dire}}\right) \times r_{alter} = r_{dire} \tag{13.43}$$

Electric resistance of a desalting cell r' and of a concentrating cell r'' in an electrodialyzer are given:

$$r' = \frac{a}{(1-\varepsilon)\kappa'} \tag{13.44}$$

$$r'' = \frac{a}{(1-\varepsilon)\kappa''} \tag{13.45}$$

ε defines an electric current screening effect of a diagonal net spacer (Fundamentals Fig. 8.26) and it is determined by the volume ratio of spacer rods in a desalting and concentrating cell as follows:

$$\varepsilon = \frac{\pi a}{8\chi \sin\theta} \tag{13.46}$$

in which χ is the distance between spacer rods, and θ is the crossing angle of the rods.

13.10 Pressure Drop in Desalting and Concentrating Cells and Slots (Section 8.13)

Hydrodynamic diameter of a desalting or a concentrating cell (electric current passing section); $d_{H,cell}$ and that of a desalting or a concentrating slot; $d_{H,slot}$ incorporated with a diagonal net spacer are expressed by the following equation (Zimmerer and Kotte, 1996; Tsiakis and Papageorgiou, 2005).

$$d_{H,cell} = \frac{8 - \pi\frac{a}{\chi}}{4\left(\frac{1}{b}+\frac{1}{a}\right) + 2\pi\left(1 - \frac{a}{4b}\right)\frac{1}{\chi}} \tag{13.47}$$

$$d_{H,slot} = \frac{8 - \pi\frac{a}{\chi}}{4\left(\frac{1}{w}+\frac{1}{a}\right) + 2\pi\left(1 - \frac{a}{4w}\right)\frac{1}{\chi}} \tag{13.48}$$

in which w is the flow-pass width in the slot (Fig. 13.3).

Pressure difference between the inlet and the outlet of the cell in a desalting or a concentrating cell; pressure drop in the cell (in the electric current passing section) ΔP_{cell} and that in a desalting or a concentrating slot; pressure drop in the slot ΔP_{slot} are (Tsiakis and Papageorgiou, 2005):

$$\Delta P_{cell}(\text{Pa}) = \frac{3.2\mu l u_{cell}}{(d_{H,cell})^2} \tag{13.49}$$

$$\Delta P_{slot}\ (\text{Pa}) = \frac{3.2\mu h u_{slot}}{(d_{H,slot})^2} \tag{13.50}$$

in which h is the flow-pass length in the slot (Fig. 13.3), u_{cell} is linear velocity in the desalting or concentrating cell, and u_{slot} is linear velocity in the slot. μ (g/cm s) is viscosity coefficient of a solution and it is expressed by the following function of temperature T (°C) and electrolyte concentration C (g salt/kg solution) of a solution (Akiyama and Tanaka, 1992):

$$\mu = 1.200 \times 10^{-2} - 1.224 \times 10^{-4}T + \left(2.107 \times 10^{-5} - 1.529 \times 10^{-7}T\right)C + \left(-1.392 \times 10^{-8} + 1.123 \times 10^{-10}T\right)C^2 + \left(5.819 \times 10^{-10} - 6.769 \times 10^{-12}T\right)C^3 \tag{13.51}$$

u_{cell} and u_{slot} are linear velocity in the cell and slot. ΔP (m) is calculated using $\Delta P\ (\text{m}) = 1.01972 \times 10^{-4}\Delta P\ (\text{Pa})$.

13.11 Current Density Distribution

13.11.1 Quadratic Current Density Distribution Equation

In an ED system, current density is decreased due to the salt concentration decrease in desalting cells along a flow-pass. This phenomenon exerts an influence to the limiting current density of an electrodialyzer $(I/S)_{\lim}$ (Section 13.13). The current density distribution is assumed to be approximated by the following quadratic equation expressed at x/l distant from the inlet of a desalting cell.

$$i = a_1 + a_2\left(\frac{x}{l}\right) + a_3\left(\frac{x}{l}\right)^2 \tag{13.52}$$

To determine a_1, a_2, and a_3 in Eqn (13.52), three simultaneous equations are set up (Tanaka, 2000, 2002) as described in the succeeding sections.

13.11.2 The First Simultaneous Equation ($V_{in} = V_{out}$) for Computing ζ_{inout}

When a large number of membrane pairs are integrated in an electrodialyzer, ohmic loss of electrodes incorporated in the electrodialyzer is negligible compared to the values between the electrodes. This phenomenon is expressed by Eqn (13.53), which means that the voltage difference between the electrodes at the entrance of desalting cells V_{in} is equal to the value at the exits V_{out}.

$$V_{in} = V_{out} \quad \text{The first equation} \tag{13.53}$$

$$V_{in} = A_1 i_{in} + A_2 \tag{13.54}$$

$$V_{out} = B_1 i_{out} + B_2 \tag{13.55}$$

A_1, A_2, B_1, and B_2 are calculated for N membrane pairs and for parameters determined in the solution velocity ratio ξ_j in $j = 0, 1, 2 \ldots n$.

Electric resistance at $x = 0$ (inlets)

$$A_1 = \left(r'_{in} + r''_{in} + r_{memb,in}\right)N \tag{13.56}$$

Membrane potential at $x = 0$ (inlets)

$$A_2 = 2(t_K + t_A - 1)\left(\frac{RT}{F}\right)N \ln\left(\frac{\gamma''_{in} C''_{in}}{\gamma'_{in} C'_{in}}\right) \tag{13.57}$$

Electric resistance at $x = l$ (outlets)

$$B_1 = \sum_{j=0}^{n} Y_j r'_{out,j} + r'' N + \sum_{j=0}^{n} Y_j r_{memb,out,j} \tag{13.58}$$

Membrane potential at $x = l$ (outlets)

$$B_2 = 2(t_K + t_A - 1)\left(\frac{RT}{F}\right)\sum_{j=0}^{n} Y_j \left(\frac{\gamma''_{out} C''_{out}}{\gamma'_{out} C'_{out}}\right) \tag{13.59}$$

$$Z_1 = \frac{B_1}{A_1} \tag{13.60}$$

$$Z_2 = -\frac{A_2 - B_2}{A_1} \tag{13.61}$$

$$\alpha_1 = \left(\frac{I}{S}\right) - Z_2 \tag{13.62}$$

$$\alpha_2 = -2\left\{3\left(\frac{I}{S}\right) - 2Z_2\right\} \tag{13.63}$$

$$\alpha_3 = 3\left\{2\left(\frac{I}{S}\right) - Z_2\right\} \tag{13.64}$$

$$\beta_1 = Z_1\left(\frac{I}{S}\right) \tag{13.65}$$

$$\beta_2 = -2(2Z_1 + 1)\left(\frac{I}{S}\right) \tag{13.66}$$

$$\beta_3 = 3(Z_1 + 1)\left(\frac{I}{S}\right) \tag{13.67}$$

$$\zeta_{inout} = \frac{\alpha_1 + \alpha_2 p + \alpha_3 p^2}{\beta_1 + \beta_2 p + \beta_3 p^2} \tag{13.68}$$

p pre-determined is substituted to Eqn (13.68) to calculate temporary ζ_{inout}.

13.11.3 The Second Simultaneous Equation ($V_{in} = V_p$) for Computing ζ_{inp}

Equation (13.69) is naturally introduced from Eqn (13.53):

$$V_{in} = V_p \quad \text{The second equation} \tag{13.69}$$

$$V_{in} = A_1 i_{in} + A_2 = \text{Eqn (13.54)} \tag{13.70}$$

$$V_p = C_1\left(\frac{I}{S}\right) + C_2 \tag{13.71}$$

$$A_1 = \text{Eqn (13.56)} \tag{13.72}$$

$$A_2 = \text{Eqn (13.57)} \tag{13.73}$$

Electric resistance at $x = pl$

$$C_1 = \sum_{j=0}^{n} Y_j r'_{p,j} + r''N + \sum_{j=0}^{n} Y_j r_{memb,p,j} \tag{13.74}$$

Membrane potential at $x = pl$

$$C_2 = 2(t_K + t_A - 1)\left(\frac{RT}{F}\right)\sum_{j=0}^{n} Y_j\left(\frac{\gamma''_p C''_p}{\gamma'_p C'_p}\right) \tag{13.75}$$

$$Z_3 = \frac{C_1}{A_1} \tag{13.76}$$

$$Z_4 = -\frac{A_2 - C_2}{A_1} \tag{13.77}$$

$$\gamma_1 = (Z_3 - 1)\left(\frac{I}{S}\right) + Z_4 \tag{13.78}$$

$$\gamma_2 = 2\left\{(3 - 2Z_3)\left(\frac{I}{S}\right) - 2Z_4\right\} \tag{13.79}$$

$$\gamma_3 = 3\left\{(Z_3 - 2)\left(\frac{I}{S}\right) + Z_4\right\} \tag{13.80}$$

$$\zeta_{inp} = \frac{\gamma_1 + \gamma_2 p + \gamma_3 p^2}{\left\{(2p - 3p^2)\left(\frac{I}{S}\right)\right\}} \tag{13.81}$$

ζ_{inp} is calculated by substituting pre-determined p, which was substituted to Eqn (13.68).

13.11.4 The Third Simultaneous Equation ($\zeta_{inout} = \zeta_{inp}$) for Computing p

p is computed by adjusting p with trial-and-error calculation to realize Eqn (13.82).

$$\zeta_{inout} = \zeta_{inp} \quad \text{The third equation} \tag{13.82}$$

13.11.5 Coefficients in the Current Density Distribution Equation and Current Density Uniformity Coefficient

Using Eqn (13.53), the coefficients in the quadratic current density distribution equation (Eqn (13.52)) are introduced as follows:

$$a_1 = Z_1\zeta_{out}\left(\frac{I}{S}\right) + Z_2 \tag{13.83}$$

$$a_2 = 2\left[\{3 - (2Z_1 + 1)\zeta_{out}\}\left(\frac{I}{S}\right) - 2Z_2\right] \tag{13.84}$$

$$a_3 = -3\left[\{2 - (Z_1 + 1)\zeta_{out}\}\left(\frac{I}{S}\right) - Z_2\right] \tag{13.85}$$

ζ_{in} and ζ_{out} included in Eqns (13.83)–(13.85) are inlet and outlet current density uniformity coefficient, respectively, defined by:

$$\zeta_{in} = \frac{i_{in}}{I/S} = \frac{a_1}{I/S} \tag{13.86}$$

$$\zeta_{out} = \frac{i_{out}}{I/S} = \frac{a_1 + a_2 + a_3}{I/S} \tag{13.87}$$

13.12 Cell Voltage, Energy Consumption, Water Recovery, and Desalting Ratio

From Eqns (13.86) and (13.87), current density at the inlet; i_{in} and outlet; i_{out} of the desalting cell are:

$$i_{in} = (I/S)\zeta_{in} \tag{13.88}$$

$$i_{out} = (I/S)\zeta_{out} \tag{13.89}$$

Ohmic voltage and membrane voltage at the inlet of the desalting cell ($V_{\Omega,in}$ and $V_{memb,in}$) and those at the outlet of the desalting cell ($V_{\Omega,out}$ and $V_{memb,out}$) are:

$$V_{\Omega,in} = \left(r'_{in} + r_{memb,in} + r''\right) i_{in} \tag{13.90}$$

$$V_{memb,in} = 2(t_K + t_A - 1) \times \left(\frac{RT}{F}\right) \ln \frac{\gamma''_{in} C''_{in}}{\gamma'_{in} C'_{in}} \tag{13.91}$$

$$V_{\Omega,out} = \left(\sum_{j=0}^{n} Y_j r'_{out,j} + \sum_{j=0}^{n} Y_j r_{memb,out,j} + r''_{out} N\right) \times i_{out} \times \left(\frac{1}{N}\right) \tag{13.92}$$

$$V_{memb,out} = 2(t_K + t_A - 1) \left(\frac{RT}{F}\right) \sum_{j=0}^{n} \ln \frac{\gamma''_{out} C''_{out}}{\gamma'_{out,j} C'_{out,j}} \times \left(\frac{1}{N}\right) \tag{13.93}$$

Cell voltage V_{cell} is introduced from Eqn (13.53) as shown in Eq. 13.94:

$$V_{cell}(\mathrm{V/pair}) = V_{\Omega,in} + V_{memb,in} = V_{\Omega,out} + V_{memb,out} \tag{13.94}$$

Energy consumption E is expressed by the following equation:

$$E\left(\mathrm{kWh/m^3}\right) = \frac{V_{cell} I}{q'_{out}} \times 3600 \times 10^3 \tag{13.95}$$

q'_{out} is the solution volume; output of the desalted solution (cm^3/s cell).
Water recovery Re is:

$$\mathrm{Re} = \frac{q'_{out}}{q'_{in} + q''_{in}} \tag{13.96}$$

Desalting ratio α is:

$$\alpha = \left(1 - \frac{C'_{out}}{C'_{in}}\right) \tag{13.97}$$

13.13 Limiting Current Density (Sections 9.5 and 9.6)

Limiting current density of a cation exchange membrane is less than that of an anion exchange membrane in inorganic strong electrolyte solutions, because the mobility of counter-ions in a solution for a cation exchange membrane is less than that for an anion exchange membrane. So, the limiting current density of an ion exchange membrane integrated in an electrodialyzer i_{lim} is given by the following empirical equation established for a cation exchange membrane (Tanaka, 2005, 2013).

$$i_{lim}(A/cm^2) = \left\{l_1 + l_2(T/25) + l_3(T/25)^2\right\}(m_1 + m_2 u'_{out}) C'^{n_1 + n_2 u'_{out}}_{out} \tag{13.98}$$

The coefficients in Eqn (13.98) were measured experimentally as follows:

$$l_1 = 0.5950,\ l_2 = 0.2731,\ l_3 = 0.1310,\ m_1 = 83.50,\ m_2 = 24.00,\ n_1 = 0.7846,\ n_2 = 8.612 \times 10^{-3} \tag{13.99}$$

When current density reaches the limit of a cation exchange membrane i_{lim} at the outlet of a desalting cell in which linear velocity becomes the least among u'_{out}; $u'^{\#}_{out}$, the average current density applied to an electrodialyzer is defined as its limiting current density $(I/S)_{lim}$, which is expressed by Eqn (13.100) introduced from Eqns (13.89) and (13.98).

$$\left(\frac{I}{S}\right)_{lim} = \frac{i_{lim}}{\zeta_{out}} = \left\{l_1 + l_2\left(\frac{T}{25}\right) + l_3\left(\frac{T}{25}\right)^2\right\} \frac{(m_1 + m_2 u'^{\#}_{out})}{\zeta_{out}} \left(C'^{\#}_{out}\right)^{n_1 + n_2 u'^{\#}_{out}} \tag{13.100}$$

in which $C'^{\#}_{out}$ is C'_{out} at $u' = u'^{\#}_{out}$.

In Eqn (13.100), $u'^{\#}_{out}$ is nearly equal to $u'^{\#}_{in}$. Substituting $u'^{\#}_{in} = u'^{\#}_{out}$ in Eqn (13.100) leads to:

$$\left(\frac{I}{S}\right)_{lim} = \left\{l_1 + l_2\left(\frac{T}{25}\right) + l_3\left(\frac{T}{25}\right)^2\right\} \frac{m_1 + m_2 u'^{\#}_{in}}{\zeta_{out}} \left(C'^{\#}_{out}\right)^{n_1 + n_2 u'^{\#}_{in}} \tag{13.101}$$

in which, $C'^{\#}_{out}$ is C'_{out} at $u' = u'^{\#}_{in}$, which is given by:

$$u'^{\#}_{in} = u'_{in}(1 - 3\,\sigma) \tag{13.102}$$

σ is the standard deviation of the normal distribution of solution velocity ratio defined by Eqn (13.14). σ scarcely affects general performance of the electrodialyzer, but it exerts an influences on the limiting current density of the electrodialyzer.

The relationship between $(I/S)_{lim}$ and $C'^{\#}_{out}$ is also introduced as follows:

$$\left(\frac{I}{S}\right)_{lim} = \left(\frac{a}{\lambda l}\right) u'^{\#}_{in}\left(C'_{in} - C'^{\#}_{out}\right) \tag{13.103}$$

Putting Eqn (13.101) = Eqn (13.103):

$$Z_1 = \left\{l_1 + l_2\left(\frac{T}{25}\right) + l_3\left(\frac{T}{25}\right)^2\right\} \frac{\left(C'^{\#}_{out}\right)^{n_1 + n_2 u'^{\#}_{in}}}{C'_{in} - C'^{\#}_{out}} \tag{13.104}$$

$$Z_2 = \left(\frac{a\zeta_{out}}{\lambda l}\right)\left(\frac{u'^{\#}_{in}}{m_1 + m_2 u'^{\#}_{in}}\right) \tag{13.105}$$

$$Z_1 = Z_2 \tag{13.106}$$

Limiting current density of an electrodialyzer $(I/S)_{lim}$ is computed with trial-and-error calculation by substituting control key $C'^{\#*}_{out}$ for $C'^{\#}_{out}$ in Eqn (13.104) to realize $Z_1 = Z_2$ (Eqn (13.106)) for determining $C'^{\#}_{out}$. $(I/S)_{lim}$ is calculated by substituting $C'^{\#}_{out}$ into Eqn (13.101).

13.14 ED Program

13.14.1 Program

The program is developed based on the equations described in Sections 13.5–13.13. It is classified to the three stand-alone programs as follows:

1. Constant current single-pass (continuous) program (Fig. 13.4)
2. Constant voltage single-pass (continuous) program (Fig. 13.5)
3. Constant salt concentration single-pass (continuous) program (Fig. 13.6)

(Remark) Computation is carried out by inputting the cell pair number N into the program. However, the computed results are not influenced by N.

13.14.2 Decision Point

Each program has the following three decision points:

1. Decision point 1 (Blue)

   C'^{*}_{p} and C''^{*}_{p} (Control key 1) are equalized to computed C'_p and C''_p for preventing circulation reference as follows:

$$C'^{*}_{p} = C'_{p} \qquad C''^{*}_{p} = C''_{p} \tag{13.107}$$

2. Decision point 2 (Purple)

   p^* (Control key 2) is adjusted to realize:

$$\zeta_{inout} = \zeta_{inp} \quad \text{The third equation; Eqn (13.82)} \tag{13.108}$$

3. Decision point 3 (Red)

   In constant voltage program:
   I/S^* (Control key 3) is adjusted to equalize V^*_{cell} (Control key 3) with computed V_{cell} as follows:

$$V^*_{cell} = V_{cell} \tag{13.109}$$

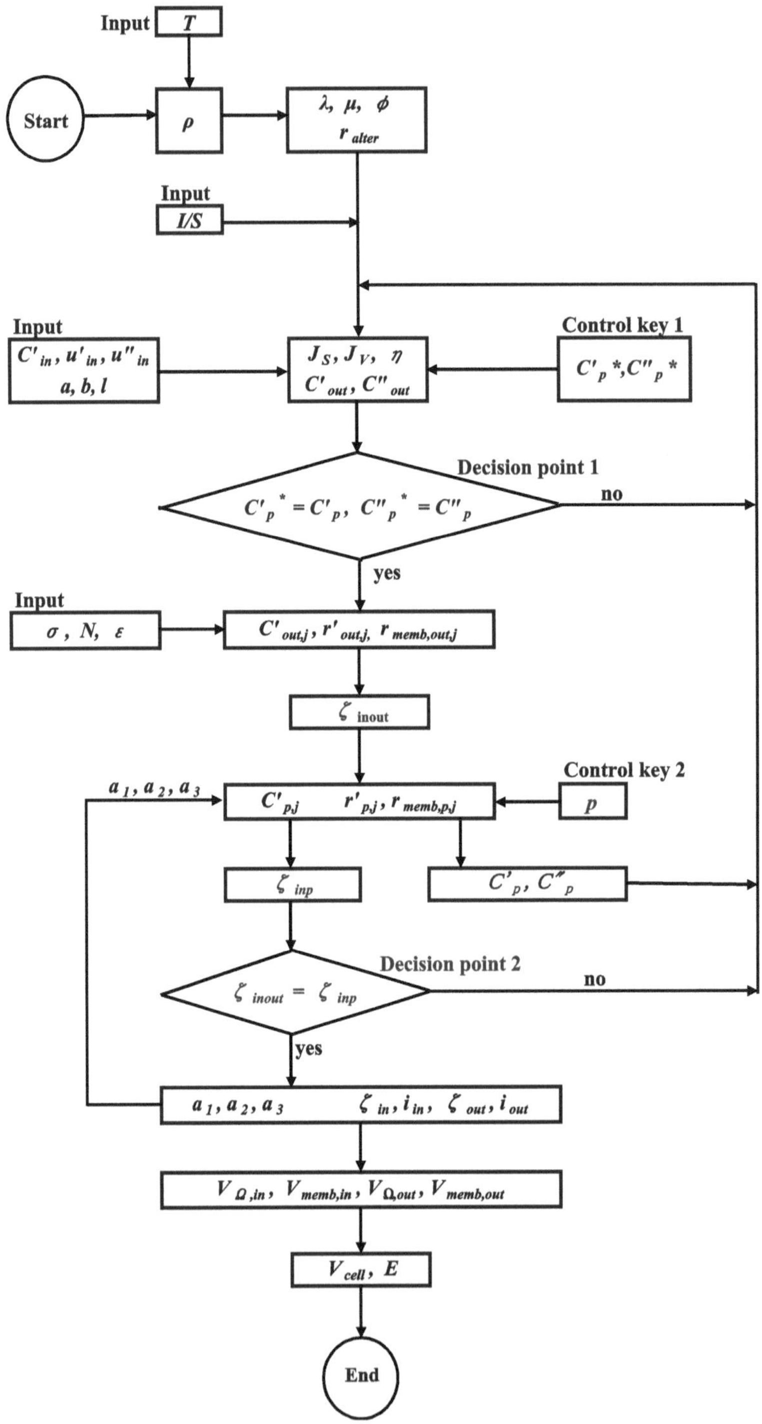

FIGURE 13.4 Constant electric current single-pass (continuous) electrodialysis program chart.

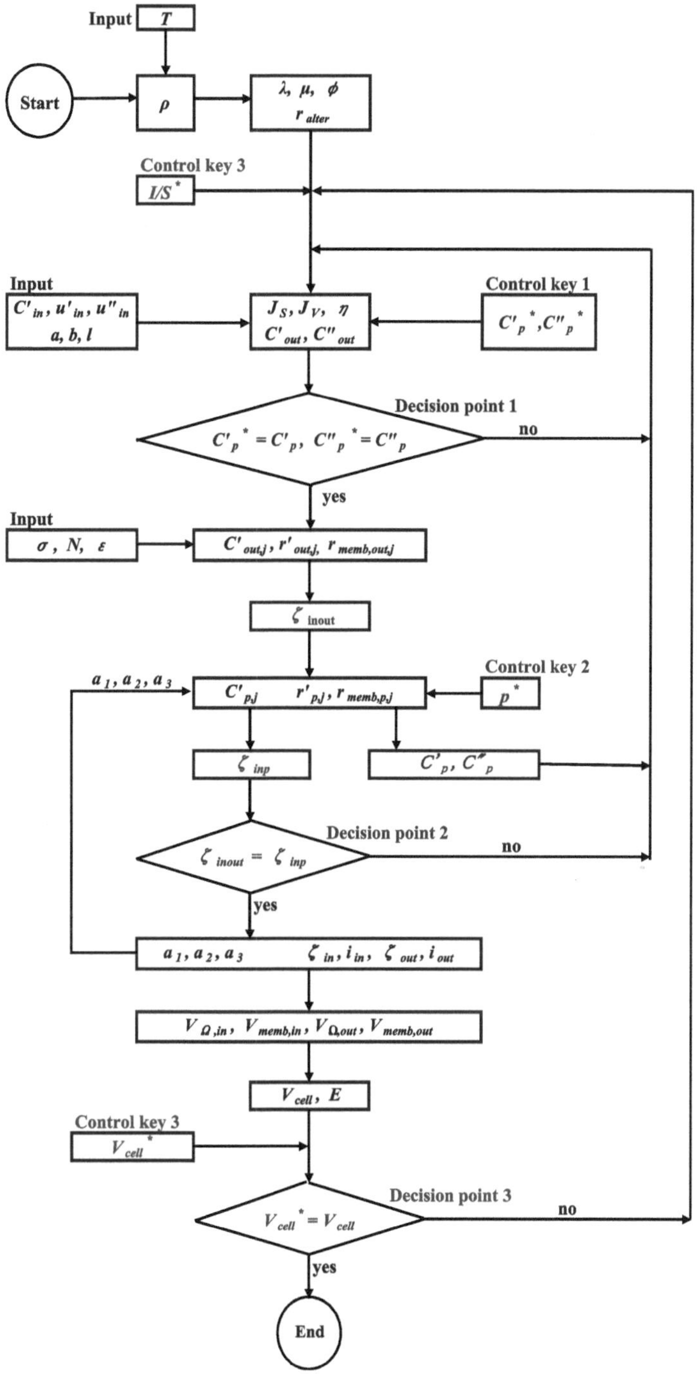

FIGURE 13.5 Constant voltage single-pass (continuous) electrodialysis program chart.

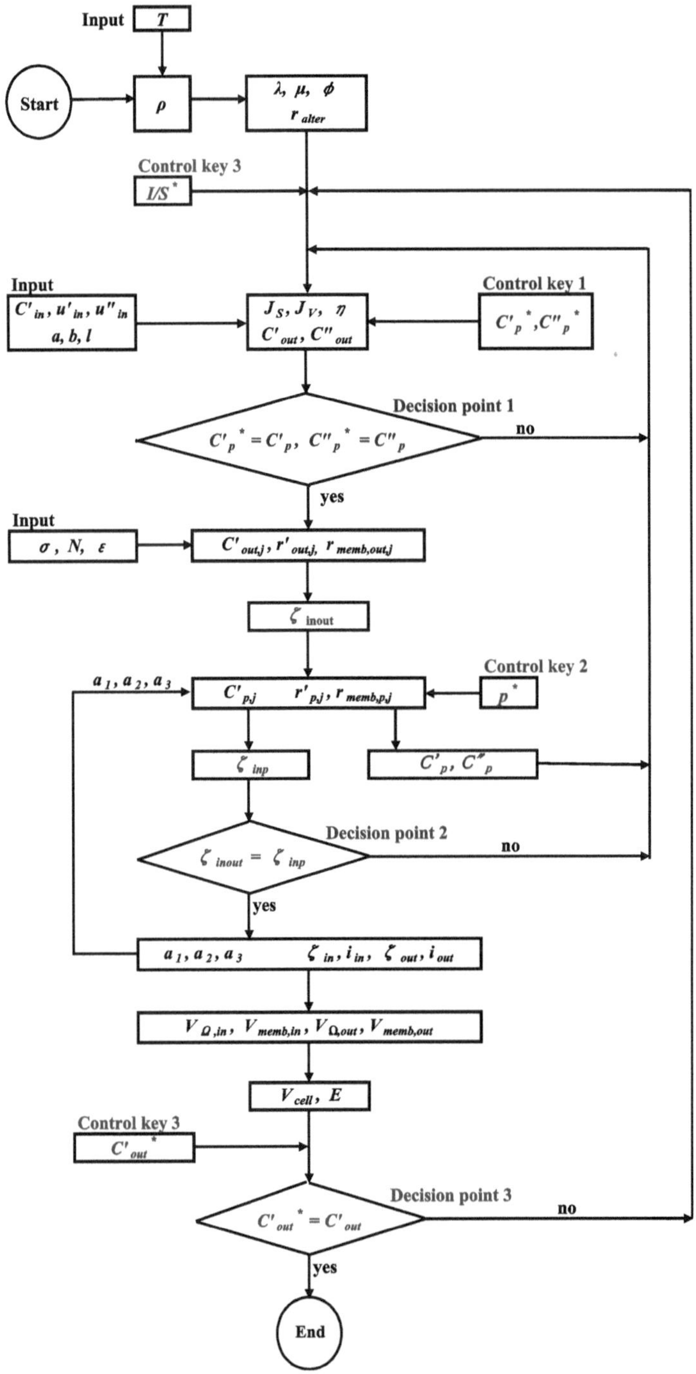

FIGURE 13.6 Constant salt concentration continuous electrodialysis program chart. *Tanaka, 2012a, p. 18, Fig. 13.5.*

In constant salt concentration program:

C'^{*}_{out} (Control key 3) is the salt concentration of a solution at the outlets of desalting cells that corresponds to the outlets of the stage. I/S^{*} (Control key 3) is adjusted to equalize C'^{*}_{out} with computed C'_{out} as follows:

$$C'^{*}_{out} = C'_{out} \tag{13.110}$$

13.14.3 Limiting Current Density

Succeeding to Figs 13.4, 13.5 or 13.6, $(I/S)_{\lim}$ is computed by adjusting control key $C'^{\#*}_{out}$, which stands for the salt concentration at the outlet of the desalting cell in which the linear velocity becomes the least. The computation is carried out to realize $Z_1 = Z_2$ (Eqn (13.106)) substituting $C'^{\#}_{out}$ into Eqn (13.101). The program is shown in Fig. 13.7.

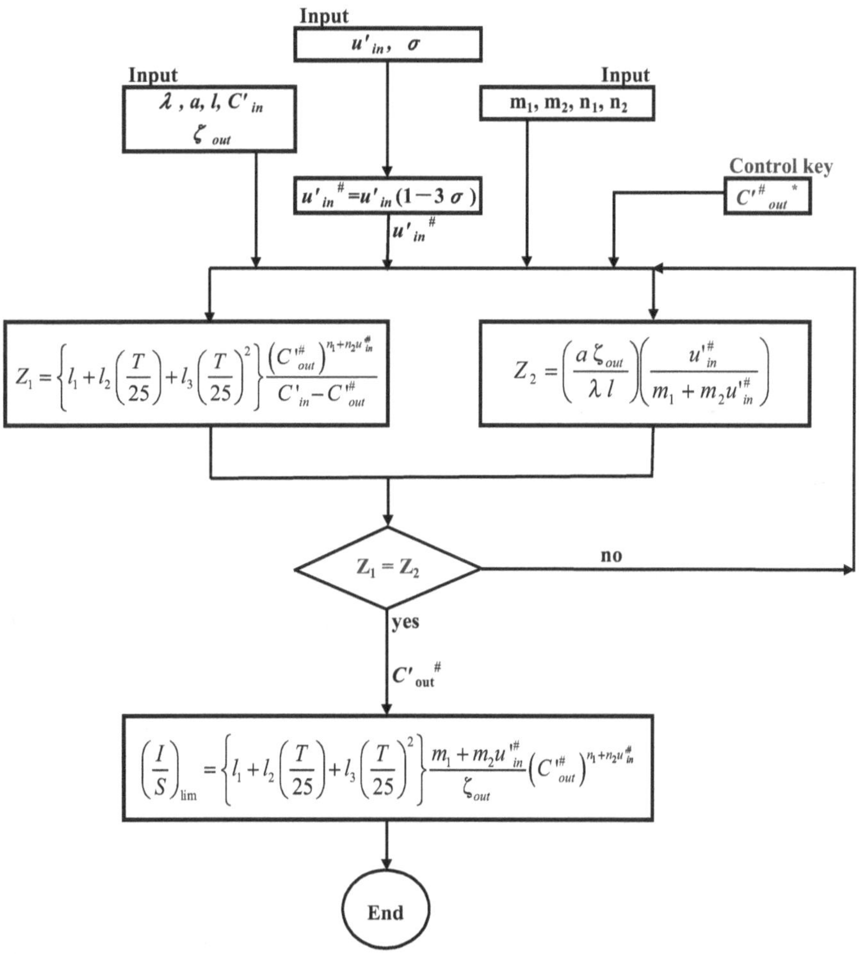

FIGURE 13.7 Limiting current density program chart. Super script # means the least value in an electrodialyzer. *Tanaka, 2012a, p. 19, Fig. 13.6.*

13.15 Companion Site (Chapter 24)

13.15.1 Companion Site

The companion site is established based on the stand alone ED program and it is classified to three kinds as follows.

1. Constant current single-pass (continuous) program (Companion site 1)
2. Constant voltage single-pass (continuous) program (Companion site 2)
3. Constant salt concentration single-pass (continuous) program (Companion site 3)

13.15.2 Steps in the Companion Site

Each companion site consists of the following 10 steps:

Step 1 Fundamental specifications of an electrodialyzer and operating conditions
Step 2 Fundamental operating conditions of an electrodialyzer
Step 3 Solution electric resistance and salt activity coefficient in concentrating cells
Step 4 Solution electric resistance and salt activity coefficient at the inlets of desalting cells
Step 5 Computation at $V_{in} = V_{out}$
Step 6 i_{in}, i_{out}, ohmic voltage, membrane potential, cell voltage, energy consumption
Step 7 Computation at $V_{in} = V_p$
Step 8 p, ζ_{in}, ζ_{out}, a_1, a_2 and a_3
Step 9 Solution viscosity and pressure difference in desalting cells and concentrating cells
Step 10 Limiting current density

13.15.3 Remarks in the Companion Site

1. Calculations in the decision points are marked with red. Inputs are marked with purple. Important values are marked with blue.
2. Readers can operate the program by inputting the optional input (purple) and carrying out the trial-and-error calculations at the decision points (red).
3. Step 10 is computed after the computations from Step 1 to Step 9 are finished. All computations are completed within about 10 min.
4. Summary is supplied at the end of the companion site for printing.

13.16 Process Specifications and ED Conditions

Table 13.1 lists the process specifications and ED conditions.

Standard deviation σ of solution velocity ratio ξ_j is influenced by the precision of part dimensions of a stack and skill of stack assembling work of an electrodialyzer.

Table 13.1 Electrodialyzer Specifications and Operating Conditions

Flow-pass thickness of desalting and concentrating cells a	0.05 cm
Flow-pass width desalting and concentrating cells b	100 cm
Flow-pass length of desalting and concentrating cells l	100 cm
Number of cell pairs N	300
Linear velocity at the inlets of desalting cells u'_{in}	10 cm/s
Linear velocity at the inlets of concentrating cells u'_{in}	1 cm/s
Temperature T	25 °C
Standard deviation of solution velocity ration in desalting cells σ	0.1
Distance between spacer rod χ	0.3 cm
Crossing angle of the rod θ	π/3 rad
Number of slots in desalting cells n'	5
Number of slots in concentrating cells n''	2
Width of slots in desalting cells w'	4 cm
Width of slots in concentrating cells w''	2 cm
Length of slots in desalting cells h'	4 cm
Length of slots in concentrating cells h''	4 cm

Tanaka, 2012a, p. 19, Table 1.

It is desirable to keep σ as low as possible but we cannot exactly control σ values in an ED operation. $\sigma = 0.1$ assumed in Table 13.1 is adopted roughly based on the experimental results. However, it is assumed that $\sigma = 0.1$ functions as a safety factor for stably operating an electrodialyzer (Tanaka, 2012b).

13.17 Computed Results

In constant voltage operation, current density is decreased with the increase of the desalting ratio. So, the constant voltage operation is more stable and widely employed than the constant current operation in saline water desalination. Because of the above reason, the computation for the constant current operation is canceled in this section.

13.17.1 Constant Voltage Operation

Computation is carried out by inputting the data in Table 13.1 and

$$C'_{in} = C''_{in} = 2000 \text{ mg/dm}^3.$$

$$V_{cell} = 0.2, 0.4, 0.6, 0.7, 0.8 \text{ V/pair}.$$

The relationships between cell voltage V_{cell} and ion flux J_S, and solution flux J_V and salt concentration at the outlets of desalting cells C'_{out} are shown in Fig. 13.8. Figure 13.9 gives the plots for current density I/S and energy consumption E. Figure 13.10 shows pressure drop in the cell (current passing sections) ΔP_{cell} and total

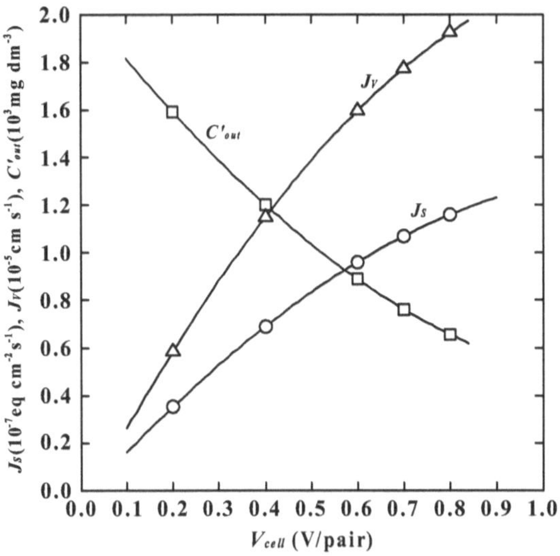

FIGURE 13.8 Ion flux, solution flux, and salt concentration at the outlets of desalting cells.

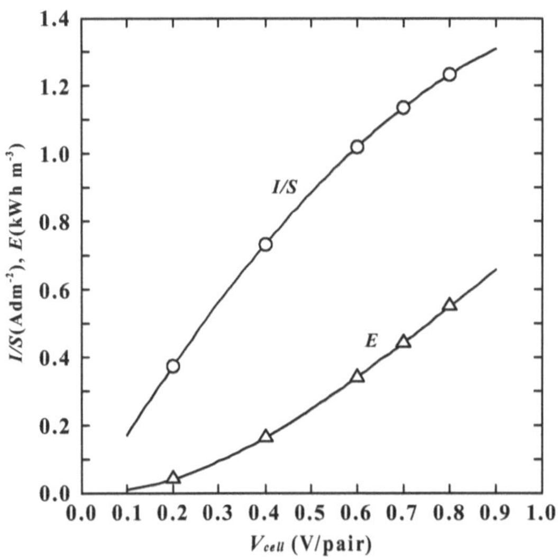

FIGURE 13.9 Current density and energy consumption.

pressure drop ΔP_{total}. The pressure drop in the slots is equivalent to $\Delta P_{total} - \Delta P_{cell}$. Superscripts, single quotation marks ($'$), and double quotation marks ($''$) refer to desalting cells and concentrating cells. Limiting current density $(I/S)_{\text{lim}}$ is plotted against current density I/S in Fig. 13.11. Real limiting current density $(I/S)_{\text{lim}} = 1.137\ \text{A/dm}^2$ is determined from the intersection between the $I/S - (I/S)_{\text{lim}}$

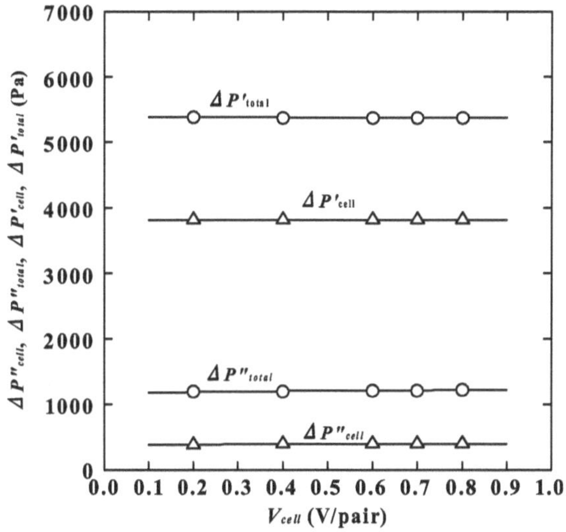

FIGURE 13.10 Pressure drop.

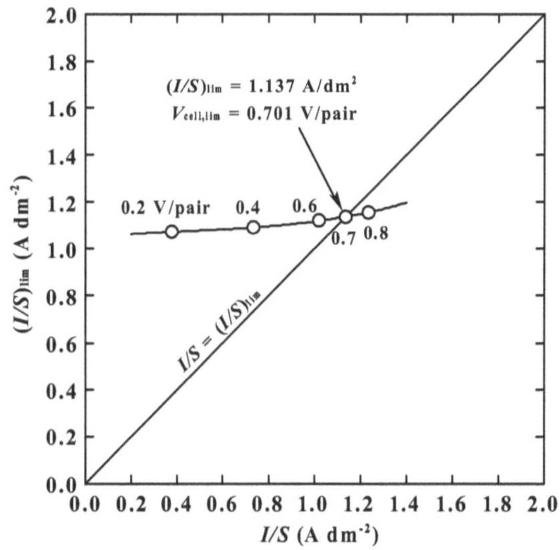

FIGURE 13.11 Limiting current density.

line and the $I/S = (I/S)_{\text{lim}}$ line. Limiting cell voltage $V_{cell,\text{lim}} = 0.701$ V/pair is the cell voltage at $(I/S)_{\text{lim}} = 1.137\ \text{A/dm}^2$.

13.17.2 Constant Salt Concentration Operation

Computation in carried out by inputting the data in Table 13.1 to produce drinking water:

$$C'_{out} = 400\ \text{mg/dm}^3\,(\text{drinking water})$$

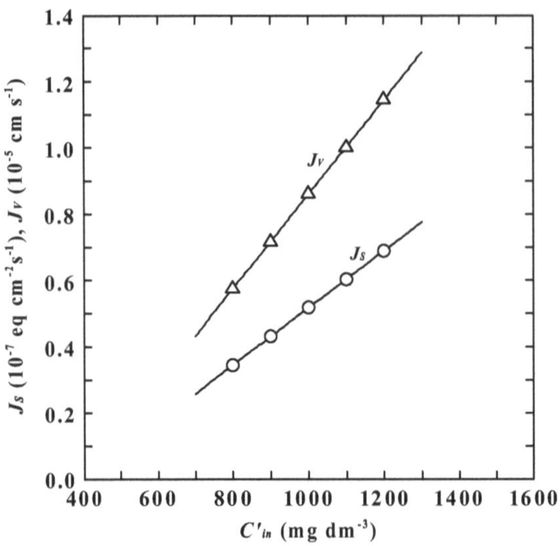

FIGURE 13.12 Ion flux and solution flux.

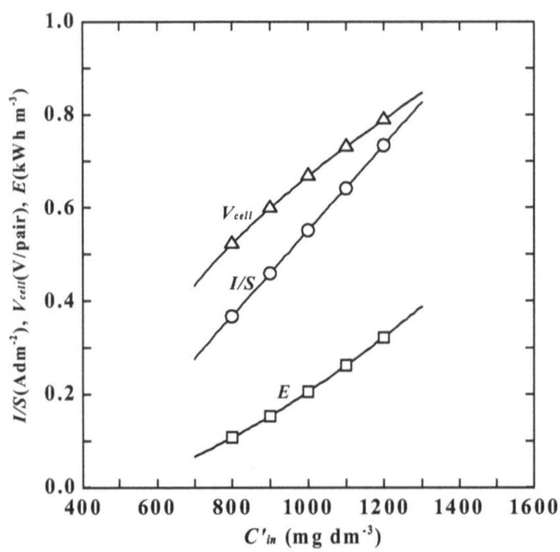

FIGURE 13.13 Current density, cell voltage, and energy consumption.

$$C'_{in} = C''_{in} = 800, 900, 1000, 1100, 1200\ \text{mg/dm}^3$$

J_S and J_V are plotted against C'_{in} and shown in Fig. 13.12. The relationships between C'_{in} and I/S, and V_{cell} and E, are shown in Fig. 13.13. Figure 13.14 shows the limiting current density. The real limiting current density is $(I/S)_{\text{lim}} = 0.629\ \text{A/dm}^2$. $V_{cell,\text{lim}} = 0.722\ \text{V/pair}$ and $C'_{in,\text{lim}} = 1086\ \text{mg/dm}^3$ are the limiting cell voltage and limiting salt concentration at $(I/S)_{\text{lim}} = 0.629\ \text{A/dm}^2$.

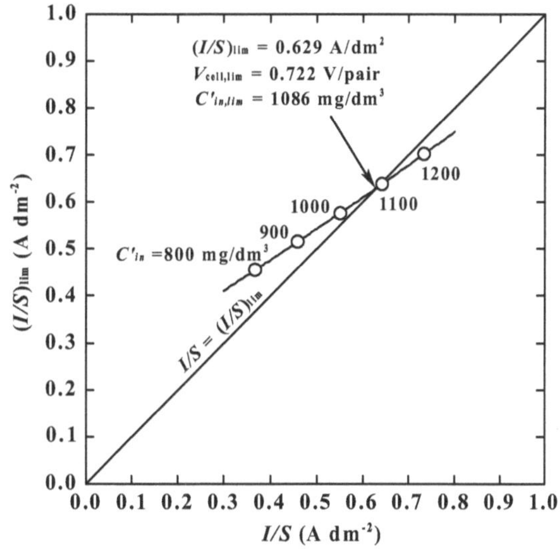

FIGURE 13.14 Limiting current density.

References

Akiyama, M., Tanaka, Y., 1992. Sea Water Science Research Laboratory. Technical Report, No. 3-22. Japan Tobacco & Salt Public Corp., p. 13.

Avriel, M., Zelighter, N., 1972. A computer method for engineering and economic evaluation of electrodialysis plant. Desalination 10, 113–146.

Belfort, G., Daly, J.A., 1970. Optimization of an electrodialysis plant. Desalination 8, 153–166.

Brauns, E., De Wide, W., Van den Bosch, B., Lens, P., Pinoy, L., Empsten, M., 2009. On the experimental verification of an electrodialysis simulation model for optimal stack configuration design through solver software. Desalination 249, 1030–1038.

Fidaleo, M., Moresi, M., 2005. Optimal strategy to model the electrodialytic recovery of a strong electrolyte. J. Membr. Sci. 260, 90–111.

Harned, H.S., Owen, B.B., 1998. Physical Chemistry of Electrolyte Solutions. Reinhold, p. 726.

Lee, H.J., Sarfert, F., Strathmann, H., Moon, S.H., 2002. Designing of an electrodialysis desalination plant. Desalination 142, 267–286.

Moon, P., Sandi, G., Stevens, D., Kizilel, R., 2004. Computational modeling of ionic transport in continuous and batch electrodialysis. Sep. Sci. Technol. 29, 2531–2555.

Nikonenko, V.V., Pismenskaya, N.D., Itoshin, A.G., Zabolotsky, V.I., Shudrenko, A.A., 2008. Description of mass transfer characteristics of ED and EDI apparatus by using the similarity theory and computation method. Chem. Eng. Process. 47, 1118–1127.

Sadrzadeh, M., Kaviani, A., Mohammadi, T., 2007. Mathematical modeling of desalination by electrodialysis. Desalination 206, 534–549.

Sato, K., Matsuo, T., 1974. Seawater handbook. Soc. Sea Water Sci. Jpn., 110.

Tanaka, Y., 2000. Current density distribution and limiting current density in ion-exchange membrane electrodialysis. J. Membr. Sci. 173, 179–190.

Tanaka, Y., 2002. Current density distribution, limiting current density and saturation current density in an ion-exchange membrane electrodialyzer. J. Membr. Sci. 210, 65–75.

Tanaka, Y., Ehara, R., Itoi, S., Goto, T., 2003. Ion-exchange membrane electrodialytic salt production using brine discharged from a reverse osmosis seawater desalination plant. J. Membr. Sci. 222, 71–86.

Tanaka, Y., 2005. Limiting current density of an ion-exchange membrane and of an electrodialyzer. J. Membr. Sci. 266, 6–17.

Tanaka, Y., 2006. Irreversible thermodynamics and overall mass transport in ion-exchange membrane electrodialysis. J. Membr. Sci. 281, 517–531.

Tanaka, Y., 2011a. Ion-exchange membrane electrodialysis of saline water and its numerical analysis. Ind. Eng. Chem. Res. 50, 10765–10777.

Tanaka, Y., 2011b. Ion-exchange membrane electrodialysis for saline water desalination and its application to seawater concentration. Ind. Eng. Chem. Res. 50, 7494–7503.

Tanaka, Y., 2012a. Ion-exchange membrane electrodialysis program and its application to multi-stage continuous saline water desalination. Desalination 301, 10–25.

Tanaka, Y., 2012b. Continuous ion-exchange membrane electrodialysis for drinking water production; numerical analysis. In: The 2012 World Congress on Advances in Civil, Environmental, and Material Research, Seoul, Korea, August 26–30, 2012, Proc, p. 141.

Tanaka, Y., 2013. Development of a computer simulation program of batch ion-exchange membrane electrodialysis for saline water desalination. Desalination 320, 118–133.

Tsiakis, P., Papageorgiou, L.G., 2005. Optimal design of an electrodialysis brackish water desalination. Desalination 173, 173–186.

Zimmerer, C.C., Kotte, V., 1996. Effects of spacer geometry on pressure drop, mass transfer, mixing behavior, and residence time distribution. Desalination 104, 129–134.

14

Computer Simulation Program: Feed-and-Bleed Program

14.1 Introduction

In a feed-and-bleed operation, a feeding solution is recycled between a circulation tank and an electrodialyzer continuously. The recycle rate is determined as desired from the targeted concentration changes of the desalted solution. This process is suitable for a middle- or a large-scale unit operation. Glancing over through the reports in literature (Kawahara, 1994; Thompson et al., 1996, 1997; Rapp and Pfromm, 1998; Ryabtsev et al., 2001; Schoeman et al., 2005; Hajdukova et al., 2011; Zhang, et al., 2011; Tran et al., 2012), the investigations of the feed-and-bleed process are directed to the process application.

In the feed-and-bleed process, an electrodialyzer and a circulation tank are the main units working in the process. The function of the electrodialyzer is computed based on the single-pass (continuous) program (Chapter 13) developed for the single-pass operation with no circulation tank. The computer program of the feed-and-bleed operation consists of (1) the program that describes the function of the electrodialyzer (Chapter 13) and (2) the program describing the function of the circulation tank (Section 14.5). In this chapter, saline water is desalinated with a use of the feed-and-bleed program. The companion site of the feed-and-bleed program is described in Chapter 24 (Companion site 4).

14.2 Feed-and-Bleed Process

The one-stage feed-and-bleed electrodialysis process (stage no. 1) is illustrated in Fig. 14.1. The electrodialyzer is incorporated with desalting cells and concentrating cells in the stack marked with gray. The number of cells is N for desalting cells, $N+1$ for concentrating cells, and N for cell pairs. The anode and cathode cells are placed at both outsides of the stack. The circulation tank is separated to the compartment I and compartment II. The salt solution is supplied to the electrodialyzer through the circulation tank and an electric current is passed through the electrodes. The electrodialyzer is operated as follows.

The raw salt solution (salt concentration, C^0; volume flow rate, Q^0) is supplied to the compartment I in the circulation tank. It is further supplied to desalting cells in the electrodialyzer (C'_{in}, Q'_{in}). Linear velocity in the desalting cells is u'_{in} at the inlets and u'_{out} at the outlets. The solution flowing out from the desalting cells is returned to the compartment II (C'_{out}, Q'_{out}). A part of the desalted solution (C'_{out}, Q''') is extracted from the compartment II to the outside of the process. Another part of the desalted solution

Ion Exchange Membranes. http://dx.doi.org/10.1016/B978-0-444-63319-4.00014-6

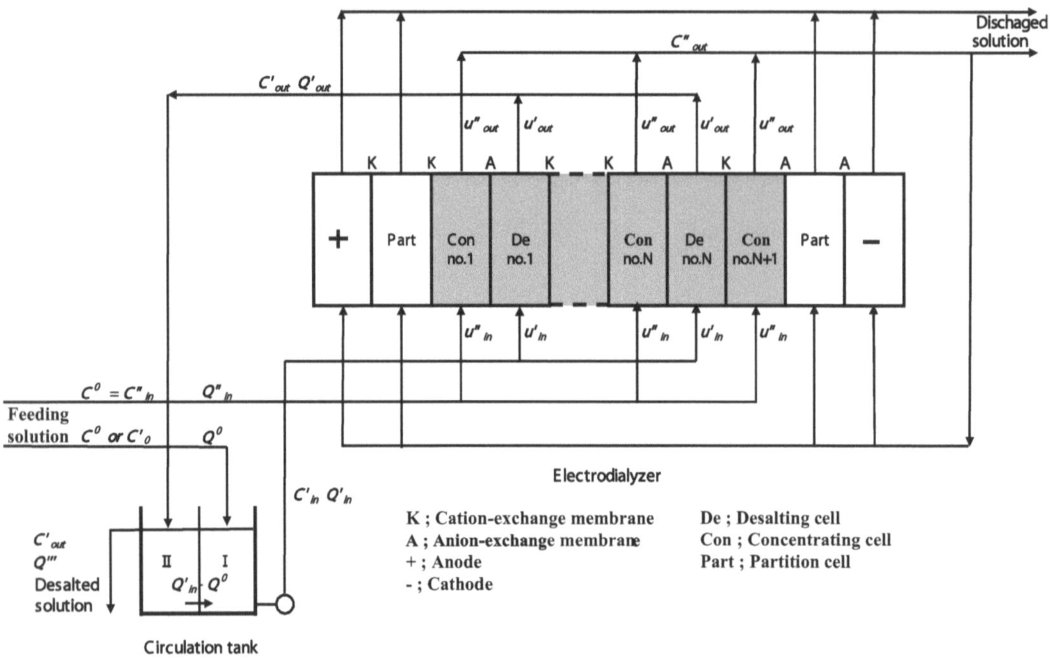

FIGURE 14.1 Feed-and-bleed electrodialysis process. *Tanaka (2014), p. 128, Fig. 1.*

(C'_{out}, $Q_{move} = Q'_{in} - Q^0$) moves from the compartment II to the compartment I. The process is operated keeping $Q_{move} > 0$.

The raw salt solution is also supplied to concentrating cells in the electrodialyzer ($C^0 = C''_{in}$, Q''_{in}) for preventing scale formation in the cells. Linear velocity in the concentrating cells is u''_{in} at the inlets and u''_{out} at the outlets. A part of the solution extracted from the outlets of concentrating cells is further supplied to the electrode cells and the partition cells (Part), which are equipped for preventing the influence of electrode reactions to the performance of the electrodialyzer. Then, the solution is discharged to the outside of the process with the residual solution flowing out from the outlets of concentrating cells.

In the multi-stage process (stage no. 2), a desalted solution (salt concentration C'_{out}) extracted from the previous stage is supplied to the compartment I in the next stage. In order to maintain the material balance in the electrodialysis system, Q^0 in the next stage must be equivalent to Q''' in the previous stage.

14.3 Specifications and Operating Conditions of an Electrodialyzer

The specifications and operating conditions are the same to the description in Section 13.4. The structure of a desalting cell and a concentrating cell is illustrated in Fig. 13.3.

14.4 Functions and Performance of an Electrodialyzer

The functions and performance are equivalent to those of the single-pass electrodialysis (Sections 13.5–13.13).

14.5 Mass Balance and Energy Consumption in the Feed-and-Bleed Process (Tanaka, 2014)

The performance of the feed-and-bleed operation is introduced from the following mass balance in the circulation tank in Fig. 14.1.

In the circulation tank, the volume rates of a solution flowing into desalting cells Q'_{in} and concentrating cells Q''_{in} are defined as:

$$Q'_{in} = abNu'_{in} \tag{14.1}$$

$$Q''_{in} = ab(N+1)u''_{in} \tag{14.2}$$

The volume rates of a solution flowing out from desalting cells Q'_{out} and that of a solution flowing out from the compartment II in the circulation tank Q''' are:

$$Q'_{out} = Q'_{in} - SNJ_V \tag{14.3}$$

$$Q''' = Q^0 - SNJ_V \tag{14.4}$$

The mass balance in the compartment I is:

$$C'_{in}Q'_{in} = C'_0Q^0 + C'_{out}(Q'_{in} - Q^0) \tag{14.5}$$

Feeding solution volume flow rate Q^0 is introduced from Eqn (14.5) as follows:

$$Q^0 = \frac{C'_{in} - C'_{out}}{C'_0 - C'_{out}}Q'_{in} \tag{14.6}$$

In the circulation tank, the volume rate $Q_{move} = Q'_{in} - Q^0$ of a solution moves from the compartment II to the compartment I. In order to operate the feed-and-bleed electrodialysis process, Q^0 must be less than Q'_{in}:

$$Q_{move} = Q'_{in} - Q^0 > 0 \tag{14.7}$$

Energy consumption E, water recovery Re, and the desalting ratio α are given as follows:

$$E = \frac{IV_{cell}N}{Q'''} = \left(\frac{1}{Q'''}\right)\left(\frac{I}{S}\right)SV_{cell}N \tag{14.8}$$

$$\mathrm{Re} = \frac{Q'''}{Q^0 + Q''_{in}} \tag{14.9}$$

$$\alpha = 1 - \frac{C'_{out}}{C^0} \tag{14.10}$$

14.6 Electrodialysis Program

14.6.1 Program

The feed-and-bleed process is operated in the constant voltage mode. The stand-alone program is developed based on the constant voltage single-pass program described in Section 13.14 and illustrated in Fig. 14.2.

14.6.2 Decision Point

The program includes the following five decision points:

1. Decision point 1 (Blue)
   $C_{in}'^*$ and $C_{out}'^*$ (Control key 1) are adjusted to realize:

$$C_{out}'^* = C_{ut}' \tag{14.11}$$

2. Decision point 2 (Orange)
   $C_p'^*$ and $C_p''^*$ (Control key 2) are equalized to computed C_p' and C_p'' for preventing circulation reference as follows:

$$C_p'^* = C_p' \quad C_p''^* = C_p'' \tag{14.12}$$

3. Decision point 3 (Purple)
   p^* (Control key 3) is adjusted to realize:

$$\xi_{inout} = \xi_{inp} \tag{14.13}$$

4. Decision point 4 (Red)
   I/S^* (Control key 4) is adjusted to equalize V_{cell}^* (Control key 4) with computed V_{cell} as follows:

$$V_{cell}^* = V_{cell} \tag{14.14}$$

5. Decision point 5 (Green)
   After the trial-and-error calculations in the Decision points 1–4, Q'''^* (Control key 5) is inputted and adjusted N^* (Control key 5) to realize:

$$Q'''^* = Q''' \tag{14.15}$$

(Remark) The performance of the electrodialyzer is not influenced by the cell pair number N inputted in the Decision points 1 – 4 (cf. Sections 13.14.1, Remark). The performance of the feed-and-bleed process is influenced by the cell pair number N^* (Control key 5) inputted in the Decision point 5 (N in Section 14.5). In the program, N (in the electrodialyzer) and N^* (or N) (in the feed-and-bleed process) are not necessary to be the same.

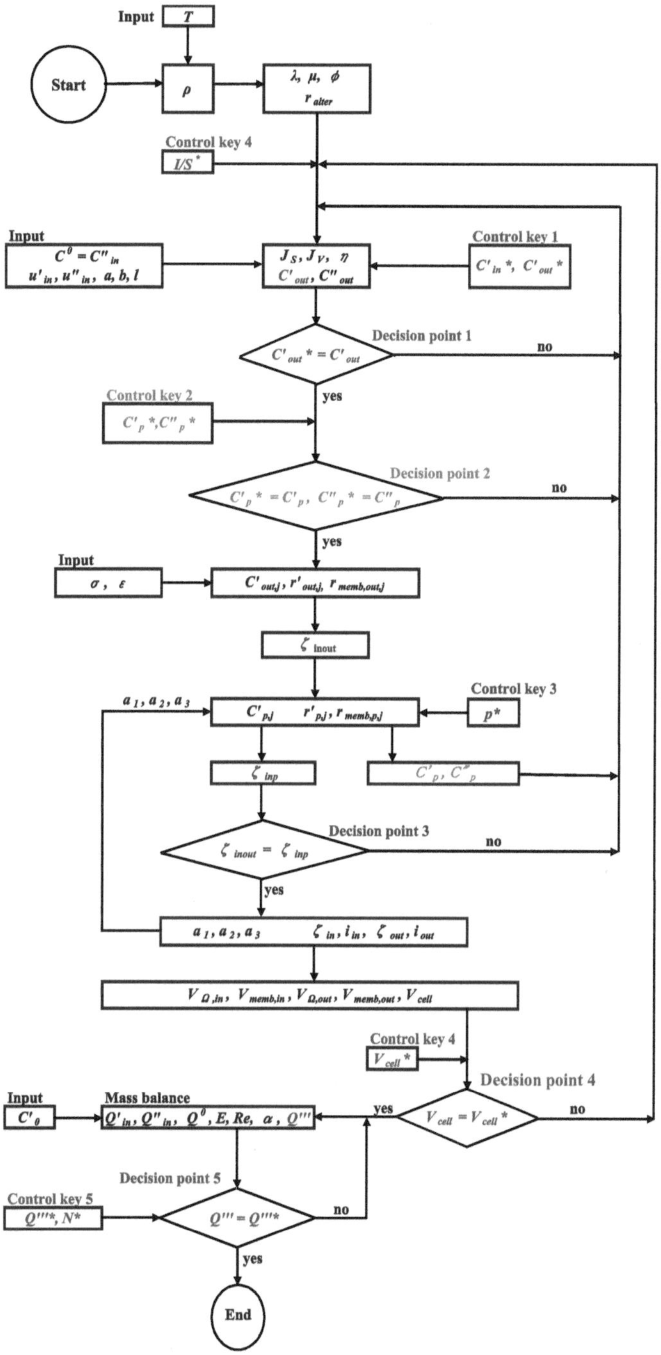

FIGURE 14.2 Feed-and-bleed electrodialysis program chart. *Tanaka (2014), p. 130, Fig. 4.*

14.6.3 Limiting Current Density

Succeeding to the above computation (Decision points 1–5), $(I/S)_{lim}$ is computed as described in Sections 13.13, 13.14.3 and Fig. 13.7. $(I/S)_{lim}$ is independent of N.

14.7 Companion Site (Chapter 24)

14.7.1 Companion Site

The Companion site 4 (Constant voltage feed-and-bleed program) is developed from the above program (Section 14.6).

14.7.2 Steps in the Companion Site

The companion site consists of the following 10 steps. Steps 1–8 and 10 are equivalent to the steps in the single-pass programs (Section 13.15.2).

Step 1 Fundamental specifications of an electrodialyzer and operating conditions
Step 2 Fundamental operating conditions of an electrodialyzer
Step 3 Solution electric resistance and salt activity coefficient in concentrating cells
Step 4 Solution electric resistance and salt activity coefficient at the inlets of desalting cells
Step 5 Computation at $V_{in} = V_{out}$
Step 6 i_{in}, i_{out}, ohmic voltage, membrane potential, cell voltage, energy consumption
Step 7 Computation at $V_{in} = V_p$
Step 8 p, ζ_{in}, ζ_{out}, a_1, a_2, and a_3
Step 9 Feed-and-bleed material balance, energy consumption, desalting ratio, water recovery
Step 10 Limiting current density

14.7.3 Remarks in the Companion Site

1. Calculations in the decision points are marked with red. Inputs are marked with purple (or red). Important values are marked with blue.
2. Readers can operate the program by inputting the optional input (purple (or red)) and carrying out the trial-and-error calculations at the decision points (red).
3. After the computations in Steps 1–8 are finished, Steps 9 and 10 are computed. All computations are completed within about 15–25 min.
4. Summary is supplied at the end of the companion site for printing.

14.8 Process Specifications and ED Conditions

Table 13.1 lists the process specifications and ED conditions. Drinking water is produced in the constant voltage one-stage feed-and-bleed operation under the following conditions:

Cell voltage $V_{cell} = 0.2, 0.4, 0.6, 0.7, 0.8$ V/pair.

Salt concentration of a feeding solution C^0 = 1000, 2000, 3000 mg/dm^3 (ionic constituent ratio is the same to that of seawater)
Salt concentration of a desalted solution $C'_{out} = 400$ mg/dm^3 (drinking water)
Output of a desalted solution $Q''' = 30\,m^3/h$.

14.9 Computed Results

Figure 14.3 shows the relationship between cell voltage V_{cell} and current density I/S taking salt concentration of a feeding solution C^0 as a parameter. Cell pair number N is increased with the increase of C^0 and decrease of V_{cell} (Fig. 14.4) (cf. Remarks in Section 14.6.2). Energy consumption E is increased with V_{cell} and C^0 (Fig. 14.5). Water recovery Re is decreased with the increase of C^0 and the decrease of V_{cell} (Fig. 14.6).

In the circulation tank, the solution moves from the compartment II to the compartment I. Thus the process must be operated in the condition defined by Eqn (14.7). Figure 14.7 gives the relationship between V_{cell} and Q_{move}. When $C^0 = 1000$ mg/dm^3, Q_{move} becomes zero at $V_{cell} = 0.67$ V/pair. So, in order to operate the process according to Eqn (14.7), cell voltage must be kept as $V_{cell} < 0.67$ V/pair. When the salt concentration of the feeding solution $C^0 = 2000$ and 3000 mg/dm^3, Q_{move} does not reach zero in the range of V_{cell} applied in Fig. 14.7, so the system is estimated to be operated stably. However, limiting cell voltage $V_{cell,lim}$ appears at about 0.7 V/pair (Fig. 14.8). Thus, the electrodialyzer should be operated at $V_{cell} < 0.7$ V/pair.

Limiting current density $(I/S)_{lim}$ is plotted against I/S and shown in Fig. 14.8. The real limiting current density $(I/S)_{lim,real}$ is determined from the intersection between the plots and the $I/S = (I/S)_{lim}$ line indicated in the figure and it is estimated as: $(I/S)_{lim,real} = 0.63$ A/dm^2. Limiting cell voltage $V_{cell,lim}$ applied under the $(I/S)_{lim,real}$ is estimated to be 0.727 V/pair ($C^0 = 1000$), 0.705 V/pair (2000 mg/dm^3) and 0.699 V/pair (3000 mg/dm^3).

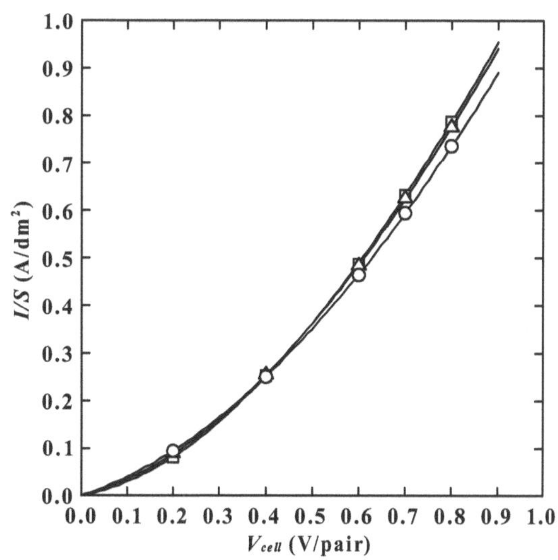

FIGURE 14.3 Cell voltage versus current density. $C^0 = 1000$ (○), 2000 (△), 3000 (□) mg/dm^3. *Tanaka (2014), p. 134, Fig. 6.*

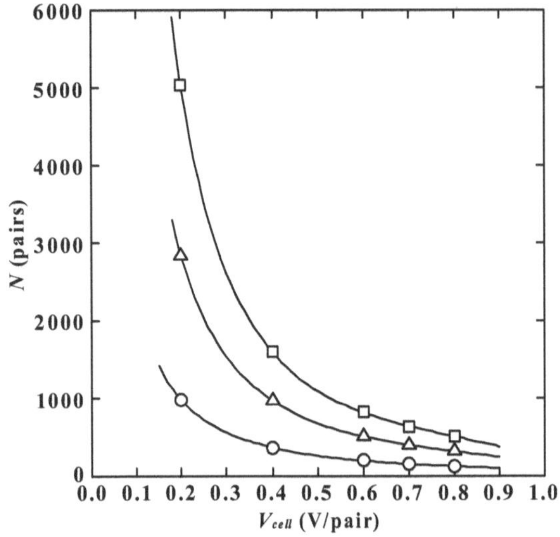

FIGURE 14.4 Cell voltage versus cell pair number. $C^0 = 1000$ (○), 2000 (△), 3000 (□) mg/dm^3. *Tanaka (2014), p. 134, Fig. 7.*

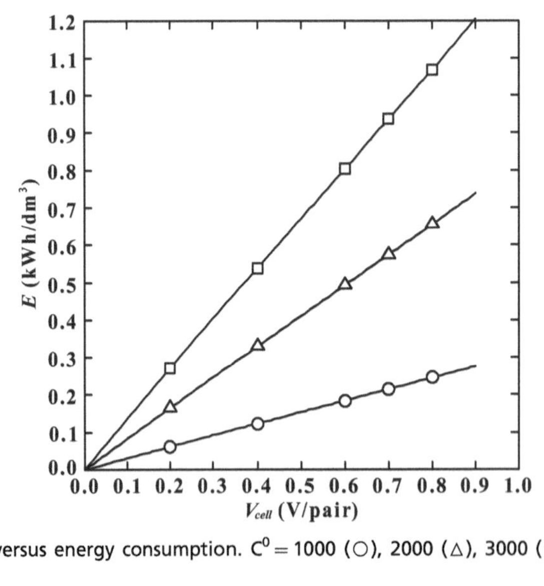

FIGURE 14.5 Cell voltage versus energy consumption. $C^0 = 1000$ (○), 2000 (△), 3000 (□) mg/dm^3. *Tanaka (2014), p.135, Fig. 8.*

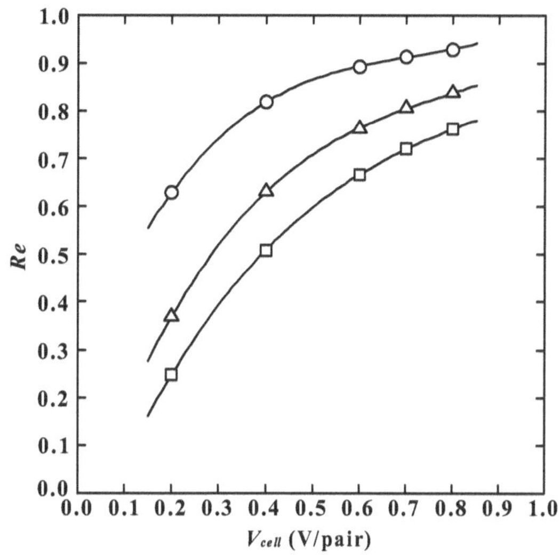

FIGURE 14.6 Cell voltage versus water recovery. $C^0 = 1000$ (○), 2000 (△), 3000 (□) mg/dm^3. *Tanaka (2014), p. 135, Fig. 9.*

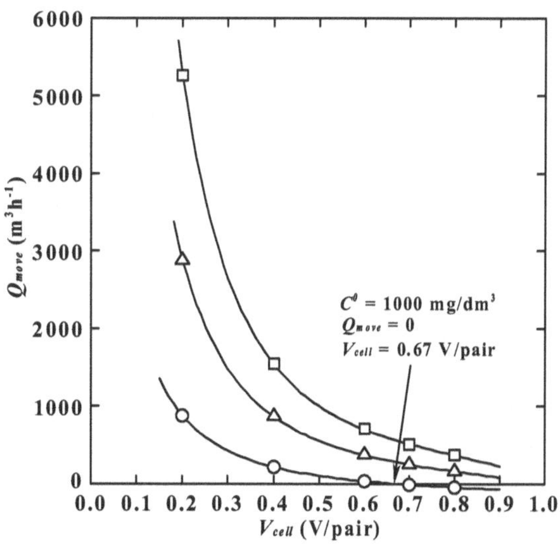

FIGURE 14.7 Cell voltage versus Q_{move}. $C^0 = 1000$ (○), 2000 (△), 3000 (□) mg/dm^3. *Tanaka (2014), p. 135, Fig. 10.*

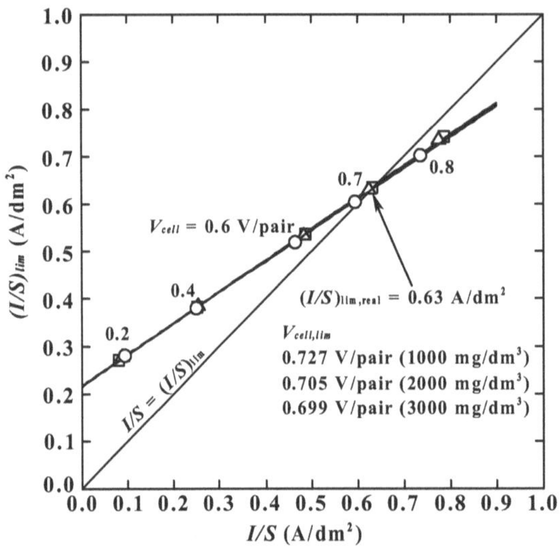

FIGURE 14.8 Limiting current density. $C^0 = 1000$ (○), 2000 (△), 3000 (□) mg/dm³. *Tanaka (2014), p. 136, Fig. 11.*

References

Hajdukova, J., Thomas, J., Malikova, P., Vidilaf, J., 2011. Demineralization of mine water by electrodialysis. Adv. Mater. Res. 378–379, 414–417.

Kawahara, T., 1994. Construction and operation experience of a large-scale electrodialysis water desalination plant. Desalination. 96, 341–348.

Rapp, H.J., Pfromm, P.H., 1998. Electrodialysis field test for elective chloride removal from the chemical recovery cycle of a kraft pulp mill. Ind. Eng. Chem. Res. 37, 4761–4767.

Ryabtsev, A.D., Kotsupalo, N.P., Tirarenko, V.I., Igmenov, I.K., Gelfond, N.V., Fedotova, N.E., Morozova, N.B., Shipachev, V.A., Tibilov, A.S., 2001. Development of a two-stage electrodialysis set-up for economical desalination of sea-type artesian and surface waters. Desalination. 137, 207–214.

Schoeman, J.J., Steyn, A., Makage, M., 2005. Evaluation of electrodialysis for the treatment of an industrial solid waste leachate. Desalination. 186, 273–289.

Tanaka, Y., 2014. Development of a computer simulation program of feed-and-bleed ion-exchange membrane electrodialysis for saline water desalination. Desalination. 342, 126–138.

Thompson, R., Paleologou, M., Jemaa, N., Wong, P.Y., Uloth, V., Van Heek, R., Berry, R.M., 1996. Electrodialysis separaters a sulfide-rich liquor from mill white liquor. Tappi J. 81, 159–164.

Thompson, R., Palaeologou, M., Wong, R.Y., Berry, R.M., 1997. Separation of sulphide from hydroxide in white liquor by electrodialysis. J. Pulp Pap. Sci. 23, J182–J187.

Tran, A.T.K., Zhang, Y., Jullok, N., Meesschaert, B., Pinoy, L., Van der Bruggen, B., 2012. RO concentrate treatment by a hybrid system of a pellet reactor and electrodialysis. Chem. Eng. Sci. 79, 228–238.

Zhang, Y., Ghyselbrecht, K., Meesschaert, B., Pinoy, L., Van der Bruggen, B., 2011. Electrodialysis on RO concentrate to improve water recovery in wastewater reclamation. J. Membr. Sci. 378, 101–110.

15

Computer Simulation Program: Batch Program

15.1 Introduction

The batch process is applicable for operating small- or middle-scale electrodialysis. The performance of the batch process has been discussed from the following viewpoints: Parulekar (1998) investigated energy consumption in a constant current and constant voltage batch operation. Demircioglu et al. (2001) introduced equations expressing the ionic mass balance around a diluate circulation tank and discussed energy consumption. Moon et al. (2004) predicted the performance of continuous and batch electrodialysis processes based on the fundamental principles of electrochemistry, transport phenomena, and thermodynamics. Ortiz et al. (2005) developed a mathematical model for a batch process and discussed mass balance, ohmic drop, and membrane potential.

In the batch process, an electrodialyzer and a circulation tank are also the main units working in the process as in the feed-and-bleed process (Chapter 14). The batch program is divided into two steps expressing the function of the electrodialyzer (step 1) and that of the circulation tank (step 2).

Step 1: Function of the Electrodialyzer

The single-pass (continuous) program is operated in the steady state. However, the batch process is operated in an unsteady state because the salt concentration in desalting cells decreases with operation time. Thus, the function of the electrodialyzer in the batch process cannot be written directly into the spreadsheet. In step 1, the function of the electrodialyzer describing the influences of salt concentration and cell voltage to the five parameters is introduced using the constant voltage single-pass program (Section 13.14) to compute 11 coefficients.

Step 2: Function of the Circulation Tank

The salt concentration in the circulation tank also decreases with operation time. The changes are usually expressed by the integral equation, which cannot be written in the spreadsheet. In step 2, the instantaneous open/shut solution feed operation is incorporated into the program and is written in the spreadsheet.

The 11 coefficients computed in step 1 (Section 15.4) are integrated into the spreadsheet with the equations developed in step 2 (Section 15.5). The computation in step 2 is carried out with the use of common software (Excel) and ordinary hardware (computer). Thus, the

Ion Exchange Membranes. http://dx.doi.org/10.1016/B978-0-444-63319-4.00015-8

overall computations are not carried out with the standalone program as developed in the single-pass and the feed-and-bleed processes. Saline water is desalinated to produce drinking water employing this batch program. The companion site of the program is described in (Chapter 24) (Companion sites 5 and 6).

15.2 Batch Electrodialysis Process

Figure 15.1 illustrates the one-stage (stage no. 1) batch electrodialysis process. In this process, valve V_1 is opened at first and a definite volume (Q_0) of a raw salt solution (salt concentration C^0) is supplied to the circulation tank. Next, V_1 is closed, V_2 is opened, and the solution is supplied to the inlets of desalting cells (De) integrated into the stack (marked gray) in the electrodialyzer. The solution flowing out from the outlets of desalting cells is returned to the circulation tank. The raw feeding solution is also supplied to the inlets of concentrating cells (Con) to prevent scale formation in the concentrating cells. A part of the solution flowing out from the outlets of concentrating cells is further supplied to electrode cells and partition cells (Part), which are integrated to prevent the influence of electrode reactions to the performance of the electrodialyzer. Then, the solution is discharged to the outside of the process with the residual solutions flowing out from the outlets of concentrating cells. Next, electrodialysis proceeds by applying constant voltage between electrodes and circulating the salt solution between the circulation tank and desalting cells. After the salt concentration in desalting cells is decreased to a targeted value, V_2 is closed, V_3 is opened, and the desalted solution (the output of the process) is extracted to the outside of the process. In the multistage process (stage no. 2), a desalted solution (salt concentration C'_{out}) produced in the previous stage is supplied to the circulation tank in the next stage to operate the next batch operation. In the multistage operation, the salt concentration supplied to the circulation tank is defined as $C'_{in,0}$ instead of C^0 defined in the one-stage operation.

15.3 Specifications and Operating Conditions of an Electrodialyzer

The description in Section 13.4 is applicable. The structure of a desalting cell and a concentrating cell is illustrated in Fig. 13.3.

15.4 Performance of an Electrodialyzer: Step 1 (Tanaka, 2013)

The performance of an electrodialyzer is defined by determining the influences of salt concentration at the inlets of desalting cells C'_{in} and cell voltage V_{cell} to the five parameters (salt concentration at the outlets of desalting cells C'_{out}, salt concentration at the outlets of concentrating cells C''_{out}, linear velocity at the outlets of desalting cells u'_{out}, current density

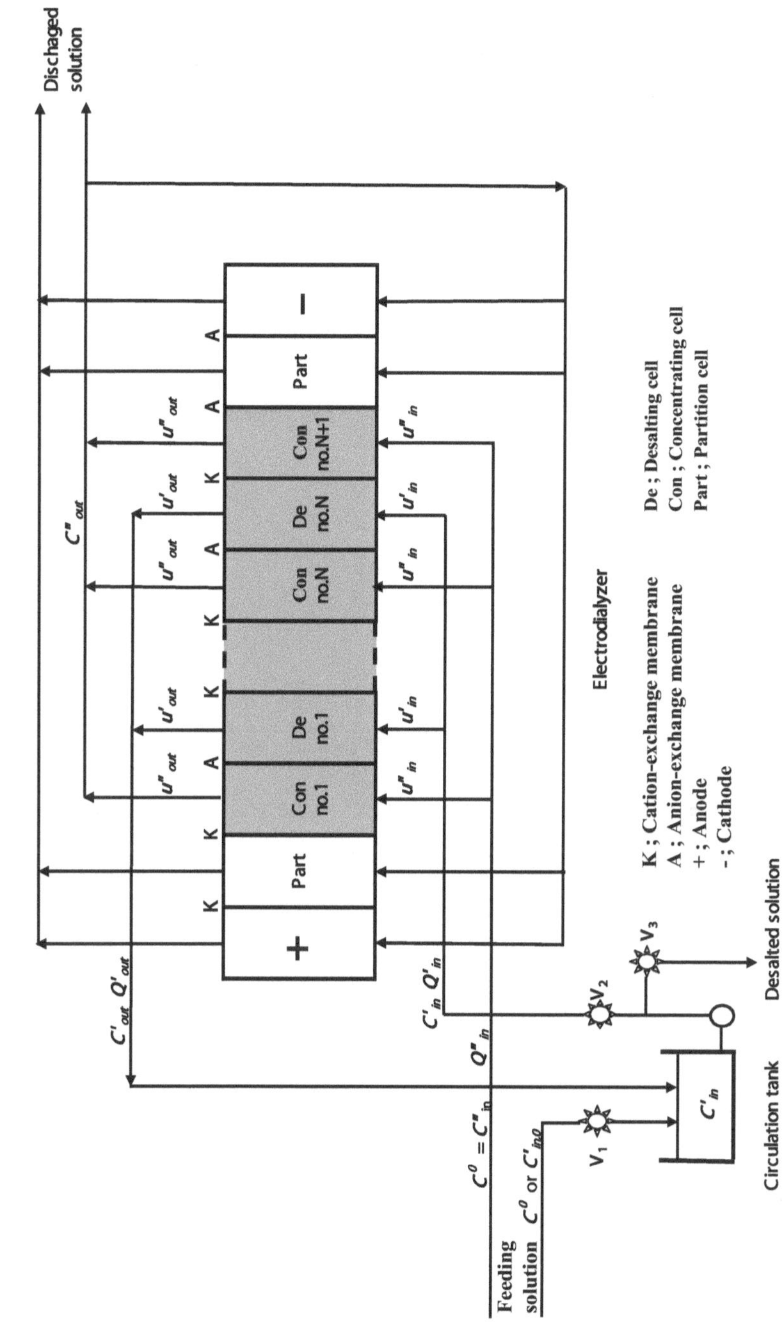

FIGURE 15.1 Batch electrodialysis process. *Tanaka (2013), p. 120, Fig. 1.*

I/S, and ion flux across a membrane pair J_S). The influences are calculated using the constant voltage single-pass (continuous) program (Section 13.14, Fig. 13.5, Companion site 2). Because of this calculation, the overall computations in the batch process are not carried out with the standalone program. The computation is carried out by inputting the electrodialyzer specifications and operating conditions listed in Table 13.1.

The relationships between C'_{in} and V_{cell} and five parameters (C'_{out}, C''_{out}, u'_{out}, I/S, and J_S) are introduced as shown in Figs 15.2–15.6 and the following equations:

$$C'_{out} = \left(\alpha_1 - \alpha_2 V_{cell}^{0.5}\right) C'_{in} \tag{15.1}$$

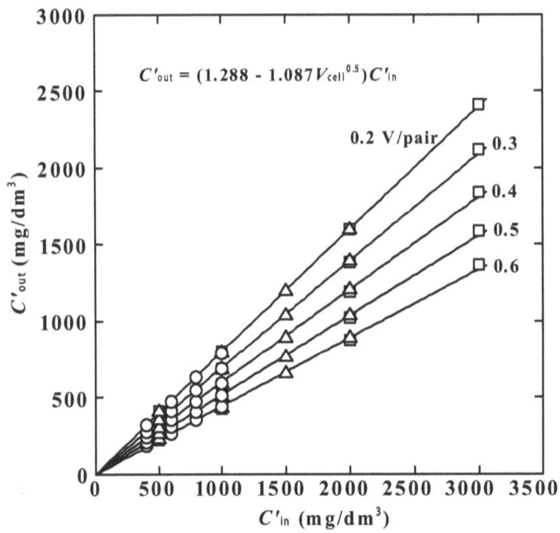

FIGURE 15.2 Salt concentration at outlets of desalting cells. $C^0 = 1000$ (○), 2000 (△), and 3000 (□) mg/dm³. *Tanaka (2013), p. 122, Fig. 5.*

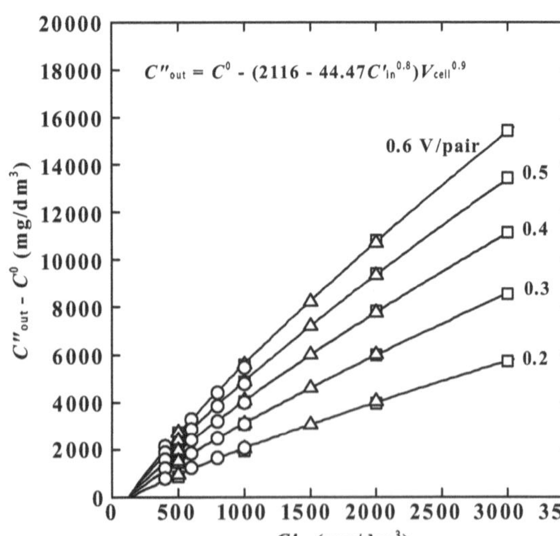

FIGURE 15.3 Salt concentration at outlets of concentrating cells. $C^0 = 1000$ (○), 2000 (△), and 3000 (□) mg/dm³. *Tanaka (2013), p. 122, Fig. 6.*

$$C''_{out} = C^0 - \left(\beta_1 - \beta_2 C'^{0.5}_{in}\right) V^{0.9}_{cell} \tag{15.2}$$

$$u'_{out} = \gamma_1 + \gamma_2 V^{1.4}_{cell} - \gamma_3 V^{0.9}_{cell} C'^{0.8}_{in} \tag{15.3}$$

$$I/S = \left(\delta_1 C'^{0.8}_{in} - \delta_2\right) V^{0.9}_{cell} \tag{15.4}$$

$$J_S = \left(\varepsilon_1 C'^{0.8}_{in} - \varepsilon_2\right) V^{0.9}_{cell} \tag{15.5}$$

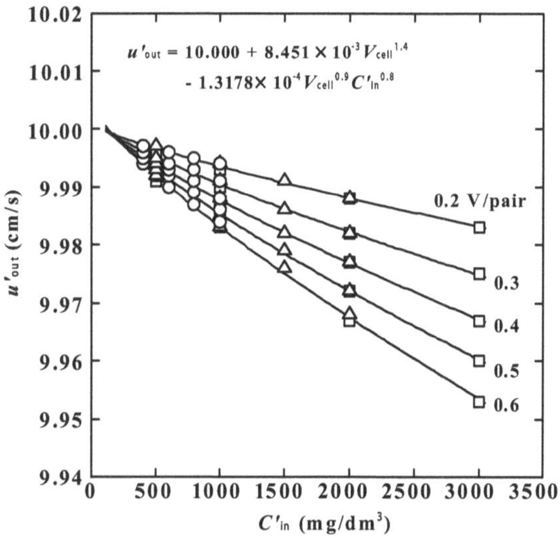

FIGURE 15.4 Linear velocity at outlets of desalting cells. $C^0 = 1000$ (○), 2000 (△), and 3000 (□) mg/dm^3. *Tanaka (2013), p. 123, Fig. 7.*

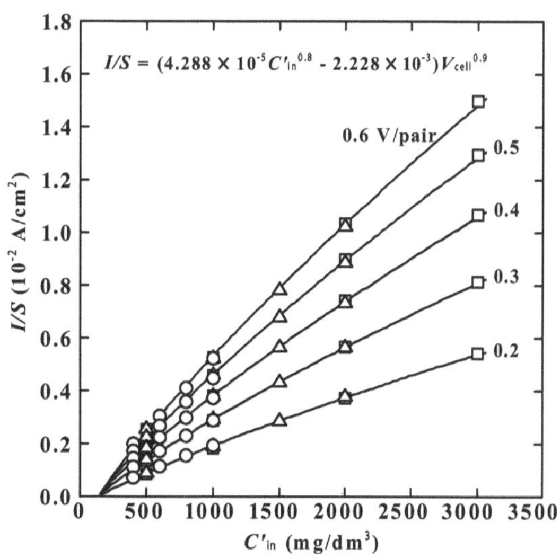

FIGURE 15.5 Current density. $C^0 = 1000$ (○), 2000 (△), and 3000 (□) mg/dm^3. *Tanaka (2013), p. 123, Fig. 8.*

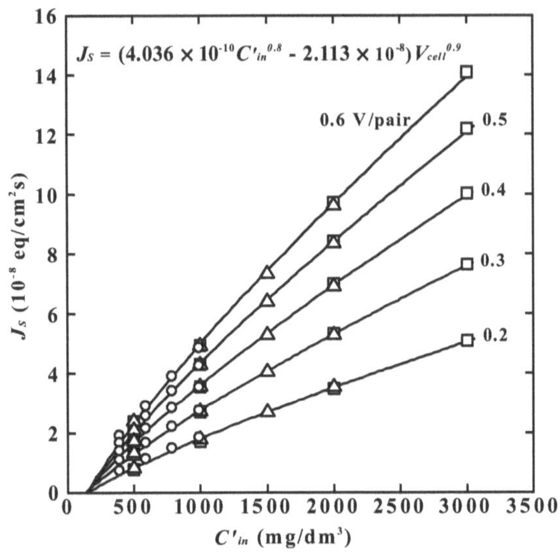

FIGURE 15.6 Ion flux. $C^0 = 1000$ (○), 2000 (△), and 3000 (□) mg/dm^3. *Tanaka (2013), p. 123, Fig. 9.*

Eleven coefficients (α_1, α_2, β_1, β_2, γ_1, γ_2, γ_3, δ_1, δ_2, ε_1, and ε_2) appearing in Eqns (15.1)–(15.5) are shown in Figs 15.2–15.6. If the electrodialyzer specifications and operating conditions in Table 13.1 are altered, the above equations must be reintroduced.

(Remark) Computation in Step 1 is carried out by inputting the cell pair number N (in the electrodialyzer) into the program. However, the computed results (11 coefficients) are not influenced by N.

15.5 Relationship between Operation Time and the Performance of an Electrodialyzer in a Batch Operation: the Open/Shut Solution Feeding Operation: Step 2 (Tanaka, 2013)

The equations are developed by incorporating the imaginary feeding cell, Fe, and shutter, Sh, at the outlet of the circulation tank, as illustrated in Fig. 15.7, in which a definite volume (Q_0 dm^3) of a solution (salt concentration: C^0 or $C'_{in,0}$ mg/dm^3) is prepared at first.

To proceed with the computation, operation time t is divided at an interval of Δt min. The shutter repeats an instantaneous imaginary open/shut operation and supplies a definite flow rate (Q'_{in}) of a solution from the circulation tank to desalting cells through the feeding cell at an interval time of Δt min.

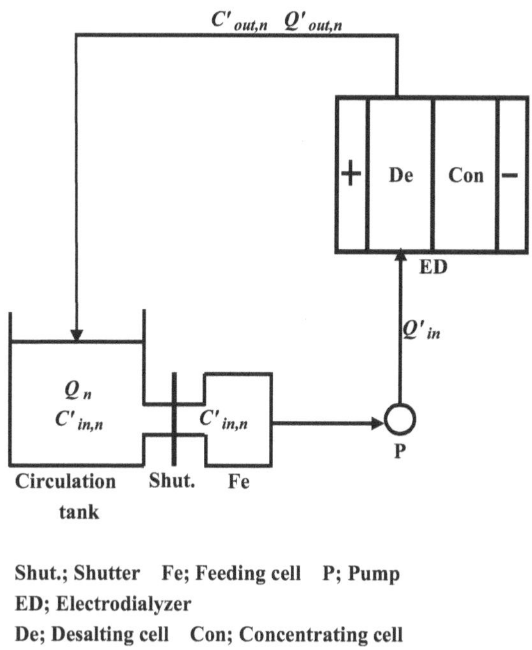

FIGURE 15.7 Instantaneous open/shut solution feed batch operation. *Tanaka (2013), p. 123, Fig. 10.*

$$Q'_{in}(\text{dm}^3/\Delta t\text{min}) = abNu'_{in} \times (60\Delta t) \times 10^{-3} \tag{15.6}$$

$$Q'_{out}(\text{dm}^3/\Delta t\text{min}) = abNu'_{out} \times (60\Delta t) \times 10^{-3} \tag{15.7}$$

$u'_{in} \fallingdotseq u'_{out}$ hold in actual electrodialysis, but this approximation is not taken into account in program development.

Before the step calculation,

$C'_{out,0}$ is computed from Eqn (15.1).

$u'_{out,0}$ is calculated from Eqn (15.3).

$Q'_{out,0}$ is computed from Eqns (15.3) and (15.7).

$C''_{out,0}$ is calculated from Eqn (15.2).

Steps during the operation time are numbered as $n=1$ ($t=\Delta t$ min), $n=2$ ($t=2\Delta t$ min), $n=3$ ($t=3\Delta t$ min), … $n=n$ ($t=n\Delta t$ min). The solution in the circulation tank is mixed vigorously and its concentration is the same as the concentration in the feeding cell and at the inlets of the electrodialyzer (desalting cells) C'_{in}. Table 15.1 gives the salt concentration changes in the circulation tank (in the feeding cell) ($C'_{in,n}$) and at the outlets of the electrodialyzer ($C'_{out,n}$) and solution volume in the circulation tank (Q_n) in each step n. The mass balance and volume balance in the circulation tank in each step are introduced as follows:

$n=1$

Table 15.1 Changes in Salt Concentration and Solution Volume in a Batch Electrodialysis Operation

Step n	Time t	Circulation Tank		Feeding Cell (ED Inlet)	ED Outlet	
	min	mg/dm^3	dm^3	mg/dm^3	mg/dm^3	dm^3/Δt min
0	0	$C'_{in,0}$	Q_0	$C'_{in,0}$	$C'_{out,0}$	$Q'_{out,0}$
1	Δt	$C'_{in,1}$	Q_1	$C'_{in,1}$	$C'_{out,1}$	$Q'_{out,1}$
2	$2\Delta t$	$C'_{in,2}$	Q_2	$C'_{in,2}$	$C'_{out,2}$	$Q'_{out,2}$
n	$n\Delta t$	$C'_{in,n}$	Q_n	$C'_{in,n}$	$C'_{out,n}$	$Q'_{out,n}$
$n+1$	$(n+1)\Delta t$	$C'_{in,n+1}$	Q_{n+1}	$C'_{in,n+1}$	$C'_{out,n+1}$	$Q'_{out,n+1}$

Tanaka, 2013, p. 124, Fig. 2.

Mass balance:

$$C'_{in,1} = \frac{C'_{in,0}Q_0 - C'_{in,0}Q'_{in} + C'_{out,0}Q'_{out,0}}{Q_1} \quad (15.8)$$

Volume balance:

$$Q_1 = Q_0 - Q'_{in} + Q'_{out,0} \quad (15.9)$$

$n = 2$

Mass balance:

$$C_{in,2} = \frac{C'_{in,1}Q_1 - C'_{in,1}Q'_{in} + C'_{out,1}Q'_{out,1}}{Q_2} \quad (15.10)$$

Volume balance:

$$Q_2 = Q_1 - Q'_{in} + Q'_{out,1} \quad (15.11)$$

$C'_{in,1}$ and Q_1 are calculated using Eqns (15.8) and (15.9). $C'_{out,1}$ and $Q'_{out,1}$ are calculated using Eqns (15.1) and (15.3), (15.7).

$n = n$

Mass balance:

$$C'_{in,n} = \frac{C'_{in,n-1}Q_{n-1} - C'_{in,n-1}Q'_{in} + C'_{out,n-1}Q'_{out,n-1}}{Q_n} \quad (15.12)$$

Volume balance:

$$Q_n = Q_{n-1} - Q'_{in} + Q'_{out,n-1} \quad (15.13)$$

$C''_{out,n}$ and I/S_n are computed substituting $C'_{in,n}$ and V_{cell} into Eqns (15.2) and (15.4). I/S_n and $J_{S,n}$ are computed using Eqns (15.4) and (15.5). $(I/S)_{average}$ is computed using the following equation:

$$\left(\frac{I}{S}\right)_{average} = \frac{\sum_{n=1}^{n} I/S_n}{n} \quad (15.14)$$

Current efficiency η_n, desalting ratio α_n, water recovery Re_n, and energy consumption E_n are computed as follows:

$$\eta_n = \frac{F\sum_{n=1}^{n} J_{S,n}}{\sum_{n=1}^{n} I/S_n} \tag{15.15}$$

$$\alpha_n = 1 - \frac{C'_{in,n}}{C'_{in,0}} \tag{15.16}$$

$$\mathrm{Re}_n = \frac{Q_n}{Q_0 + Q''_{in}(\mathrm{dm^3/min}) \times t(\mathrm{min})} \tag{15.17}$$

$$Q''_{in}(\mathrm{dm^3/min}) = ab(N+1)u''_{in} \times (60\ \mathrm{s/min}) \times 10^{-3} \tag{15.18}$$

$$E_n(\mathrm{kWh/m^3}) = \frac{S(\mathrm{cm^2})NV_{cell}(I/S)_{average}(\mathrm{A/cm^2}) \times 10^{-3}}{\frac{Q_n(\mathrm{dm^3})}{t(\mathrm{min})} \times 60\,(\mathrm{min/h}) \times 10^{-3}} \tag{15.19}$$

(Remark) In the program, it is not necessary for N (in the batch process) inputted in Step 2 (Section 15.5) to be the same as N (in the electrodialyzer) inputted in Step 1 (Section 15.4). $(I/S)_{\mathrm{lim}}$ is independent of N.

15.6 Program and Computation with Companion Site (Chapter 24)

15.6.1 Saline Water Desalination (Companion Site 5)

The performance of one-stage batch electrodialysis for saline water desalination is computed by inputting C^0 $(=C'_{in,0}) = 2000\ \mathrm{mg/dm^3}$, $V_{cell} = 0.4$ V/pair, $N = 300$ pairs, $Q_0 = 30\ \mathrm{m^3}$, and $\Delta t = 4$ min (marked with purple). The fundamental parameters are listed in Table 13.1. In Figs 15.8–15.10, the salt concentration at the inlets of desalting cells C'_{in}, current density I/S, and energy consumption E are plotted against operation time t taking cell voltage V_{cell} as a parameter. Readers can operate the program by inputting the optional input (purple). If the electrodialyzer specifications and operating conditions in Table 13.1 are altered, the 11 coefficients must be reintroduced in the Step 1 (Section 15.4). Operation time t can be changed by copying and pasting.

15.6.2 Drinking Water Production (Companion Site 6)

The threshold of acceptable aesthetic criteria for human drinking water is 500 mg/dm^3. Thus, drinking water (salt concentration, 400 mg/dm^3) is produced using the program. The performance of the one-stage electrodialysis spreadsheet is computed inputting C^0 $(=C'_{in,0}) = 2000\ \mathrm{mg/dm^3}$, $V_{cell} = 0.4$ V/pair, $Q_0 = 30\ \mathrm{m^3}$, $\Delta t = 4$ min (marked with purple), and operation time $t = 60$ min. The fundamental parameters are also listed in Table 13.1. Computation is carried out by adjusting cell pair number (desalting cell number)

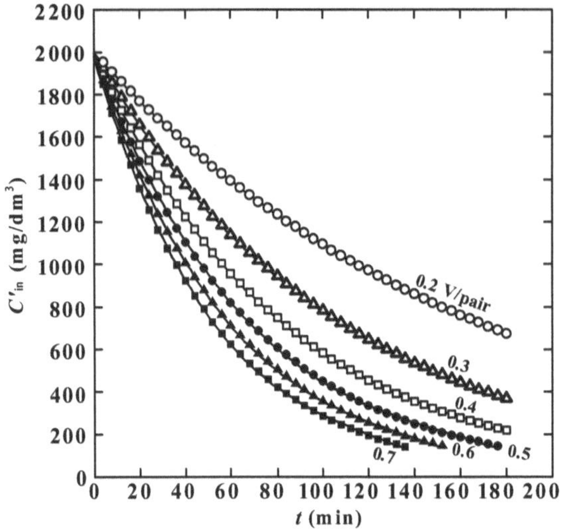

FIGURE 15.8 Salt concentration at inlets of desalting cells. $C^0 = 2000$ mg/dm^3, $Q_0 = 30$ m^3, and $N = 300$ pairs.

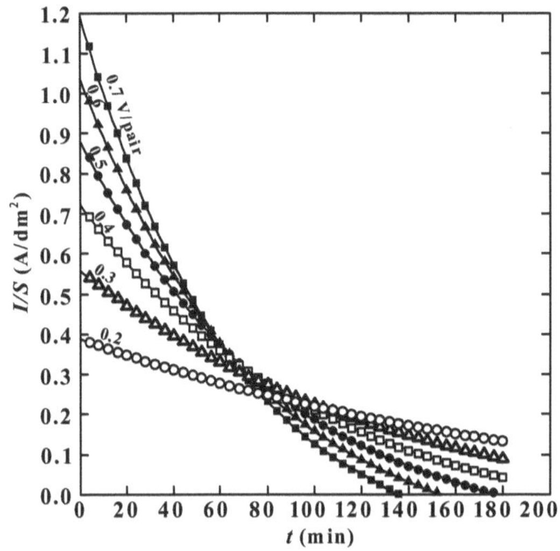

FIGURE 15.9 Current density. $C^0 = 2000$ mg/dm^3, $Q_0 = 30$ m^3, and $N = 300$ pairs.

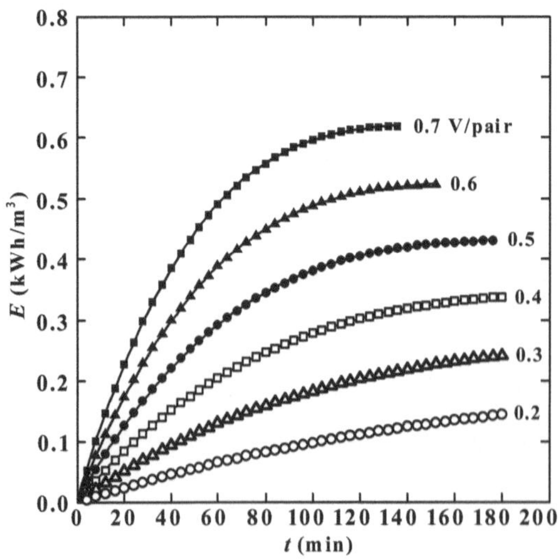

FIGURE 15.10 Energy consumption. $C^0 = 2000$ mg/dm^3, $Q_0 = 30$ m^3, $N = 300$ pairs.

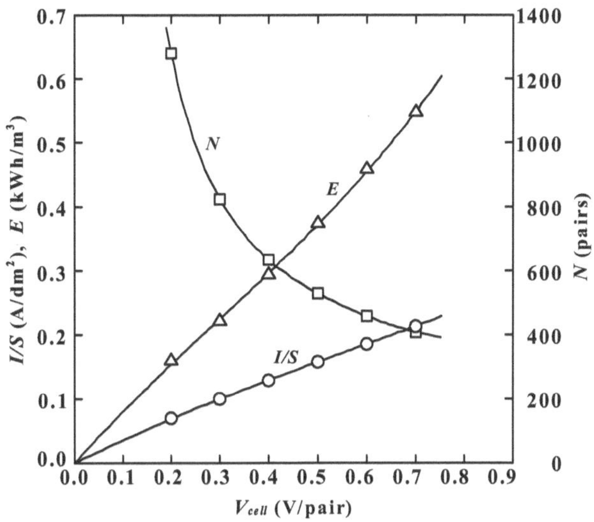

FIGURE 15.11 Current density, energy consumption, and cell pair number. $C^0 = 2000$ mg/dm^3, $C'_{in} = 400$ mg/dm^3, $Q_0 = 30$ m^3, and $t = 60$ min.

N^*(control key, marked with red) to realize $C'_{in} = 400$ mg/dm^3 at $t = 60$ min (marked with blue) (cf. Remark in Section 15.5). Figure 15.11 shows cell voltage V_{cell} versus current density I/S, energy consumption E, and cell pair number N. Readers can operate the program by inputting the optional input (purple). If the electrodialyzer specifications and operating conditions in Table 13.1 are altered, the 11 coefficients must be reintroduced in the Step 1 (Section 15.4).

References

Demircioglu, M., Kabay, N., Ersoz, E., Kurucaovali, I., Safak, S., Gizli, N., 2001. Cost comparison and efficiency modeling in the electrodialysis of brine. Desalination 136, 317–323.

Moon, P., Sandi, G., Stevens, D., Kizilel, R., 2004. Computational modeling of ionic transport in continuous and batch electrodialysis. Sep. Sci. Technol. 39, 2531–2555.

Oritz, J.M., Sotoca, J.A., Exposito, E., Gallud, F., Garcia-Garcia, V., Montiel, V., Aldaz, A., 2005. Brackish water desalination by elrctrodialysis: batch recirculation operation modeling. J. Membr. Sci. 252, 65–75.

Parulekar, S.J., 1998. Optimal current and voltage trajectories for minimum energy consumption in batch electrodialysis. J. Membr. Sci. 148, 91–103.

Tanaka, Y., 2013. Development of a computer simulation program of batch ion-exchange membrane electrodialysis for saline water desalination. Desalination 320, 118–133.

16

Electrodialysis Reversal

16.1 Overview of Technology

Electrodialysis reversal (EDR) was developed from electrodialysis (ED) technology by Ionics in 1974 (Reahl, 2004). The membranes were used to provide a continuous self-cleaning electrodialysis process with periodic reversal of the direct current (DC) polarity. The DC electric field was alternately reversed to dissolve salt scale deposited on the membranes. DC field reversal eliminated the need to feed either acid or anti-scalant chemicals into the desalination process. Ionics further improved membrane spacers, system design, and operating efficiency of the EDR system in 1997. On lower total dissolved solids (TDS) waters (less than 1500 ppm), EDR electrical power consumption can be less than reverse osmosis (RO). This chapter introduces the development of EDR and its operating performance.

16.2 Spacer

Ion flux across the membrane is strongly influenced by the boundary layer formed at the membrane–solution interface. The boundary layer thickness is a function of velocity and flow spacer geometry, and it may be minimized by promoting turbulent flow, thus increasing the limiting current density (Belfort and Guter, 1972; Chiapello and Bernard, 1993; Zhong et al., 1983). The operating velocity in an ED stack is limited owing to the pressure drop along the flow spacer and through the stack. There is a maximum inlet pressure at which to operate a membrane stack to prevent external leakage. If the pressure drop is too great, the number of stages is limited. Thus, it is desirable to incorporate a turbulence promoter that optimizes performance by maximizing the limiting current density and minimizing pressure drop.

The type of flow spacer that has had the most commercial success in terms of the total installed capacity of ED/EDR plants is the tortuous path spacer Mark III. This spacer is manufactured by using two sheets of low-density polyethylene with a die-cut flow channel. The two sheets of polyethylene are glued together to form an under/over flow path that creates turbulence. Because the under/over straps that create the turbulence are infrequent, the amount of turbulence promotion is limited. At higher velocities, however, the turbulence is sufficient to obtain reasonable limiting current densities. The pressure drop per unit length of this design is low, so a long flow path can be used without having excessive pressure drops. By using a long tortuous flow path and operating at velocities in the 20–40 cm/s range, this design optimizes turbulence

Ion Exchange Membranes. http://dx.doi.org/10.1016/B978-0-444-63319-4.00016-X

promotion and stack pressure drop. The disadvantage of this approach is that the tortuous design has significant membrane area wasted as a sealing area. In a typical commercial design, 36% of the membrane area is shadowed by the sealed area, so only 64% of the membrane area is usable for desalting. Figure 16.1 illustrates the Mark III.

For practical long-term operation of large electrodialysis systems, the thickness tolerances of flow spacers are critical. Because 1000 or more spacers are piled on top of each other in an electrodialysis stack, even slight variations in component thickness can lead to large variations in the stack heights. The thickness of the sealing area around the flow path must be close to the thickness of the flow spacer. If the sealing area is thicker than the turbulence promoters, the turbulence promoters do not touch the membrane surface. This can cause laminar flow at the membrane surface, which in turn reduces limiting current density significantly.

Several manufacturers of electrodialysis equipment have employed screen spacers to promote turbulence. In these spacers, woven or non-woven netting is used in the flow path to create turbulence, and the performance of the screen spacer is better than that of the tortuous path spacer Mark III. In the 1990s, Ionics developed bench scale screen

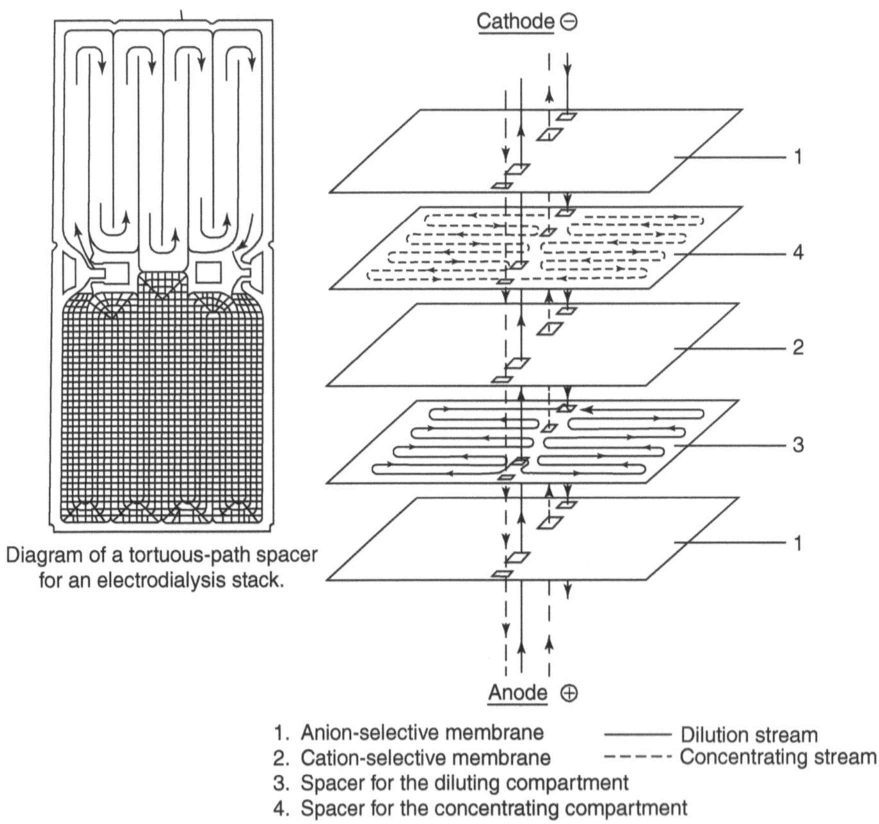

FIGURE 16.1 Tortuous flow path spacer (Mark III). *Lacey and Loeb (1972).*

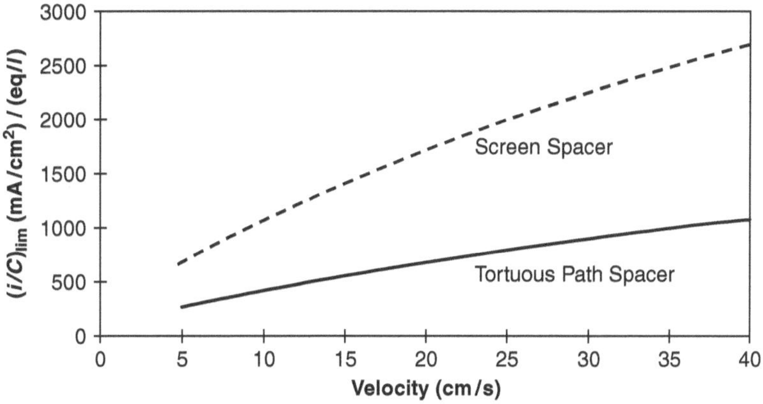

FIGURE 16.2 $(i/C)_{\text{lim}}$ versus solution velocity. 70 °F. *von Gottberg (1998), p. 5, Fig. 4.*

spacer Mark IV; performance in terms of limiting current density and pressure drop was measured in the laboratory for a sodium chloride solution. Figure 16.2 shows a graph of limiting current density versus solution velocity for conventional Mark III and the Mark IV manufactured with non-woven netting. For a given velocity, the limiting current density for the Mark IV is about three times that of the Mark III. The pressure drop per unit length is much greater for the Mark IV than the Mark III (Figure 16.3). To minimize the pressure drop, the optimum velocity for the Mark IV is in the 6–12 cm/s range, and a short flow-path length is selected.

The outside dimensions of 18″ × 40″ for the high-performance Mark IV were based on the dimensions of preexisting ion exchange membrane production lines and the requirement for membranes and spacers to be easy to handle by one person for stack maintenance. A *U*-shaped flow path was developed to fit the optimum flow-path length and width into the 18″ × 40″ configuration. The Mark IV has about 74% usable area in

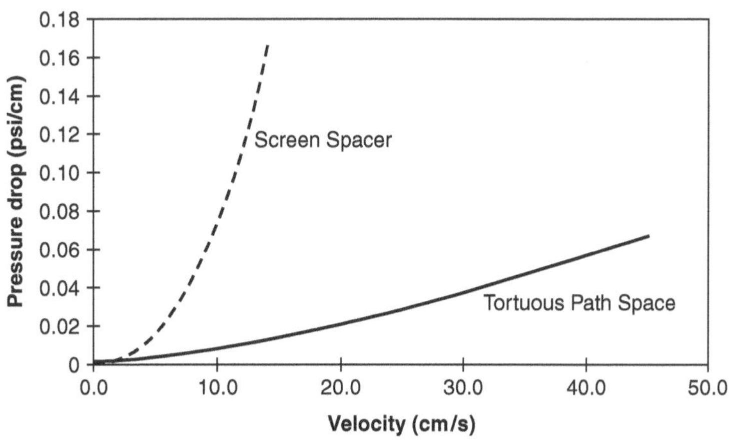

FIGURE 16.3 Pressure drop per unit length versus solution velocity. *von Gottberg (1998), p. 5, Fig. 6.*

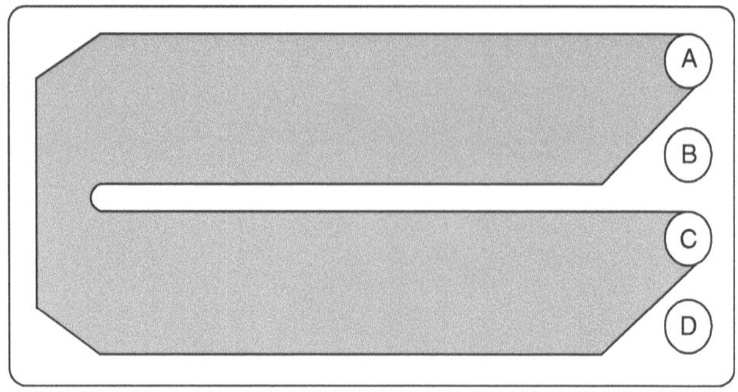

FIGURE 16.4 *U*-shaped flow path spacer (Mark IV). *von Gottberg (1998), p. 6, Fig. 7.*

contrast to 64% for the conventional Mark III (Figure 16.4). The Mark IV spacer thickness is 0.030″, versus 0.040″ for the Mark III spacers. The maximum stack height of the Mark III is 500 cell pairs. This height is determined by the maximum safe DC voltage that can be applied to a stack and the dimensions of a standard shipping container in which the stack can fit. With the Mark IV, the maximum stack height is 600 cell pairs. The combination of more usable area per membrane and more cell pairs per stack has led to an increase of 38% usable membrane area in an electrodialysis stack. The increase in membrane area per stack combined with the increased current density that can be applied to a stack means that fewer stacks are required to desalt a given volume of water. This leads to capital and building cost savings for the customer.

In electrodialysis plant design, there is a tradeoff between power consumption and capital cost. For any particular spacer, the greater the number of cell pairs, the lower the DC power consumption, and vice versa (Figure 16.5). In developing the Mark IV, it was important to make sure that capital cost savings resulting from to fewer stacks and cell pairs were not accompanied by increases in power consumption. The use of the thinner spacer means that the electrical resistance of each cell pair is reduced, so that the overall DC power consumption for a given number of cell pairs decreases. Therefore, at

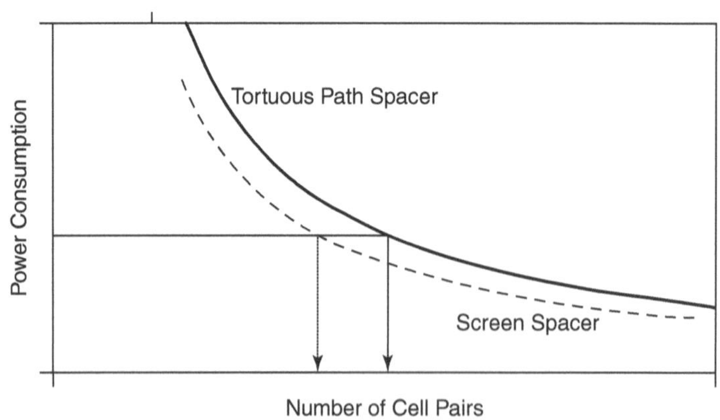

FIGURE 16.5 Power consumption versus number of cell pairs. *von Gottberg (1998), p. 6, Fig. 8.*

constant DC power, fewer cell pairs can be achieved with a use of the Mark IV, providing capital cost savings without increases in operating costs. The overall system pressure drop is lower because fewer stages are required; this in turn reduces pumping power. Use of the Mark IV offers significant savings in capital costs and can reduce the overall power consumption of an EDR system (von Gottberg, 1998).

16.3 Water Recovery

The source water feed to an EDR system is divided into three streams (Figure 16.6). The largest flow is the dilute feed, which passes through the stacks once to produce a desalted product. The second stream, concentrate makeup, is fed into a concentrate stream being re-circulated through the stacks and generates overflow waste at the exit from the stacks. The third stream is electrode compartment flush water, which on most system is 1% of the dilute flow. Normally this water is directed to waste after it passes through the stacks. This system enhances water recovery in the process as follows (Allison, 1993).

A small amount of the dilute stream water flow is transferred to the concentrate stream as it passes through the stack. First, water is transferred through the membrane with the passage of ions. This water transfer amounts to 0.15–0.45% of the dilute flow per 1000 mg/l of salts removed. Internal leakage also occurs between the dilute stream and the concentrate stream, which is approximately 0.25–0.50% per stage. The concentrate stream is normally operated about 1 psi lower in pressure than the dilute stream, so this leakage will be from the dilute stream to the concentrate stream. Water transfer and cross-leakage do not reduce water recovery. These flows are calculated as part of the concentrate makeup flow in the plant design.

Periodic reversal of the DC power polarity interchanges the solutions in the individual spacers. Spacers carrying the dilute flow in negative polarity carry concentrate in the positive polarity. Automatic inlet and outlet diversion valves direct the dilute and concentrate flows to the correct spacers for each polarity.

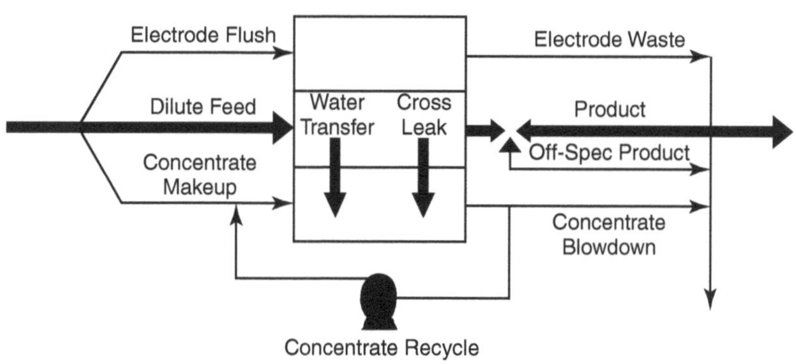

FIGURE 16.6 Electrodialysis reversal flow diagram. *Allison (1993), p. 1, Fig. 1.*

While the concentrate is being flushed out during reversal, poor-quality brine comes out. This is diverted to waste by a conductivity-controlled automatic valve. By timing the operation of the automatic diversion valves and the DC polarity reversal of each stack to the flow through the system, high-conductivity brine is produced for only 36 s. This phased polarity reversal occurs every 15–30 min. The high conductivity or off specification brine from the phased reversal represents 2–4% waste. This waste is high TDS concentrate and cannot be recycled to the feed or concentrate makeup.

The highest water recovery EDR plant in operation is the 3.76 million gallons per day (mgd) facility operated by the City of Suffolk, Virginia (Section 16.8.1). The plant has three stages and achieves 94% water recovery. Table 16.1 shows a flow summary for one of the three EDR units (Suffolk) broken down to show where the waste is generated. The units have no continuous chemical feeds. A 14.9 gallon per minute (gpm) concentrate (brine) makeup flow is used to control calcium carbonate scaling potential in the concentrate. If the units did not require this controlled concentrate makeup flow, waste could be reduced to 43.1 gpm, giving an ultimate recovery capacity of 95.3%.

A 5 mgd EDR plant located on Grand Canary Island operates on 5000– 7000-mg/l feed water at 85% water recovery. The unit flow summary is shown also in Table 16.1. The controlled concentrate (brine) makeup flow is approximately half of the total waste. The ultimate water recovery capability based on flows of this plant is only 91.8%.

16.4 Prevention of Scale Formation

The water recovery capability of nearly all EDR plants is limited by the potential for salts of limited solubility to precipitate from the concentrate stream as scale. The polarity reversal of EDR alternately exposes membrane surfaces and the water flow paths to concentrate that tends to precipitate scale, and desalted water, which tends to dissolve scale. This allows the process to operate with super-saturated concentrate streams

Table 16.1 Electrodialysis Reversal Unit Flow Summary

	Suffolk, Virginia	Grand Canary Island	
Total EDR feed	926	409.1	gpm
Total EDR product	870.4	347.7	gpm
Total water recovery	94	85	%
Waste sources			
Off-specification product	17.4	7.8	gpm
Brine makeup flow	14.9	27.8	gpm
Brine makeup (from electrode flush)	9.3		gpm
Electrode flush (sent to waste)		4.1	gpm
Brine makeup (from cross-leak)	12.5	11.4	gpm
Brine makeup (from water transfer)	1.5	10.3	gpm
Total waste	55.6	61.4	gpm

Allison (1993), p. 2, Table 1; Table 2.

without chemical additions to prevent scale formation, as follows (Allison, 1993). If chemicals are added to the concentrate, higher levels of super-saturation can be achieved.

The design limit for $CaCO_3$ without anti-scalants is a Langelier Saturation Index (LSI) (Section 12.3.3.1) of +1.8. Actual scale precipitation starts at an LSI near +2.2. This limit is sufficiently high that few EDR units need chemical additions to prevent $CaCO_3$ scaling from high concentrate LSI. Where chemical addition is used, acid or carbon dioxide addition to the concentrate is employed.

The design limit for $CaSO_4$ saturation without anti-scalant addition is an ion product of $2.25 \times K_{sp}$ (K_{sp}: ion product specific limit). The real limit at which precipitation starts is near $4 \times K_{sp}$. Dell City, Texas has operated a 150,000 gpd EDR system since 1975 with the concentrate $CaSO_4$ ion product at $3.5 \times K_{sp}$ without adding anti-scalant. The limit cannot be increased by adding acid or CO_2 because pH adjustment does not increase the solubility of $CaSO_4$. The only way to increase water recovery is to use anti-scalants (Section 12.3.3.2). In 1981 and 1982, the EDR was operated at high $CaSO_4$ saturation with sodium hexameta phosphate anti-scalant addition at Roswell, New Mexico. The final test in this program was operation of a 50,000 gpd EDR unit with the concentrate $CaSO_4$ ion product at an average of $12.5 \times K_{sp}$ for over 5000 operating hours.

It is known that the threshold point for $BaSO_4$ is an ion product of $100 \times K_{sp}$ when no anti-scalants are used. The limit with anti-scalant addition is currently unknown. Several plants are operating at a $BaSO_4$ ion product of $150 \times K_{sp}$ and two plants have reached an ion product of $225 \times K_{sp}$ with a polymer anti-scalant dosage of 2–5 mg/l to the concentrate.

The $SrSO_4$ scale has not been detected in any EDR plants and CaF_2 precipitation has never occurred in an EDR plant. Plants are operating with ion products $500 \times K_{sp}$, and much higher levels were realized in two industrial pilot studies with no CaF_2 precipitation. Silica is essentially nonionic at pH < 9.5 and is not removed or concentrated by EDR. EDR plant tests show that silica levels are equal in the feed, product, and brine. High water recoveries can be obtained on high silica waters because silica does not limit EDR recovery.

Combining EDR with anti-scalant addition increases the allowable concentration of these scale-forming entities even further. EDR is not affected by as many feed water constituents as RO, which limit the process performance, as illustrated in Table 16.2 (Reahl, 2004).

16.5 Anti-Organic Fouling

In late 1979, a 100,000 gpd industrial EDR system was installed in Texas to desalt municipal water before ion exchange demineralization for boiler feed. The municipality uses a surface water source. The original styrene divinyl benzene–based anion exchange membranes irreversibly fouled and failed after 14 months of use. New acrylic-based anion exchange membranes had been under field testing for several years at that time,

Table 16.2 Feed Water Constituents That Limit Process Performance

Influent Feed Water Factor	EDR Limits	RO Limits
Well or surface supply	Flow through system is unaffected by source	Reduced flux on surface water of wastewater
Silt density index	15 (5 min SDI)	5 or less (15 min SDI)
TOC	Up to 15 ppm	Up to 2–5 ppm
Silica	Unlimited to saturation	Depends on water recovery
Oil and grease	Up to 1 ppm	0
Turbidity	Up to 2 NTU	0.1–1 NTU
Iron (Fe^{2+})	0.3 ppm	0.3 ppm
Mn^{2+}	0.1 ppm	0.1 ppm
H_2S	Up to 1 ppm	No limit if not mixed with O_2
Normal temperature range	40–100 F	40–100 F
Free chlorine	0.5 ppm continuous with Spikes to 15–20 ppm+	None; Cl_2 will destroy TFC membranes
Feed *pH*	2.0–11.0	2.0–12.0

Reahl (2004), p. 3.

and their improved organic fouling resistance was becoming evident. This was an opportunity to put the new membrane to test. The failed membranes were replaced with the new membranes and were operated for 9 years. The new membranes experienced some organic fouling, but the fouling could be fully controlled by circulating a 5% sodium chloride salt solution through the stacks at about 4-month intervals (Allison, 2001).

Analytical results of new acrylic-based anion membrane samples were obtained from an industrial plant in the late 1980s (Table 16.3). Samples from the edge gasket area and the active flow path area were typically analyzed. In the received condition of the sample membranes, the active flow path area had an electrical resistance two times higher than the edge area that was not exposed to the water in the plant operation. The procedures for analysis involve treating the samples in 2 N HCl and 2 N NaCl solutions to recover membrane performance. For the edge gasket area, the recovered electrical resistance is 57.1% of the received value. For flow path area, recovered resistance is only 32% of the received value and close to the edge area value. During recovering treatment of this

Table 16.3 Analysis of Electric Resistance of Anion Exchange Membrane with Organic Fouling

		Acrylic-Based Membrane		Styrene Divinyl Benzene Membrane	
Property		Edge Area	Flow Path Area	Edge Area	Flow Path Area
Received resistance	Ω cm^2	21.9	43.5	25.1	97.0
Recovered resistance	Ω cm^2	12.5	14.1	19.1	72.0
Ion exchange capacity	mEq/dry g	2.29	2.25	1.55	0.75
Water content	%	47.0	47.3	44.0	41.6

Allison (2001), p. 2, Table 1; Table 2.

sample, the NaCl solution changed from clear to about the color of a cup of black coffee, indicating extraction of organic foulants. The ion exchange capacity and percent water content of both areas were nearly the same. The analysis showed that the membranes could easily be cleaned.

The same analysis was achieved for styrene divinyl benzene–based membrane (Table 16.3). The received electrical resistance of the flow path area is high and it is also high after acid and salt solution treatments (recovered resistance). The ion exchange capacity of the flow path area decreases to less than half of the edge area value. The ion exchange sites are still present in the membrane, but they are occupied by organic anions (foulants) that inactivate the sites. Because the organic foulants are not removed by the salt solution recovering treatment, the fouling is permanent. The water content is also lower in the flow path. The organic anion foulants occupy spaces in the resin pores and displace some of the water volume in the membrane. Because the membrane performances could not be recovered with cleaning in this plant, they had to be replaced.

Benzene rings in the styrene divinyl benzene membranes are assumed be fouled owing to bonding between benzene ring and ionic sites in the foulants (organic anions). The intensity of the ionic bond is strong enough to make removal impossible. The acrylic-based membranes have no benzene ring structures to provide the extra bonding, and therefore can be cleaned with a salt solution. The acrylic anion exchange membranes have chlorine tolerance, and now are commonly applied to maintain a disinfectant level when they treat biologically active water.

16.6 Colloidal Deposit Formation on the Membrane Surface and Its Removal

The driving force for colloidal deposit formation on an RO membrane is the water flow toward and through the membrane. Water flow carries the colloids to the surface where they are blocked and deposited. In an EDR system, species that carry an electrical charge are carried toward the membranes. Species such as dissolved salts and small organic ions with molecular weights below about 200 pass freely through the membrane to the concentrate. Particles in water nearly always have a negative electrical charge. In an ED or EDR system, the applied DC power is the driving force that moves the particles toward the anion exchange membrane, where the organic fouling is induced (Fundamentals Section 11.5). Figure 16.7 shows the deposition force and how in EDR the periodic reversal of the applied DC power reverses the driving force for deposition into a driving force for deposit removal (Allison, 2001).

With over 1500 ED and EDR plants around the world, it is inevitable that a few have been operated with feed turbidities over 0.5 NTU (Nephelometric Turbidity Unit). EDR will survive short-term operation at higher turbidities but will foul if operated for an extended time over this limit. As with RO systems, severe deposits are difficult to remove with chemical solution cleaning. EDR stacks are made to be disassembled for

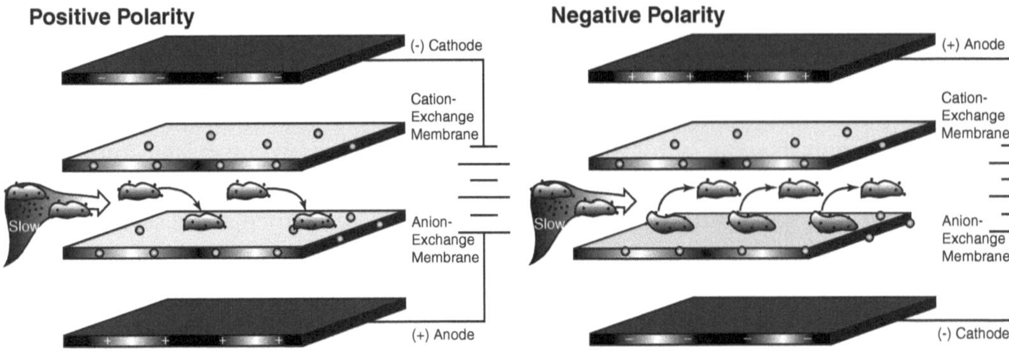

FIGURE 16.7 Colloid deposition and removal forces in EDR. *Allison (2001), p. 5, Fig. 3.*

maintenance if needed. It takes two people about 8 h to disassemble, hand clean the components, and reassemble a full-size ERD stack (cf. Section 12.3.4). This is a job that can be done by unskilled workers.

16.7 Nitrate and Nitrite Removal

Nitrate contamination of drinking water is a widespread problem. It has long been known that levels of nitrates exceeding the 10-mg/l (as N (nitrogen)) limit are associated with certain health problems. Although high nitrogen concentrations in drinking water are found mainly in regions of intensive agricultural use, there are sources of nitrate contamination other than agricultural. Nitrates and nitrites are removed efficiently and economically using EDR, as follows (Prato and Parent, 1993).

Fertilizer runoff, farm animal wastes, and septic tank discharge are percolated through the soil into groundwater aquifers and ultimately into water supplies. Agricultural sources of nitrates are the most common. Regions of the country where corn is grown experience peak levels of nitrates in groundwater from heavy fertilization. Other sources of nitrate and nitrite contamination are natural and industrial in origin. Standards for maximum levels of nitrates in drinking water were established by the federal government in 1975 with passage of the Safe Drinking Water Act (SDWA). As of the May 1990 SDWA regulations, some major allowable inorganic contaminants are as follows:

Contaminant	Maximum or Secondary Maximum Contaminant level (mg/l)
Chloride	250
Fluoride	2
Nitrate (as N)	10
(as NO_3)	45
Nitrite (as N)	1
(as NO_2)	3.3
TDS	500

Before the SDWA, there was no requirement or practical, affordable method to remove nitrates from drinking water. Since that time a number of demineralization technologies have been given best available technology (BAT) status for nitrate removal. These BAT processes include EDR, RO, and ion exchange (IE). The EDR process can effectively reduce nitrate concentrations to the maximum contaminant level (MCL).

Table 16.4 presents operating data on all of the various EDR units, including three public drinking water installations and one industrial application. All of the waters contain high levels of nitrate, and the industrial feed water contains exceptionally high

Table 16.4 Electrodialysis Reversal Plant Data: Bermuda, Delaware, Industrial, Italian

Plant specifications	Bermuda	Delaware	Industrial	Italian
Model	2 AquamiteXX	Aquamite XX Aquamite X	Aquamite X	Aquamite XX
Production	300,000 gpd each 600,000 gpd total	300,000 gpd 100,000 gpd 400,000 gpd total	100,000 gpd	300,000 gpd
Recovery	90%	90%	80%	90%
Product purity	TDS 278 ppm NO_3 8.8 ppm	TDS 11 ppm NO_3 4.5 ppm	TDS 534 ppm NO_3 128 ppm NO_3 21 ppm	TDS 474 ppm NO_3 37 ppm
Raw water	TDS 1614 ppm NO_3 66 ppm	TDS 114 ppm NO_3 61 ppm	TDS 1753 ppm NO_3 655 ppm NO_3 64 ppm	TDS 1012 ppm NO_3 120 ppm
Desalting stages	3	3	3	2
Percent removal	TDS 81% NO_3 86.7%	TDS 88% NO_3 92.6%	TDS 66% NO_3 80.4% NO_3 67.2%	TDS 53% NO_3 69.2%

Water quality

	Feed	Product	Feed	Product	Feed	Product	Feed	Product
Constituent	**(mg/l)**	**(mg/l)**	**(mg/l)**	**(mg/l)**	**(mg/l)**	**(mg/l)**	**(mg/l)**	**(mg/l)**
Sodium	349	72	12	1.6	24	14	73	49
Calcium	138	13	9	0.5	141	28	127	63
Magnesium	40	4	8	0.6	34	8	34	13
Potassium	19	2					4	1.7
Chloride	656	92	15	1.2	35	11	120	44
Bicarbonate	259	75	9	2.4	514	235	449	240
Sulfate	85	10			113	23	85	25
Nitrate	66	8.8	61	4.5	655	128	120	37
Nitrite					64	21		
pH	7.9	7.0	6.2	5.4	7.3	7.0	7.3	7.1
TDS	1614	278	114	11	1753	534	1012	474

Prato and Parent (1993), p. 4, Table 4; Table 5.

levels of nitrite. Operating data demonstrate the practicality of the EDR demineralization process for removing nitrates and nitrites as well as TDS.

16.8 Practice

16.8.1 Electrodialysis Reversal Operation for Production of Drinking Water in the City of Suffolk

The City of Suffolk (population approximately 55,000) is located in southeast Virginia, an area that experienced rapid growth over the past 2 decades. In the late 1980s the city evaluated a number of source alternatives for expanding its water supply to meet the increasing demand for water. Well water has moderate levels (689 mg/l). However, a fluoride level of 4.6 mg/l is higher than the primary maximum contaminant level of 4.0 mg/l. Also, the sodium concentration of 191 mg/l exceeds Suffolk's water quality guidelines of 50 mg/l for potable water. Hence, a membrane desalination treatment process was required to raise well water quality to meet drinking water standards. A typical feed water analysis is shown in Table 16.5. Ionics succeeded in treating this feeding water using EDR as follows (Werner and von Gottberg, 1998).

Based on the results of a feasibility study, RO and EDR were identified as the most appropriate technologies to reduce the levels of fluoride and sodium, and pilot testing was undertaken in 1987 to evaluate the technologies. EDR was eventually selected for full-scale application in Suffolk because it would result in higher water recovery rates and lower operational costs (Thompson et al., 1991). Water recovery was important for Suffolk because concentrate disposal is a major issue. Because EDR could produce 94% water recovery compared with 85% water recovery for RO, the volume of concentrate discharge was lower for EDR.

Table 16.5 Feed Water Quality, Suffolk

Parameter	Concentration (mg/l)
Bicarbonate	453
Calcium	2.27
Carbonate	1.2
Chloride	21.1
Fluoride	4.57
Magnesium	0.931
pH	8.15
Phosphate	2.79
Potassium	5.05
Sodium	191
Sulfate	7.04
TDS	689

Werner and von Gottberg (1998), p. 2, Table 1.

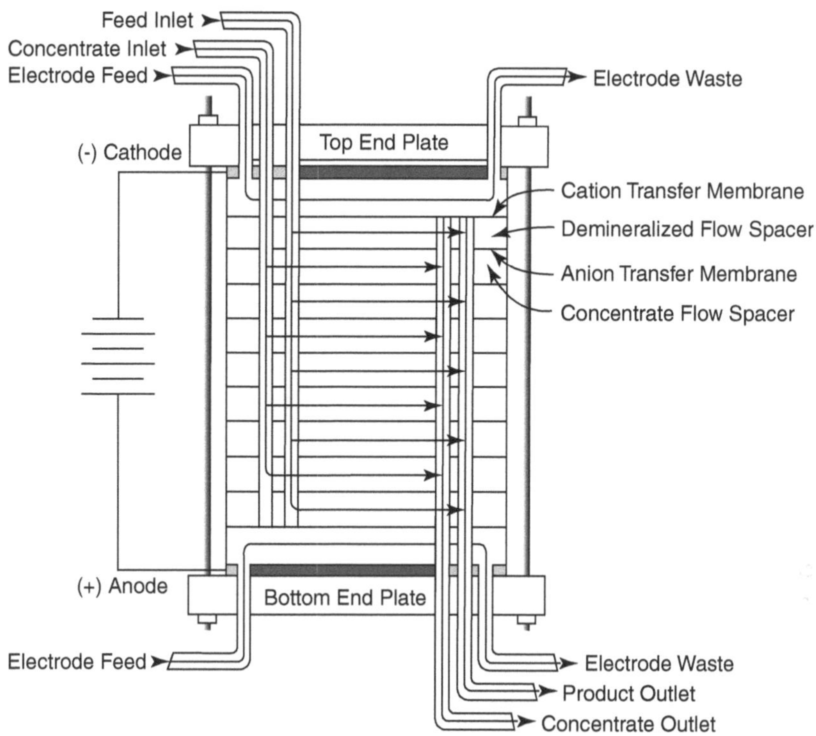

FIGURE 16.8 Electrodialysis membrane stack. *Werner and von Gottberg (1998), p. 2, Fig. 1.*

Figure 16.8 shows a typical electrodialysis membrane stack. EDR is an automatic self-cleaning version in which the polarity of the DC voltage is reversed two to four times per hour (Siwak, 1993). A 3.76 mgd EDR facility started operation at Suffolk in August 1990. Since its commissioning, the plant has produced over 5 billion gallons of potable water. The plant has three separate units (Ionics Aquamite 120's), each of which produces 1.25 mgd. Each unit contains eight parallel lines of membrane stacks, and each line has stages of stacks in series. Operating lines in parallel increases the production capacity of the unit and operating stacks in series increases the salt removal capacity of the plant. Three stages were required to reduce the fluoride level to 1.4 mg/l.

Suffolk needed a membrane process that could operate at supersaturated levels of calcium fluoride because high recovery was important to minimize concentrate disposal. Although the concentration of calcium in the feed is low, the fluoride concentration is high, and at 94% water recovery the concentrate stream is supersaturated with calcium fluoride. Reversing the DC polarity switches the dilute and concentrate streams every 30 min. Any calcium fluoride that has started to precipitate in the concentrate stream is dissolved by the dilute stream. Operation in this unsteady state mode can continue up to levels of calcium fluoride saturation of more than 500 × K_{sp} (Allison, 1993).

Table 16.6 Electrodialysis Reversal Product Water Quality, Suffolk

Parameter	Concentration	
Chloride	7.3	mg/l
Conductivity	280	μMho
Fluoride	1.43	mg/l
pH	7.3	
Sodium	6.1	mg/l
TDS	117	mg/l

Werner and von Gottberg (1998), p. 3, Table 2.

Table 16.7 Operation and Maintenance Costs, Suffolk

Cost Category	Cost ($1000/gal)
Fixed	0.72
Professional services	0.06
Chemicals	0.02
Utilities ($0.05/kWh)	0.21
Maintenance	0.17
EDR stack replacement	0.23
Total	1.41

Production in 1997: 827,339,440 gal.
Werner and von Gottberg (1998), p. 4, Table 4.

With no change in feed water quality, EDR product water quality has remained constant. Table 16.6 shows typical values for several major chemical parameters. This product water is then blended with the product from the conventional surface water treatment plant before its delivery into the city's distribution system. Table 16.7 represents recent cost data related to operation of the EDR plant. Fixed costs are associated with such items as wages and salaries, vehicle cost, telephone, and other items that apply equally to both treatment plants. Professional services represent cost associated with service obtained from outside the department, such as sludge removal, laboratory testing, and other items that apply to each of the plants on an individual basis. EDR stack replacement figures represent costs associated with replacement of stack membranes, spacers, and electrodes. Electrodes are considered a consumable item whereas the membranes and spacers have a theoretical life expectancy. Costs not represented here are treatment plant replacement costs and depreciation or debt service on the original plants.

16.8.2 EDR, Nanofiltration, and RO at a Brackish Water Reclamation

In Port Hueneme, California, a new state-of-art desalination facility uses three brackish water desalination technologies: RO, nanofiltration (NF), and EDR, operated side by side to produce over 3 mgd of high-quality drinking water. The Brackish Water Reclamation

Demonstration Facility (BWRDF) is the cornerstone of the Port Hueneme Water Agency's (PHWA) Water Quality Improvement Program. In addition to providing desalted water for local use, the BWRDF also serves as a full-scale research and demonstration facility.

It is usually a difficult task to compare the long-term performance and operating costs of three technologies owing to variables in source water quality, plant capacities, labor, power, and chemical costs. Operating three full-scale desalination technologies in parallel at the same site has made direct comparison possible. During the course of the plant's operation, the PHWA will collect data on long-term costs and performance characteristics of the three membrane systems. This will provide a realistic comparison of the desalination technologies that can then be used by water purveyors across the country to help determine which technology best suits their specific needs. The PHWA's Water Quality Improvement Program was implemented over a 6-year period starting in 1993. Design of the BWRDF was completed in late 1996 and facility construction was completed in late 1998. During the period of the BWRDF construction, the PHWA also constructed several major pipelines to deliver raw and treated surface water to the facility and deliver treated and blended water to the customers. The BWRDF has been in continuous operation since January 1999. Because the BWRDF also serves as a full-scale brackish membrane research and demonstration facility, the United States Bureau of Reclamation funded approximately 25% of the cost of the facility. The demonstration research was proceeded using the facility as follows (Passanisi et al., 2000).

The three membrane treatment processes, RO, NF, and EDR, operate side by side to produce a total of 3 mgd of treated and blended water, as shown in Table 16.8. The source water is chlorinated groundwater from inland and upper aquifer wells that are under the influence of surface water. These wells are recharged with surface water from the Santa Clara River through spreading basins. Typical source water characteristics are presented in Table 16.9.

The RO system is a two-stage process with 14 first-stage vessels and 7 second-stage vessels, each with six elements per vessel. The concentrated reject stream from the first-stage membranes is the feed water to the second-stage membranes. The RO membrane elements are thin film composite, Filmtec BW40LE-440 elements. The product recovery for the RO system, defined as the product water out of the system divided by the source water entering the system, is approximately 75%. The RO pressure

Table 16.8 Plant Water Flow Rates

Stream	Flow Rate (mgd)
Raw	3.843
Reject	0.780
Bypass	0.648
Product	2.415
Total treated	3.063

Passanisi et al. (2000), p. 2, Table 1.

Table 16.9 Typical Raw Water Parameters

Parameter	Measurement	
Conductivity	1385	μS/cm
pH	7.4	
ORP	<400	mV
Cl_2 residual	1.6	ppm free
Ammonia	0	ppm
SDI	<0.5	
Temperature	64	F
Turbidity	0.02	NTU

Passanisi et al. (2000), p. 3, Table 2.

required to desalt source water of approximately 1000 mg/l TDS is about 160 psi. The RO product water has a TDS of about 15 mg/l.

The NF system is a two-stage process with 15 first-stage vessels and 7 second-stage vessels, each with six elements per vessel. The concentrated reject stream from the first-stage membranes is the feed water to the second-stage membranes. The NF membrane elements are thin film composite, Filmtec NF90-400 elements. The product recovery for the NF system is approximately 73%. NF pressure required to desalt source water of approximately 1000 mg/l TDS is about 140 psi. The NF product water has a TDS of about 20 mg/l.

The EDR system is an Ionics EDR 2020 with five parallel lines of three stages of Mark IV membrane stacks (cf. Section 16.2). Each membrane stack contains 600 cell pairs of ion exchange membranes and flow spacers. The product recovery for the EDR system is approximately 85% (some source water is added to the reject water loop to keep the dissolved ion concentrations low enough to prevent mineral scale formation). The EDR system, unlike the RO and NF systems, uses no filtered raw water to blend with the product water.

Operating costs characteristics of the three systems were monitored over the first year of operation, from February 1999 to January 2000. EDR, RO, and NF can be fairly compared using these data as follows (Passanisi et al., 2000).

1. Water quality and plant flow rates

The average TDS of the raw water feed to the plant is 1000 mg/l. The system is designed to produce water of 370 mg/l TDS and 150 mg/l of hardness, which matches the water quality of the state water imported from the Calleguas Municipal Water District facility. Table 16.10 gives the average water analysis for the raw water and final blended product. It also shows the product and reject quality from the three-membrane system. Representative plant flows are shown in Table 16.11.

2. Labor

Total manpower cost was $168,000, which includes two operators for 9 h/day at a productive hourly rate of $32/h. Labor was fairly evenly split between the technologies,

Table 16.10 Membrane System Water Quality

Ion		Raw Water	Final Blend	Product			Reject		
				EDR	NF	RO	EDR	NF	RO
Calcium	ppm	140	46	27	ND	ND	951	518	515
Magnesium	ppm	52	16	11	ND	ND	341	184	193
Sodium	ppm	94	53	67	4	3	237	310	318
Potassium	ppm	5	2	3	ND	ND	19	18	18
Alkalinity	ppm	220	123	148	8	5	553	713	740
Bicarbonate	ppm	270	150	178	15	13	670	868	905
Sulfate	ppm	480	138	69	2	2	3200	1725	1750
Chloride	ppm	56	25	29	2	1	239	185	185
Nitrate	ppm	20	8	8	4	2	70	52	56
pH		7.2	8.1	7.2	6.0	5.8	7.2	7.6	7.6
Conductivity		1390	595	533	32	23	5725	4225	4350
TDS	ppm	1000	370	320	20	15	5825	3575	3675
Hardness		560	179	114	ND	ND	3803	2065	2063
Langelier index		0.2	0.2	−0.6	ND	ND	1.3	1.7	1.6
Iron	μg/l	ND-300	38	ND	ND	ND	ND-490	ND-263	ND-83
SDI		<0.5	NA	NA	NA	NA	NA	2.3	2–4.5

Passanisi et al. (2000), p. 5, Table 4; Table 5.

with 16% for EDR, 15% for RO, 16% for NF, and 53% for treatment plant operations including general and preventative maintenance, laboratory analyses, and reports. Labor related to the EDR system included biweekly wash-downs and stack probing. Cleaning in place (CIP) was performed every 1000 h of runtime. Daily SDI (Silt Density Index) measurement on the NF and RO systems, as well as CIPs every 2–3 weeks because of particulate and biofouling, were the major labor requirements for these systems.

3. Chemical consumption

The daily cost for EDR chemicals was $30.87, which included hydrochloric acid, anti-scalant (Argo AS120), and sodium bisulfite. The daily chemical cost for NF was $34.89. This included anti-scalant (Argo AS120, Permacare 191 and Argo AF200) sodium bisulfite. Cost for RO was $31.06, including anti-scalant (Argo AF200) and sodium

Table 16.11 Plant Flow Rates

Technology	Raw gpm	Product gpm	Reject gpm	Bypass gpm	Blended Product gpm
EDR	782	694	88	0	694
NF	726	534	178	160	694
RO	683	506	167	188	694

Passanisi et al. (2000), p. 6, Table 6.

bisulfite. Cleaning chemical costs were $1302 for EDR, $12,635 for NF, and $841 for RO. The NF chemical costs are highest because proprietary cleaners were used for CIPs related to particulate and biofouling, and the cost for NF anti-scalant pretreatment is about double per gallon compared with RO.

4. Power consumption

Power consumption was 1275 kWh/day for EDR, 1460 kWh/day for NF, and 1690 kWh/day for RO. Power cost is $0.069/kWh. The VFD pumps on the EDR system helped to save power consumption. Considerable energy saving could be realized if the NF and RO pressurizing pumps were converted to VFD operation. The current RO and NF systems require an orifice plate and motorized valve to control inlet pressure and flow. If VFDs were used on the NF and RO, the power reduction would be 40%.

5. Downtime

The total downtime for the EDR was 236 h for the year for CIPs, stack probing, stack wash-downs, and valve maintenance. The downtime for RO was 600 h, and for NF was 530 h. This included moving membrane elements, CIPs, and valve repair. Therefore, EDR had the largest production for the year at 342 million gallons (mg), followed by NF with 239 mg and RO with 224 mg.

6. Annual operating costs

Table 16.12 summarizes the costs per 1000 gallons of water produced ($/kgal) facility.

16.8.3 Brackish Water Desalination to Produce Drinking Water in Barcelona

The 4.5 million inhabitants of Barcelona and the surrounding area are mainly supplied with surface water from the Llobregat and Ter river basins. The Llobregat river water shows important concentrations of parameters associated with salinity (Na^+, K^+, Cl^-, and Br^-) owing to both natural and anthropogenic processes. Furthermore, many

Table 16.12 Annual Operation and Maintenance Cost Comparison ($/kgal)

Process	Labor	CIP Chemicals	Pretreatment Chemicals	Power	Total
EDR	0.13	0.01	0.03	0.09	0.23
RO	0.14	0.01	0.03	0.12	0.27
NF	0.14	0.04	0.04	0.1	0.29
Overall[a]	0.14	0.015	0.035	0.11	0.30[a]

[a]Costs based on actual annual water production figures.
Passanisi et al. (2000), p. 7, Table 7.

problems are associated with the increase in micropollutant and microbiological levels caused by both urban and industrial sewage.

The Llobregat–Abrera Drinking Water Treatment Plant (DWTP) can produce up to 4 m^3/s with conventional treatment including pre-oxidation with potassium permanganate, coagulation, flocculation, oxidation with chlorine dioxide, sand filtration, granular activated carbon (GAC), and chlorination using NaClO. This process does not affect the salinity of the water, and high levels of trihalomethanes (THMs) are formed after outlet chlorination in the distribution system. Total THMs are regulated to a maximum level of 100 μg/l.

To minimize the THM problem, in 2009 the water utility Aigues Ter–Llobregat (ATLL) introduced a new EDR step in the DWTP. After GAC filtration, it was possible to process up to 2.3 m^3/s (58 mg/day) by means of an EDR brackish water desalination plant, as illustrated in Figure 16.9 (Valero and Arbos, 2010). The EDR step operates discontinuously to optimize the energy consumption of the whole process, according to expected levels of THM in the outlet plant water. Then, EDR-treated water can be blended with conventional product or be distributed directly to the system.

The EDR process includes a cartridge filtration (18 filters with 170 cartridge each, 50″ long and 5 μm in normal size) and nine modules of EDR with two hydraulic EDR stages, equipped with a total of 576 stacks of 600 cell pairs, each of which integrates homogeneous AR204 anion exchange membranes and CR67 cation exchange membranes. Each module is provided with reversing systems of flow for the change of polarity, automatic valves, and pumps that allow a fully automated system. The EDR process is operated according to the levels of THMs expected in the final drinking water. Then, one to nine modules are operated to blend the obtained water with conventional treatment product to maintain THMs levels at the lowest possible cost. Energetic average consumption for the EDR process (stacks and pumps) has been lower than 0.6 kWh/m^3. Photo 16.1 shows the EDR units at the DWTP.

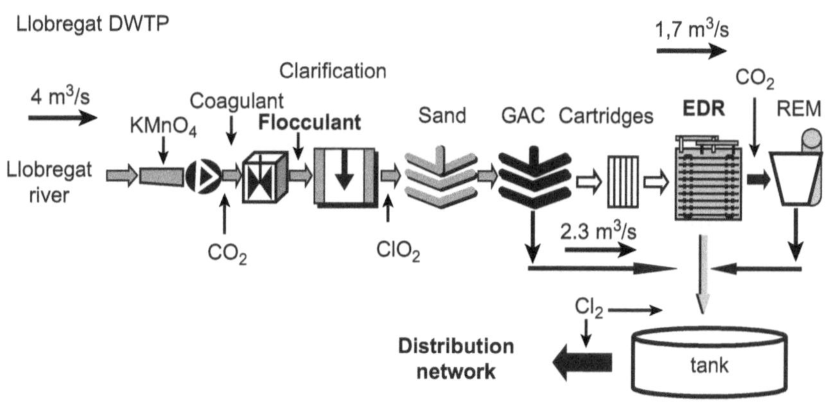

FIGURE 16.9 Drinking water treatment process. *Valero and Arbos (2010), p. 171, Fig. 1.*

PHOTO 16.1 Electrodialysis reversal units in the drinking water treatment plant.

Each EDR module is equipped with 165 automatic valves, 86 mechanical valves, 64 flow meters, 1 pH meter, 1 thermometer, 21 pressure controls, 4 conductivimeters, 23 module sampling points, 128 stack sampling points, eight rectifiers to convert alternating current power to DC power and 4 ampere meters. In addition, each of the 64 stacks of each module has two specific sampling points. Data are collected or logged for the different streams and during positive or negative polarity. In addition, the operation team is involved in chemical operation including continuous hydrochloric acid dosing at the electrode compartment and periodically for the CIP procedures. Other chemicals included are anti-scalant, NaCl, and NaClO.

The EDR step reduces the contents of significant parameters from GAC filtered water. The reduction values were (monthly analysis 2009–2011 expressed in percentage as average ± standard deviation): bromide: 77 ± 8%, chloride: 76 ± 8%, NO_3^-: 74 ± 6%, sulfate: 72 ± 8%, Ca^{2+}: >84 ± 10%, K^+: 74 ± 9%, Mg^{2+}: 82 ± 7%, Na^+: 56 ± 12%, conductivity: 69 ± 9%, and TOC: 30 ± 12%. The average values of THM in the water product ranged between 40 and 60 μg/l.

ATLL is concerned about the impact of workload on employees and managers, mainly because of the plant size and its discontinuous operation. Thus, company efforts have been directed to minimize low-value work especially related to EDR technology. In this way, some work of the research and development department has focused on design and development of special devices to use in the following three areas involved in operational maintenance (Valero et al., 2013).

1. Stack probing

The most frequent manual operation is to check the inter-membrane voltage to prevent hot spots or current leakage. Excess current can melt or burn the membranes

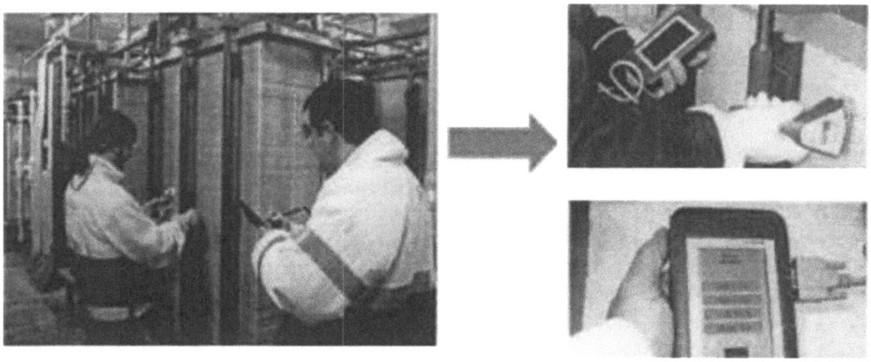

PHOTO 16.2 Stack proving: from manual to semi-automated work. *Valero et al. (2013), p. 1140, Fig. 1.*

and spacers. Normal design practices limit this voltage to 80% of the current that would cause burning. The limit is determined by the water temperature, conductivity of the source water, and membrane stack size. When operators find increases in current in a located point, they have to check the voltage for some days to prevent a hot spot. To facilitate the work, the stack probing is now improved with a new device (Photo 16.2) that identifies each position with a radiofrequency identification label and measures three points each time.

2. Cleaning membranes

One advantage of EDR technology is its ability to troubleshoot by opening and disassembling the stack. In the case of problems such as biofouling or scale, each unit (membrane or spacer) can be washed manually, brushing with diluted chemical products. Usually, if no problems are found, each first-stage stack is washed once a year as preventive maintenance work that includes an inspection and washing process. This involves washing 2400 units for each of the 288 stacks, only in the first stage. Now, a specific washing machine (Photo 16.3) is available to wash the membranes and spacers (2.6 units/min).

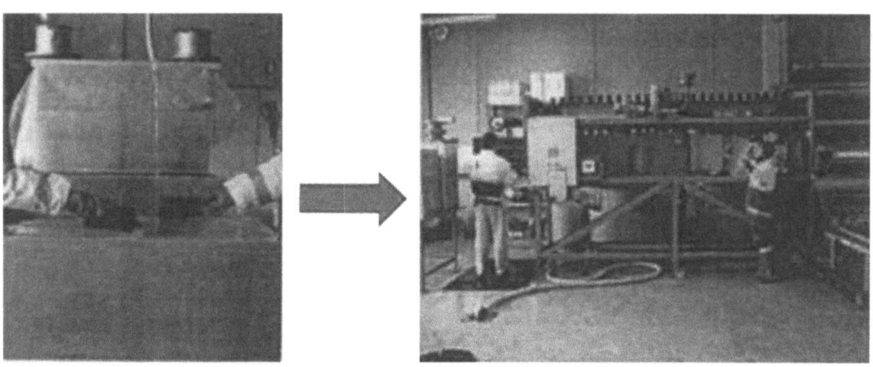

PHOTO 16.3 New membrane washing machine. *Valero et al. (2013), p. 1141, Fig. 2.*

PHOTO 16.4 Robotic device for assembling and disassembling the stack. *Valero et al. (2013), p. 1141, Fig. 3.*

3. Stack repair and disassembly

During maintenance procedures or when a problem arises, it might be necessary to disassembly the stack for wash procedures or repairs. Generally, disassembly requires that each piece be removed separately, which can be replaced without the removal of any of the membranes. In some cases, the operator finds a hot spot and an area of the stack can be burned. It is then necessary to disassembly the stack to reach to the affected units and wash or replace them. In this case, ATLL has designed a robotic device (Photo 16.4) that can assemble and disassemble a stack (2.5 units/min). A computer program allows operation until a pre-programmed position of the stack. The operator can then check the problem and repair it. Then, the machine can assemble the stack in the correct position. Normally, the manual operation requires two people.

References

Allison, R.P., 1993. High water recovery with electrodialysis reversal. Proceedings of 1993 American Water Works Association Membrane Conference, Baltimore, Maryland, USA, August 1–4.

Allison, R.P., 2001. Electrodialysis Treatment of Surface and Waste Water. Proceedings of 2001 American Water Works Association Annual Conference.

Belfort, G., Guter, G.A., 1972. An experimental study of electrodialysis hydrodynamics. Desalination 10, 221–262.

Chiapello, J.M., Bernard, M., 1993. Improved spacer design and cost reduction in an electrodialysis system. J.Membr. Sci. 80, 251–256.

Lacey, R.E., Loeb, S. (Eds.), 1972. Industrial Processing with Membranes, p. 348. Wiley-Interscience, New York.

Passanisi, J., Persechino, J., Reynolds, T.K., 2000. EDR, NF and RO at a brackish water reclamation facility. Proceedings of 2000 American Water Works Association Annual Conference.

Prato, T., Parent, R.G., 1993. Nitrate and nitrite removal from municipal drinking water supplies with electrodialysis reversal. Proceedings of 1993 American Water Works Association Membrane Conference, Baltimore, Maryland, USA, August 1–4.

Reahl, E.R., 2004. Half a Century of Desalination with Electrodialysis, Ionics Technical Paper.

Siwak, L.R., 1993. Here's how electrodialysis reverse...and why EDR works. Int. Desalin. Water Reuse Quarterly. Vol. 2/4.

Thompson, M.A., Robinson Jr., M.P., 1991. Suffolk introduce EDR to Virginia. Proceedings, American Water Works Assoc. Membrane Conference, Orlando, Fl., USA.

Valero, F., Arbos, R., 2010. Desalination of brackish river water using electrodialysis reversal (EDR). Control of the THMs formation in the Barcelona (NE Spain) area. Desalination 253, 170–174.

Valero, F., Barcelo, A., Medina, M.E., Arbos, R., 2013. Barcelona, three years of experience in brackish water desalination using EDR to improve quality. New O&M procedures to reduce low-value work and increase productivity. Desalin. Water Treat. 51, 1137–1142.

von Gottberg, A., 1998. New high-performance spacers in electrodialysis reversal (EDR) systems. Proceedings of 1998 American Water Works Association Annual Conference, Dallas, Texas, USA, June 21–25.

Werner, T.E., von Gottberg, A.J.M., 1998. Five billion gallons later-operating experience at city of Suffolk EDR plant. The American Desalting Association 1998 North American Biennial Conference and Exposition, August 2–6, Williamsburg, Verginia.

Zhong, K.W., Zhang, W.R., Hu, Z.Y., Li, H.C., 1983. Effect of characterization of spacer in electrodialysis cells on mass transfer. Desalination 46, 243–252.

17 Bipolar Membrane Electrodialysis

17.1 Overview of Technology

The bipolar membrane (BPM) is a composite membrane composed of three parts: a cation exchange layer, an anion exchange layer, and the interface formed between the two layers. With a direct current passed across the BPM, the salt concentration in the interface decreases. When all salt ions are removed from the interface, electrical conduction is achieved by the transport of H^+ and OH^- ions that are produced from the dissociation of water (water splitting). The water-splitting behavior of the BPM was first studied by Frilette (1956) using a multi-chamber cell integrated with Permutit 1373 cation/1374 anion exchange membranes and with their BPMs in an NaCl solution (Fig. 17.1). In this experiment, after an electric current was passed, the individual chambers were drained. A series of experiments with filled 0.2 M NaCl at various current densities i is summarized in Table 17.1. At low current densities (experiment 1), the initial low voltage rose rapidly and attained voltage E within less than half a minute, which is presumably the time required for salt depletion from the interface. At higher current densities (experiments 2–4), salt depletion appeared faster accompanied by an increase in coulomb efficiency for H^+ ion transfer: η_H caused by the water splitting in the BPMs. Experiment 5 showed the apparent loss in η_H during increased operation time owing to the loss of H^+ ions by competitive electromigration with Na^+ ions across the cation exchange membranes. Experiment 6 shows the data obtained at electrodes reversed, indicating a decrease in η_H owing to extinction of the water-splitting reaction.

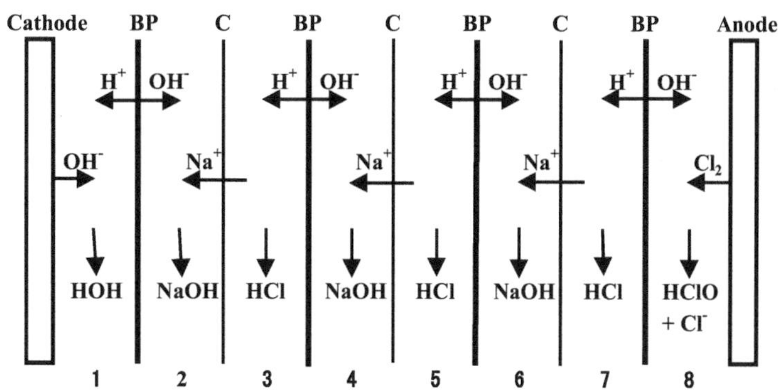

FIGURE 17.1 Multi-chamber salt-hydrolyzing cell. BP, bipolar membrane; C, cation exchange membrane. *Frilette (1956), p. 437, Fig. 1.*

Ion Exchange Membranes. http://dx.doi.org/10.1016/B978-0-444-63319-4.00017-1

Table 17.1 Proton Transport Efficiency

Exp. No	E^a (V)	Current Density i (A/cm^2 × 10^2)	Operation Time (min)	Approx. Concentration[b] H^+, OH^- (eq/dm^3)	Coulomb Efficiency for Proton Transfer η_H (%) From OH^- Chambers 2, 4, 6	From H^+ Chambers 3, 5, 7
1	9.8	3.8	10	0.006	34	31
2	21.6	7.5	5	0.008	48	46
3	33.3	15.2	5	0.02	55	52
4	55.3	30.2	2.5	0.02	63	60
5	51.6	30.2	12.5	0.06	34	30
6[c]	18.7	30.2	12.5	0.003	<2	<2

[a]Applied emf in steady state, less 2.7 V for terminal electrodes.
[b]Determined from titration of solutions collected from chambers, 2, 4, 6 and 3, 5, 7.
[c]Electrodes reversed.
Frilette (1956), p. 437, Table III.

Electrodialytic water splitting is an energy-efficient means for converting salts into their acids and bases. The technology uses ion exchange membranes to concentrate the ions in solution and is driven by an electrical potential; in this respect, it is closely related to the electrodialytic concentration. The water-splitting reaction is generated in the system in Fig. 17.2, exhibiting three chambers including acid, salt, and base bounded by the bipolar and anion/cation exchange membranes. The salt MX, e.g., sodium sulfate, is fed into the chamber between the cation and anion exchange membranes. When an electric current is passed through the electrodes, the cations Na^+ and anions SO_4^{2-} move across the monopolar membranes and combine with the OH^- and H^+ ions generated at the BPM to form the base and acid. Figure 17.3 illustrates the construction and operation of the BPM working in Fig. 17.2.

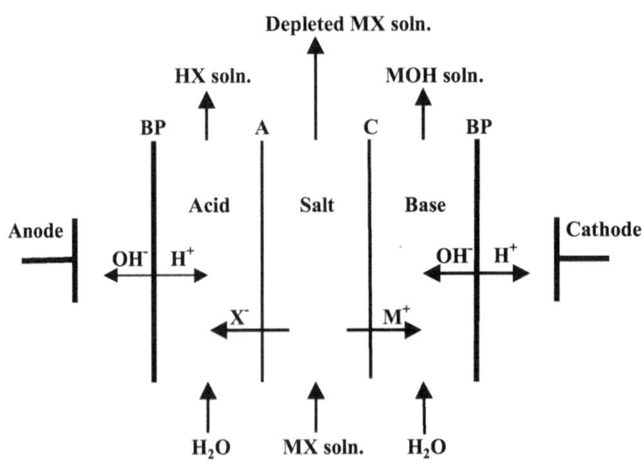

FIGURE 17.2 Three-compartment cell. BP, bipolar membrane; C, cation exchange membrane; A, anion exchange membrane. *Mani (1991), p. 118, Fig. 1.*

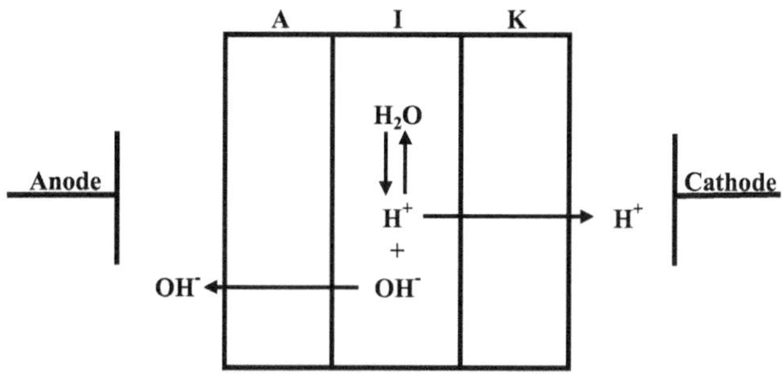

FIGURE 17.3 Bipolar membrane construction and operation. A, anion exchange layer; K, cation exchange layer; I, interface layer. *Mani (1991), p. 119, Fig. 2.*

17.2 Free Energy Changes in BMP Electrodialysis Process

The water-splitting process is electrodialytic in nature because the process merely involves changing the concentration of ions that are already present in a solution. The theoretical energy for concentrating H^+ and OH^- ions from their concentration in the interface layer in the BPM (approximately 10^{-7} M at 25 °C) to the acid and base concentrations at the outer surface of the membrane is expressed by the free energy change $-\Delta G$ in going from the interior of the membrane to the outside:

$$-\Delta G = nFE = RT \ln \frac{a_{\mathrm{H}}^{i} a_{\mathrm{OH}}^{i}}{a_{\mathrm{H}}^{o} a_{\mathrm{OH}}^{o}} \tag{17.1}$$

where a's are the activity of the H^+ and OH^- ions, and superscripts i and o refer to the interface layer and the outer surfaces of the membrane, respectively, F is the Faraday constant, 96,500 C/eq., E is the reversible electro-motive force (V), R is the gas constant, n represents the number of equivalents/mole of reactants, and T is the absolute temperature.

For generating a 1 N ideal product solution, Eqn (17.1) reduces to (because $n = 1$):

$$\Delta G = -FE = -RT \ln\left(a_{\mathrm{H}}^{i} a_{\mathrm{OH}}^{i}\right) \quad or \quad E = -\frac{RT}{F} \ln K_W \tag{17.2}$$

where K_W is the dissociation constant of water.

To overcome this potential, a positive potential $E_0 = -E$ must be applied across the membrane. Using the data on free energy for the dissociation of water $\Delta G = 0.0222$ kWh $= 22.2 \times 3600$ VAs (International Critical Tables, 1930), one can calculate the theoretical potential for generating acid and base for an ideal (i.e., perfectly permselective) BPM as $E_0 = \Delta G/F = 22.2 \times 3600/96{,}500 = 0.828$ V at 25 °C. Figure 17.4 shows the actual potential drop behavior of a BPM. In the practically applicable range of current density; 50–150 mA/cm^2, the membrane has a potential drop E of 0.9–1.1 V in a 0.5 M Na_2SO_4 solution slightly higher than E_0 (Mani, 1991). The difference between

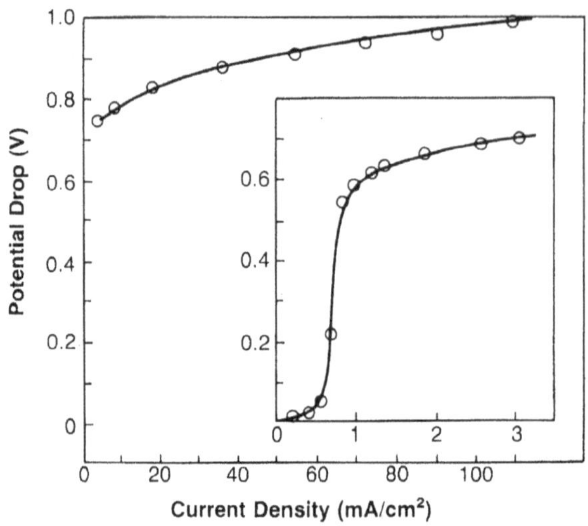

FIGURE 17.4 Potential drop across a bipolar membrane. *Mani (1991), p. 120, Fig. 3.*

E and E_0 is caused by the electric resistance of the cation exchange and anion exchange layers and the resistance of the interface layer.

17.3 Interface Layer

With the BPM placed in a salt solution, the water dissociation (splitting) reaction is suggested to be generated in the interface (transition) layer formed between the cation exchange layer and the anion exchange layer, as illustrated in Fig. 17.3. The interface layer is also referred to as the space charge region (Mauro, 1962), depletion layer (Simons, 1972), abrupt junction (Coster, 1965), or neutral region (Simons and Khanarian, 1978). The thickness of the interface layer is a few nanometers (Maletzki et al., 1992; Atkins, 1994), which roughly approaches the values in the order of the electric double-layer thickness.

The structure of the BPM is illustrated in Fig. 17.5 (Strathmann et al., 1997).

In the case of no applied external electric potential field and the symmetrical fixed charge densities in the cation exchange layer and anion exchange layer, the potential difference across the interface layer $\Delta\psi$ is given by doubling the Donnan potential ψ_{Don} as follows:

$$\Delta\psi = 2|\psi_{Don}| = 2\frac{RT}{F}\ln\frac{X}{C_0} \tag{17.3}$$

The thickness of the interface layer 2λ is expressed by the following equation (Mauro, 1962):

$$2\lambda = 2\sqrt{\frac{\Delta\psi\varepsilon_0\varepsilon_r}{XF}} \tag{17.4}$$

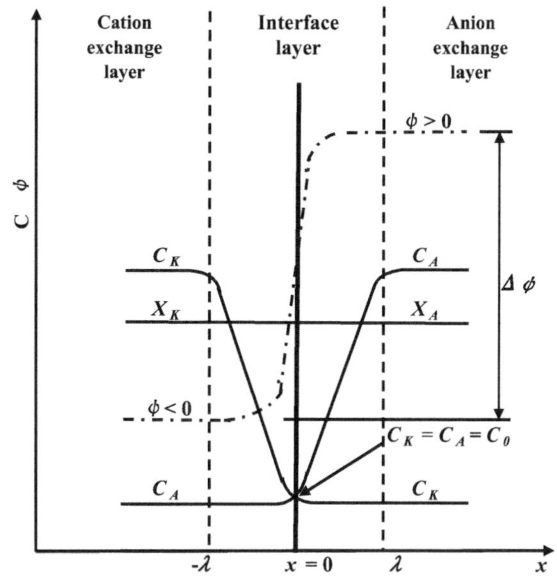

FIGURE 17.5 Concentration and potential profiles in a bipolar membrane. *Strathmann et al. (1997), p. 127, Fig. 5.*

where ε_0 is the permittivity of free space (vacuum) and ε_r is the relative permittivity. X $(=X_K = X_A)$ is the fixed charge density in the cation exchange layer X_K and the anion exchange layer X_A. $\Delta\psi$, 2λ, and potential gradient in the interface layer $d\psi/dx$ are calculated by substituting: $T = 293.16\ \text{K} = 20\ °\text{C}$, $C_0 = 1 \times 10^{-4}\ \text{mol/m}^3 = 1 \times 10^{-7}\ \text{M}$ and $X = 1000\ \text{mol/m}^3 = 1\ \text{M}$ into Eqns (17.3) and (17.4) as follows:

$$\Delta\psi = 0.814\ \text{V}$$
$$2\lambda = 4.84 \times 10^{-9}\text{m} = 4.84\ \text{nm}$$
$$d\psi/dx = \Delta\psi/2\lambda = 1.68 \times 10^{8}\ \text{V/m}$$

With external voltage ΔV applied across the BMP $\Delta\psi$ and 2λ are given as follows:

$$\Delta\psi = 2|\psi_{Don}| + \Delta V \tag{17.5}$$

$$2\lambda = 2\sqrt{\frac{(2|\psi_{\text{Don}}| + \Delta V)\varepsilon_0\varepsilon_r}{XF}} \tag{17.6}$$

The water dissociation rate in the BPM is several orders of magnitude faster than in a free solution. The reaction is generally understood to be accelerated in the thin interface layer under the increased potential gradient $d\psi/dx$, as described in the above calculation (Strathmann et al., 1997).

17.4 Structural Heterogeneity of the Membrane Surface

The interface layer is estimated to be created as the result of the structural heterogeneity of the membrane surface. Sata et al. (1995) observed the membrane surface structure before and after introducing anion exchange groups to a

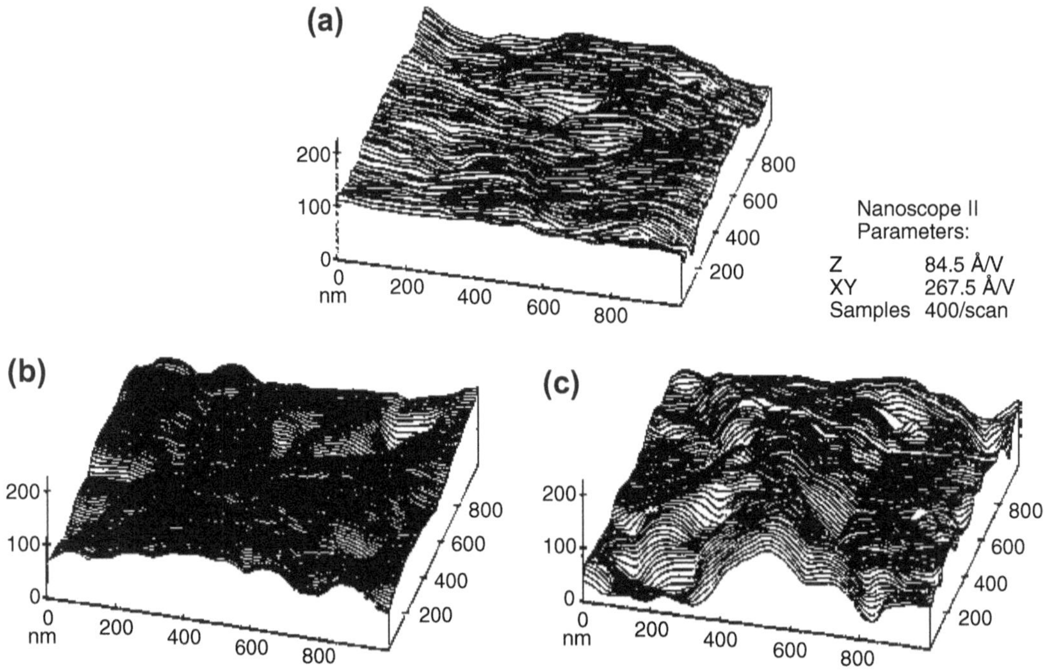

FIGURE 17.6 AFM images of the base membrane and anion exchange membranes. (a) Base membrane. (b) Amination with trimethylamine. (c) Amination with tributylamine. *Sata et al. (1995), p. 12881, Fig. 12.*

chloromethylstyrene/divinyl benzene copolymer membrane by means of an atomic force microscope (AFM). Figure 17.6(a–c) shows AFM images of three different membranes: (a) copolymer membrane (base membrane), (b) anion exchange membrane with trimethylbenzylammonium groups, and (c) anion exchange membrane with tri-*n*-butylbenzylammonium groups. Before introducing anion exchange groups, the copolymer membrane has a relatively flat surface. However, after introducing anion exchange groups, the membrane surface is uneven (b and c). From the graded scale on the perpendicular axis, the dimension of the spaces formed on the membrane is estimated to be on the order of 100 nm, which corresponds to the thickness of the interface layer.

17.5 Water Dissociation Reaction

The special difference between the BPM and the monopolar membrane is the existence of the interface layer. The mechanism of the water dissociation (splitting) reaction in the BPMs discussed in Section 17.3 seems to be different from the reaction in the monopolar membranes. The following is an understanding of the mechanism introduced from the viewpoints of the monopolar membrane (Fundamentals Section 7.8, Tanaka, 2007, 2010).

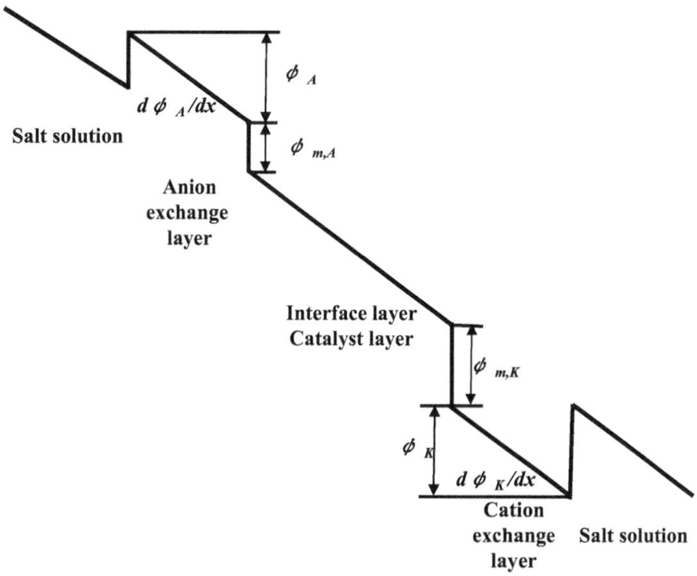

FIGURE 17.7 Potential profile in the bipolar membrane.

The BPM consists of three layers: (1) cation exchange, (2) anion exchange, and (3) interface (catalysis). A potential profile in the BPM is illustrated in Fig. 17.7, which is substantially similar to the profile in the monopolar membrane (Fundamentals Fig. 7.23). Based on the conceptual mechanism of the water dissociation reaction in the monopolar membranes, the following dissociation reaction in the BPM is assumed to be generated in the three layers in principle.

Equilibrium water dissociation reaction (Fundamentals Eqns (7.1) and (7.20)) (Eigen, 1954):

$$H_2O \underset{k_b}{\overset{k_a}{\Leftrightarrow}} H^+ + OH^- \tag{17.7}$$

Current efficiency of the water dissociation reaction (Fundamentals Eqn (7.28)) (Tanaka, 2002):

$$\left(\frac{i}{F}\right)\eta = \sigma_H + \sigma_{OH} = k_a C_{H_2O} l - k_b \int_0^l C_H C_{OH} dx = (k_a C_{H_2O} - k_b K_W) l \tag{17.8}$$

in which l is the thickness of a cation exchange layer, an anion exchange layer, or an interface layer (on the order of 100 nm thickness; cf. Section 17.4).

The forward reaction rate constant k_a is given by the following equation (Fundamentals Eqn (7.51); Timashev and Kirganova, 1981; Mafe et al., 1998):

$$k_a\left(\frac{d\psi}{dx}\right) = k_a(0)\alpha \exp\frac{F}{RT}\left(l\frac{d\psi}{dx}\right) \tag{17.9}$$

in which $d\psi/dx$ is the potential gradient in the cation exchange layer, anion exchange layer, or interface layer interface layer.

k_a in the interface layer $k_{a,inter}$ is much lower than k_a in the cation exchange layer $k_{a,cation}$, which is much lower than k_a in the anion exchange layer $k_{a,anion}$ (Fundamentals Fig. 7.22):

$$k_{a,inter} \ll k_{a,cation} \ll k_{a,anion} \tag{17.10}$$

To accelerate the water dissociation reaction in the BPM, appropriate catalytic components (Fundamentals Table 1.1) are supplied into the interface layer to form a catalysis layer. The catalysis components increase $k_{a,inter}$ and induce a drastic water dissociation reaction (Fundamentals Sections 7.7.4 and 7.8.6) caused by metallic hydroxides (Fundamentals Eqns (7.59) and (7.60); Ganych et al., 1992):

$$\mathrm{R}-\left[\mathrm{Me(H_2O)_2}\right]^{z+} + \mathrm{H-OH} \underset{k_{-1}}{\overset{k_1}{\Leftrightarrow}} \mathrm{R}-\left[\mathrm{Me(H_2O)OH}\right]^{(z-1)+} + \mathrm{H_3O^+} \tag{17.11}$$

$$\mathrm{R}-\left[\mathrm{Me(H_2O)OH}\right]^{(z-1)+} + \mathrm{H-OH} \underset{k_{-2}}{\overset{k_2}{\Leftrightarrow}} \mathrm{R}-\left[\mathrm{Me(H_2O)_2}\right]^{z+} + \mathrm{OH^-} \tag{17.12}$$

17.6 Current Efficiency

When BPM consists of the cation exchange, anion exchange, and interface layers, the water dissociation reaction is generated in the cation exchange layer and the anion exchange layer ($k_{a,inter}$ in Eqn (17.10) is so low that the reaction in the interface layer is negligible). According to suggestions described in Section 17.5, the electric current efficiency of the BPM η_B is defined as (Tanaka, 2007):

$$1-\eta_B = (1-\eta_K)(1-\eta_A) \tag{17.13}$$

where η_K and η_A are the current efficiency for the cation exchange layer and the anion exchange layer, respectively.

Rearranging Eqn (17.13):

$$\eta_B = \eta_K + \eta_A - \eta_K\eta_A \tag{17.14}$$

From the experimental observation, the intensity of the water dissociation is much stronger in the anion exchange layer than in the cation exchange layer (Fundamentals Sections 7.7.1 and 7.7.2):

$$\eta_A \gg \eta_K \tag{17.15}$$

Thus,

$$\eta_B \fallingdotseq \eta_A \tag{17.16}$$

When catalytic materials are supplied into the interface layer, the BMP includes the cation exchange, anion exchange, and catalysis layers and η_B is expressed by (Tanaka, 2007):

$$1-\eta_B = (1-\eta_K)(1-\eta_A)(1-\eta_{cat}) \tag{17.17}$$

where η_{cat} is the current efficiency for the catalysis layer.

Rearranging Eqn (17.17):

$$\eta_B = \eta_K + \eta_A + \eta_{cat} - \eta_K\eta_A - \eta_K\eta_{cat} - \eta_A\eta_{cat} + \eta_K\eta_A\eta_{cat} \tag{17.18}$$

η_{cat} is increased drastically owing to the catalytic acceleration of the water dissociation reaction (Eqns (17.11) and (17.12)) and increase $k_{a,inter}$ in Eqn (17.10):

$$\eta_{cat} \gg \eta_A \gg \eta_K \tag{17.19}$$

Thus,

$$\eta_B \fallingdotseq \eta_{cat} \tag{17.20}$$

Figure 17.8 illustrates mass transfer in the two-compartment splitting system generating H^+ ions and OH^- ions from salt MX (Mani, 1991). The ionic flux J transported into the acid chamber owing to electromigration is expressed by the following equation.

For cations,

$$J_+ = J_{\mathrm{H}}^{\mathrm{BP}} + J_{\mathrm{M}}^{\mathrm{BP}} - J_{\mathrm{H}}^{\mathrm{A}} \tag{17.21}$$

For anions,

$$J_- = J_{\mathrm{X}}^{\mathrm{A}} - J_{\mathrm{X}}^{\mathrm{BP}} \tag{17.22}$$

Here, subscripts H, M, and X refer to H^+, M^+, and X ions, and superscripts BP and A refer to the BMP and anion exchange membrane, respectively. $J_{\mathrm{M}}^{\mathrm{BP}}$, $J_{\mathrm{H}}^{\mathrm{A}}$ (Eqn (17.21)), and $J_{\mathrm{X}}^{\mathrm{BP}}$ (Eqn (17.22)) are caused by the imperfection of permselectivity of the membranes. The current efficiency in the acid chamber η_{acid} is introduced from Eqns (17.21) and (17.22) as:

$$\eta_{\mathrm{acid}} = \frac{J_{\mathrm{H}}^{\mathrm{BM}} - J_{\mathrm{H}}^{\mathrm{A}}}{i/F} = \frac{J_{\mathrm{X}}^{\mathrm{A}} - J_{\mathrm{X}}^{\mathrm{BP}}}{i/F} \tag{17.23}$$

The ionic transfer and current efficiency in the base chamber are introduced in the same way:

$$J_+ = J_{\mathrm{M}}^{\mathrm{C}} - J_{\mathrm{M}}^{\mathrm{BP}} \tag{17.24}$$

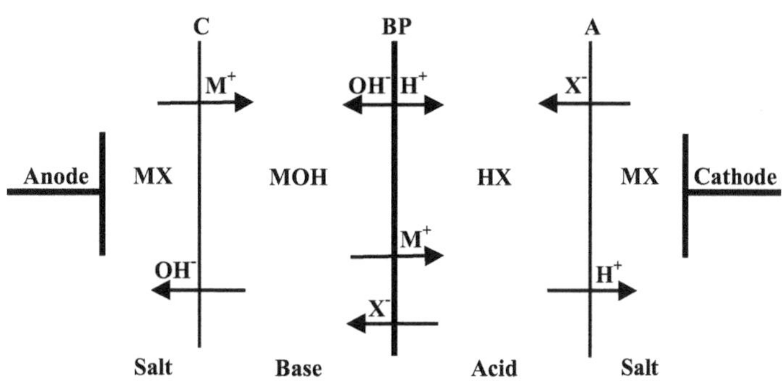

FIGURE 17.8 Mass transfer in the two-compartment cell. BP, bipolar membrane; C, cation exchange membrane; A, anion exchange membrane. *Mani (1991), p. 125, Fig. 6.*

$$J_{-} = J_{OH}^{BP} + J_{X}^{BP} - J_{OH}^{C} \tag{17.25}$$

$$\eta_{base} = \frac{J_{OH}^{BP} - J_{OH}^{C}}{i/F} = \frac{J_{M}^{K} - J_{M}^{BP}}{i/F} \tag{17.26}$$

where subscript OH and superscript C refer to OH^- ions and the cation exchange membrane, respectively. J_M^{BP} (Eqn (17.24)), J_X^{BP}, and J_{OH}^{C} (Eqn (17.25)) are caused by the imperfection of permselectivity of the membrane.

Today, most BPMs are prepared by incorporating the catalysis layer between the cation exchange layer and the anion exchange layer (Fundamentals Section 1.6), as described below.

Hanada et al. (1991, 1993, 1996) introduced inorganic catalytic substances (Fe, Ti, Sn, Zr, Pa, and Ru) into BPMs to enhance the water dissociation reaction. Kang et al. (2004) investigated the optimal contents of inorganic substances such as iron hydroxides and silicon compounds to accelerate the water dissociation reaction, and they developed a method for immobilizing the inorganic materials in the BPM.

17.7 Energy Consumption and Production Capacity

The total costs of the water-splitting process are broken down into costs related to the membrane area and to power consumption (Nagasubramanian et al., 1977).

The mass of electrolyte transported across the membrane for unit time J (eq/s) is:

$$J = \left(\frac{I}{F}\right)\eta = \left(\frac{iS}{F}\right)\eta \tag{17.27}$$

where I (A) is the electric current, i (A/m^2) is the current density, S (m^2) is the membrane area, and η is the current efficiency.

Equation (17.27) may be written as:

$$S = \frac{JF}{\eta i} \tag{17.28}$$

The voltage drop across a unit cell of a water-splitting stack E_{cell} (V/cell) is:

$$E_{cell} = iR_{cell} + E_{bp} \tag{17.29}$$

R_{cell} (Ωm^2/cell) is the resistance of a unit area of the unit cell and may be obtained from the sum of the individual solution (acid, base, and salt chamber) and membrane (anion and cation exchange membrane) resistance (cf. Fig. 17.2):

$$R_{cell} = R_{acid} + R_{base} + R_{salt} + R_{an.m} + R_{cat.m} \tag{17.30}$$

R_{bp} (V/cell) is the potential drop across the BPM and can be written as:

$$E_{bp} = i\left(R_{bp,an} + R_{bp,cat} + R_{bp,int}\right) + E + E_{irr} \tag{17.31}$$

where $R_{bp,an}$, $R_{bp,cat}$, and $R_{bp,int}$ are, respectively, the electric resistance of the anion exchange layer, cation exchange layer, and interface layer. E approaches the thermodynamic water-splitting potential E_0, and E_{irr} refers to the potential drop caused by the irreversible characteristics of the membrane.

The power consumption for unit cell P (W/cell) is thus given by:

$$P = \frac{JF}{\eta}(iR_{\text{cell}} + E_{\text{bp}}) = i^2 SR_{\text{cell}} + iSR_{\text{bp}} \qquad (17.32)$$

As be seen in Eqn (17.27), the membrane area S is inversely proportional to the current density i for constant product rate J and current efficiency η, so in the process design, the number of membranes integrated in the process decreases with the increase in current density. However, this situation results in an increase in power consumption P, as is seen in Eqn (17.32). Thus, the reduction in fixed capital brought about by reducing the membrane area by increasing current density must be weighed against the increase in fixed capital brought about by higher power requirements (for rectifiers and possibly cooling) and the increased operating costs owing to the higher consumption of electricity.

17.8 Water Transfer

In the BPM system, almost all of the water dissociation reactions are generated in the membrane in which H_2O molecules split into H^+ and OH^- ions. The H_2O must be supplied from the external solution into the BPM. The water transfer in the BPM is the basis of the water dissociation reaction. However, the mechanism of the water transfer is not yet clear and it is necessary to be understood in order to establish the basis for BPM technology. This section introduces two suggestions:

1. Maxwell pressure

The electrolyte concentration in feeding solutions in acid and base compartments is high whereas that in the interface (catalysis) layer is nearly zero. Thus, an osmotic pressure difference exceeding 100 atm must be generated between the interface layer and the external solution. Simons (1993) suggested based on Eqn (17.33) that the Maxwell pressure difference ΔP_M between the outside (external solution) and the inside (interface layer) of the BPM exceeds the osmotic pressure and supplies the water.

$$\Delta P_M = \varepsilon_0 \int_0^{E_{inter}} \varepsilon_W(E) E dE \qquad (17.33)$$

in which E is electric field, E_{inter} is the electric field in the interface layer, ε_W is dielectric constant of water, and ε_0 is the dielectric constant in a vacuum.

The theory is developed on the basis of the ion concentration profile illustrated in Fig. 17.5. To estimate ΔP_M, the thickness of the interface layer 2λ and potential difference across the interface layer $\Delta\psi$ are calculated as 1.79 nm and 0.80 V, respectively; thus, the electric field in the interface layer $E_{inter} = \Delta\psi/2\lambda = 4.5 \times 10^8$ V/m. ε_W is expressed by the function of E as shown in Eqn (17.34) and is predicted as $\varepsilon_W = 80$ at $E = 0$ and $\varepsilon_W \fallingdotseq 60$ at $E_{inter} = 3 \times 10^8$ V/m. Hence, ΔP_M is estimated as:

$$\Delta P_M > \varepsilon_0 \int_0^{4.5\times10^8} \varepsilon_W(E) E dE > \varepsilon_0 \int_0^{3\times10^8} \varepsilon_W(E) E dE > \varepsilon_0 \int_0^{3\times10^8} 60 E dE = 236 \text{ atm} > 100 \text{ atm} \qquad (17.34)$$

In this theory, the thickness of the interface layer $2\lambda = l$ is assumed to be 1.79 nm, and water splitting is assumed to occur in the interface layer under the potential difference of 3.0×10^8 V/m.

2. Swelling pressure

The swelling pressure of a dried ion exchange membrane supposedly reaches several hundred atmospheres and it splits a rock (Yamabe and Seno, 1964); in ancient Egypt, rocks were split with the swelling pressure of timber in the construction of pyramids. In the water dissociation reaction, water molecules are supplied from an external solution into the ion exchange layer with the influence of the swelling pressure, where the molecules split into H^+ ions OH^- ions.

The ion exchange layers swell to only a limited degree under normal conditions; an equilibrium is attained beyond which swelling does not proceed (Helfferich, 1962). However, the water dissociation reaction consumes H_2O molecules to produce H^+ ions and OH^- ions in the ion exchange layer, so the equilibrium is disturbed. With an applied electric current, the swelling pressure works to supply water molecules from the external solution into the ion exchange layer.

The swelling pressure is caused by the following phenomena (Helfferich, 1962):

1. There is a tendency of the fixed ion exchange groups and mobile ions to form solvation shells.
2. The interior of the ion exchange layer, as a highly concentrated solution of ions, has a tendency to be diluted by taking up additional water molecules.
3. Neighboring fixed ion exchange groups repel one another electrostatically and thus tend to stretch the chain of the matrix.

These phenomena induce water transfer from the external solution toward the ion exchange layer and in which the above

1. generates hydration water of fixed groups,
2. generates free water, and
3. generates free water and hydration water.

With an applied potential field, the free water and hydration water in the ion exchange layer split into H^+ ions and OH^- ions, which are transported to the cathode and anode, respectively.

In the case of an anion exchange layer immersed in an NaCl solution, the phenomena are illustrated in Fig. 17.9, where the flux of H_2O, $-J_{H_2O}$, is equivalent to the flux of H^+ ions, $-J_H$ and OH^- ions, J_{OH} as follows:

$$-J_{H_2O}(dP/dx)(\mathrm{mol/cm^2s}) = -J_H(d\psi/dx) = J_{OH}(d\psi/dx) \tag{17.35}$$

where dP/dx (swelling pressure gradient) and $d\psi/dx$(potential gradient) are the driving forces of the transport.

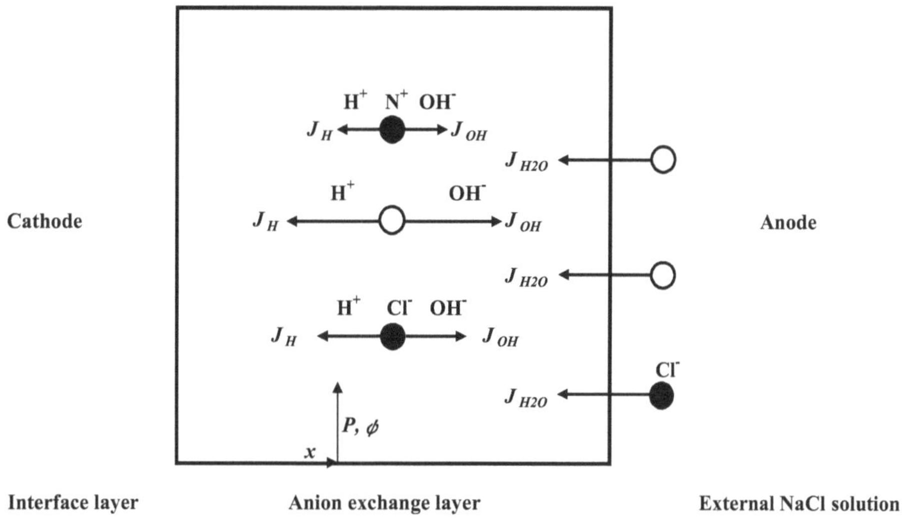

FIGURE 17.9 Water transfer and water dissociation in a bipolar membrane system. ○, Free water; ●, hydration water; N^+, quartanary ammonium group; Cl^-, Cl^- ion.

The water-splitting reaction in Fig. 17.9 is suppressed owing to the repulsive forces generated between ion exchange groups (quaternary ammonium group N^+) and co-ion (OH^-) (cf. Fundamentals Fig. 7.25). The intensity of the water dissociation reaction in the anion exchange layer is stronger than that in the cation exchange layer (Fundamentals Sections 7.7.1 and 7.7.2). Thus, the swelling pressure in the anion exchange layer is predominant compared with the pressure in the cation exchange layer. Violent water dissociation is generated in the catalysis layer formed between the cation exchange layer and anion exchange layer. This phenomenon is also assumed to be caused by the swelling pressure of the catalytic components, in which free water and hydration water are included.

3. Current–voltage relationship

Aritomi et al. (1996) measured an inflection point in current–voltage curves at high current density, as shown in Fig. 17.10, and indicated that at high current densities water supply from the outsides of the BPM to the interface layer becomes the rate determining step. The inflection points are increased with temperature and decreased with an increase in the salt concentration of an external solution (outsides of the BMP). In the bipolar electrodialysis system, water molecules in the interface layer are changed to H^+ and OH^- ions generated by the water dissociation reaction. Thus, the concentration of the water molecules C_{H_2O} is decreased and that of H^+ ions; C_H and OH^- ions; C_{OH} are increased in the interface layer. At high current density, C_{H_2O} is highly decreased and this phenomenon causes the inflection point corresponding to the rate determining

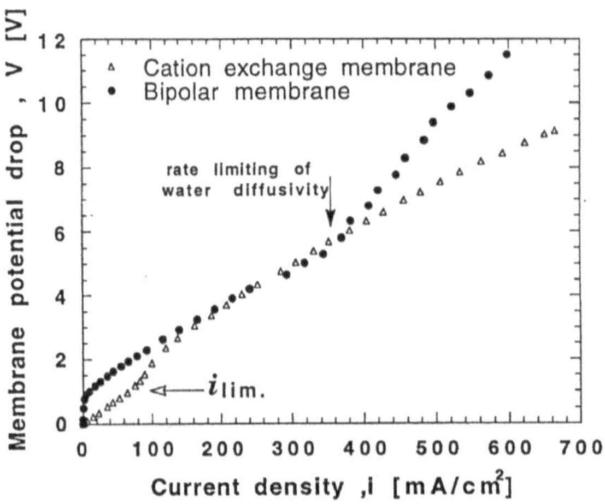

FIGURE 17.10 Current–voltage curve of a bipolar membrane. Bipolar membrane; Neosepta BP-1 0.5 M NaCl, 20 °C. *Aritomi et al. (1996), p. 16, Fig. 3.*

step. The rate limiting of water diffusivity in Fig. 17.10 corresponds to the inflection point and is much larger than the limiting current density of the monopolar membrane (Neosepta CL-25T cation exchange membrane).

4. Deterioration of BMP owing to water transfer insufficiency

As the current density increases, the rate of water removal from the interface layer increases. If the water transfer J_{water} is insufficient, the numbers of water molecules in the interface layer reaches zero and a part of the BPM dries and becomes non-conducting, resulting in membrane damage (Nagasubramanian et al., 1977).

17.9 Rectification Effect

Ohki (1965) discussed the rectification effect based on the Donnan equilibrium and the Nernst–Planck equation, and indicated that it is strongly affected by the fixed charge densities in the cation and anion exchange layers. Tanioka et al. (1996), Tanioka and Shimizu (1997) discussed experimentally the effect of the polymer material and interface structure on the rectification phenomena of the BPMs. Figure 17.11 shows the typical current–voltage curves obtained in this experiment, and demonstrates that in all sample membranes, an electric current is increased when positive potential ($V > 0$) is supplied, whereas it is suppressed when negative potential ($V < 0$) is supplied. Thus, the p–n junction theory applied to a semiconductor is also applicable to the behavior of ions in a BPM. When the voltage is less than −3 V, the reverse electric current flows to some extent. This phenomenon is presumably due to (1) the imperfectness of

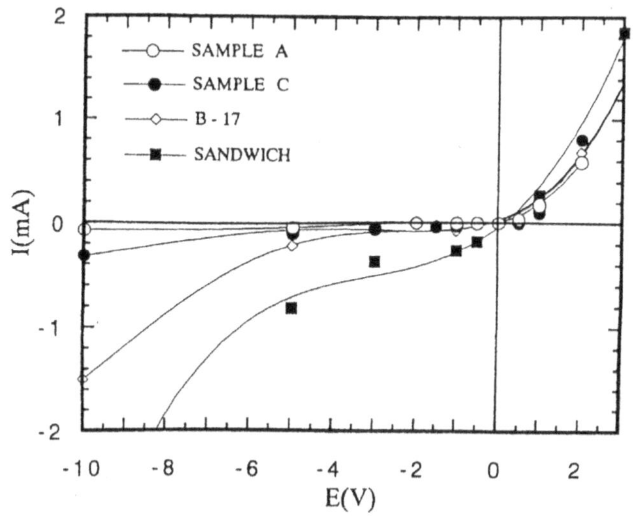

FIGURE 17.11 Current–voltage curves of a bipolar membrane. *Tanioka et al. (1996), p. 1889, Fig. 9.*

permselectivity of the membranes, and (2) H^+ and OH^- ion generation owing to the water-splitting reaction.

17.10 Desirable Properties and Operational Problems in BMP Electrodialysis Process

Desirable characteristics of the BPM electrodialysis process are (Nagasubramanian et al., 1977):

1. Low power consumption
2. Ability to operate at high current densities
3. High acid and base concentrations
4. High current efficiency
5. Low maintenance costs, i.e., use of low cost and long-life membrane

A number of these factors are interrelated and optimization is needed to arrive at a proper balance of the various factors.

A major problem in the practical application of BPM is assumed to be as follows (Strathmann, 2004):

1. The permselectivity of the membrane is decreased drastically at high product concentrations owing to a decrease in the Donnan exclusion. It results in significant salt leakage through the membrane into the product.

2. Di- or multivalent cations precipitate in the alkaline solution and must therefore be removed from the solution. The same is true for organic components that can precipitate at extreme pH values.
3. At high acid and base concentrations, retention of the anion exchange layer for H^+ ions and that of the cation exchange layer for OH^- ions are poor; thus, current use reaches an economically unacceptable low value.

17.11 Practice

Substantial efforts have been made to apply BPM technology in the fields of pollution control, resource recovery, and chemical processing, as shown in Table 17.2 (Xu, 2002). This section introduces the process flow, pretreatment requirement, and process economics presented in the literature.

17.11.1 Recovery of Mixed Acids from Stainless Pickling (Mani et al., 1988)

In the production of stainless steel, the surface oxide film is removed via a chemical pickling (de-scaling) step that uses a mixed-acid solution composed of 2–5 wt% HF, 8–15 wt% HNO_3. The spent acid from the pickling step containing metal fluorides and nitrates is conventionally limited, filtered, and discharged to landfills and waterways. The disposal of these effluents is becoming increasingly difficult and expensive.

Aquatech technology commercialized by Allied-Signal, Inc., can be used to recover the acids at the proper strength and purity for recycling back to the pickling operation. Figure 17.12 shows a simplified flow sheet of the process. The process was fully piloted at the facilities of Washington Steel in Western Pennsylvania. In the process, the spent pickle liquor is neutralized with recycle potassium hydroxide to precipitate the metal values in the form of hydroxides while the fluoride and nitrate are retained in the solution as soluble salts. The neutralized mixture is then filtered and washed with recycle water to recover substantially all of the fluoride and nitrate values. The hydroxide cake from the filtration step, containing 38–45% solids, can then be dried and returned to the steel smelter.

The KF/KNO_3 solution from the filtration step is then admitted to the salt loop of the three-compartment bipolar cell (Aquatech) (cf. Fig. 17.2). Here, through use of the direct current driving force, KOH is generated in the base loop while an HF/HNO_3 mixture is generated (recovered) in the acid loop. The product from the base loop is typically 1.5 N KOH/0.4 N KF and is forwarded to the waste acid neutralization step.

During the water-splitting step salt (KF/KNO_3) concentration decreases from 1.1–1.5 to 0.3–0.5 N in the salt loop. Salt (and water) recovery from the dilute salt solution from the unit was achieved using a conventional electrodialyzer (ED) containing Allied's cation membranes and Asahi Glass's AMV anion membranes. The concentrate from the

Table 17.2 Application of Bipolar Membrane Process

Application Field	Scale	Process Characteristics	Economics Estimation
Recovery of HF and HNO_3 (Pourcelly and Gavach, 2000)	Industrial plant Aqualytics system	Three-Compartment cell, membrane area: 3×10^5 m^2 BPM life time: 2 years, recovery ratio: 90% for HF, 95% for HNO_3, production: 8000 hour per year	Total investment: 2,950,000 US$ Operating profits: 1,620,000 US$ Total operating costs: 750,000 US$ Profits per year: 870,000 US$
Recovery of NaOH from a stream containing Na_2SO_4 (Pourcelly and Gavach, 2000)	Semi-industrial pilot	Membrane area: 0.5 m^2, feed rate: 5 l/h, feed concentration Na: 22 g/l, current applied: 900 A/m^2, current yield: 82%, product NaOH concentration: 1 M	Energy consumption: 5.0 kWh/kg NaOH
Recovery of NH_3 from HNO_3 containing NH_4NO_3 (Pourcelly and Gavach, 2000; Graillon et al., 1996)	Semi-industrial pilot	Membrane area: 120 m^2, feed concentration NH_4NO_3: 250 g/l, current applied: 1000 A/m^2, desalination rate: 97%, production: 8000 hour per year	Total cost: 0.34 US$/kg$NaNO_3$
Recycling of dimethyl isopropyl amine in production of aluminum casting molds (Pourcelly and Gavach, 2000)	Semi-industrial pilot	Membrane area: 0.3 m^2, feed concentration amine sulfate: 1 M, current applied: 800 A/m^2, current yield: 30–70%, operating time: 8000 hour per year	Energy consumption: 2.5–5.0 kWh/kg amine
Flue gas desulphuration (Mani, 1991; Pourcelly and Gavach, 2000; Liu et al., 1978)	Industrial plant Soxal™ process	Three-compartment cell, membrane area: 560 m^2, cell voltage: 2.0 V at 1000 A/m^2, current efficiency: 86%, operating time: 7200 hour per year Two-compartment cell, membrane area: 5000 m^2, cell voltage: 1.7 V at 1000 A/m^2, current efficiency: 92%, operating time: 7200 hour per year	Three-compartment cell Energy consumption: 1400 kWh/ton NaOH Two-compartment cell Energy consumption: 1120 kWh/ton NaOH
Recovery of gluconic acid from sodium gluconate (Pourcelly and Gavach, 2000; Grib et al., 1998)	Pilot scale	Two-compartment cell, membrane area: 0.19 m^2, elementary cell voltage: 2.2 V at 415 A/m^2, conversion ratio: 98.3%, Na faradic yield: 85.4%	For a plant of 10,000 ton per year: total 2.5 MUS$ Membrane replacement: 0.03 US$/Kg Na-gluconate Recovery chemicals: NaOH 0.5 MUS$ Gluconic acid unknown
Recovery of methane-sulphonic acid (MTA) from sodium methane-sulphonate (MTS) solutions (Pourcelly and Gavach, 2000)	Industrial plant (Italy)	Three-compartment cell, membrane area: 64 m^2, elementary cell voltage: 2.26 V at 800 A/m^2, methane sulphonic conversion ratio: 95%, concentration: MTS 250, MTA 100, NaOH: 80 g/l	Total investment: 700,000 US$ Total costs per ton MTA: 354 US$ Market price of MTA: 5500 US$/ton

(Continued)

Table 17.2 Application of Bipolar Membrane Process—cont'd

Application Field	Scale	Process Characteristics	Economics Estimation
Recovery of amino acid from fermentation broth (Pourcelly and Gavach, 2000)	Industrial plant Aqualytics system	Three-compartment cell, membrane area: 3 × 180 m^2, BPM lifetime: 2 years, organic acid concentration: 4–6 M, production: 8000 hour per year	No information available
Production of lactic acid from fermentation broth (Pourcelly and Gavach, 2000; Lee et al., 1998; Siebolder et al., 1995)	Industrial plant	Two-compartment cell, membrane area: 280 m^2, current yield: 60%, conversion ratio: 96%	Bipolar membrane costs: 0.12 US$/kg lactic acid, energy consumption 1 kWh/kg lactic acid
Regeneration of camphor-sulphonic acid (Pourcelly and Gavach, 2000)	Pilot scale (France)	Three-compartment cell, BPM area: 0.14 m^2, current density: 500 A/m^2, Faradic yield: 7%, salt conversion: 98.5%, final acid concentration: 0.8 M	Energy consumption: 3000 kWh/ton of products
Production of vitamin C (ascorbic acid HAsc) from sodium ascorbate (NaAsc) (Pourcelly and Gavach, 2000; Yu et al., 2000b; Novalic et al., 2000)	Labo scale and semi-industrial pilot	Two-compartment cell, current density: 1000 A/m^2, current efficiency: 75%, acid concentration: 1 M	Energy consumption: 1.4–2.3 kWh/kg HAsc
Production of citric acid (Xu and Yang, 2001, 2002; Novalic and Kulbe, 1998)	Pilot scale (China)	Two-compartment cell, BPM area: 0.004 m^2, current density: 1000 A/m^2, current efficiency: 70%, acid concentration: 30 g/l	Energy consumption: 2–5 kWh/kg citric acid
Production of silicic acid (Lin, 2001)	Industrial pilot	Two-compartment cell, elementary cell voltage 2.5–4 V at 100–200 A/m^2, current efficiency: 55–75%, acid concentration: 6–10%	Energy consumption: 0.6 kWh/kg product (6–10%)
Production of salicylic acid (Alvarez et al., 1997)	Laboratory pilot	Three-compartment cell, Tokuyama BPM, elementary cell voltage: 30 V at 750 A/m^2, current efficiency: 80–90% at 40 °C, acid concentration: 4.5 (maximum) g/l	Energy consumption: 15–20 kWh/kg product
Conversion of sodium acetate to acetic acid (Trivedi et al., 1997; Yu et al., 2000a)	Pilot scale	Five-compartment cell, BPM area: 0.008 m^2, (0.5 M sodium acetate) current efficiency: 99.9%, acid concentration: 1 M, (1.0 M sodium acetate) current efficiency: 96.8%, acid concentration: 1.5 M	Energy consumption: (0.5 M sodium acetate) 1.3–2.0 kWh/kg product, (1.0 M sodium acetate) 1.5–2.5 kWh/kg product

Xu (2002), p. 17–19, Table 1.

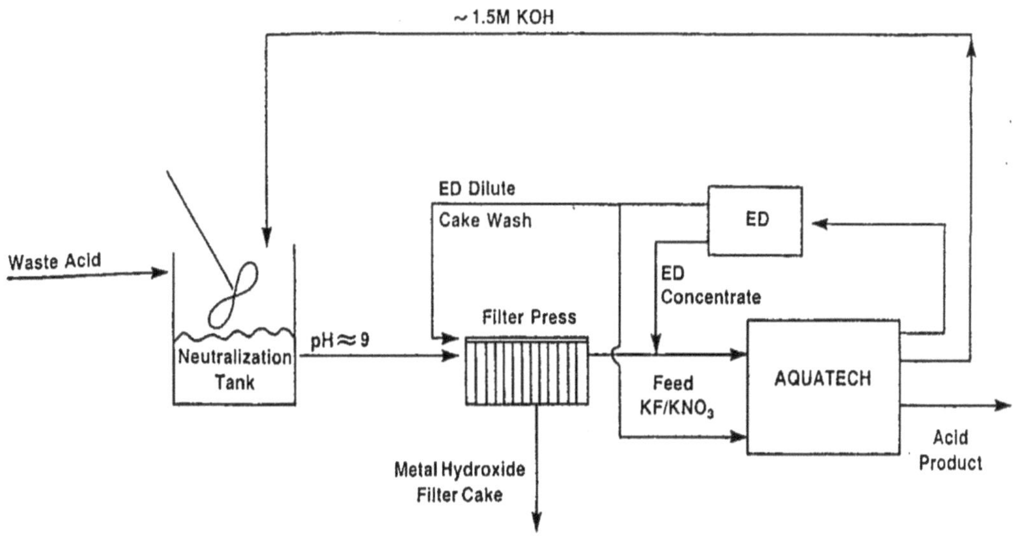

FIGURE 17.12 Aquatech pickle-liquor recovery process. *Mani et al. (1988), p. 155, Fig. 6.*

ED was returned to the salt loop while the dilute containing 0.02–0.04 N salt was used to wash the filter cake.

Table 17.3 shows an overall material balance for the 100-day run. Recovery of the key materials—namely fluoride, nitrate, and potassium—was high. Obviously the neutralization step is effective in breaking down fluoride complexes in the waste pickle liquor. The results of the pilot run are summarized in Table 17.4. All of the membranes retained

Table 17.3 100-Day Pilot Plant Run: Overall Balance

	Rate		Material Handled (kg/d)		
Stream	**l/h**	**kg/h**	F^-	NO_3^-	K^+
Waste acid	10		15	22	
Recycle base	30		5		55
Acid product	18		14	21.8	
Cake		3.2	1	0.2	1.8
Recovery (%)			93	99	96

Mani et al. (1988), p. 156, Table III.

Table 17.4 Washington Steel Corporation Pilot Test Review: Membrane Analysis

Type	Manufacturer	Units	Nominal (When New)	100 days	170 days
Bipolar	Allied	V at 100 mA/cm^2	1.1	1.29	1.4
Cation	Allied	$\Omega\ cm^2$	6.0	6.9	6.8
Anion	RAI 4035	$\Omega\ cm^2$	2.3	2.7	2.2

Mani et al. (1988), p. 158, Table IV.

Table 17.5 Aquatech Pickle-Liquor Recovery Process: Capital and Operating Costs

Basis	6×10^6 l per year of spent pickle liquor containing 5 wt% F^-, 10 wt%
	8000 h per year
	1 year life for cell assembly
	Acids recovered (net): 320 Mt per year of HF and 675 Mt per year of HNO_3
Total battery limits capital for plant \$1.8 × 10^6 (January 1986 costs (\$US per year))	
Operating Cost Summary	
Raw materials	
Makeup KOH at \$600/ton	36,000
Filter aid at \$400/ton	18,000
Cell maintenance	300,000
Electricity at \$0.05/kWh	128,000
Other maintenance	78,000
Labor and supervision	40,000
Depreciation at 10%	180,000
Taxes and insurance at 2%	36,000
Total operating costs	816,000
Product credits	
HF at \$1500/ton	480,000
HNO_3 at \$225/ton	152,000
Waste disposal at \$0.08/l	480,000
Metals credit	150,000
Total credits	1,262,000
Savings	446,000
Return on investment	25%

Mani et al. (1988), p. 158, Table V.

their selectivity and mechanical properties and showed no evidence of fouling. The electrical resistance of the monopolar membranes was stable whereas the BPMs had an acceptably low potential increase of <0.08 V per month. Table 17.5 shows an estimated capital and operating cost summary for a plant processing 6×10^6 l per year of spent pickle liquor. The battery limits investment covers the entire plant including filtration, neutralization, tankage, BP system, and ED water recovery unit. A considerable part of the total investment is for hardware and equipment other than the Aquatech unit itself.

17.11.2 Recovery of Caustic Soda and Sulfuric Acid from Sodium Sulfate (Mani et al., 1988)

Sodium sulfate is generated as a waste or byproduct in many operations such as waste acid neutralization, ion exchange regeneration, flue-gas desulfurization, and rayon manufacture. Sulfate concentration in a given stream is low, such as in waste acid neutralization, or high, such as in rayon manufacture. Water-splitting technology is applied to these sulfate streams to upgrade the salt to acid and base, thus reducing or eliminating a pollution problem.

Applicability of the water-splitting technology to sulfate conversion is illustrated with reference to the manufacture of rayon. The process flow sheet is shown in Fig. 17.13 and uses a two-compartment bipolar cell. In this process, sodium sulfate values from the spin bath are recovered via an evaporative crystallization step. The mother liquor from the evaporation step is returned back to the spin bath. Solid sulfate from the crystallizer, typically Glauber salt, is then dissolved in water and purified to remove metals such as Ca, Mg, and Zn. The purified sulfate solution is then fed to the acid compartments of a two-compartment bipolar Aquatech unit. Here, under a direct current driving force, sodium hydroxide is generated in the base loop while the sulfate is acidified in the acid loop. The base is forwarded to the cellulose dissolving step while the acidified sulfate is returned to the spin bath.

Table 17.6 shows the estimated economics of the Aquatech technology for a generalized sulfate conversion application. The plant has been sized to produce 10,000 Mt per

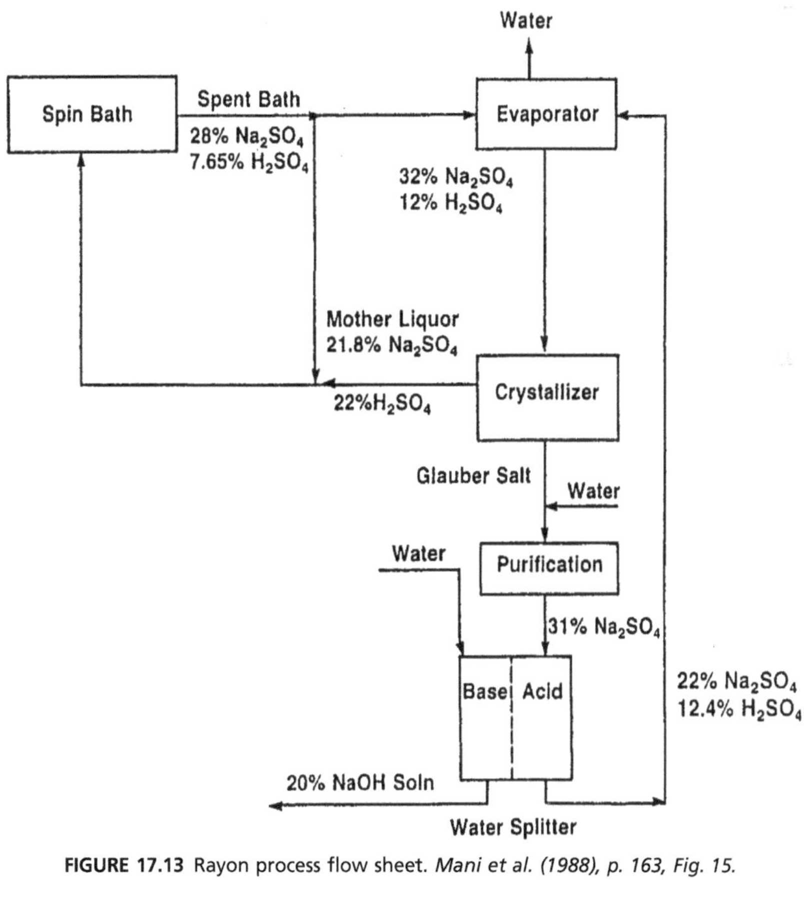

FIGURE 17.13 Rayon process flow sheet. *Mani et al. (1988), p. 163, Fig. 15.*

Table 17.6 Sodium Sulfate Conversion via Aquatech: Operating Cost Summary

Basis	10,000 Mt per year NaOH 12,250 Mt per year H_2SO_4 8000 h per year 2-year life for cell stacks	
Battery limits capital: \$3.7 × 10^6 (two-compartment cell) and \$5.2 × 10^6 (three-compartment cell) (January 1986 costs (\$US))		
Items	Two-Compartment Cell	Three-Compartment Cell
Cell maintenance	360,000	612,000
Electricity at \$0.05/kWh	730,000	850,000
Other maintenance	150,000	180,000
Labor and supervision	140,000	160,000
Depreciation at 10%	370,000	520,000
Taxes and insurance at 2%	74,000	104,000
Total operating cost	1,824,000	2,426,000
Product credits		
NaOH at \$250/Mt	2,500,000	2,500,000
H_2SO_4 at \$80/Mt	980,000	980,000
Total credits	3,480,000	3,480,000
Profit	1,656,000	1,054,000
Return on investment	45%	20%

Mani et al. (1988), p. 164, Table VI.

year of sodium hydroxide (100% basis) from an aqueous sulfate feed. The total capital equipment for the plant includes the cell stacks, power supply, tanks, plumbing, etc., to convert a clean sulfate solution to its aqueous acid/base constituents. Cell stack life is estimated at 2 years based on laboratory and pilot studies. The major variable cost items are for electricity and cell maintenance.

References

Alvarez, F., Alvarez, R., Coca, J., Sandeaux, J., Sadeaux, R., Gavach, C., 1997. Salicylic acid production by electrodialysis with bipolar membranes. J. Membr. Sci. 123, 61–69.

Aritomi, T., van den Boomgarrd, Th., Strathmann, H., 1996. Current-voltage curve of a bipolar membrane at high current density. Desalination 104, 13–18.

Atkins, P.W., 1994. Physical Chemistry, fifth ed. Oxford University Press, Oxford, UK.

Coster, H.G.L., 1965. A quantitative analysis of the voltage-current relationships of fixed charge membranes and the associated properties of "punch-through". Biophys. J. 5, 669–686.

Eigen, M., 1954. Method for investigation of ionic reaction in aqueous solutions with half-time as short as 10^{-9} s, application to neutralization and hydrolysis reaction. Discuss. Faraday Soc. 17, 194–205.

Frilette, V.J., 1956. Preparation and characterization of bipolar ion-exchange membranes. J. Phys. Chem. 60, 435–439.

Ganych, V.V., Zabolotskii, V.I., Shel'deshov, N.V., 1992. Electrolytic dissociation of water molecules in systems comprising solutions and MA-40 anion-exchange membranes modified with transition metal ions. Sov. Electrochem. 28, 1138–1143.

Graillon, S., Persin, F., Pourcelly, G., et al., 1996. Development of electrodialysis with bipolar membrane for the treatment of concentrated nitrate effluents. Desalination 107, 159–169.

Grib, H., Bonnal, L., Sandeaux, J., et al., 1998. Extraction of amphoteric amino acids by an electromembrane process. pH and electrical state control by electrodialysis with bipolar membranes. J. Chem. Technol. Biotechnol. 73, 64–70.

Hanada, F., Hirayama, K., Ohmura, N., Tanaka, S., 1991. Bipolar Membrane and Method for its Production, European Patent 0459820.

Hanada, F., Hirayama, K., Ohmura, N., Tanaka, S., 1993. Bipolar Membrane and Method for its Production, US Patent 5,221,455.

Hanada, F., Hirayama, K., Tanaka, S., Ohmura, N., 1996. Bipolar Membrane and Method for its Production, JP Patent 2,524,012.

Helfferich, F.G., 1962. Ion Exchange. McGraw-Hill, New York, p. 100.

International Critical Tables, vol. 7, 1930. McGraw-Hill, p. 232.

Kang, M.S., Choi, Y.J., Moon, S.H., 2004. Effects of inorganic substances on water splitting in ion-exchange membranes, II. Optimal contents of inorganic substances in preparing bipolar membranes. J. Colloid Interface Sci. 273, 533–539.

Lee, E.G., Moon, S.H., Chang, Y.K., et al., 1998. Lactic acid recovery using two-stage electrodialysis and its modeling. J. Membr. Sci. 145, 53–66.

Lin, A.G., 2001. Application of bipolar membrane electrodialysis in production of silicic acid. Membr. Sci. Technol. 21, 66–69.

Liu, K.J., Chlanda, F.P., Nagsubramanian, K., 1978. Application of bipolar membrane technology: a novel process for control of sulfur dioxide from flue gases. J. Membr. Sci. 3, 57–70.

Mafe, S., Ramirez, P., Alcaraz, A., 1998. Electric field associated proton transfer and water dissociation at the junction of a fixed-charge bipolar membrane. Chem. Phys. Lett. 294, 406–412.

Maletzki, F., Roster, H.W., Staude, E., 1992. Ion transfer across electrodialysis membranes in the overlimiting current range: stationary voltage current characteristics and current noise power spectra under different conditions of free convection. J. Membr. Sci. 71, 105–115.

Mani, K.N., Chlanda, F.P., Byszewski, C.H., 1988. Aquatech membrane technology for recovery of acid/base values from salt streams. Desalination 68, 149–166.

Mani, K.N., 1991. Electrodialysis water splitting technology. J. Membr. Sci. 58, 117–138.

Mauro, A., 1962. Space charge regions in fixed charge membranes and the associated property of capacitance. Biophys. J. 2, 179–198.

Nagasubramanian, K., Chlanda, F.P., Liu, K.J., 1977. Use of bipolar membrane for generation of acid and base – an engineering and economic analysis. J. Membr. Sci. 2, 109–124.

Novalic, S., Kulbe, K.D., 1998. Separation and concentration of citric acid by means of electrodialysis bipolar membrane technology. Food Technol. Biotechnol. 36, 193–195.

Novalic, S., Kongbangkerd, T., Kulbe, K.D., 2000. Recovery of organic acids with high molecular weight using a combined electrodialysis process. J. Membr. Sci. 166, 99–104.

Ohki, S., 1965. Rectification by a double membrane. J. Phys. Soc. Jpn. 20, 1674–1985.

Pourcelly, G., Gavach, C., 2000. Electrodialysis water splitting – application of electrodialysis with bipolar membranes. In: Kempermann, A.J.B. (Ed.), Handbook – Bipolar Membrane Technology. Twente University Press, Enschede, Holland, pp. 17–46.

Sata, T., Yamaguchi, T., Matsusaki, K., 1995. Effect of hydrophobicity of ion exchange groups of anion exchange membranes on permeability between two anions. J. Phys. Chem. 99, 12875–12882.

Siebolder, M., Freling, P.V., Joppien, R., Rindfleisch, D., Schugeri, K., Roper, H., 1995. Comparison of the production of lactic acid by three different lactobacilli and its recovery by extraction and electrodialysis. Proc. Biochem. 30, 81–95.

Simons, R., 1972. The steady and non-steady state properties of bipolar membranes. Biochim. Biophys. Acta 274, 1–14.

Simons, R., Khanarian, G., 1978. Water dissociation in bipolar membranes: experiments and theory. J. Membr. Biol. 38, 11–30.

Simons, R., 1993. A mechanism for water flow in bipolar membranes. J. Membr. Sci. 82, 65–73.

Strathmann, H., Krol, J.J., Rapp, H.J., Eigenberger, G., 1997. Limiting current density and water dissociation in bipolar membranes. J. Membr. Sci. 125, 123–142.

Strathmann, H., 2004. Ion Exchange Membrane Separation Process. In: Membrane Science and Technology Series, vol. 9. Elsevier, Amsterdam, p. 272.

Tanaka, Y., 2002. Water dissociation in ion-exchange membrane electrodialysis. J. Membr. Sci. 203, 227–244.

Tanaka, 2007. Ion Exchange Membranes; Fundamentals and Applications. In: Membrane Science and Technology Series, vol. 12. Elsevier, Amsterdam, pp. 419–421.

Tanaka, Y., 2010. Water dissociation reaction generated in an ion exchange membrane. J. Membr. Sci. 350, 347–360.

Tanioka, A., Shimizu, K., Miyasaka, K., Zimmer, J., Minoura, N., 1996. Effect of polymer materials on membrane potential, rectification and water splitting in bipolar membranes. Polymer 37, 1883–1889.

Tanioka, A., Shimizu, K., 1997. Effect of interface structure and amino groups on water splitting and rectification effects in bipolar membranes. Polymer 38, 5441–5446.

Timashev, S.F., Kirganova, E.V., 1981. Mechanism of the electrodialytic decomposition of water molecules in bipolar ion-exchange membranes. Sov. Electrochem. 17, 366–369.

Trivedi, G.S., Shah, B.G., Adhikary, S.K., Indusckhar, V.K., Rangarajan, R., 1997. Studies on bipolar membranes. Part II – conversion of sodium acetate to acetic and sodium hydroxide. React. Funct. Polym. 32, 209–215.

Xu, T.W., Yang, W.H., 2001. Citric acid production by electrodialysis with bipolar membranes. Chem. Eng. Proc. 41, 519–524.

Xu, T.W., Yang, W.H., 2002. Effect of cell configuration on the performance of citric acid production by bipolar membrane electrodialysis. J. Membr. Sci. 203, 145–153.

Xu, T.W., 2002. Electrodialysis process with bipolar membranes (EDBM) in environmental protection – a review. Resour. Conserv. Recycl. 37, 1–22.

Yamabe, T., Seno, M., 1964. Ion Exchange Resin Membranes. Gihodo, Tokyo, p. 29.

Yu, L.X., Guo, Q.F., Hao, J.H., et al., 2000a. Recovery of acetic acid from dilute wastewater by means of bipolar membrane electrodialysis. Desalination 129, 283–288.

Yu, L.X., Lin, A.G., Zhang, L.P., Chen, C.X., Jiang, W.J., 2000b. Application of electrodialysis to the production of vitamin C. Chem. Eng. J. 78, 153–157.

18

Electro-Deionization

18.1 Overview of Technology

Electrodeionozation (EDI) was applied to concentrate radioactive aqueous wastes at first by Walters et al. (1955). Kollsman (1957) performed some experimental work for deionizing water. However, the EDI technology was not commercialized until the late 1980s. It is now applied to the supplies of purified water (PW) in power generation, electronics, pharmaceutical, and food and beverage industries. EDI is essentially a mixed-bed (MB) deionization process built into an electrodialysis (ED) system that continuously regenerates the ion exchange resin using electrical power. The process produces high-purity water in the range of 8–17 MΩ from feed water that contains 1–100 mg/l of total dissolved solid (TDS). The electrically generated hydrogen and hydroxyl ions recombine to form water in the concentrate and produce no extra salts. EDI has the advantage of being a continuous process with constant stable product quality, which is able to produce high-purity water without the need for acid or caustic regeneration. In a process train in which reverse osmosis (RO) is followed by EDI, water can be produced that is comparable to MB ion exchange treated water.

Ganzi et al. (1987) show a schematic of the EDI process as in Fig. 18.1. In this figure, an applied DC potential is denoted by (+) and (−). Here, ion exchange membrane sheets are represented by the vertical lines labeled in terms of their ionic permeability. These membranes, used as barriers to bulk water flow, define district compartments, through which liquid streams containing ions (Na^+ and Cl^- ions) can flow tangentially. A DC electrical potential, applied by an external power supply, causes the transfer of ions to occur. In the diluting compartment, the space between membranes is filled with cation and anion exchange resins (or resin fibers).

The transfer of ions is represented schematically by arrows. Ions entering the diluting compartment react with the ion exchange resin. Ions then transfer through the resin in the direction of the potential gradient. Ions simultaneously transfer across the membranes, maintaining neutrality in all compartments. Because of the permselective properties of the membranes and the directionality of the electrical potential gradient, ions in the solution become depleted in the diluting compartment and become concentrated in the adjacent (concentrating) compartment.

The use of ion exchange resins in the diluting compartment is the key to the process. One reason for this is that without the ionic conductivity afforded by the resin, ion transfer does not occur at a practical rate for most fresh water sources. Without the ion exchange resins, the solution in the diluting compartment is deionized, and its electrical

Ion Exchange Membranes. http://dx.doi.org/10.1016/B978-0-444-63319-4.00018-3

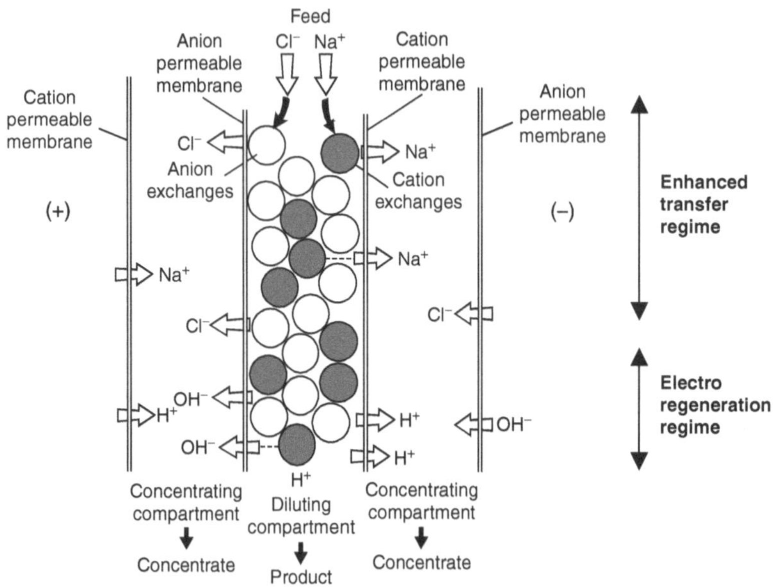

FIGURE 18.1 Schematic of ionpure process. *Ganzi et al. (1987), p. 43, Fig. 1.*

resistance increases to such an extent that useful electrical current transfer is outweighed by membrane inefficiencies, resulting in back-diffusion of ions from the concentrating to diluting compartments. By filling the ion exchange resins in the diluting compartment, the conductivity of the compartment is increased and enhanced ionic transport rates are achieved in the enhanced transfer regime depicted in Fig. 18.1. In this regime, the resins remain in the salt forms, and the rate-limiting step is most often film-diffusion of the ions in the bulk solution to the ion exchange resin surfaces. The film-diffusion leads to limiting current density, under which the electric current is limited by the diffusion of counter-ions to a resin surface and co-ions away from a resin surface.

A second reason for the need for resin in the diluting compartment relates to water dissociation reaction generated in the electro-regeneration regime in Fig. 18.1. In this regime, high-potential gradients are created at the interface between the cation exchange resins and the anion exchange resins or at the interface between the cation (anion) exchange resins and the anion (cation) exchange membranes. Significant amounts of H^+ and OH^- ions are produced due to the water dissociation (water splitting), and the resins become electrochemically regenerated to the H^+ and OH^- forms. The H^+ ions and OH^- ions produced in this regime are transported toward the electrodes under the applied potential field. They recombine in the desalting or the concentrating compartments and regenerate water. Under these conditions, the resin acts as a continuously regenerated MB ion exchange column, exchanging H^+ ions and OH^- ions in stoichiometric amounts with salts in the solution. EDI devices operating in the electro-regeneration regime can remove weakly ionized compounds. These phenomena account for the ability of the EDI process to efficiently deionize water to the 10–0.1 μS/cm region of electrical conductivity.

Allison (1996) suggests that the feed water to the EDI process must have very low levels of fine suspending matter because the ion exchange resin essentially acts as a media filter and there is no developed method to backwash the media. This means extensive pretreatment such as ultrafiltration or RO is needed to ensure continuous low levels of the suspending matter. Stack repairs are also more difficult. Stacks are filled with resin after assembly, but there is no effective means to unload the resin prior to disassembly. Each compartment must be cleaned to remove the resin beads (or resin fibers) prior to reassembly. There are no real advantages, and there are these disadvantages over ED when non-water splitting operation is considered. When EDI is operated in the water splitting mode, the efficiency of using power for desalting is low. Typically, only 10–20% of the applied DC current transports ionic salts. The rest of the current splits water. With the low power use efficiency, the process is really practical only on low TDS waters, such as RO permeate with a TDS in the range of 100 mg/l or less.

18.2 Mixed-Bed, Layered-Bed, and Separated-Bed

The deionization process is classified to the (a) mixed-bed (MB), (b) layered-bed (LB), and (c) separated-bed (SB), as illustrated in Fig. 18.2. MB is general process applied widely and shown also in Fig. 18.1.

The idea of LB is presented in relatively old time and presently commercially applied (Salem and Tessier, 1998; Liang and Wang, 2001). Bipolar contacts between ion exchange resins in LB is less than those in MB, while the electrical resistance of the diluate compartment is decreased and the deionization performance is improved for the removal of weakly dissociated electrolytes. Jha and Gifford (2004) achieved a diluate resistivity of about 18 MΩ cm and more than 99.9% silica removal. Drawbacks of LB can result from the different conductance of anion and cation exchange resin beds and from the different intensity of water dissociation. This can lead to local difference in the

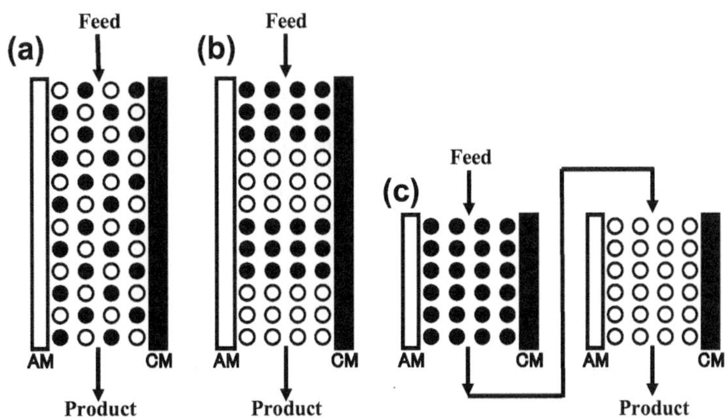

FIGURE 18.2 Mixed-bed, layered-bed, and separated-bed. (a) Mixed-bed (b) Layered-bed (c) Separated-bed. AM, Anion exchange membrane; CM, Cation exchange membrane; ○, Anion exchange resin; ●, Cation exchange resin.

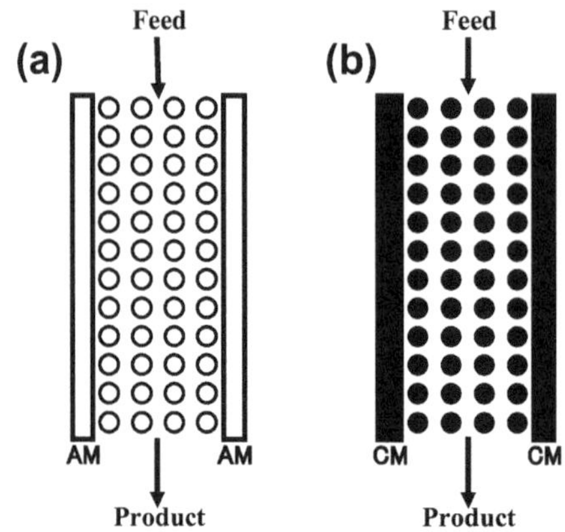

FIGURE 18.3 EDI process integrated with unipolar exchangers. (a) Anion exchanger bed (b) Cation exchanger bed. AM, Anion exchange membrane; CM, Cation exchange membrane; ○, Anion exchange resin; ●, Cation exchange resin.

current passing through the bed, which leads to differences in the regeneration of both resins (Grabowski, 2010).

The problems of unequal current distribution can be overcome by the concept of SB, which was proposed in patents (Kollsman, 1957; McRay and Rigopulos, 1967; Parsi, 1964, 1966). Korngold and Selegny (1972) have proposed a concept of the SB process illustrated in Fig. 18.3, where (a) the anion exchange resins are fixed between two

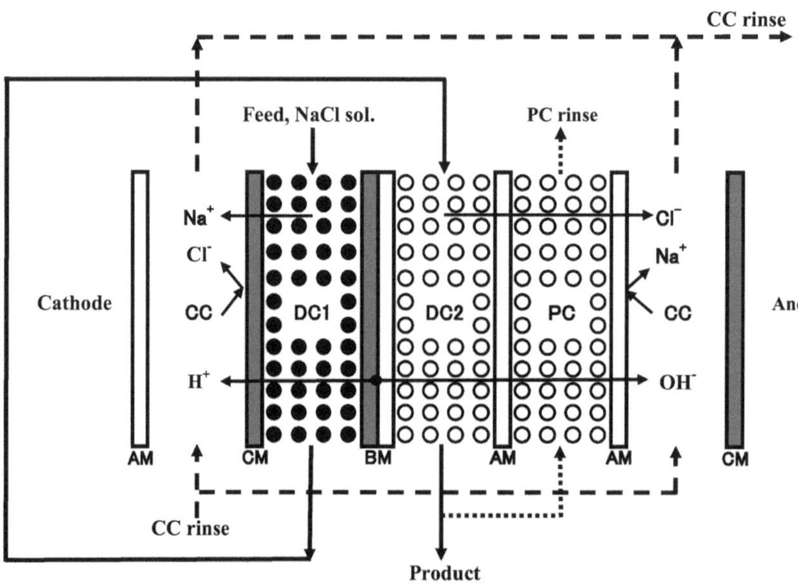

FIGURE 18.4 EDI process integrated with bipolar membranes. AM, Anion exchange membrane; CM, Cation exchange membrane; ○, Anion exchange resin; ●, Cation exchange resin; DC, Dialysate compartment; CC, Concentrate compartment; PC, Protection compartment. *Grabowski et al. (2006), p. 300, Fig. 3.*

anion exchange membranes and (b) the cation exchange resins are fixed between two cation exchange membranes. A similar set-up was also described in patents (Kunz, 1983, 1987). In this concept there are no bipolar contacts. They assumed that regeneration of ion exchange groups in the resins is performed by H^+ and OH^- ions supplied from electrodes. To supply H^+ and OH^- ions effectively, Thate (2002) and Grabowski et al. (2006) developed the SB process incorporated with bipolar membranes, as illustrated in Fig. 18.4. In this figure, the protection compartment (PC) filled with anion exchange resins is incorporated between the concentrate compartment (CC) and diluate compartment (DC2) filled with anion exchange resins. PC functions to prevent cation leakage from CC through the anion exchange membranes into the DC2.

18.3 Structure of the Electrodeionization Unit and Energy Consumption

The structure of the EDI unit is illustrated in Fig. 18.5 (Deguchi and Karibe, 1998). Based on the principle of electrodialysis, cation exchange membranes and anion exchange membranes are arranged alternately. Desalting cells and concentrating cells incorporated with spacers are arranged between the membranes. Cation exchange resins (or resin fibers) and/or anion exchange resins (or resin fibers) are filled in the desalting cells. A cathode chamber and an anode chamber are placed at both outsides of the cell arrangement, and a direct current is passed. The electric current is expressed as:

$$I = \frac{QF(N_t - N_p)}{n\eta} \tag{18.1}$$

in which, I is a direct current (A), Q is feeding solution flow rate (ml/s), N_t is ion concentration in the feeding solution (eq/l), N_p is ion concentration in the product solution (eq/l), n is the constant, F is the Faraday constant, and η is current efficiency.

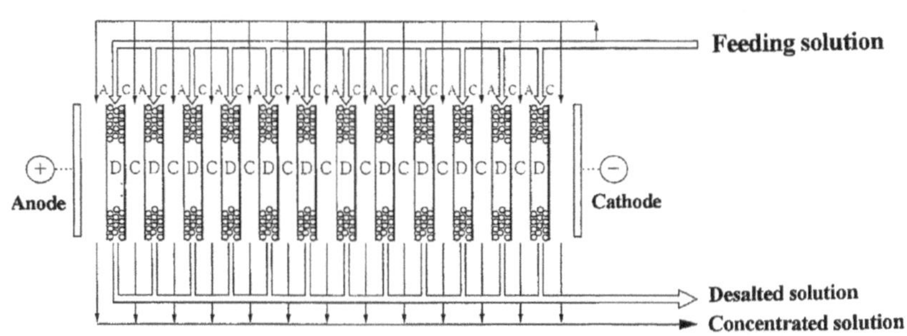

FIGURE 18.5 EDI cell arrangement. A, Anion exchange membrane; C, Cation exchange; membrane; D, Desalting cell; C, Concentrating cell. *Deguchi and Karibe (1998), p. 220, Fig. 3.68.*

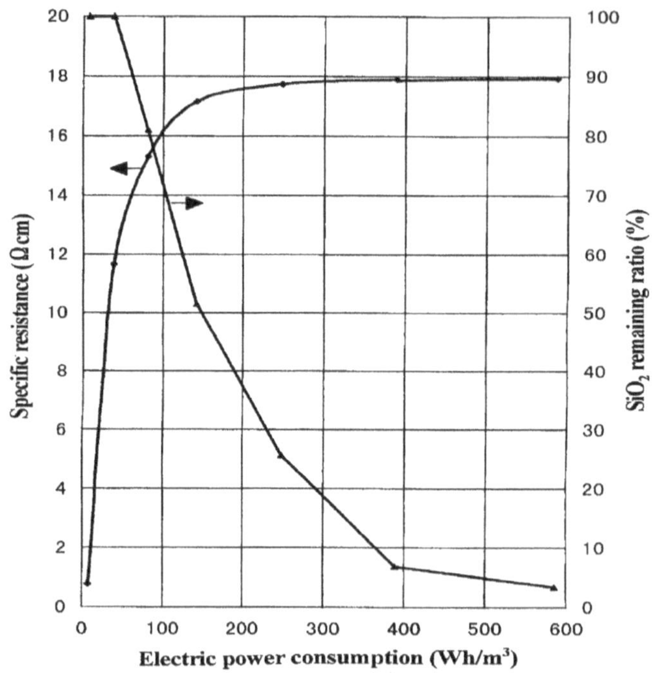

FIGURE 18.6 Energy consumption versus electric resistance and silica removal in the GDI process. *Akahori and Konishi (1999), p. 64, Fig. 5.*

Energy consumption is:

$$P = I^2 \times R_m \tag{18.2}$$

in which, P is direct current electric power (W), and R_m is electric resistance of the system (Ω).

Ebara Corporation developed an EDI (GDI) apparatus with ion exchange nonwoven fabric (IEN) and ion conducting spacer (ICS) (Akahori and Konishi, 1999). IEN and ICS are prepared by radiation graft polymerization and placed between ion exchange membranes. They reduce cell pair resistance and promote water dissociation. Figure 18.6 gives electric resistance and silica removal plotted against energy consumption in the GDI process, showing that ultrapure water (more than 17.5 MΩ cm) is produced at an energy consumption of 250 Wh/m^3.

18.4 Mass Transport and Water Dissociation

To understand the mass transport phenomena arising in the EDI system, Glueckauf (1959) discussed the two-stage ion transport process in the desalting compartment. The first concerns the diffusion transfer from the flowing solution into the ion exchange resin particles. The second concerns the transfer of the ions along the ion exchange particle chains from the interior of the compartment to the ion exchange membranes. The theory is developed on the assumption that the point of the water dissociation lies at the contact between the membrane and exchanger of opposite type, bipolar contact. Verbeek et al. (1998) proposed the concept expressing the mass transport in the EDI system, in which the anode compartment and cathode compartment are filled with cation exchange resins and anion exchange resins, respectively, to form separated beds. The water dissociation reaction is assumed to be generated only at the interface between the electrodes and ion exchange resins. Matejka (1971) reported the experimental and theoretical work that advanced the theory of ionic transport in the EDI system. The mechanism of the mass transport described in the above investigations are developed on the assumption that the water dissociation reaction is generated at the interface.

DiMascio and Ganzi (1999) suggest that the surfaces of ion exchange resins and ion exchange membranes are in contact with each other and form the resin/membrane interfaces and resin/resin interfaces at which water splitting reactions arise at over limiting current density. They discussed the catalytic influence of functional groups (quaternary ammonium groups and tertiary alkyl amine groups) in an anion exchange membrane and anion exchange resin to the intensities of the water dissociation reaction. The feature of the water dissociation generated in the EDI system considerably resembles that in the bipolar membrane electrodialysis system described in Chapter 17.

As described above, the water dissociation in the EDI process is generally discussed by assuming that the reaction is generated at the resin/membrane interfaces and resin/resin interfaces, i.e., bipolar contacts. The bipolar contact seems to be an essential requirement to achieve an EDI operation. However, from this view point, the phenomenon illustrated in Fig. 18.3 is not explained. Figure 18.3 does not include bipolar contacts, although it is operated successfully. The questions are resolved by assuming that the water dissociation is generated in ion exchange membranes and resins. The concept described below is developed from electrodialysis integrated with monopolar ion exchange membrane (Fundamentals, Section 7.8).

A NaCl solution extracted from an RO process (NaCl concentration; C_{feed}) is assumed to be supplied to the diluate compartment of the EDI process integrated with cation exchange membranes (resins) and anion exchange membranes (resins). The NaCl solution is purified to produce ultrapure water ($C_{product} \fallingdotseq 0$) through the three steps illustrated in Fig. 18.7. In the figure, the intensity of the water dissociation reaction in the cation exchange membrane (resin) is suppressed (shown with short arrows) comparing to the phenomenon in the anion exchange membrane (shown with long arrows). This phenomenon is observed experimentally in monopolar ion exchange membranes (Fundamentals, Section 7.7) and its mechanism is discussed (Tanaka, 2010) (Fundamentals, Section 7.8).

Step 1

NaCl concentration C_1 is so high that water splitting does not occur. Supplied Na^+ ions and Cl^- ions are transferred from dilute compartment toward the concentrate

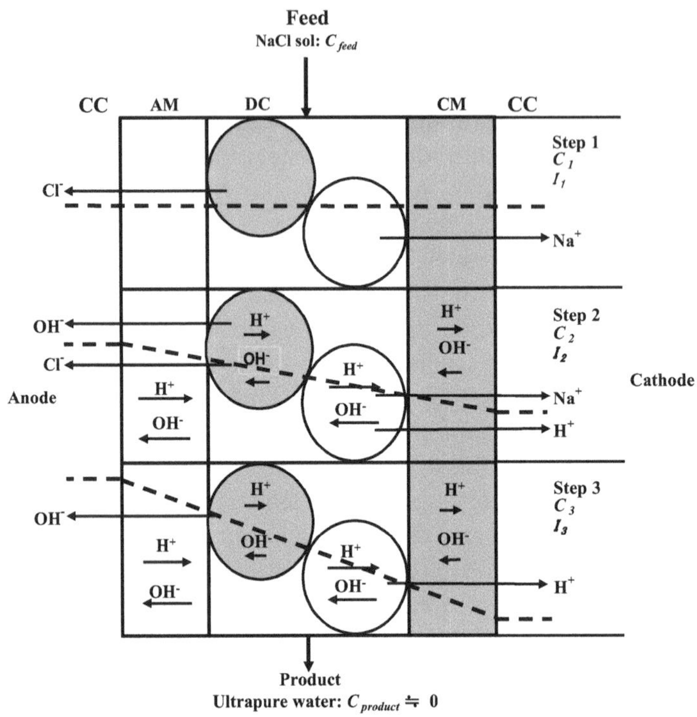

FIGURE 18.7 Mass transport and water dissociation in the EDI process. AM, Anion exchange membrane; CM, Cation exchange membrane; ○, Anion exchange resin; ●, Cation exchange resin; DC, Diluate compartment; CC, Concentrate compartment; ----, pH change.

compartment through the ion exchange membrane (resin) with electro-migration and carry an electric current I_1. pH is unchanged.

$$I_1/F = J_{\mathrm{Na}} + J_{\mathrm{Cl}} \tag{18.3}$$

Step 2

NaCl concentration C_2 is moderate; not so high and not so low. Water splitting takes place to generate H^+ ions and OH^- ions (Eqn (18.4))and carry an electric current I_2 ($<I_1$). A part of generated H^+ ions and OH^- ions is recombined (neutralization) to form water molecules (water regeneration; Eqn. (18.5)), and at the same time Na^+ and Cl^- ion transport (desalination) is accelerated to support on the electric current I_2. pH declines to alkaline in the anode side and acidic in the cathode side (pH distribution).

Water splitting:

$$H_2O \rightarrow H^+ + OH^- \tag{18.4}$$

Water regeneration (neutralization):

$$H^+ + OH^- \rightarrow H_2O \tag{18.5}$$

$$I_2/F = J_{Na} + J_{Cl} + J_H + J_{OH} \tag{18.6}$$

Step 3

NaCl concentration C_3 is very low. Water splitting (Eqn (18.4)) takes place to generate H^+ and OH^- ions to carry an electric current I_3 ($<I_2$). J_{Na} and J_{Cl} is so low (negligible) that the electric current is carried mainly by H^+ ions and OH^- ions. pH declines to alkaline in the anode side and acidic in the cathode side (pH distribution). Ultrapure water is produced in the diluate compartment.

$$J_{Na} + J_{Cl} \fallingdotseq 0 \tag{18.7}$$

$$I_3/F \fallingdotseq J_H + J_{OH} \tag{18.8}$$

Yoshida et al. (2002) observed pH changes in the concentrate compartment and pH distribution in the diluate compartment using an electrodialysis cell integrated with ion exchange fibers adsorbed with a phenolphthalein indicator. From the color changes of the indicator, they confirmed the pH distribution in the compartment and the neutralization caused by the water regeneration. They suggest that the current density in the EDI unit is smaller than the limiting current density by two orders of magnitude and the water dissociation reaction is not caused by the concentration polarization. Thus the water dissociation is also not assumed to be caused by the salt concentration decrease in the bipolar contact.

18.5 Removal of Weakly Ionized Species

Weakly ionized species such as silica, carbon dioxide, boron, and ammonia are difficult to remove through a membrane process such as RO, electrodialysis reversal (EDR), etc. However, EDI removes these species effectively. Hernon et al. (1998) discussed how the EDI process successfully removes various weakly ionized species as follows.

18.5.1 Effect of Water Dissociation

Figure 18.8 illustrates the basic flow scheme of EDI process. In EDI, the ion exchange resins facilitate mass transfer of weakly ionized species mainly due to water splitting. Namely, in the diluting cell, the DC electrical field splits water at the surface of the ion exchange resin beads, producing hydrogen and hydroxyl ions that act as continuous regenerants of the ion exchange resin. This allows a portion of the resins in the EDI to always be in the fully generated state. In turn, the fully generated resins are able to ionize weakly ionized species. Once ionized, these species are quickly removed under the influence of the DC electrical field. Figure 18.8 roughly depicts how different ions are removed as water travels through the EDI diluting cell, strongly ionized ions being removed first in the flow-path and weakly ionized species removed as the water moves

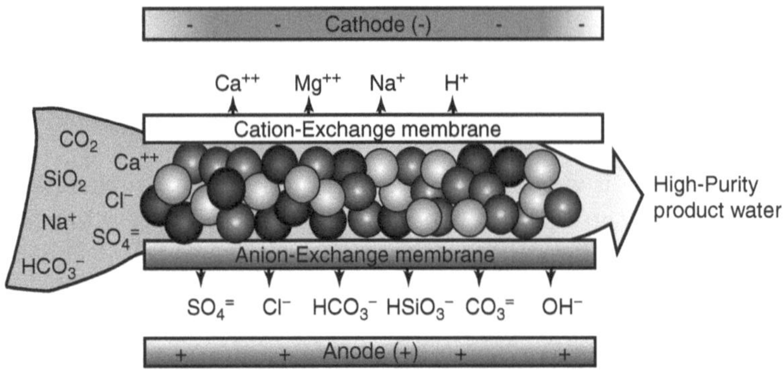

FIGURE 18.8 Basic flow scheme of the EDI process. *Hernon et al. (1998), p. 17, Fig. 1.*

down the flow-path. EDI's performance in removing these weakly ionized species thus assuredly will lead to greater use of EDI in the production of ultrapure water.

18.5.2 Silica Removal

To further explore the performance limits of EDI on weak ion removal, Hernon et al. (1998) conducted the following laboratory study, focusing on the effect of flow rate and amperage on silica removal in a full-scale, 50 gpm unit. EDI feed water for this test contained 20 μS/cm of NaCl and 0.5 ppm of CO_2 when the feed water was treated with RO. As shown in Fig. 18.9, silica removal decreases with increasing flow rate for constant amperage. Increasing flow rate results in lower residence time, more use of available current for strong ion transfer, and less water splitting available for weak ion removal. However, even at velocities that exceed nominal velocities by 50%, silica removal remained above 95%. Figure 18.10 shows the effect of amperage on silica removal for a constant flow. Initially, silica removal increases with increasing amperage due to higher extent of resin/membrane regeneration; however, it levels off at a certain current, I_0, which is a function of flow rate and feed composition. Operation at higher current densities than I_0 does not impact silica removal any further, as the limiting factor at this point for silica removal is the residence time.

The above studies confirm that the degree of weak ion removal in an EDI unit is tied to the degree of water splitting occurring within it. The above studies also provide a basis for calculating design conditions for the removal of weakly ionized species.

18.5.3 Carbon Dioxide Removal

Carbon dioxide removal by EDI is an important facet of its performance. When carbon dioxide is present in an ion exchange feed stream, the carbon dioxide competes with silica for ion exchange sites on an anion resin. Carbon dioxide often represents the largest anion load on an ion exchange system, especially when the ion exchange unit is preceded by an RO unit. The presence of excess amounts of carbon dioxide both limits

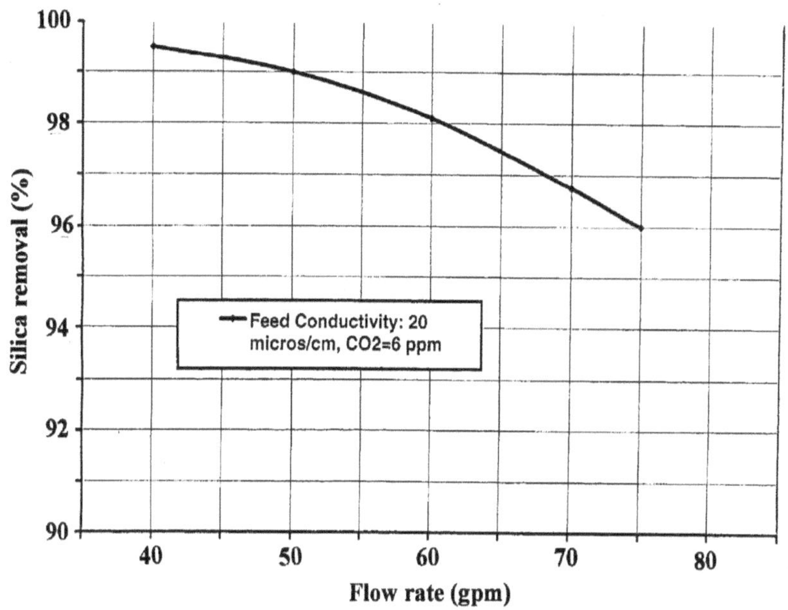

FIGURE 18.9 Silica removal. *Hernon et al. (1998), p. 18, Fig. 2.*

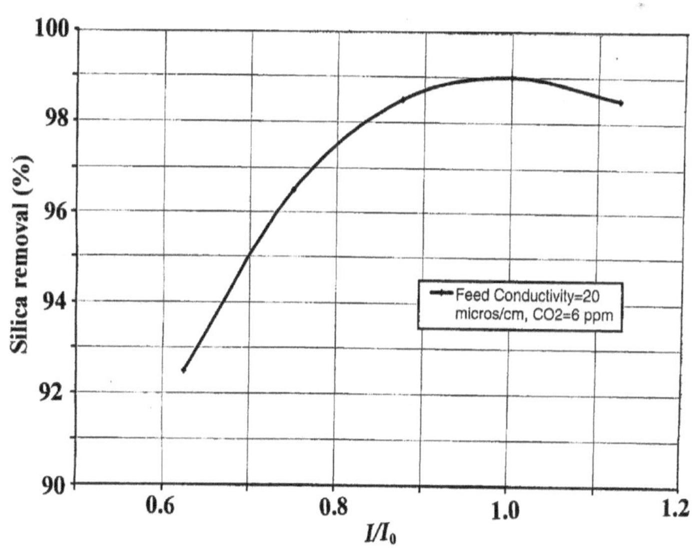

FIGURE 18.10 Silica removal. *Hernon et al. (1998), p. 19, Fig. 3.*

the capacity of ion exchange resin to remove silica and limits the efficiency of silica removal by the ion exchange bed. As discussed in Section 18.6.1, silica removal is critical in both power generation and in semiconductor production. The presence of an excess amount of carbon dioxide in effect exposes a plant to a potentially serious silica problem. Carbon dioxide cannot normally be removed via other membrane demineralization processes such as RO and EDR unless chemical adjustments are made to change alkalinity levels. EDI, on the other hand, is a membrane demineralization process that routinely reduces CO_2 levels by more than 99% in most applications (cf. Table 18.4).

18.5.4 Boron Removal

Boron is often present in water and is problematic because it can cause defects in semiconductor chip manufacturing. Boron is not being removed well by the ion exchange resin process due to its poor ionization and low selectivity. Table 18.1 shows removal of boron by both RO and EDI at four plants. The RO units typically remove only a small portion of the boron from the feed water. However, the EDI units consistently remove more than 96% of the boron in the feed water.

18.5.5 Ammonia Removal

The ammonia study was part of a largest project addressing the development of total water-recycle systems for the Space Station. In a self-contained life-support system, such as the future-planned Space Station, the ability to efficiently recycle water is key to long-term mission success. It is projected that a four-person team will use a supply of 225 lbs of water for general use during a space mission to Mars and back. All water used will need to be recycled through an intricate recovery system and treated through a sequence of processes to produce drinking water. Ammonium ions are generated as a decomposition product of urea and over time will build up in concentration, thus rendering the water unsuitable for human consumption. EDI, being one of the purification steps, is required to reduce the level of ammonia from 200 ppm down to the range of 0.5–1 ppm.

Table 18.1 Boron Levels and Removals at Various Sites

	Power Plant#2	Semiconductor Plant#2	Semiconductor Plant#3	Semiconductor Plant#4
RO Feed Boron (ppb)	22	280	110	85
RO Product Boron (ppb)	14	170	83.5	64.5
EDI Feed Boron (ppb)	14	71	83.5	64.4
EDI Product Boron (ppb)	0.45	2.75	2.8	0.74
RO Boron Removal (%)	36.3	39.3	24.1	24.1
EDI Boron Removal (%)	96.8	96.1	96.6	98.9
RO Feed pH	6.5	7.7	7.8	7.9
EDI Feed pH	5.7	6.5	6.4	6.4

Hernon et al. (1998), p. 13, Table 4.

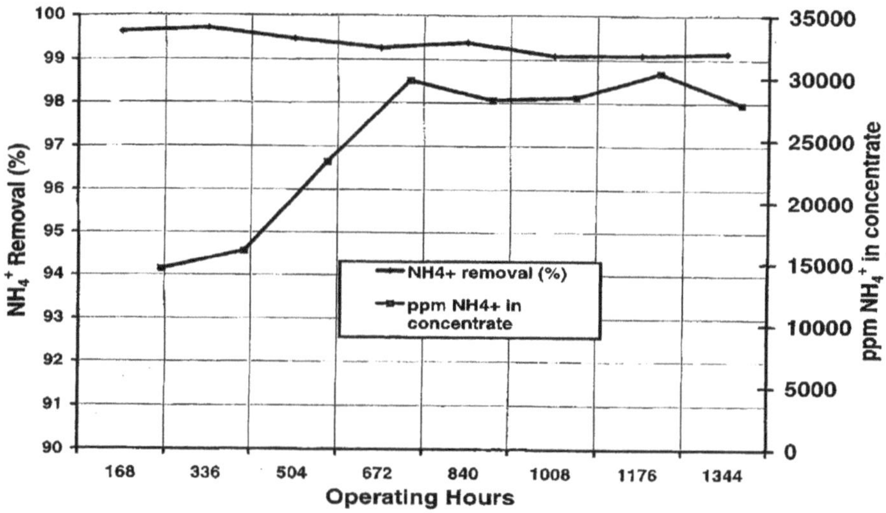

FIGURE 18.11 NH_4^+ removal. pH = 10.3, 99.9% water recovery. *Hernon et al. (1998), p. 23, Fig. 7.*

Figure 18.11 shows the percent ammonia removal by EDI. EDI operation was optimized in terms of flow rate and current. The system was operated in a semicontinuous mode with a continuous 200 ppm ammonia feed at 160 ml/min. The concentrate stream in turn was continuously recycled from a separate waste tank. The waste tank volume was kept constant by periodically removing a small amount of water that would make up for any water transfer from the dilute to the concentrate through the membranes. The amount of periodic waste blowdown corresponds to a water recovery value of 99.9%. Due to such continuous recycle, the waste concentration increased over time, as pictured in Fig. 18.11. Such a high concentration resulted in a slight decrease in ammonia removal due to back-diffusion.

18.6 Practice

18.6.1 Ultrapure Water Production in Electric Power Generation and Semiconductor Manufacturing Processes

In the nuclear power plants, ultrapure water (UPW) is used for high-pressure stream boiler. In the semiconductor plants, the UPW is used for rinsing semiconductor wafers after various processing steps. Ionic Inc. applied the EDI process operating in the following locations (Hernon et al., 1995):

1. Grand Gulf Nuclear Station (GGNS)
2. Arkansas Nuclear ONE (ANO)
3. Seabrook Nuclear Station (SNS)

Table 18.2 EDI Installations; Raw Feedwater Sources and Pretreatment Steps

Location	Feed Source	Feed TDS (ppm)	Temp. (°C)	Flowrate (gpm)	Stream Flowsheet
GGNS	Surface	400	17–25	50	MMF-ACF-UF-EDR-RO-EDI
ANO	Surface	100	10–30	200	MMF-ACF-UF-EDR-RO-EDI
SNS	Well	475	25	150	UF-RO-EDI
Semi 1	Surface	230	20	65	CF-UF-RO-EDI
Semi 2	Well	900	15–25	100	MMF-DEGAS-UF-RO-EDI
Semi 3	Surface	700	20–25	200	UF-RO-EDI

Hernon et al. (1995), Table 1.

4. A New England semiconductor manufacturing plant (Semi 1)
5. A Midwest semiconductor manufacturing plant (Semi 2)
6. An Arizona semiconductor manufacturing plant (Semi 3)

In all these installations, EDI is one building block of a multistep treatment process. The use of various treatment steps in the treatment process depends largely on differences in feed water characteristics such as TDS, pH, organic load, dissolved gases, temperature, and the presence of various undesirable constituents, such as trihalomethanes. Table 18.2 summarizes some of the feed water variables to each of these units, as well as the unit processes that are used in each installation.

Figure 18.12 shows conductivity reductions that are directly related to TDS removed across the EDI unit at GGNS. In the figure, 100% reduction corresponds to the minimum conductivity achievable due to water dissociation. During all of the operating terms, the overall conductivity reduction averages more than 99.5%. The values in the other stations and plants have been steady in the range of 99%–99.5+%.

EDI feed water and product water samples were analyzed for sodium, calcium, magnesium, chloride, and sulfate. Rates of removal range from one to three orders of

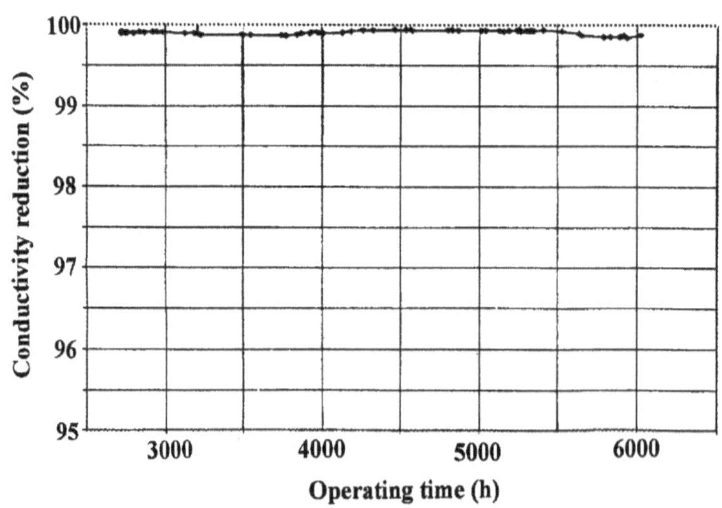

FIGURE 18.12 EDI conductivity reduction, GGNS. *Hernon et al. (1995), Fig. 1.*

Table 18.3 EDI Feed and Product Ion Levels

Ion		Na^+(ppb)	Ca^{2+}(ppb)	Mg^{2+}(ppb)	Cl^-(ppb)	SO_4^{2-} (ppb)
GGNS	Feed	1560	10	4	58	70
	Prod	2	<0.5	<0.5	<2	<4
ANO	Feed	871	42	7	594	88
	Prod	3	0.7	<0.5	<2	<4
SNS	Feed	1210	165	nd	528	157
	Prod	2.4	2.5	nd	<2	<2
Semi 1	Feed	289	158	nd	140	245
	Prod	<2	<0.5	nd	<2	<4
Semi 2	Feed	1710	320	160	2200	1110
	Prod	<2	<0.5	<5	<2	<4
Semi 3	Feed	437	35	25	78	31
	Prod	4	<3	<5	<2	<4

Hernon et al. (1994), Table 2; Hernon et al. (1995), Table 2.

magnitude, depending on the specific ion and its feeding water concentration. EDI product water quality approaches MB ion exchange quality. Table 18.3 summarizes the results of these analyses.

In power plants, high silica levels in the UPW supply to steam generators can lead to silica deposition on the electrical generator turbine blades. CO_2 competes with silica ions for exchange sites, so the amount of CO_2 in a feed stream affects both bed life and silica removal efficiency in the ion exchange bed. Table 18.4 presents a summary of silica and carbon dioxide feed levels and the corresponding percent removals achieved with EDI at the six sites mentioned above.

18.6.2 Ultrapure Water Production in Pharmaceuticals

Various grades of water are used by the pharmaceutical industry for a number of applications. Definitions of each water type and quality are provided in the United States Pharmacopoeia (USP). Of this water, water for injection (WFI) and PW are of

Table 18.4 Silica and CO_2 Levels and Removal

	Silica			CO2		
Location	**Feed SiO_2 (ppb)**	**Product SiO_2 (ppb)**	**Rejection (%)**	**Feed Total CO_2 (ppb)**	**Product Total CO_2 (ppb)**	**Rejection (%)**
GGNS	640	3	>99.5	2920	17	99.4
ANO	3830	138	96.4	4780	8	99.8
SNS	170	8	95.2	5910	<10	>99.8
Semi 1	208	<2	>99.0	2800	<10	>99.6
Semi 2	86	<2	>97.7	3260	<10	>99.7
Semi 3	165	<2	>98.8	5340	<10	>99.8

Hernon et al. (1995), Tables 3 and 4.

Table 18.5 Water Quality Standard (Numerical Interpretation of USPC Standard)

Constituent	Purified Water	Water for Injection
pH	5.0–7.0	5.0–7.0
Chloride	≦0.5 mg/l	≦0.5 mg/l
Sulfate	≦1.0 mg/l	≦1.0 mg/l
Ammonia	≦0.1 mg/l	≦0.1 mg/l
Calcium	≦1.0 mg/l	≦1.0 mg/l
Carbon dioxide	≦5.0 mg/l	≦5.0 mg/l
Heavy metal	≦0.1 mg/l as Cu[a]	≦0.1 mg/l as Cu[a]
Oxidizable substances[b]	Passes USP Permanganate test	Passes USP Permanganate test
Total solids	≦10 mg/l	≦10 mg/l
Total bacterial count	≦50 CFU/ml	≦10 CFU/100 ml
Pyrogen	Non specified	≦0.25 EU/ml

[a]Limits for other heavy metals may be determined.
[b]Limits for specific oxidizable substances may be determined.
Parise et al. (1990), p. 14, Table 1.

primary interest to water treatment engineers and equipment suppliers. Table 18.5 provides a summary of the numerical interpretation of the USP chemistry limits for this water.

A leading manufacturer of ophthalmic products (eye drops, contact lens cleaner, etc.) planned to increase its capacity by establishing a new production facility in the continental United States. As water is the main ingredient in ophthalmic solutions, the choice of a design for the new plant's central water purification system was a primary concern. The operational and performance requirements for the water purification system were as follows (Parise et al., 1990):

1. Product quality equal to or better than that specified for USP Purified Water.
2. Product quantity of 16 gpm, to be available on a continuous basis, 24 h per day, 6 days per week.
3. Minimize the use of hazardous chemicals and chemical waste discharge.
4. Minimize the opportunity for bacterial growth in the system.
5. Achieve consistent product quality and minimize equipment downtime.

After evaluation of the available water purification technologies, the following RO/CDI system developed by Ionpure Technologies Corporation was selected. The continuous deionization (CDI) module is the EDI system developed by Millipore Corporation. The analysis of the municipal feed water in Table 18.6 indicates that a single-pass RO system, followed by a single-pass CDI unit would meet the client's final product quality specification.

The system design is shown in block format in Fig. 18.13. Raw municipal water entered the system at 60 psig, and was pumped through dual online multimedia filters to remove medium- and large-size particles down to a 10 μm level. The next process in the

Table 18.6 Purified Water System Feed Quality (Raw Water Analysis)

Cations	ppm	Anions	ppm
Ca	119.1	OH	0.0
Mg	20.4	CO_3	0.0
Na	31.1	HCO_3	92.0
K	4.9	SO_4	35.1
Fe	0.1	Cl	37.1
Cu	0.1	NO_3	2.2
Ba	0.0	F	2.4
Sr	0.3		
Al	0.4		

Dissolved SiO_2 4.5 ppm, pH 8.2
Parise et al. (1990), p. 17, Table B.

system was filtration by activated carbon to remove organic contaminants and chlorine. Following the carbon beds were dual online cation exchange softeners to reduce the incoming total hardness to less than 1 ppm. The last step in the pretreatment train was filtration to a nominal 1 μm level with polyethylene cartridge filters. The pretreatment train had one more design feature to minimize bacterial growth. Whenever the RO system was idle, a pump was activated to recycle water around the pretreatment train,

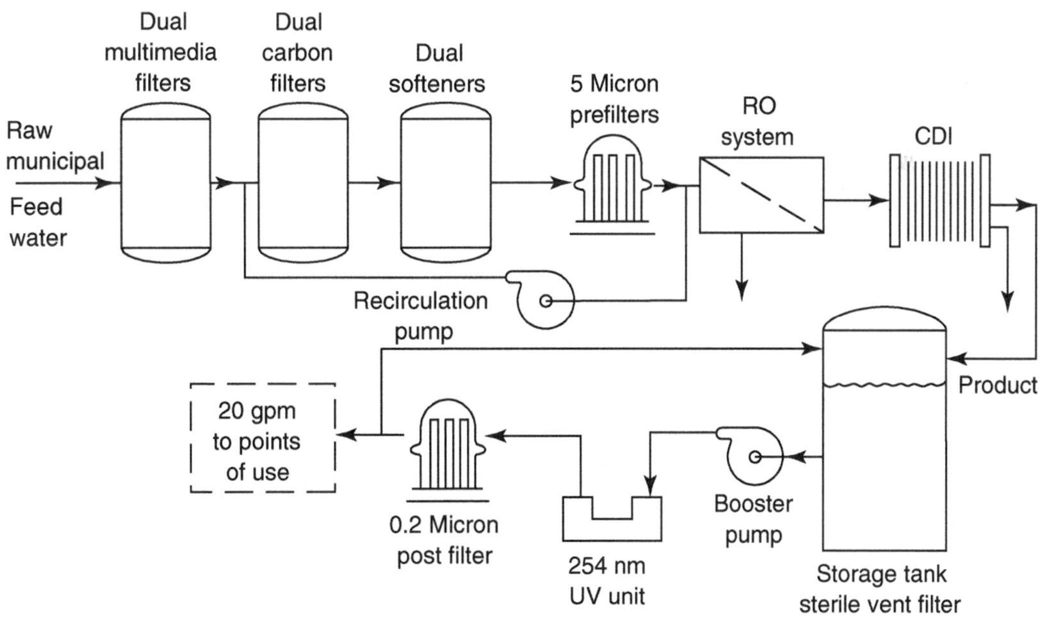

FIGURE 18.13 EDI system to produce purified water meeting the United States Pharmacopoeia. *Parise et al. (1990), p. 18, Fig. 8.*

from just before the RO inlet to the carbon filter. This eliminated the stagnant condition that would otherwise occur when the RO was inoperative.

The RO system currently uses 1500 square feet of Hi-Flux CP polysulfone membrane to remove suspended and dissolved contaminants and is operated at 400 psig to produce 20 to 32 gpm of product water. The RO is currently operating at 50% recovery, but is capable of operating at up to 70% recovery.

The 20 gpm RO product is next fed to a CDI installation consisting of two cell-pair CDI modules. The CDI modules remove dissolved ionic contaminants to polish the RO product water up to the required final production resistivity of 0.50 MΩcm. The CDI system is producing 16 gpm of final product water. The 60 cell-pair stacks are running at 220 V and 1.1 A each. The final product quality from the CDI system varies with the RO product conductivity. With an RO product TDS of 17.7 ppm, the CDI product resistivity is 1 MΩ cm. With an RO product TDS of 10.6 ppm, the CDI product resistivity stays between 3 and 4 MΩ cm. The CDI modules operate at a recovery rate of 80%.

The CDI product water is fed to a storage tank, from which water is recirculated at 20 gpm through a 254 nm ultraviolet unit and final 0.2 μm final filters before being sent to the distribution loop.

The multimedia filters are back-washed twice per week. The carbon filters are steam sterilized and back-washed every 2 weeks. The softeners are regenerated automatically based on total flow. The RO system is sanitized monthly with 100 ppm of sodium hypochlorite, and cleaned as needed to maintain flux, the frequency of which depends on the SDI (Silt Density Index) of the incoming raw water. No periodic maintenance is performed on the CDI modules.

18.6.3 Economic Comparison between Electrodeionization and Mixed-Bed Ion Exchange

EDI is an effective technology alternative to MMB ion exchange polishing after RO. Edmonds and Salem (1998) provided the following economic comparison of EDI and MB for new plant installation. In this study, a comparison of three product flow cases (50, 200 and 600 gpm, see Table 18.7) and three feed water total dissolved solids (TDS) cases (low,

Table 18.7 EDI/MB Feed Water Quality

Constituent	Low TDS (ppm)	Medium TDS (ppm)	High TDS (ppm)
Na^+	1.2	5.27	9.25
HCO_3^-	2.68	3.15	3.78
Cl^-	0.28	6.32	11.9
CO_2	0.9	1.0	1.2
SiO	0.5	0.5	0.5
TDS	4.16	14.74	24.9

pH = 6.7.

Edmonds and Salem (1998), p. 49, Table A.

Table 18.8 Capital Equipment Cost and System Annualized Cost

		Capital Equipment Cost (1000$)			System Annualized Cost (1000$)		
Flow	gpm	50	200	600	50	200	600
High TDS	EDI	60	150	385	30.8	83.6	177.8
	MB	152	216	397	73.8	120.9	240.2
	EDI/MB	0.39	0.69	0.97	0.42	0.69	0.74
Medium TDS	EDI	60	150	385	28.7	75.6	153.5
	MB	152	216	354	60.2	94.2	176.7
	EDI/MB	0.39	0.69	1.09	0.48	0.80	0.87
Low TDS	EDI	60	150	385	27.6	71.1	140.3
	MB	151	216	352	49.6	72.6	126.2
	EDI/MB	0.40	0.69	1.09	0.56	0.98	1.11

Edmonds and Salem (1998), p. 49, Tables B and C.

medium, and high) were considered. The deionized water quality required was normally defined as greater than or equal to 17 MΩ cm and silica reduction to less than 20 ppb.

Capital costs for the EDI and the MB equipment were obtained and given in Table 18.8. Installation costs were estimated to be percentage of the total capital cost. A factor of 0.2 was used for the EDI systems, and a factor of 0.4 was used for the MB systems. The major operating costs considered for each process were as follows: Both systems had labor and water use/wastewater costs. Electricity and replacement EDI stacks were other operation costs for the EDI. MB system costs also included chemicals and replacement resins. The following cost factors were used: operating labor, $40/h; electricity, $0.07/kWh; water, $1/kgal; wastewater, $2/kgal; sulfuric acid, $0.05/lb of 100%; sodium hydroxide, $0.15/lb of 100%; cation resin, $55/ft^3; anion resin, $150/ft^3; and stack replacement cost, $6300/12.5 gpm product.

A summary of the annualized costs for each case is given also in Table 18.8. The total system annualized costs were developed from the capital equipment costs, the operating cost items, and installation factors as described above. As the RO product water TDS increases, EDI becomes increasingly cost-effective. At the high flow, low TDS case (600 gpm, 4.16 ppm), EDI is slightly (~10%) more expensive than MB system. This is primarily the result of the replacement frequency of the ED stacks. EDI demonstrates a clear capital cost advantage in the 50 to 200 gpm range over MB polishing for new plants. MB polishers require considerable ancillary equipment (chemical storage and delivery systems, waste neutralization system) that negatively impacts their cost effectiveness in low-to-moderate flow range. The 600 gpm case shows that the capital cost of EDI and MB systems are approximately equal.

It was found that EDI, in low-to moderate flow-rate system cases, is significantly less expensive, on an annualized cost basis, than MB. In addition, EDI offers other cost benefits over MB that was not quantified in this study. These include minimization of wastewater monitoring and discharge activity, significantly reduced space requirements, reduction in spill protection plan designs and tasks, and less chemical handling for plant operators. EDI also has the ability to respond to feed-water TDS changes without compromising deionized water production and quality, with minimal impact on operation costs.

References

Akahori, M., Konishi, S., 1999. Development of electrodeionization apparatus containing novel ion-exchange materials. J. Ion. Exch. 10, 60–69.

Allison, R.P., 1996. The continuous electro-deionization process. In: American Desalting Association 1996 Biennial Conference & Exposition, Monterey, CA, USA, August 4–8.

Deguchi, T., Karibe, T., 1998. Electro deionization. In: Water Treatment Handbook. Maruzen Publishing Co., Tokyo, pp. 220–225.

DiMascio, F., Ganzi, G.C., 1999. Electrodeionization Apparatus and Method. US 5858191.

Edmonds, C., Salem, E., November 1998. Demineralization. An economic comparison between EDI and mixed-bed ion exchange. Ultrapure Water. November, pp. 43-49.

Ganzi, G.C., Egozy, Y., Giuffrida, A.J., Jha, A.D., 1987. High purity water by electrodeionization performance of the ionpure continuous deionization system. Ultrapure Water 4 (3), 43–50.

Glueckauf, E., 1959. Electro-deionization through a packed bed. Brit. Chem. Eng. 4, 646–651.

Grabowski, A., Zhang, G., Strathmenn, H., Eigenberger, G., 2006. The production of high purity water by continuous electrodeionization with bipolar membranes: Influence of the anion-exchange membrane permselectivity. J. Membr. Sci. 281, 297–306.

Grabowski, A., 2010. Electromembrane Desalination Processes for Production of Low Conductivity Water. Logos Verlag Berlin GmbH, Berlin, pp. 135–138.

Hernon, B.P., Zanapalldou, R.H., Li, Zhang, Siwak, L.R., Schoepke, E.J., 1994. Application of electro-deionization in ultrapurewater production: Performance and theory. 55th Annual Meeting International Water Conference. Pittsburgh, PA. Oct. 30–Nov. 2.

Hernon, B.P., Zhang, Li, Siwak, L.R., Schoepke, E.J., 1995. Process report, application of electro-deionization in ultrapure water production. 56th Annual Meeting International Water Conference. Pittsburgh, PA. Oct. 29–Nov. 2.

Hernon, B.P., Zanapalidou, H., Prato, T., Zhang, Li, 1998. Removal of weakly ionized species by EDI. IONICS ESD 3770, pp. 9–28.

Jha, A.D., Gifford, J.D., 2004. Ultrapure CEDI for microelectronics applications: a cost effective alternative to mixed-bed polishers. Proceedings for Ultrapure Water Asia 2004, Singapore, March 10–11, 2004.

Kollsman, P., 1957. Method and Apparatus for Treating Ionic Fluids by Dialysis. US 2815320.

Korngold, E., Selegny, E., 1972. Method of Separation of Ions from a Solution. US 2366089.

Kunz, G., 1983. Verhahren und Vorrichtung zum Behandeln von Flussigkeiten, insbesondere Entsalzen wassriger Losungen. DE 3217990.

Kunz, G.K., 1987. Process and Apparatus for Treatment of Fluid, Particularly Desalination of Aqueous Solutions. US 4636296.

Liang, L., Wang, L., 2001. Continuous electrodeionization processes for production of ultrapure water. Semicond. Pure Water Chem. Conf.

Matejka, Z., 1971. Continuous production of high-purity water by electrodeionization. J. Appl. Chem. Biotech. 21, 117–120.

McRay, W.A., Rigopulos, P.N., 1967. Removal of Gases by Electrode-ionization. US 3330750.

Parise, P.L., Parekh, B.S., Waddington, G., 1990. The use of ionpure continuous deionization for the production of pharmaceutical and semiconductor grades of water. Ultrapure Water 7 (8), 14–27.

Parsi, E., 1964. Removal of Dissolved Salts and Silica from Liquids. US 3149061.

Parsi, E.J., 1966. Removal of Weakly Basic Substances from Solution by Electrodeionization. US 3291713.

Salem, E., Tessier, D.F., 1998. Advances in Electrodeionization. International Water Conference, Pittsburgh, PA. Oct. 1998.

Tanaka, Y., 2010. Water dissociation reaction generated in an ion exchange membrane. J. Membr. Sci. 350, 347–360.

Thate, S., February 2002. Untersuchung der elektrochemischen Deionisation zur Reinstwasserherstellung (Ph. D. thesis). Universitat Stuttgart, Institut fur Chemische Verfahrenstechnik, Stuttgart, Germany.

Verbeek, H.M., Furst, L., Neumeister, H., 1998. Digital simulation of an electrodeionization process. Comput. Chem. Engng. 22 (Suppl.), S913–S916.

Walters, W.R., Weiser, D.W., Marek, L.J., 1955. Concentration of radioactive aqueous wastes. Ind. Eng. Chem. 47 (1), 61–67.

Yoshida, S., Kanazawa, N., Qiu, L., Umeda, M., Uchino, H., Fukuda, J., Aoyagi, M., Watanabe, T., 2002. Regeneration mechanism of ion exchange materials in electrodeionization system. Electrochemistry 70, 784–788.

19

Electrolysis

19.1 Overview of Technology

A feature of the sodium chloride electrolysis industry is to produce chlorine and caustic soda, i.e., chlor-alkali, in parallel. Because of this, serious problems occur when the production–consumption difference of chlorine is not consistent with that of caustic soda. Chlor-alkali production has been influenced by a chlor-alkali consumption difference change between too much and too little (Takarada, 2004). In the first half of the twentieth century, caustic soda consumption increased for the production of rayon, pulp, glass, etc. but consumption of chlorine was limited to the fields such as disinfectants and bleaches, so the chlor-alkali process was restricted and alkali production by means of an ammonia soda process was developed. In and after the middle of the twentieth century, the chlor-alkali industry expanded owing to an increase in chlorine consumption in the synthetic chemical, petrochemical, and synthetic fiber industries, for example. However, in the next stage, chlorine consumption decreased largely owing to environmental protection and the chlor-alkali process was again restricted.

The chlor-alkali process is classified into the processes using mercury, asbestos diaphragms, and ion exchange membranes. Here, we have to pay attention to poisoning caused by the mercury process and the process conversion policy enacted by the Japanese government. Figure 19.1 gives the production capacity change of the chlor-alkali process in Japan; the mercury method fell into disuse by 1986, and the processes were unified to the ion exchange membrane method until 1999 (Furuya, 2000). This

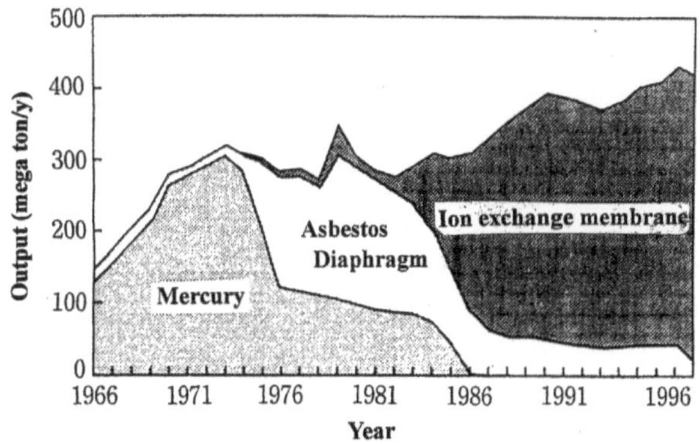

FIGURE 19.1 Production capacity change of chlor-alkali processes in Japan. *Furuya (2000), p. 380, Fig. 11.11.*

Ion Exchange Membranes. http://dx.doi.org/10.1016/B978-0-444-63319-4.00019-5

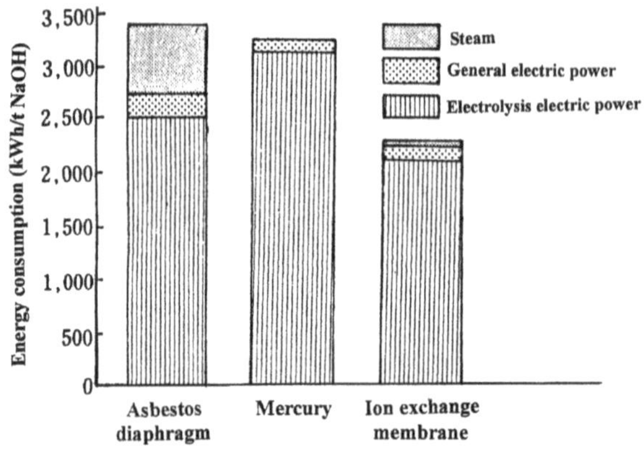

FIGURE 19.2 Energy consumption in chlor-alkali processes. *Asawa (1991), p. 514, Fig. 8.*

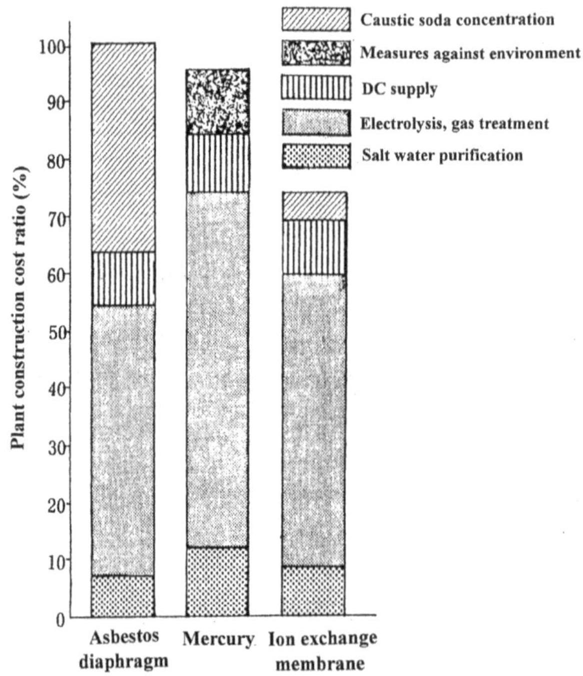

FIGURE 19.3 Construction cost of chlor-alkali processes. *Asawa (1991), p. 515, Fig. 9.*

conversion was needed to achieve lower energy consumption and lower capital costs in the ion exchange membrane process compared with the others, as indicated in Figs 19.2 and 19.3 (Asawa, 1991). Because of the lower energy consumption, the caustic soda production capacity of the ion exchange membrane method in the world increased: 12.8 Mt; 29% of the total caustic soda production capacity (1994) → 33.0 Mt; 56% (2004) → 50.0 Mt; 64% (2007) (Kado, 2009).

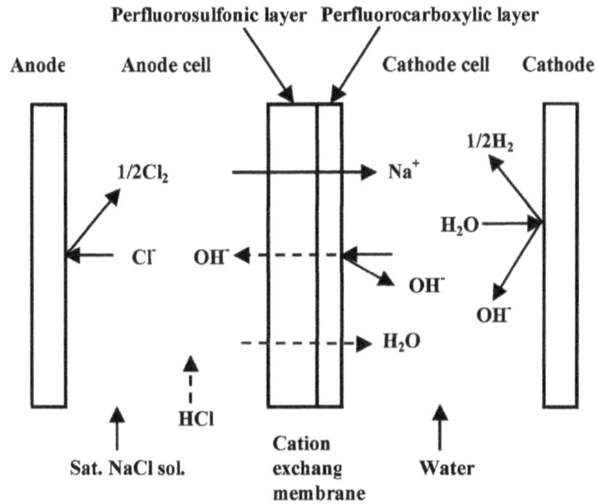

FIGURE 19.4 Principle of ion exchange membrane electrolysis of sodium chloride.

Figure 19.4 illustrates the principle of the ion exchange membrane electrolysis system, in which a saturated NaCl solution and pure water are supplied to an anode cell and a cathode cell, respectively. When an electric current is passed, the following reactions occur.

At an anode:

$$2Cl^- \rightarrow Cl_2 + 2e^- \quad E^0 = 1.36\ V \tag{19.1}$$

At a cathode:

$$2H_2O + 2e^- \rightarrow H_2 + 2OH^- \quad E^0 = -0.828\ V \tag{19.2}$$

Total reaction:

$$2NaCl + 2H_2O \rightarrow 2NaOH + H_2 + Cl_2 \quad E^0 = 2.19\ V \tag{19.3}$$

Here, Cl_2 gas is generated at the anode; Na^+ ions are transferred from the anode cell to the cathode cell through the cation exchange membrane and combine with OH^- ions generated at the cathode to produce NaOH. The cation exchange membrane indicated in the figure is a double-layer membrane consisting of a perfluorocarboxylic acid layer and a perfluorosulfonic acid layer (Fundamentals, Section 1.7.2).

19.2 Ion Exchange Membrane

19.2.1 Perfluorosulfonic Acid Membrane

In the electrolysis system, cation exchange membranes have to exhibit both high chlorine resistance and high current efficiency in highly concentrated caustic soda. In the

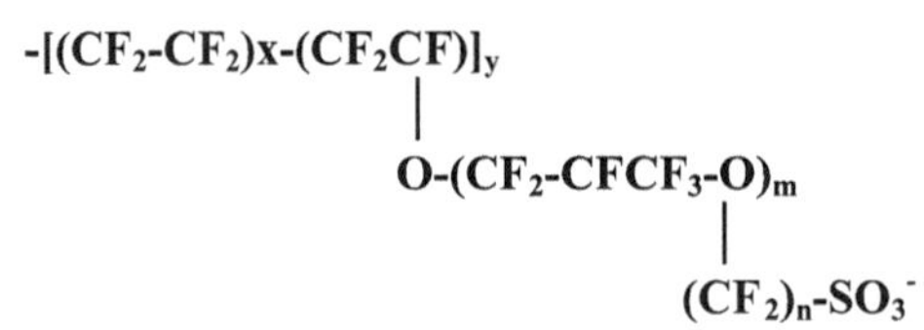

Nafion	**: m ≧ 1; n = 2, x = 5 – 13.5; y = 1000**
Dow membrane	**: m = 0; n = 2**
Aciplex	**: m = 0, 3; n = 2.5; x = 1.5 – 1.4**
Flemion	**: m = 0, 1; n = 1 – 5**

FIGURE 19.5 Structural formula of a perfluorosulfonic acid membrane.

system in Figure 19.4, the perfluorosulfonic acid membrane presented in Figure 19.5 was used at a first developing stage because of chemical and thermal stability. This membrane is based on perfluorosulfonyl fluoride monomer copolymerized with tetrafluoroethylene and was developed by Du Pont (Putnam, 1967; Connally, 1966). Ion exchange capacity can be adjusted by changing *x*, *y*, *m*, and *n* for the commercially available membrane shown in the figure.

Gierke (1977) investigated the microstructure of a Nafion membrane by means of x-ray diffractometry and suggested the ion-cluster channel model illustrated in Figure 19.6 (cf. Fig. 22.6), which shows that the distribution of fixed ions is not uniform. The clusters are connected through narrow channels and the distance between the clusters is nearly 50 Å. An electron microscopic observation revealed the shape of the cluster diameter to be about 40 Å. The diameter of the channel between the clusters was estimated to be nearly 10 Å from measuring the velocity of the water passing through the membrane. Kusumocahyo and Sudoh (1999) measured the cluster diameter of Nafion membranes combined with long chain counter-ions by means of the water passing experiment and obtained the following values.

Counter Ions	Cluster Diameter (nm) (Å)
H^+	38
$(CH_3)_4N^+$	30.93
$(C_4H_9)_4N^+$	15.09
$(C_8H_{17})_4N^+$	17.32

The cluster size is controlled by the ion exchange capacity, which corresponds to equivalent weight: EW (g-dry/Eq) = 1000/ion exchange capacity (mEq/g). The cluster size decreases with an increase in EW owing to the decrease in water content in the membrane. Furthermore, the cluster size is influenced by hydrolysis conditions of precursors to perfluorosulfonic acid or perfluorocarbolic acid polymers in the synthesis of the membranes because the water content is affected by hydrolysis conditions. With the decrease in cluster size, ion permselectivity is increased but it induces a decrease in

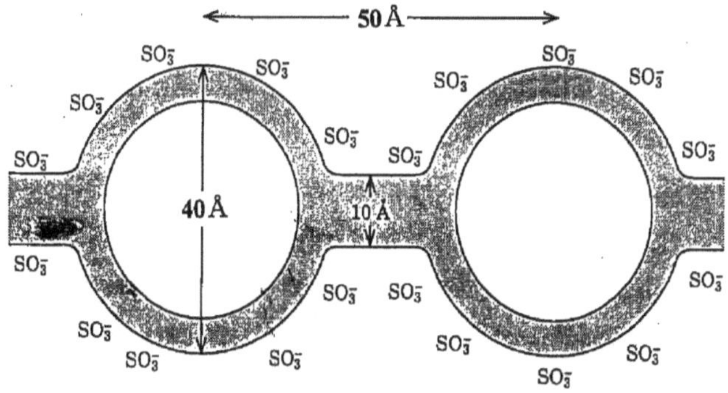

FIGURE 19.6 Ion-cluster channel model. *Gierke (1977).*

electric conductivity. Thus, the influence of the cluster size on ion permeability and electric conductivity is a tradeoff (Kado, 2009).

The Nafion persulfonic acid membrane is usually reinforced with polytetrafluoroethylene (PTFE) fabrics and is useful as a separator in a wide range of applications (DuPont, 2002). However, the reinforcement partially screens an electric current and increases the electric resistance of the membrane (Kado, 2009). To reduce electric resistance, fine channels are prepared in the membrane for passing an electrolyte solution by embedding polyethylene (PET) and PTFE fabrics into the membrane and dissolving PET by hydrolysis (Kameyama et al., 2010).

19.2.2 Sulfonic Acid/Carboxylic Acid Double-Layer Membrane

The electric conductivity of the perfluorosulfonic acid membrane is high; however, it cannot prevent back migration of OH^- ions generated at a cathode toward an anode cell. A perfluorocarboxylic acid membrane effectively prevents the back migration of OH^- ions. Currently, a double-layer membrane consisting of sulfonic acid and carboxylic acid layers is generally used (Grot et al., 1972) (Fundamentals, Section 1.7.2).

Table 19.1 shows the performance of sulfonic acid type Rf-SO_3^-, carboxylic acid type Rf-COO^-, and double-layer type (Rf-SO_3^-/Rf-COO^-) membranes. The sulfonic acid type membrane has lower electric resistance and lower ohmic loss in the electrolysis process, but it has larger hydrophilicity and larger OH^- ion penetrability and so does not achieve higher current efficiency. On the contrary, the carboxylic acid membrane shows lower hydrophilicity and higher current efficiency, but it has defects such as higher electric resistance and larger ohmic loss. Furthermore, although it is a carboxylic acid membrane, it cannot prevent the back migration of OH^- ions from the cathode cell to the anode cell. The back-migrated OH^- ions are discharged at the anode (cf. Section 19.3.3) and generate O_2, ClO_3^- ions, etc. decrease Cl_2 generation efficiency, lower the product Cl_2 purity, and shorten the anode life. To prevent OH^- ion discharge at the anode, OH^- ions are usually neutralized by injecting HCl into the anode cell. However, HCl injection

Table 19.1 Comparison of Ion Exchange Groups and Membrane Characteristics

Ion Exchange Group	Rf-SO_3^{2-}	Rf-COO^-	Rf-COO^-/Rf-SO_3^{2-}
pKa	<1	2–3	2–3/<1
Hydrophilicity	High	Low	Low/high
Current efficiency	<80%	>93%	>93%
Electric resistance	Low	High	Low
Chemical stability	High	High	High
pH range	>1	>3	>1
HCl neutralization	Possible	Impossible	Possible
O_2/Cl_2	<0.5%	>2%	<0.5%
Current density	High	Low	High

Ohmura (1984), p. 646, Table 1.

converts Rf-COO^- groups to Rf-COOH groups, which increases the electric resistance of the carboxylic acid membrane (Ohmura, 1984).

To prevent these problems, a double-layer membrane consisting of a perfluorocarboxylic layer and a perfluorosulfonic acid layer was prepared. Aciplex membranes developed by Asahi Chemical Co. were prepared by coating perfluorocarboxylic acid polymers on the perfluorosulfonic acid membrane or laminating two different polymer films containing carboxylic and sulfonic acid, or by chemical conversion of perfluorosulfonic acid to perfluorocarboxylic acid (Fundamentals, Section 1.7.2). A comparison of these membranes is shown in Table 19.2. Chemical treatment was considered the best method for preparation of the membrane for chlor-alkali process. With this method, the thickness of the carboxylic acid layer formed on the surface of the perfluorosulfonic acid membrane can be easily reduced to less than 10 μm, resulting in a corresponding reduction in the electric resistance of the membrane. When the

Table 19.2 Comparison of Monolayer and Double-Layer Perfluoro-Ionomer Membranes

Type of Membrane	Monolayer Perfluorocarboxylic Acid Membrane COOH	Double-Layer Membrane of Perfluorocarboxylic and Sulfonic Acid by Lamination or Coating COOH/SO_3H	Double-Layer Membrane of Perfluorocarboxylic and Sulfonic Acid by Chemical Treatment COOH/SO_3H
Thickness of COOH layer (microns)	>125	20–50	<10
Current efficiency (%) 8 N NaOH	96	96	96
Electrical resistance	Heigh	Medium	Low
Necessary number of functional monomers	1	2	1
Membrane fabrication	Extrusion	Extrusion and lamination or coating	Extrusion and chemical treatment

Seko et al. (1982), p. 472, Table 3.

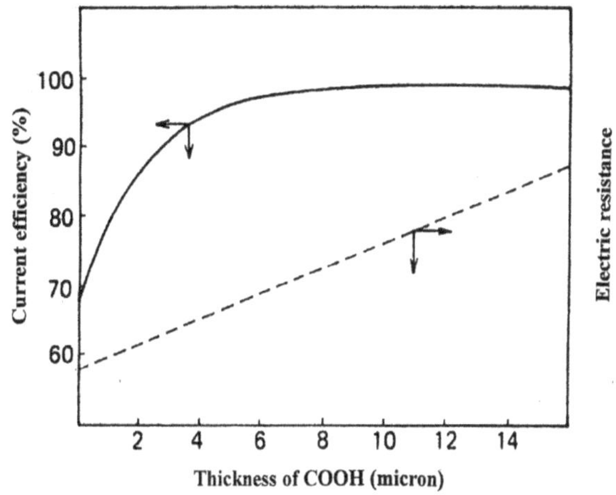

FIGURE 19.7 Thickness of COOH layer, current efficiency, and electric resistance of a multilayer perfluorocarboxylic and sulfonic acid membrane prepared by chemical treatment. *Seko et al. (1982), p. 472, Fig. 19.3.*

carboxylic acid and sulfonic acid layers are faced on the cathode and anode sides, respectively, as shown in Figure 19.4, it becomes possible to inject HCl into the anode cell (Seko et al., 1982).

The current efficiency and the electric resistance are shown in Figure 19.7 for the double-layer membrane prepared by chemical treatment. As shown in this figure, the electric resistance increases linearly with the thickness of the carboxylic acid layer. The current efficiency increases greatly until the thickness of the carboxylic acid layer reaches 7 μm, when saturation tends to occur. From the point of view of power consumption, the optimal thickness of the carboxylic acid layer is in the range of 2–10 μm. Thus, chemical treatment was assumed to be the most appropriate method for preparation of the carboxylic acid layer in the Aciplex membrane, because the highest current efficiency was achieved with a minimal increase in electric resistance.

Figure 19.8 shows the relationship between the current efficiency, caustic soda concentration, and ion exchange capacity of the double-layer Aciplex membrane prepared by chemical treatment of the perfluorosulfonic acid membrane. Each membrane shows maximal current efficiency at a certain caustic soda concentration. This is because with increasing concentration of the external caustic solution up to a certain value, the membrane tends to shrink and the COO^- fixed ion concentration in the membrane tends to increase. However, if the concentration of the caustic soda increases beyond this value, hydroxide ion diffusion increases owing to increased electrolyte sorption; thus, current efficiency is decreased. An increase in the ion exchange capacity results in the occurrence of maximal current efficiency at a higher caustic soda concentration. Therefore, to achieve a high NaOH concentration of catholyte, the membrane should have carboxylic acid groups of a higher exchange capacity.

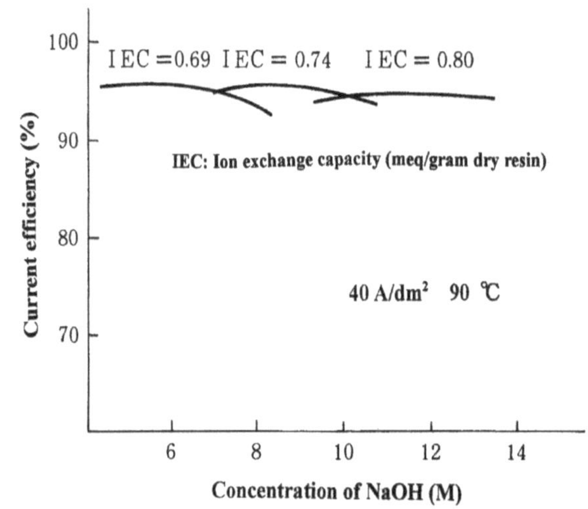

FIGURE 19.8 Current efficiency, ion exchange capacity, and NaOH concentration for double-layer membranes prepared by chemical treatment. *Ohmura (1984), p. 647, Fig. 5.*

19.3 Mass Transport and Electrode Reactions in an Electrolysis System

19.3.1 Mass Transport

The mass transport in an electrolysis system illustrated in Figure 19.9 is discussed here according to the explanation by Yawataya (1986).

A salt solution (C'_{NaCl} (mol/dm^3), V'_A (dm^3/h)) is assumed be supplied to an anode cell at the flux $J'_{NaCl} = C'_{NaCl} \times V'_A$ (mol/h). Impurities in the salt solution are assumed to be strictly removed through pretreatment. C'_{HCl} is the HCl concentration in the feeding salt solution caused by a small amount of HCl added to neutralize OH^- ions transferred from the cathode. C''_i s are the concentrations of components at the exit of the anode cell involving NaCl and every kind of byproducts, and V''_A is the amount of this solution.

Cl_2 and O_2 are generated at the anode and create the flux J'_{Cl_2} and J'_{O_2}, respectively. J_{Cl_2} is the amount of product Cl_2. J_{O_2} is the amount of O_2 mixed with the product Cl_2. A part of the Cl_2 is dissolved into the anode solution and carried outside the system. However, the Cl_2 is recovered in the succeeding process.

J'_{H_2} and J'_{OH} are, respectively, the flux of H_2 and OH^- generated by the reaction at the cathode. V'_C is the amount of water supplied into the cathode cell to maintain the NaOH concentration C''_{NaOH} of the product extracted from the system to be constant. V''_C and J''_{NaOH} are, respectively, the amount (dm^3/h) and flux (mol/h) of NaOH produced in this system; $J''_{NaOH} = C''_{NaOH} \times V''_C$.

J_{Na} is the flux of Na^+ transported from the anode cell to the cathode cell across the cation exchange membrane. J_{H_2O} (mol/h) is the flux of H_2O accompanying the Na^+.

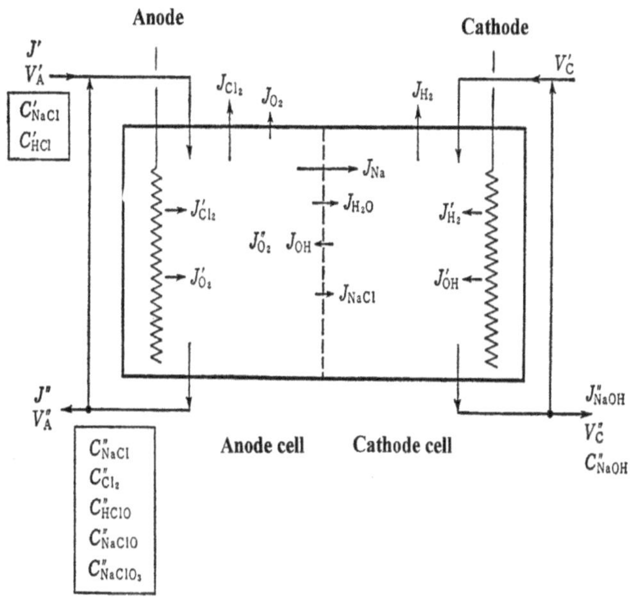

FIGURE 19.9 Mass transport in an electrolysis system. *Yawataya (1986), p. 152, Fig. 8.6.*

A small amount of OH^-, J_{OH} is transferred from the cathode cell to the anode cell across the membrane. J_{NaCl} is the flux of NaCl across the membrane owing to the diffusion.

19.3.2 Reaction at a Cathode

OH^- ions are generated by the following reactions at the cathode.

Primary reaction: the same as Eqn (19.2):

$$2H_2O + 2e^- \rightarrow H_2 + 2OH^- \tag{19.4}$$

Secondary reaction:

$$1/2O_2 + H_2O + 2e^- \rightarrow 2OH^- \tag{19.5}$$

$$ClO^- + H_2O + 2e^- \rightarrow Cl^- + 2OH^- \tag{19.6}$$

Current efficiency for OH^- generation ε'_{OH} totaled by these reactions becomes 1, so that OH^- flux at the cathode J'_{OH} is expressed by the following equation:

$$J'_{OH} = \varepsilon'_{OH}(I/F) = I/F \tag{19.7}$$

in which I is an electric current and F is the Faraday constant.

The flux of NaOH in the product taken from the cathode cell J''_{NaOH} is given by the following equation subtracting J_{OH} from J'_{OH}:

$$J''_{NaOH} = J'_{OH} - J_{OH} = I/F - J_{OH} \tag{19.8}$$

Accordingly, current efficiency for NaOH in the product ε_{OH} is:

$$\varepsilon_{OH} = \frac{J''_{NaOH}}{I/F} = 1 - \frac{J_{OH}}{I/F} = \frac{J_{Na}}{I/F} \tag{19.9}$$

Equation (19.9) shows that NaOH current efficiency depends on the Na^+ ion transport number of the cation exchange membrane.

Water molecules are converted into J'_{OH} (mol/h) by the following electrolysis reaction at the cathode:

$$NaCl + H_2O \rightarrow NaOH + 1/2H_2 + 1/2O_2 \tag{19.10}$$

The flux of water transferred from the anode cell into the cathode cell is J_{H_2O} (mol/h) $= 18 \times 10^{-3} J_{H_2O}$ (dm^3/h). The amount of water supplied to the cathode is V'_C (dm^3/h). Accordingly, the NaOH concentration extracted from the cathode cell C''_{NaOH} (mol/dm^3) is introduced as follows:

$$C''_{NaOH} = \frac{J''_{NaOH}}{V''_C} = \frac{J'_{OH} - J_{OH}}{V'_C + 18\left(J_{H_2O} - J'_{OH}\right) \times 10^{-3}} = \frac{(I/F)\varepsilon_{OH}}{V'_C + 18\left(J_{H_2O} - J'_{OH}\right) \times 10^{-3}} \tag{19.11}$$

19.3.3 Reaction at an Anode

The primary and the secondary reactions at the anode are, respectively, the generation of Cl_2, J'_{Cl_2} and O_2, J'_{O_2}, as follows:

$$J'_{Cl_2} = \varepsilon'_{Cl_2}(I/2F) \tag{19.12}$$

ε'_{Cl_2} is current efficiency for Cl_2 generation at the anode. An electric current generated at the anode is expressed by the following equation:

$$I = 2FJ'_{Cl_2} + 4FJ'_{O_2} \tag{19.13}$$

From Eqns (19.12) and (19.13), Eqn (19.14) is introduced:

$$J'_{O_2} = (1 - \varepsilon'_{Cl_2})(I/4F) \tag{19.14}$$

O_2 is assumed to be generated by another chemical reaction J''_{O_2} (Eqn (19.21)), except for the above reactions, so that the flux of total O_2; O_2 discharged to the outside of the anode cell, J_{O_2} is given by the following equation:

$$J_{O_2} = J'_{O_2} + J''_{O_2} \tag{19.15}$$

Now, Cl_2 generated at the anode undergoes changes as follows.

1. Dissolving into the anode solution:

$$Cl_2(g) \rightarrow Cl_2(aq) \tag{19.16}$$

2. Decomposition to form HClO:

$$Cl_2 + NaOH = NaCl + HCLO \tag{19.17}$$

The rate of this reaction is $V''_A C''_{HClO}$ (mol/h). Equation (19.17) shows that the reaction consumes 1 mol NaOH and 1 mol Cl_2 to generate 1 mol HClO.

3. Decomposition to form NaClO:

$$Cl_2 + 2NaOH = NaCl + NaClO + H_2O \tag{19.18}$$

Equation (19.18) shows that the reaction consumes 2 mol NaOH and 1 mol Cl_2 to generate 1 mol NaClO.

4. Decomposition of HClO to form $NaClO_3$:

$$2HClO + ClO^- = ClO_3^- + 2Cl^- + 2H^+ \tag{19.19}$$

Taking account of Cl_2 consumption generating HClO and ClO^-:

$$3Cl_2 + 6NaOH = NaClO_3 + 5NaCl + 3H_2O \tag{19.20}$$

Equation (19.20) shows that the reaction consumes 6 mol NaOH and 3 mol Cl_2 to generate 1 mol $NaClO_3$.

5. Decomposition of HClO to form O_2

A part of HClO formed in the solution generates $O_2(J''_{O_2})$ according to the following reaction:

$$2HClO + 2NaOH \rightarrow 2NaCl + 2H_2O + O_2 \tag{19.21}$$

Based on the reactions described above, the destination of OH^- ions transferred from the cathode to the anode across the membrane is introduced as shown in the following equation:

$$J_{OH} = V'_A \cdot C'_{HCl} + 4J'_{O_2} + V''_A \cdot C''_{HClO} + 2V''_A \cdot C''_{NaClO} + 6V''_A \cdot C''_{NaClO_3} + 4J''_{O_2} \tag{19.22}$$

The first term in Eqn (19.22) shows OH^- flux for neutralizing HCl added into the feeding salt solution. The second term corresponds to OH^- flux to generate O_2 at the anode by the following equation:

$$4OH^- \rightarrow O_2 + 2H_2O + 4e^- \tag{19.23}$$

Finally, the current efficiency for product Cl_2, ε_{Cl_2} is introduced as the following equation:

$$\begin{aligned}\varepsilon_{Cl_2} &= \frac{J_{Cl_2}}{I/2F} = \frac{J'_{Cl_2} - 2J''_{O_2} - \left(J''_{Cl_2} + J''_{HClO} + J''_{NaClO} + 3J''_{NaClO_3}\right)}{I/2F} \\ &= 1 - \frac{2F}{I}\left(2J'_{O_2} + 2J''_{O_2}\right) - \frac{2F}{I}\left(J''_{Cl_2} + J''_{HClO} + J''_{NaClO} + 3J''_{NaClO_3}\right)\end{aligned} \tag{19.24}$$

Here, J''_{Cl_2} is the Cl_2 flux discharged from the anode cell dissolving into the anode solution. This flux is recovered at the succeeding process. J'_{O_2} is the electric current loss caused by O_2 generation at the anode (Eqn (19.13)). J''_{O_2} is another chemical reaction rate generating O_2. $\left(J''_{HClO} + J''_{NaClO} + 3J''_{NaClO_3}\right)$ shows the electric current loss owing to the consumption of Cl_2.

When HCl is added into the salt solution supplied to the anode, ε_{Cl_2} is increased because every side reaction is suppressed. However, if a part of product Cl_2 is consumed to obtain HCl, ε_{Cl_2} is substantially decreased and is comprehensively consistent with ε_{OH} (Eqn (19.9)).

19.4 Electrolyzer and Its Performance

19.4.1 Monopolar and Bipolar System

The electrolyzer is classified into the monopolar and bipolar systems. The structure of both systems is illustrated in Figure 19.10. The electrical connections between the electrolyzers are shown in Figure 19.11.

In the monopolar electrolyzer:

1. Unit cells are arranged in parallel.
2. An electric current is supplied to each unit cell from monopolar electrodes.
3. The total electric current supplied to an electrolyzer corresponds to the current supplied to a unit cell multiplied by the number of unit cells.

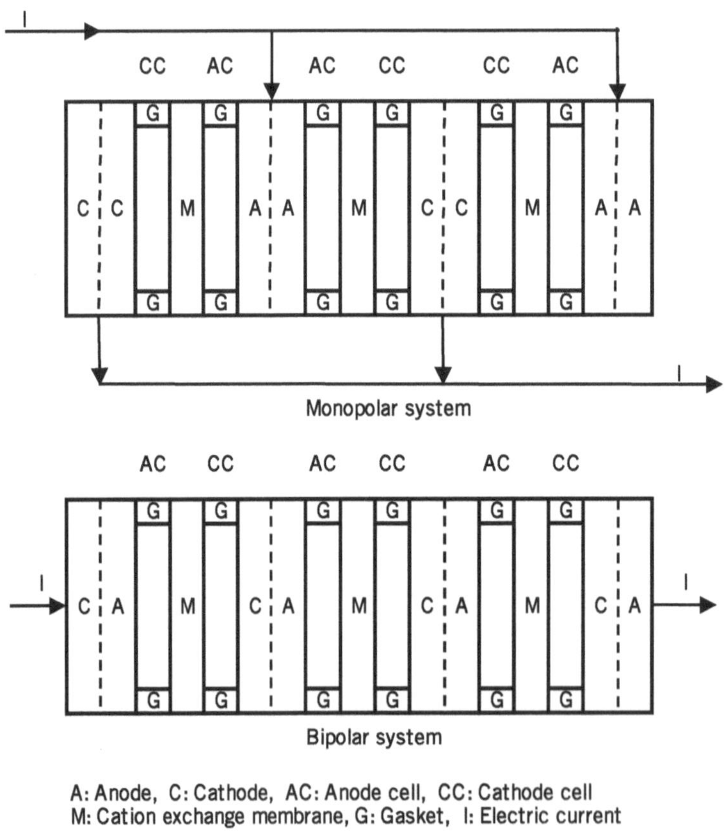

FIGURE 19.10 Structure of monopolar and bipolar electrolyzers.

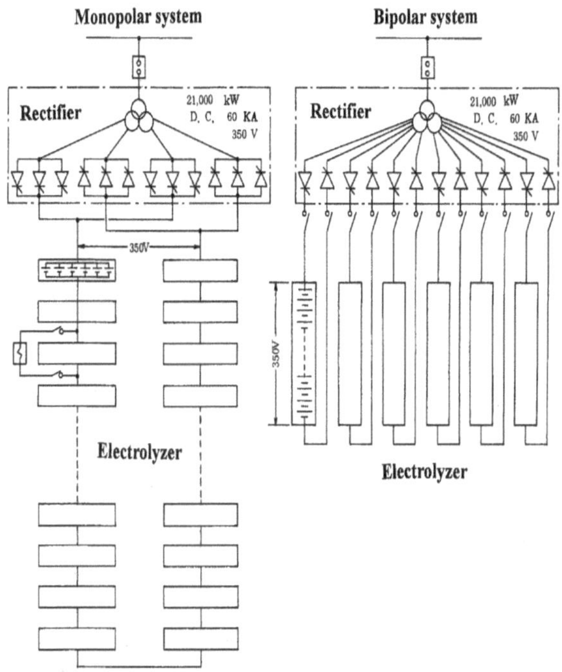

FIGURE 19.11 Electrical connections between the electrolyzers. *Ohmura (1984), p. 649, Fig. 11.*

4. The total voltage applied to the electrolyzer is equivalent to the voltage applied to the unit cell.
5. A high electric current, low-voltage rectifier supplies an electric current to many small-scale electrolyzers.

In the bipolar electrolyzer:

1. The unit cells are arranged in series.
2. An electric current is supplied to each unit cell from bipolar electrodes.
3. An electric current passing through the electrolyzer is equivalent to the current flowing through the unit cell.
4. Voltage applied to the electrolyzer is equal to the voltage applied to the unit cell multiplied by the number of the unit cells.
5. A low electric current, high-voltage rectifier supplies an electric current to few large-scale electrolyzers.

19.4.2 Decrease in Energy Consumption

Table 19.3 shows the specifications of commercially available bipolar electrolyzers. Figure 19.12 shows the capacity changes of the monopolar and bipolar systems in the world. The increase in monopolar systems during 1990–1995 was caused by an increase

Table 19.3 Specifications of Commercially Available Bipolar Electrolyzers

	Asahi Chemical	Toso-Chlorine Engineers	Asahi Glass	Krupp Uhde	DeNora
Denomination	ML32NCH	BiTAC873	AZEC-B1	HU	DN350
Rating current density (kA/m^2)	6	5	5	5	5
Maximum current density (kA/m^2)	8	6	6	6	8
Effective membrane area (m^2)	2.70	3.28	2.88	2.72	3.50
Unit cell number	100	73	80	160	90
Voltage (at rating) (V)	3.18	3.08–3.15	3.08	3.12	3.15
Current efficiency (%)	96	96–96.5	96	96	96.5
Energy consumption (kWh/t NaOH)	2220	2150–2187	2150	2178	2188
Main Materials					
Anode	DSA	DSA	DSA	Metallic	DSA
Anode cell	Ti	Ti	Ti	Ti	Ti
Cathode	Oxidated Ni	Ni system	Raney Ni	RuO_2	RuO_2
Cathode cell	Ni	Ni	Ni	Ni	Ni

Takarada (2004), p. 112, Table 4.2.2.

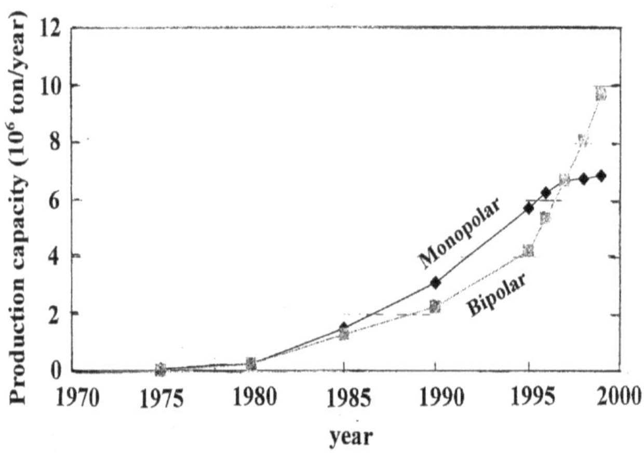

FIGURE 19.12 Production capacity changes of the monopolar and bipolar systems. *Takarada (2004), p. 111, Fig. 4.2.5.*

in process conversion from the mercury to ion exchange membrane process, because rectifiers in the mercury process were applicable to the ion exchange membrane process. The increase in bipolar systems after 1995 is due to the increase of current density, because ohmic loss of the bipolar system is less than that of monopolar systems.

A further decrease in unit cell voltage is achieved owing to the following technical developments.

1. Cut down air bubble electric resistance.
2. Make zero gap space between the membrane and the electrodes.

3. Decrease in overvoltage of the electrodes.
4. Decrease in electric resistance of the membrane.

Air bubbles increase the unit cell voltage, so it is strongly expected to prevent bubble attachment to the membrane surface. Asahi Chemical Co. prevented the bubble attachment by making the Aciplex membrane surface structure coarse (Ohmura, 1984). Asahi Glass Co. coated a porous layer consisting of nonconductive inorganic materials on the Flemion membrane (Suzuki and Shirogami, 1984).

To achieve zero gap space between the membrane and electrode, the structure and materials of the electrodes and electrode chambers must be developed. Furthermore, technology must be developed to avoid damaging the membrane structure. Bipolar electrode chambers developed by Asahi Chemical Co. are manufactured by joining Ti (anode cell) and Ni (cathode cell) between which a Ti/Fe plate is placed. Figure 19.13 shows the electrode configuration developed by Chlorine engineers and Toso Corp. (Chlorine Engineers, 2006). An electric current flows from the cathode to anode through a diaphragm between the electrodes. For the bipolar electrolyzer, an anode chamber made of Ti has six times larger resistance than a cathode chamber made of Ni. In this electrode chamber, the electric current route through Ti is shortened by making the diaphragm a wave shape and keeping enough capacity to exhaust Cl_2 gas. The grooves and projections of the wave shape diaphragm are arranged alternately (Fig. 19.14). An electrolyte solution supplied from the bottom of the electrolyzer is repeatedly mixed and separated at the edge of the grooves and projects. It is evenly distributed on the electrolysis surface.

The following is typical unit cell voltage (Takarada, 2004):

Theoretical electrolysis voltage	2.21 V
Anode overvoltage	0.05 V
Cathode overvoltage	0.12 V
Membrane electric resistance	0.33 V
Electric resistance owing to air/solution bubbles and unit structure	0.21 V
Total	2.92 V

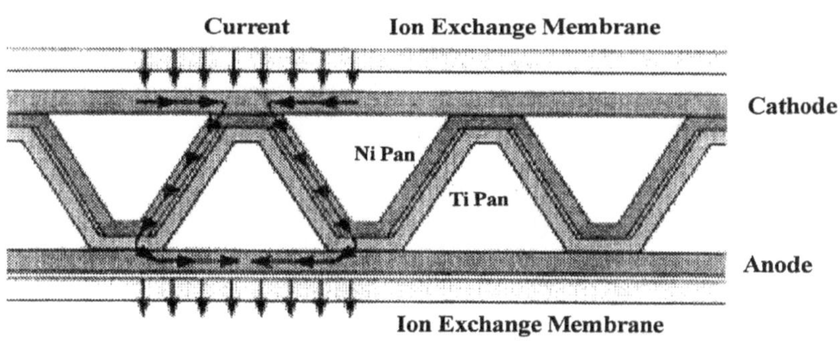

FIGURE 19.13 Electrode configuration of a bipolar electrolyzer. Chlorine Engineers Corp. BiTAC. *Chlorine Engineers Corp (2006).*

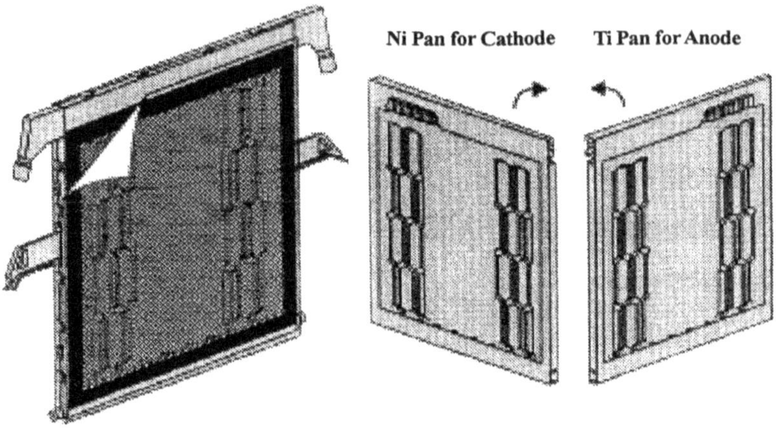

FIGURE 19.14 Bipolar element assembly. Chlorine Engineers Corp. BiTAC. *Chlorine Engineers Corp (2006).*

Here, the following operating conditions are assumed: current density: 4 kA/m^2, NaOH concentration in the product: 32%, concentration of feeding salt solution: 205 g/l, temperature: 90 °C. Extremely low anode overvoltage, 0.05 V, results from a metallic DSA electrode. Cathode overvoltage, 0.12 V, was achieved with catalytic active electrodes.

19.4.3 Electrolysis Process

Figure 19.15 (Bergner, 1982) shows the chlor-alkali process. First, saturated saltwater prepared by dissolving raw salt into water is supplied to the primary refining process in which Ca^{2+} and Mg^{2+} ions are removed by Na_2CO_3 and NaOH adding and settling. Further SO_4^{2-} ions are removed by adding $BaCl_2$ or $BaCO_3$. In the secondary refining process, the remaining small amount of Mg^{2+} and Ca^{2+} ions is removed using chelate resins to a concentration on the order of parts per billion. The refined saltwater is supplied to the anode cells and electrolyzed. The saltwater discharged from the anode cells is fed to a chlorate ($NaClO_3$) decomposition process after product Cl_2 gas is separated and returned to a raw salt dissolving tank.

The cathode solution discharged from the cathode cells is separated from H_2 gas and circulated. In this process, a part of the solution is fed to the chlorate decomposition tank and water is added to keep a definite NaOH concentration. A high-concentration NaOH solution or solid NaOH is produced by evaporating the overflow from the cathode solution circulating tank.

19.5 Purification of Saltwater

19.5.1 Primary Purification

Chemical reactions occurring in the primary refining process are:

$$Ca^{2+} + Na_2CO_3 \rightarrow CaCO_3 + 2Na^+ \quad (19.25)$$

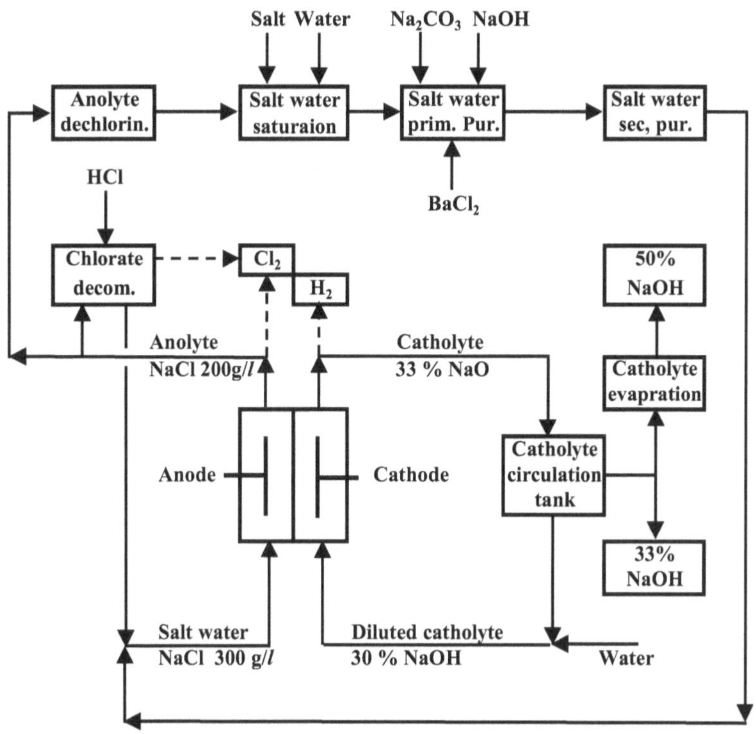

FIGURE 19.15 Chlor-alkali process. *Bergner (1982), p. 564, Abb 5.*

$$Mg^{2+} + NaOH \rightarrow Mg(OH)_2 + 2Na^+ \quad (19.26)$$

$$SO_4^{2-} + BaCl_2(\text{or } BaCO_3) \rightarrow BaSO_4 + 2Cl^-(\text{or } CO_3^{2-}) \quad (19.27)$$

$CaCO_3$, $Mg(OH)_2$, and $BaSO_4$ are separated using a sedimentation separator (thickener or circulator) and filtration (sand filter, ceramic filter, or leaf filter).

Ca^{2+}, Mg^{2+}, and SO_4^{2-} concentrations in purified saltwater are controlled as (Aikawa, 1994):

Ca^{2+} ions	1–3 ppm
Mg^{2+} ions	<1 ppm
SO_4^{2-} ions	<4 g/l

19.5.2 Secondary Purification

Two or three columns filled with imino-2-acetic acid ($-CH_2-N(CH_2CO-OH)_2$) or aminophosphoric acid ($-CH_2-NH-PO(OH)_2$) chelate resins are operated as a merry-go-round into which primarily purified saltwater is supplied. Table 19.4 shows the saltwater quality after the second purification (Chlorine Engineers Corp, 1987).

Table 19.4 Saltwater Quality after Secondary Purification

		Standard	Measured Values
NaCl	g/l	300–305	305
Na_2SO_4	g/l	<5	4
NaClO	g/l	<20	9
Ca + Mg	ppb	<20 as Ca	Ca < 10, Mg < 10
Sr	ppm	<0.06	0.01
Ba	ppm	<0.5	0.13
Hg	ppm	<15	<0.01
Al	ppm	<0.1	<0.05
SiO_2	ppm	<5	<1.0
Fe	ppm	<0.2	<0.03
Ni	ppm	<0.01	<0.01
TOC	ppm	<10	8.5
I	ppm	<0.2	<0.2

Aikawa (1994), p. 444, Table 4.

19.5.3 Influence of Impurities in Saltwater on the Performance of an Electrolyzer

Atmosphere in the membrane is alkaline because OH^- ions back migrate through the membrane. pH values in the membrane becomes higher near the cathode side because an NaOH solution flows in the cathode cell. Cationic impurities supplied into the anode side transfer across the membrane toward the cathode side under an electromotive force. In this circumstance, the cationic impurities combine with OH^- ions and precipitate as hydroxides in the membrane. The precipitating points depend on the solubility of the hydroxide. $Mg(OH)_2$ precipitates near the anode side, whereas $Ca(OH)_2$, $Sr(OH)_2$, and $Ba(OH)_2$ are precipitated near the cathode side. These phenomena increase the voltage and decrease the current efficiency during the electrolysis operation. Table 19.5 shows the effects of impurities on the performance of electrolysis (Aikawa, 1994).

Figure 19.16 shows the influence of added Ca^{2+} and Mg^{2+} ions on cell voltage and current efficiency, respectively, in laboratory cells integrated with a Nafion 90209 membrane (Momose et al., 1991). The cell voltage increases by adding both Mg^{2+} and Ca^{2+} ions. The addition of Mg^{2+} ions induces an apparent larger voltage increase. Current efficiency is decreased after the addition of Ca^{2+} ions, whereas the addition of Mg^{2+} ions has no such effect. To clarify these variations, the location of impurity precipitation in the membrane was observed by electron probe X-ray microanalysis. Nafion 90209 includes a carboxylate layer with high selectivity. The figure indicates that the Ca^{2+} ions pass through the membrane from the anode side and are precipitated in carboxylate layer in the cathode side (Photo 19.1 (a)) to form of particles (Photo 19.1 (b)).

Table 19.5 Effects of Impurities on Performance of Electrolysis

	Influence on Membrane		Influence on Electrodes		
	η	V	Ano	Cath	Remark
Cation					
Ca	○				$Ca(OH)_2$ deposit on membrane
Mg	Δ	○			$Mg(OH)_2$ deposit on membrane
Sr	○				$Sr(OH)_2$, $Sr + SiO_2$ deposit on membrane
Ba			○		I + Ba deposit on membrane, adhered to anode
Ni		○			Deposit on membrane
Fe			○	○	Indistinct
Anion					
SiO_2		○			Composite compound deposit on membrane
F			○		Influence on anode
I	○				Ba + I deposit on membrane
SO_4	○		○		Na_2SO_4 deposit on membrane
ClO_3					NaCl solubility decrease
Composite Compound					
$Ca + SiO_2$	○				Deposit on membrane
$Al + SiO_2$	○		○		Deposit on membrane, adhered to anode
$Ca + Mg + SiO_2$	○				Deposit on membrane
Ba–I	○				Deposit on membrane
Organic compound			○		Adhered to anode

η, Current efficiency; V, voltage; Ano, anode; Cath, cathode; ○, influence; Δ, influence slightly.
Aikawa (1994), p. 447, Table 6.

Mg^{2+} ions are precipitated in the sulfonate layer on the anode side (Photo 19.1 (c)) to the form a layer (Photo 19.1 (d)).

The difference in the location of calcium and magnesium precipitation in the membrane is understandable from the differences in their hydroxide solubility. The solubility products of $Mg(OH)_2$ and $Ca(OH)_2$ are 4.11×10^{-12} and 9.33×10^{-7} at 90 °C, respectively; the solubility product of $Mg(OH)_2$ is much lower than that of $Ca(OH)_2$. pH distribution in the Nafion membrane is 9.1–10.5 near the anode side (sulfonate layer) and sufficiently high (>14) near the cathode side (carboxylate layer). In this situation, the Mg^{2+} ions supplied to the membrane from the anode side immediately precipitate as the hydroxide because the pH is high enough to cause the cell voltage increase, whereas Ca^{2+} ions travel to the carboxylate layer where the pH is high enough to cause the precipitation and current efficiency (selectivity) decrease (Fig. 19.16).

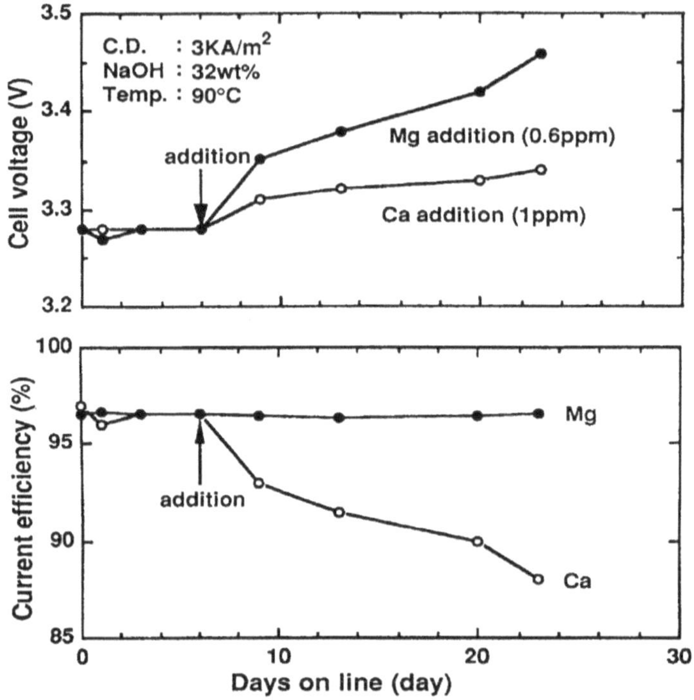

FIGURE 19.16 Changes in cell voltage and current efficiency by adding Ca^{2+} and Mg^{2+} ions to feeding saltwater. *Momose et al. (1991), p. 736, Fig. 1.*

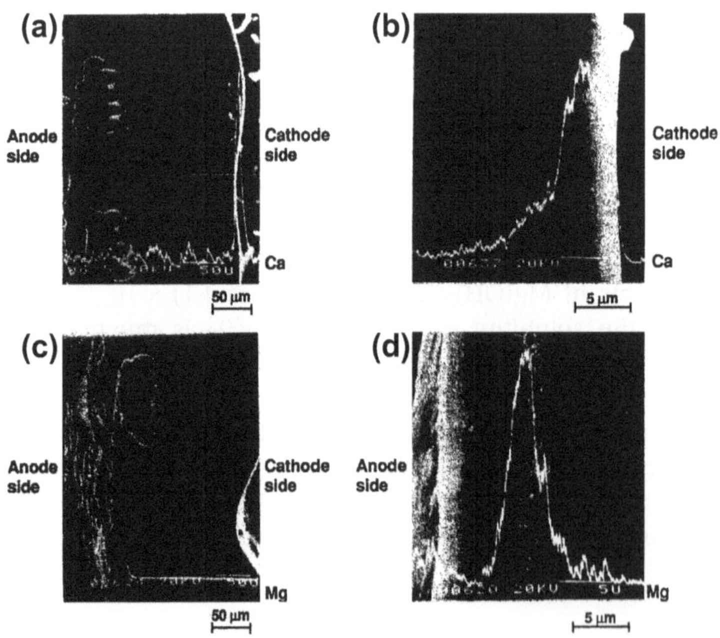

PHOTO 19.1 Distribution of precipitate in the membrane analyzed by EPMA. (a) Calcium line profile, (b) calcium line profile near the cathode side surface, (c) Mg line profile, (d) Mg line profile near the anode side surface. *Momose et al. (1991), p. 738, Fig. 5.*

19.6 Research and Development

19.6.1 Gas Diffusion Electrodes

Replacement of conventional hydrogen evolution cathodes by gas diffusion electrodes (cf. Section 22.3.1) has been found to lead to lower energy consumption in the chlor-alkali electrolysis process. Electrochemical reactions generated in the gas diffusion electrodes are as follows.

At an anode:

$$2Cl^- \rightarrow Cl_2 + 2e^- \quad E^0 = 1.36\ V \tag{19.28}$$

At a cathode:

$$O_2 + 2H_2O + 4e^- \rightarrow 4OH^- \quad E^0 = 0.40\ V \tag{19.29}$$

Total reaction:

$$NaCl + 1/2H_2O + 1/4O_2 \rightarrow NaOH + 1/2Cl_2 \quad E^0 = 0.96\ V \tag{19.30}$$

The theoretical decomposition voltage in the gas diffusion electrode $E^0 = 0.96$ V (Eqn (19.30)) is less than that in the traditional electrode $E^0 = 2.19$ V (Eqn (19.3)). Thus, electrolytic voltage is assumed to decrease from 3.20 V/unit in a traditional electrolyzer to 2.00 V/unit in a gas diffusion type electrolyzer (current density 3 kA/m^2, NaOH concentration 32 wt%, and 90 °C). From 1999 to 2002, the Association of the Progress of New Chemistry investigated the gas diffusion electrode system for practical application as part of a national project supported by the New Energy and Industrial Technology Development Organization (NEDO, 2004). The system was introduced in electrolyzers and commercial operations were started (Toagosei, 2013).

19.6.2 Fuel Cell

Hydrogen gas produced in a chlor-alkali process (Fig. 19.9) can be supplied to a fuel cell (Chapter 22). Electricity generated in the fuel cell can be returned to the chlor-alkali process and therefore save on the consumption of electrical energy by up to 20% (Bebelis et al., 2013). In 2011, Nedstack installed a large fuel cell system with 12,600 polymer electrolyte fuel cells and generated 1 MW of electrical energy at Solvay's chlor-alkali plant in Lillo, Belgium. The annual production is estimated to be 5475 GWh/year.

References

Aikawa, H., 1994. Brine purification for ion exchange membrane chlor-alkali process. Bull. Soc. Sea Water Sci. Jpn. 48, 439–450.

Asawa, T., 1991. Sodium chloride electrolysis, Asahi Glass process. In: Nakagaki, S., Shimizu, H. (Eds.), Membrane Treatment Technology System, vol. 1. Fuji Techno System, Tokyo, pp. 510–515.

Bebelis, S., Bouzek, K., Cornell, A., Ferreira, M.G.S., Kelsall, G.H., Lapicque, F., Ponce de Leon, C., Rodrigo, M.A., Walsh, F.C., 2013. Highlights during the development of electrochemical engineering. Chem. Eng. Res. Des. 91, 1998–2020. http://www.cleanenergyactionproject.com.

Bergner, D., 1982. Alkalichlorid-Elektrolyse nach dem membranverfahren. Chem. Ing. Tech. 54 (6), 562–570.

Chlorine Engineers Corp., 1987. Technical report, Tokyo.

Chlorine Engineers Corp., 2006. Product Bulletin, Tokyo.

Connally, D.J., 1966. US. 3 282 875.

DuPont, 2002. Nafion Perfluorinated Membranes. Product information (3/18/02).

Furuya, N., 2000. Sodium chloride electrolysis. In: Soc. Electrochem., Jpn. Electrochemistry Handbook, vol. 5, pp. 380–386.

Gierke, T.D., October 1977. In: Am. Electrochem. Soc. Meeting, Atranta.

Grot, W.G.F., Munn, G.E., Walmsley, P.N., 1972. Perfluorinated Ion Exchange Membranes. Presented at 141st National Meeting of Electrochem. Soc., Houston Texas, May 7–11.

Kado, Y., 2009. Soda industry. In: Kanzaki, Y. (Ed.), Ion Exchange Technology. Kogyo Chosakai Publishing Co. Ltd., Tokyo, pp. 98–103.

Kameyama, H., Sugimoto, M., Kado, Y., 2010. A Cation Exchange Membrane and Manufacturing Method of an Electrolyzer and the Cation Exchange Membrane. JP2010-068855.

Kusumocahyo, S.P., Sudo, M., 1999. Dehydration of acetic acid by pervaporation with charged membranes. J. Membr. Sci. 161, 77–83.

Momose, T., Higuchi, N., Arimoto, O., Yamaguchi, K., Walton, C.W., 1991. Effects of low concentration levels of calcium and magnesium in the feed brine on the performance of a membrane chlor-alkali cell. J. Electrochem. Soc. 138, 735–741. http://www.jes.ecsdl.org/content/138/3/735.abstract - aff-2.

NEDO, 2004. http://www.nedo.go.jp/content/100089530.pdf.

Ohmura, J., 1984. Sodium chloride electrolysis using ion exchange membranes – Asahi Chemical process. In: Shimizu, H., Nishimura, M. (Eds.), Advanced Membrane Treatment Technology and Its Applications. Fuji Techno System, Tokyo, pp. 645–649.

Putnam, P.E., 1967. (to E. I. du Pont de Nemours and Co.) US 3 301 893.

Seko, M., Miyauchi, H., Ohmura, J., Kimoto, K., 1982. Ion-exchange membrane for the chlor-alkali process. J. Electrochem. Soc. Jpn. 50, 470–476.

Suzuki, K., Shirogami, O., 1984. Sodium chloride electrolysis using ion exchange membranes – Asahi Glass Flemion process. In: Shimizu, H., Nishimura, M. (Eds.), Advanced Membrane Treatment Technology and Its Applications. Fuji Techno System, Tokyo, pp. 656–660.

Takarada, H., 2004. Perfluorocarbon ion exchange membrane, production of sodium hydroxide using ion exchange membranes. In: Seno, M., Tanioka, A., Itoi, S., Yamauchi, A., Yoshida, S. (Eds.), Functions and Applications of Ion Exchange Membranes. Industrial Publishing & Consulting Inc., Tokyo, pp. 106–118.

Toagosei, 2013. http://toagosei.co.jp/whatsnew/news/n130806.pdf.

Yawataya, T., 1986. Ion Exchange Membrane for Engineers, second ed. Kyritu Shuppan Co., Tokyo. pp.149–157.

20

Diffusion Dialysis

20.1 Overview of Technology

Diffusion dialysis is a separation process using the ionic diffusion due to the concentration difference across a membrane. The phenomenon is governed by Fick's law, and the diffusion velocity is generally low. To promote the process efficiency, it becomes necessary to decrease membrane thickness and increase the membrane area. The process uses high mobility of H^+ ions across an anion exchange membrane and is applied to recover acid from an electrolyte solution in the following instances (Itoi and Mochida, 1985):

- Treatment of waste solutions from an aluminum foil etching process,
- Composition control in an aluminum anodizing bath,
- Acid separation in a metallic rust removing process,
- Acid separation in a chemical reaction process,
- Acid concentration control in a metal surface treatment process,
- Purification of crude acid,
- Treatment of waste acid in a stainless steel washing process.

20.2 Transport Phenomena

In the process illustrated in Fig. 20.1, a high concentration salt solution including acid (Feed) is supplied to the bottom of the feeding cell, flows upward in the cell, and flows out at the top of the cell (Deacid). A low-concentration solution (water) is supplied to the top of the recovering cell, flows down in the cell, and flows out at the bottom of the cell (recovery). The acid transfers across the anion exchange membrane because of its high mobility in the membrane due to the Grotthuss proton conduction mechanism (Fundamentals Section 7.6). However, the salt transfer is restricted because of the Donnan exclusion (Fundamentals, Section 2.2.1) due to the interaction between the salt cations and functional groups (quaternary ammonium groups) in the membrane. In the steady state the flux of acid or salt J (mol/h) is defined by the following Fick's law (Fundamentals Section 2.4):

$$J = US\Delta C_{av} \tag{20.1}$$

where S (m^2) is membrane area.

Ion Exchange Membranes. http://dx.doi.org/10.1016/B978-0-444-63319-4.00020-1

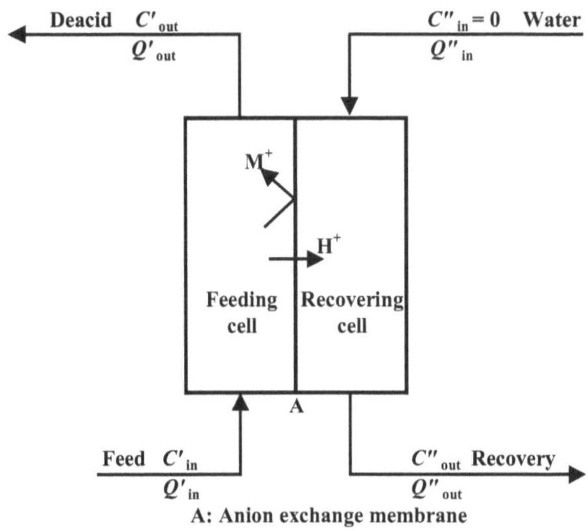

FIGURE 20.1 Mass transport in diffusion dialysis.

U ((mol/(h m^2))/(mol/l)) in Eqn (20.1) is the overall dialysis coefficient of solutes (acid or salt) defined by:

$$\frac{1}{U} = \frac{1}{K} + \frac{1}{k'} + \frac{1}{k''} \tag{20.2}$$

in which K is the diffusion coefficient for the anion exchange membrane, and k' and k'' are the diffusion coefficients for the boundary layer formed on the feeding and recovering surfaces of the membrane, respectively.

Table 20.1 shows the overall diffusion coefficient U measured for Neosepta AFN (Noma, 1991). U is influenced by temperature and solute concentration in a feeding solution. Membranes having larger U_{acid} and smaller U_{salt} show excellent performance in diffusion dialysis. The degrees of acid permeabilities are arranged as HCl· $HNO_3 > H_2SO_4 > HF > H_3PO_4$. U_{salt} is decreased with the increase of charge numbers. U_{salt} for HNO_3–$Cu(NO_3)_2$ and HNO_3–$Zn(NO_3)_2$ systems take larger values because they form anionic complex salt.

ΔC_{av} (mol/l) in Eqn (20.1) is the average concentration difference of solutes between the feeding cell and recovering cell.

$$\Delta C_{av} = \frac{(C'_{in} - C''_{out}) - C'_{out}}{\ln\left(\frac{C'_{in} - C''_{out}}{C'_{out}}\right)} \tag{20.3}$$

The solute concentrations in the deacid solution C'_{out} and in the recovered solution C''_{out} are expressed by the following equations:

$$C'_{out} = \left\{1 - \frac{Q''_{in}}{Q'_{in}} \times \frac{US\left(\frac{1}{Q'_{in}} - \frac{1}{Q''_{in}}\right) - 1}{\frac{Q''_{in}}{Q'_{in}} \exp US\left(\frac{1}{Q'_{in}} - \frac{1}{Q''_{in}}\right) - 1}\right\} \times C'_{in} \tag{20.4}$$

Table 20.1 Overall Diffusion Coefficient of Neosepta AFN

Acid·salt mixture	Acid concentration N	Salt concentration N	U_{acid} (mol/hm^2)/(mol/l) (25 °C)	U_{salt} (mol/hm^2)/(mol/l) (25 °C)	U_{salt}/U_{acid} (25 °C)
$HCl \cdot NaCl$	2.0	1.0	8.6	0.47	5.5×10^{-2}
$HCl \cdot FeCl_2$	2.0	1.0	8.6	0.17	2.0×10^{-2}
$HCl \cdot FeCl_3$	2.0	1.0	8.5	0.055	6.5×10^{-3}
$H_2SO_4 \cdot Na_2SO_4$	2.0	1.0	3.5	0.14	4.0×10^{-2}
$H_2SO_4 \cdot FeSO_4$	2.0	1.0	3.6	0.037	1.03×10^{-3}
$H_2SO_4 \cdot ZnSO_4$	2.0	1.0	3.6	0.053	1.5×10^{-2}
$H_2SO_4 \cdot Al_2(SO_4)_3$	2.0	1.0	3.6	0.004	1.1×10^{-3}
$HNO_3 \cdot Al(NO_2)_3$	1.5	1.5	9.3	0.048	5.2×10^{-3}
$HNO_3 \cdot Zn(NO_3)_2$	1.5	1.5	9.8	0.14	1.4×10^{-2}
$HNO_3 \cdot Cu(NO_3)_2$	1.5	1.6	9.6	0.17	1.8×10^{-2}
$H_3PO_4 \cdot MgHPO_4$	3.0	0.2	0.85	0.018	2.1×10^{-2}

Noma (1991), p. 176, Table 1.

$$C''_{out} = \frac{\exp US\left(\frac{1}{Q'_{in}} - \frac{1}{Q''_{in}}\right) - 1}{\frac{Q''_{in}}{Q'_{in}} \exp US\left(\frac{1}{Q'_{in}} - \frac{1}{Q''_{in}}\right) - 1} \times C'_{in} \tag{20.5}$$

Membrane area S is introduced from Eqn (20.5) as follows:

$$S = \frac{1}{U\left(\frac{1}{Q'_{in}} - \frac{1}{Q''_{in}}\right)} \times \ln \frac{1 - Y}{1 - \frac{Q''_{in}}{Q'_{in}} Y} \tag{20.6}$$

in which $Y = C''_{out}/C'_{in}$ (acid recovering ratio).

The performance of diffusion dialysis is given by the following equations:

$$\text{Acid remaining ratio or salt remaining ratio} = \frac{C'_{out} Q'_{out}}{C'_{in} Q'_{in}} \tag{20.7}$$

$$\text{Acid recovering ratio or salt moving ratio} = \frac{C''_{out} Q''_{out}}{C'_{in} Q'_{in}} \tag{20.8}$$

20.3 Diffusion Dialyzer and Its Operation

A practical scale diffusion dialyzer consists of gaskets (feeding cells and recovering cells) and anion exchange membranes (10-1000 sheets), as shown in Fig. 20.2. Figure 20.3 gives the process flow, which consists of mainly a diffusion dialyzer itself and a filter to remove sludge or oil slicks in the feeding solution (Kawahara, 1984). The velocities of the feeding solution and water are controlled using valves and by adjusting the head tank level to about 2.5 m. Linear velocity is in the range of 50–300 cm/h, so flow resistance is low. In

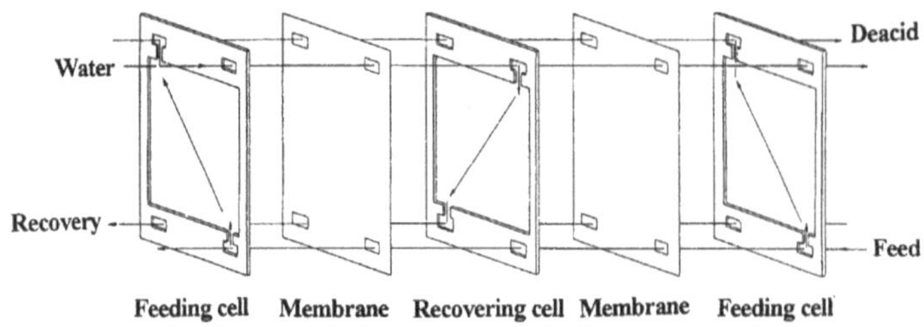

FIGURE 20.2 Cell arrangement in a diffusion dialyzer. *Kawahara (1984), p. 251, Fig. 7.*

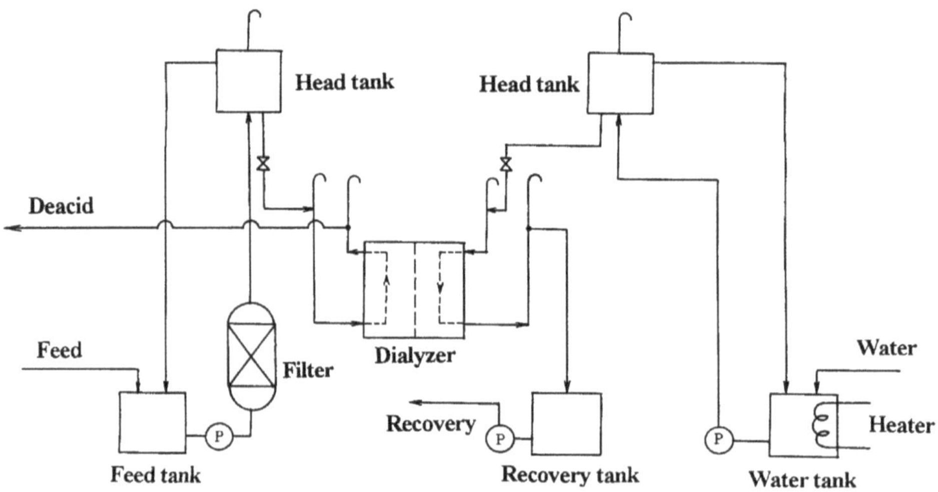

FIGURE 20.3 Diffusion dialysis process flow. *Kawahara (1984), p. 251, Fig. 8.*

winter, water is heated to prevent acid diffusivity lowering. When substances in the feeding solution are precipitated on the membrane surface, the stack is disassembled and the membranes are washed about one to two times in a year. The dialyzer is operated continuously and stably. Energy consumption is very low because electric energy consumption is only for pumping the solutions through the stack. The operating process is simple, and operating costs are low. The main cost factor is charges for the capital investment, which is considerably high because a large membrane area is required due to low acid diffusivity. The useful membrane life under operating condition in strong acids is relatively short. The process costs are, therefore, governed by costs and life of the membrane.

The diffusion dialysis process for sulfuric acid recovery is effective because it reduces neutralization/sludge treatment cost in the traditional process. The processes for nitric acid and hydrofluoric acid are prospective because unit prices of both acids are relatively high. The processes for phosphoric acid and organic acid are not yet practically operated because the diffusion coefficients of both acids are low. Chromic acid deteriorates membrane performance and reduces membrane life time due to high oxidizability (Katayama, 2004).

20.4 Practice

20.4.1 Composition Control in an Anodized Aluminum Processing Bath

Figure 20.4 gives an aluminum sash manufacturing process consisting of the aluminum surface treatment by alkali etching and anodic oxidation. The solution in the aluminum anodizing bath is taken out and treated in a diffusion dialyzer to separate Al and H_2SO_4. Figure 20.5 shows the material balance in the process operating in the aluminum sash manufacturing plant. In this process, a part of an anodic solution in the bath is supplied continuously to the diffusion dialyzer. 75–85% of H_2SO_4 in the feeding solution is recovered into the recovery, which is returned to the anodizing bath with new H_2SO_4 supplied to maintain constant acid concentration in the bath. The process was developed by Asahi Glass Co, and it improves the product quality and decreases the quantities of new H_2SO_4 addition and alkali supplement for neutralization of discharged acid (Utsunomiya, 1984).

Running costs of the process are:

1. Ion exchange renewal	880,000 yen/year
2. Electric power	104,000 yen/year
3. Others	72,000 yen/year
Total	1,056,000 yen/year

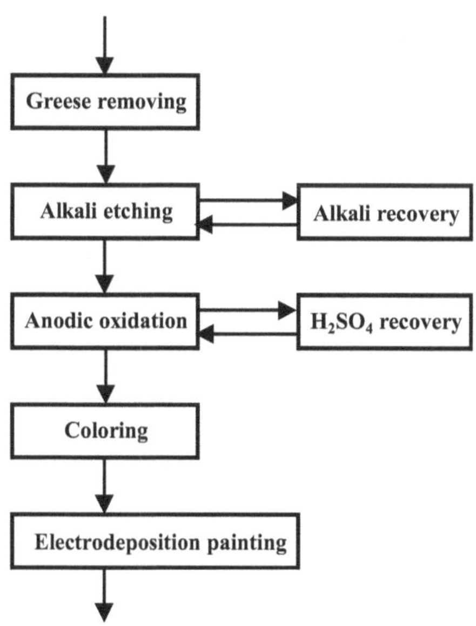

FIGURE 20.4 Aluminum sash manufacturing process. *Utsunomiya (1984), p. 456, Fig. 2.*

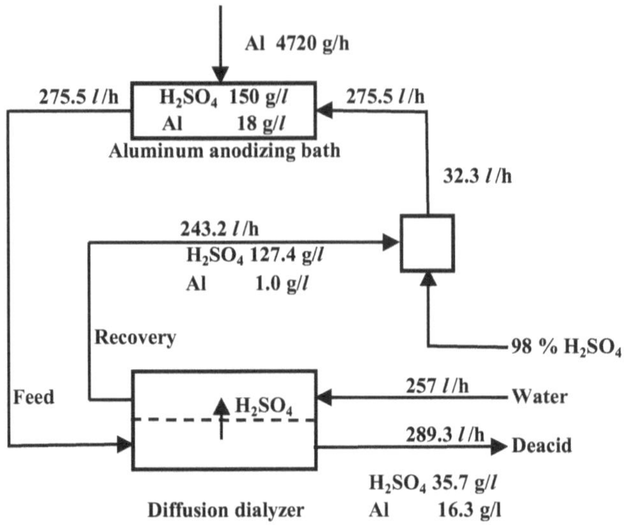

FIGURE 20.5 Material balance in a H_2SO_4 recovering process in a 1,000 t/month aluminum sash manufacturing factory. *Utsunomiya (1984), p. 457, Fig. 3.*

Cost-saving merits in chemical reagent consumption are:

1. 98% H_2SO_4	160 l/year	2,400,000 yen/year
2. 48% NaOH	265 l/year	8,500,000 yen/year
Total		10,900,000 yen/year

20.4.2 Recovery of Nitric Acid in an Acid Washing Process

Pretreatment in a plating process, surface treatment of stainless steel, or etching treatment of electronic parts includes an acid washing process using acid such as H_2SO_4, HCl, HNO_3, HF, etc. or their mixed acid. In these processes, metal is dissolved into the acid solution, and its washing performance is gradually lowered. To prevent such a problem, Tokuyama Co. developed diffusion dialysis technology for recovering HNO_3 from the acid washing solution, as shown in Fig. 20.6. The specifications and performance of the process are enumerated as follows (Motomura, 1986):

1. Diffusion dialysis
   a. Feeding solution
      - 6.8 m^3/day, HNO_3 100 g/l, $Al(NO_3)_3$ 100 g/l, SS 200–500 ppm.
   b. Diffusion dialyzer
      - Neosepta TSD-50-400
   c. Process performance
      - See Fig. 20.7.

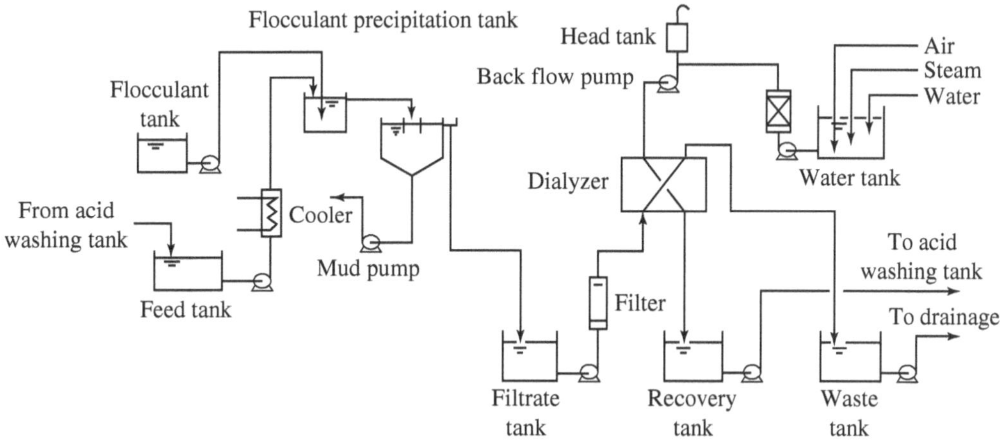

FIGURE 20.6 Diffusion dialysis process for recovering HNO_3. *Motomura (1986), p. 228, Fig. 1.*

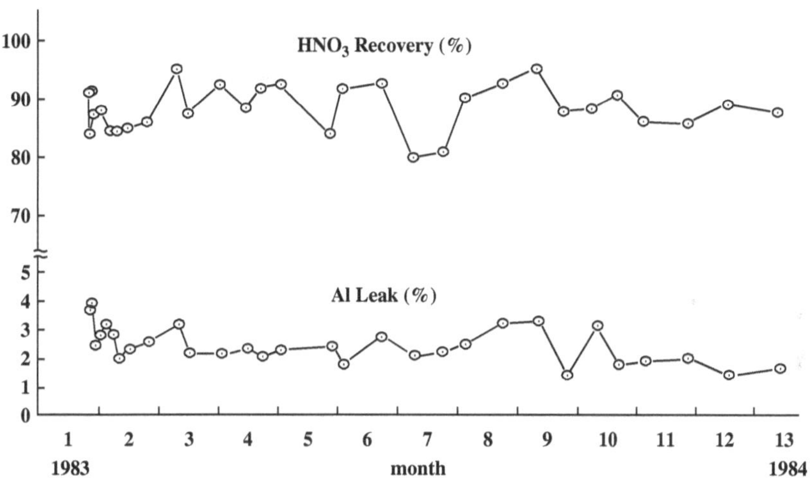

FIGURE 20.7 Performance of HNO_3 diffusion dialysis. *Motomura (1986), p. 231, Fig. 4.*

2. Running cost and merits

a. Running cost

Ion exchange membrane renewal	75,000 yen/month
Filter	33,000 yen/month
Electric power	20,000 yen/month
Water supply	12,000 yen/month
Flocculant	4,000 yen/month
Total	144,000 yen/month

b. Merits

HNO_3 Consumption Per One Ton of Steel Material	
Before diffusion dialysis adoption	365 kg/ton
After diffusion dialysis adoption	213 kg/ton
Steal material treated	100 ton/month
HNO_3 unit cost (as 65% HNO_3)	55 yen/kg

Gain for recovering HNO_3

(365 − 213) (kg/ton) × 100 (ton/month) × 55 (yen/kg) = 836,000 (yen/month)+ 58,000 (yen/month) (cost saving for neutralization agent $Ca(OH)_2$) = 894,000 yen/month.

Accordingly

Cost merit = (894,000 − 144,000) (yen/month) × 12 (month/year) = 9,000,000 (yen/year).

3. Maintenance
 a. Renewal of filter materials: one time during 4–5 months (acid washing solution feeding side), one time in a month (water supplying side)
 b. Heat exchanger washing: one time during 5–6 months
 c. Diffusion dialyzer: no disassembly and no washing during 2 years.

References

Itoi, S., Mochida, M., 1985. Present Status of the Dialysis Technique, Ionics. No. 120, pp.171–176. Ionics Co., Tokyo.

Katayama, S., 2004. Acid recovery from waste acid. In: Seno, M., Tanioka, A., Itoi, S., Yamauchi, A., Yoshida, S. (Eds.), Functions and Applications of Ion Exchange Membranes, pp. 151–162.

Kawahara, T., 1984. Industrial diffusion dialyzer. In: Shimizu, H., Nishimura, M. (Eds.), The Latest Membrane Treatment Technology and its Applications. Fuji Techno System Co., Tokyo, pp. 248–252.

Motomura, H., 1986. Recovery of nitric acid and fluoric nitric acid by diffusion dialysis. In: Industrial Application of Ion Exchange Membranes. Research Group of Electrodialysis and Membrane Separation Technology, Soc. Sea Water Sci., Jpn., vol. 1, pp. 223–233.

Noma, Y., 1991. Diffusion dialysis membrane. In: Nakagaki, M., Shimizu, H. (Eds.), Membrane Treatment Technology, Part I, Fundamentals. Fuji Techno System Co., Tokyo, pp. 175–179.

Utsunomiya, H., 1984. Recovery of waste acid by diffusion dialysis. In: Shimizu, H., Nishimura, M. (Eds.), The Latest Membrane Treatment Technology and its Applications. Fuji Techno System Co., Tokyo, pp. 455–463.

21

Donnan Dialysis

21.1 Overview of Technology

Figure 21.1 illustrates the principle of Donnan dialysis (DD) incorporated with an anion exchange membrane. Feed and strip dissolving monovalent electrolytes AX and AY are assumed to be supplied into the feeding and extracting cells, respectively. Here, electrolyte concentrations at the entrance of the feeding and extracting cell are $C'_{in.X}, C'_{in.Y} = 0$ and $C''_{in,X} = 0, C''_{in,Y}$, respectively. In this system, monovalent anion Y (designated as driving ion) in the extracting cell transfer into the feeding cell by diffusion due to the concentration difference ($C''_{in.Y} > C'_{in.Y} = 0$), which is the primary driving force of the system. When the permselectivity of the anion exchange membrane is perfect, the cations A do not permeate the membrane. Accordingly, the solution phase in the feeding cell is charged negatively (potential ψ'), that in the extracting cell is charged positively ($\psi'' > \psi'$), and the electric neutrality is disturbed. To maintain the electric neutrality in the system, anion X in the feeding cell is transferred with H_2O across the anion exchange membrane under the potential gradient $d\psi/dx > 0$ and extracted into the extracting cell. In a steady state, X ion concentration comes to be decreased to C'_X in the feeding cell and comes to be increased to C''_X in the extracting cell. In this circumstance, the movement of X ions is against the concentration difference ($C'_X < C''_X$) and exhibits the uphill transport. The uphill transport movement of the anions X is caused by the diffusion of driving ions Y and resultant potential difference $\psi'' - \psi'$ to maintain the electric neutrality. This phenomenon leads to the exchange of anion X and anion Y between the two cells through the membrane. The principle of the DD described above is also achieved in the system incorporated with a cation exchange membrane. The process is applicable to concentrate or desalinate X ions dissolving in feed and was called the Donnan dialysis by Wallace (1967).

The DD process does not use an electric current, so it is energy-saving compared to electrodialysis. Compared to a mixed-bed ion exchange process, the DD has the advantages that it does not need regeneration and is operated continuously. However, the system is not economical because the effect of the driving force (concentration difference $C''_Y - C'_Y$ and resultant potential difference $\psi'' - \psi'$) is weak, and it becomes necessary to use a larger membrane area (higher capital cost) to maintain the output of the process. The system had been investigated in laboratories in the following instances; however, the industrial scale application of the process has not been seen up to now:

1. Concentration of uranyl ions (Wallace, 1967; Davis et al., 1971)
2. Removal of fluoride (Hichour et al., 1999,2000; Durmaz et al., 2005)

Ion Exchange Membranes. http://dx.doi.org/10.1016/B978-0-444-63319-4.00021-3

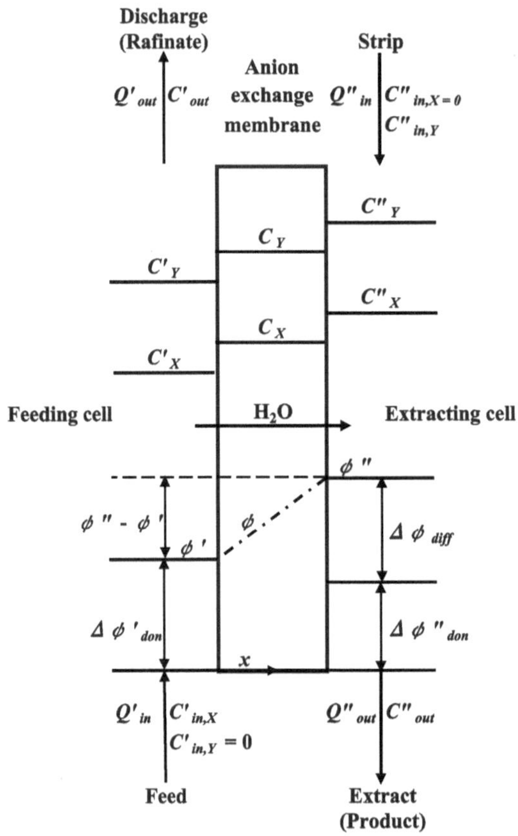

FIGURE 21.1 Principle of Donnan dialysis.

3. Removal of chromium (Tor et al., 2004)
4. Removal of metals (Na, Fe, Al, Ti) from red mud (Cengeloglu et al., 2003)
5. Removal of electrolyte ions (Mg, Ca, SO_4, HCO_3) in saline water (Wisniewski and Rozanska, 2007)
6. Separation of copper, gold and silver cyanide (Akretche and Kerdjoudj, 2000)
7. Removal of ammonium from waste water (Takeuchi et al., 1987)
8. Separation of amino acid (Sato, 2002)

21.2 Mass Transport

21.2.1 Principle

In Figure 21.1, the potential difference is written as follows (Fundamentals Figure 3.1, Eqn (3.1)):

$$\Delta\psi'_{Don} + (\psi'' - \psi') = \Delta\psi''_{Don} + \Delta\psi_{diff} \tag{21.1}$$

$\Delta\psi'_{Don}$ and $\Delta\psi''_{Don}$ are the Donnan potential (Fundamentals Eqn (2.7)). $\Delta\psi_{diff}$ is the diffusion potential (Fundamentals Eq. (4.27)). Potential difference $\psi'' - \psi'$causing the driving force in the DD process is obtained from Eqn (21.1) as follows:

$$\psi'' - \psi' = \Delta\psi''_{Don} - \Delta\psi'_{Don} + \Delta\psi_{diff} \tag{21.2}$$

In an equilibrium state, the electrochemical potentials in both solutions $\eta\prime$ and η'' in Figure 21.1 become equal:

$$\eta' = RT\ln a'_i + z_i F\psi' = \eta'' = RT\ln a''_i + z_i F\psi'' \tag{21.3}$$

in which a_i is the activity of i ions. The driving force in the DD process $\psi'' - \psi'$ is introduced from Eqn (21.3) (cf. Fundamentals Eq. (2.31)):

$$\psi'' - \psi' = \frac{RT}{z_i F}\ln\left(\frac{a''_i}{a\prime_i}\right) \tag{21.4}$$

According to Eqn (21.4), $(a''_i/a'_i)^{1/zi}$ is expressed as follows (Donnan, 1924; Helferrich, 1962a):

$$\left(\frac{a''_i}{a'_i}\right)^{\frac{1}{z_i}} = const. \tag{21.5}$$

Cwirko and Carbonell (1990) discussed the DD phenomena across charged porous membranes based on the capillary tube model. This model takes into account the effects of concentration, electrical potential, and pressure driving forces on ion flux across charged cylindrical pores and determines the effect of membrane properties (average pore radius and surface charged density) and solution concentration on fluxes of ions and a solution in a steady state.

Ruiz et al. (2003) developed the theory of the ion exchange kinetics in the DD phenomenon, and deduced the forward reaction rate constant of the ion exchange equilibrium reaction (Section 21.2.2). Further, ion transport in the cells is discussed based on the diffusion–convective equation. Mass balance at the membrane/solution interface is written by Fick's first law. Based on this theory, the ion concentration profile in the cells and ion concentration changes at the outlet of a feeding compartment are simulated.

Wen and Hamil (1981) defined the effective rate constant of ion concentration decrease with time in the feeding cell and material balance of ions for the whole feed solution in the batch mode process. Integrating the material balance equation, the ion depletion rate constant is deduced (Section 21.2.3).

Ktari and Auclair (1987) derived a flux equation in which the reciprocal of the flux is directly related to the quotient of the concentrations of the counter-ions in the feed compartment. This equation is based on a mechanism that considers the ion exchange reactions at the membrane/solution interface in addition to the diffusional processes within the membrane.

The driving force in the DD is not a simple chemical potential gradient, but an electrochemical potential gradient. Such a phenomenon is discussed with the Nernst-Planck equation. Sudoh et al. (1987) and Ktari et al. (1993) discussed the transport phenomena in the DD system based on the Nernst-Planck equation, with the mass transport coefficient in the boundary layer.

Higa et al. (1988) presented the uphill transport theory across ion exchange membranes in multiple component ionic systems. The theory is formulated on the basis of both the Donnan equilibrium and the Nernst-Planck equation, and it enables to simulate uphill transport of ions in a nonstationary dialysis by numerical computation.

Miyoshi (1998) measured the diffusion coefficients of ions through ion exchange membranes in DD using cations of different valence in the feed and extract solutions. The experiment shows that it is better to use monovalent driving ions to obtain a larger ion flux because monovalent ions can move more easily inside the membrane than bivalent ions, which interact more strongly with the ionized sites of the membrane. Moreover, the electrolyte leakage through the membrane from the extract to feed solution is 2.5 times higher with NaCl than Na_2SO_4.

21.2.2 Ion Exchange Kinetics

Ruiz et al. (2003) developed the following theory of the ion exchange kinetics. A DD phenomenon in Figure 21.1 is assumed to be operated in the batch mode. When the membrane equilibrated in monovalent Y anion form is in contact with a solution containing monovalent anions X and monovalent cations A, an ion exchange reaction occurs as follows:

$$X + \bar{Y} \underset{k_2}{\overset{k_1}{\rightleftarrows}} \bar{X} + Y \tag{21.6}$$

where the bars refer to the membrane.

Instead of an equilibrium constant, a selectivity coefficient, K_B^A, is often used for describing ion exchange equilibrium (Helfferich, 1962b).

$$K_Y^X = \frac{\left(\bar{C}_X\right)_{eq} (C_Y)_{eq}}{(C_X)_{eq}\left(\bar{C}_Y\right)_{eq}} \tag{21.7}$$

In Eqn (21.7), the anion concentrations are equilibrium concentrations.

Before the equilibrium is established in the above process, an instantaneous selectivity coefficient, $U_Y^X(\mathrm{t})$, is defined as follows:

$$U_Y^X(t) = \frac{\bar{C}_X C_Y}{C_X \bar{C}_Y} \tag{21.8}$$

In Eqn (21.8), the anion concentrations are concentrations at elapse time t. Thus, the deviation of the unsteady system at t from equilibrium state can be described by the ratio $U_Y^X(\mathrm{t})/K_Y^X$.

Ion exchange related to the Eqn (21.6) is formally treated as a following chemical reaction:

$$-\frac{QdC_X}{Sdt} = k_1 C_X \bar{C}_Y - k_2 \bar{C}_X C_Y \tag{21.9}$$

where Q is the volume of the AX solution, S is the membrane area, and k_1 and k_2 are the rate constants for the forward and reverse reactions of Eqn (21.6), respectively.

Equation (21.9) can be written:

$$-\frac{QdC_X}{Sdt} = -k_1 C_X \bar{C}_Y \left(1 - \frac{k_2 \bar{C}_X C_Y}{k_1 C_X \bar{C}_Y}\right) \quad (21.10)$$

At equilibrium, $dC_X/dt = 0$, and therefore, the combination of Eqns (21.7) and (21.9) gives the following relation between the rate constants and selectivity:

$$\frac{k_1}{k_2} = K_Y^X \quad (21.11)$$

Thus, combining Eqns (21.8), (21.10) and (21.11), the exchange rate can be written:

$$-\frac{QdC_X}{Sdt} = k_1 Z \quad (21.12)$$

$$Z = C_X \bar{C}_Y \left(1 - \frac{U_Y^X(t)}{K_Y^X}\right) \quad (21.13)$$

The dC_X/dt variation as a function of Z is linear and the k_1 value can be deduced from the gradient of the straight line.

Experiments were performed in the batch mode in the two compartments separated by the Selemion DSV anion exchange membrane (Asahi Glass Co.). A NaF solution and a NaCl solution were supplied to the feeding cell and extracting cell. Substituting $X = F$ and $Y = Cl$ in Eqns (21.12) and (21.13) and assuming $Na = A$, the relationship between dC_F/dt and $Z = C_F\bar{C}_{Cl}(1 - U_{Cl}^F(t)/K_{Cl}^F)$ is plotted in Figure 21.2, in which the initial fluoride concentration is 10^{-3} M = 19 mg/l. The gradient obtained with the various concentration studies gave for k_1 an average value of 0.008 ± 0.001 mg/mol/s.

21.2.3 Donnan Dialysis Kinetics

The DD unit described in this section is assumed to be incorporated with a cation exchange membrane and operated supplying an AX solution and a BX solution into the

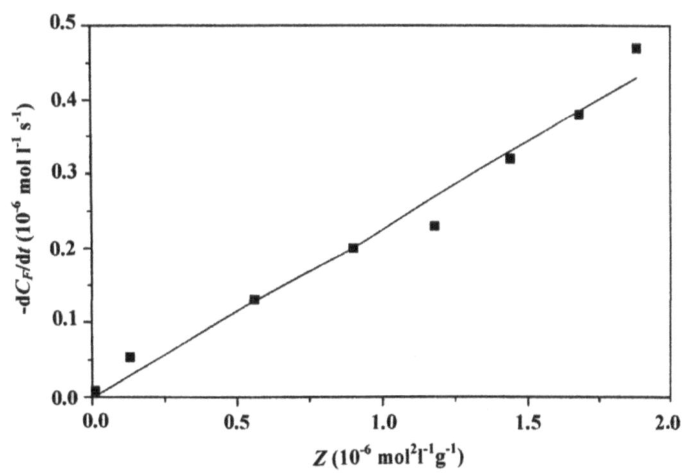

FIGURE 21.2 Variation of the ion exchange rate as a function of $Z = C_F\bar{C}_{Cl}(1 - U_{Cl}^F(t)K_{Cl}^F)$. Initial fluoride concentration: 19°mg/l. *Ruiz et al. (2003), p. 117, Fig. 3.*

feeding cell and extracting cell, respectively, in the batch mode. The operating condition is assumed to be reached to ion exchange equilibrium, as discussed in Section 21.2.2. Figure 21.3 illustrates the batch mode DD process in the feeding side. Here, C and C' are A ion (cation) concentration in the reservoir and feeding cell respectively. $C'_{in}(\fallingdotseq C)$ and C'_{out} are A ion concentration at the inlet and the outlet of the feeding cell. u is the linear velocity in the feeding cell. Q, Q' and Q_t are the solution volume in the reservoir, feeding cell and connected tubes ($Q >> Q'$ and Q_t).

Wen and Hamil (1981) defined the rate of concentration decrease of A ions with time t in the feeding cell in Figure 21.3 as:

$$\frac{dC}{dt} = -kC \tag{21.14}$$

in which k is the effective rate constant of ion concentration decrease (depletion) in the solution. In this theory, the osmotic water flow is assumed to be negligibly small compared to the feed recirculation rate. Therefore, the volume of the solution is essentially constant, and the amount of A ions in the solution is proportional to ion concentration. Solutions in the cells are sufficiently mixed, so that the influence of the concentration polarization to the ion transport phenomena is negligible. Integration of Eqn (21.14) from the inlet to the outlet of feeding cell gives:

$$C'_{out} = C'_{in} \exp(-kQ'/u) \tag{21.15}$$

Since $C'_{in} \fallingdotseq C$, the material balance of A ions for the whole feed solution is given as:

$$\frac{d(QC)}{dt} = -uC\left\{1 - \exp\left(\frac{-kQ'}{u}\right)\right\} \tag{21.16}$$

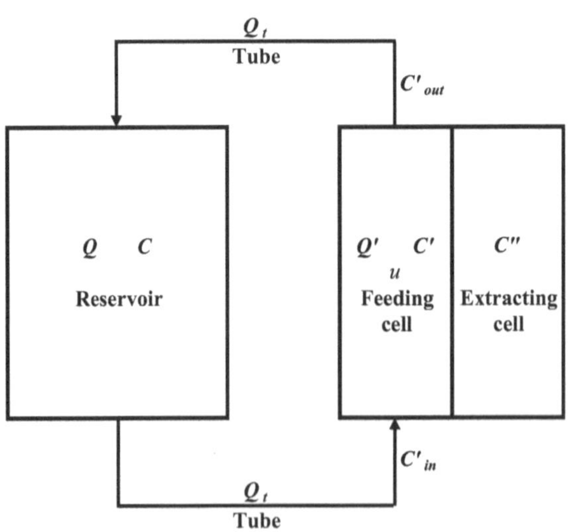

FIGURE 21.3 Donnan dialysis batch process.

Because Q is much larger than $Q' + Q_t$, and C'_{out} is smaller than C, Eqn (21.16) can be written as:

$$dC/dt = -(u/Q)\{1 - \exp(-kQ'/u)\}C \tag{21.17}$$

Integrating Eqn (21.17) from C_0 (at $t = 0$) to C (at anytime t) gives:

$$\ln(C_0/C) = (u/Q)\{1 - \exp(-kQ'/u)\}t \tag{21.18}$$

Here G is defined as the gradient of the linear plot $\ln(C_0/C)$ versus t then the A ion depletion rate constant k can be calculated as:

$$k = (u/Q')\ln\{u/(u - QG)\} \tag{21.19}$$

The experiments were performed using a DD unit incorporated with a cation exchange membrane prepared from an irradiation-grafted polyethylene film. The batch mode operation was carried out supplying a $CuSO_4$ (AX) solution and a Na_2SO_4 (BX) solution into a feeding cell and an extracting cell, respectively. Figure 21.4 exemplifies $\ln(C_0/C)$ versus t plots and G calculated from the gradient of the lines.

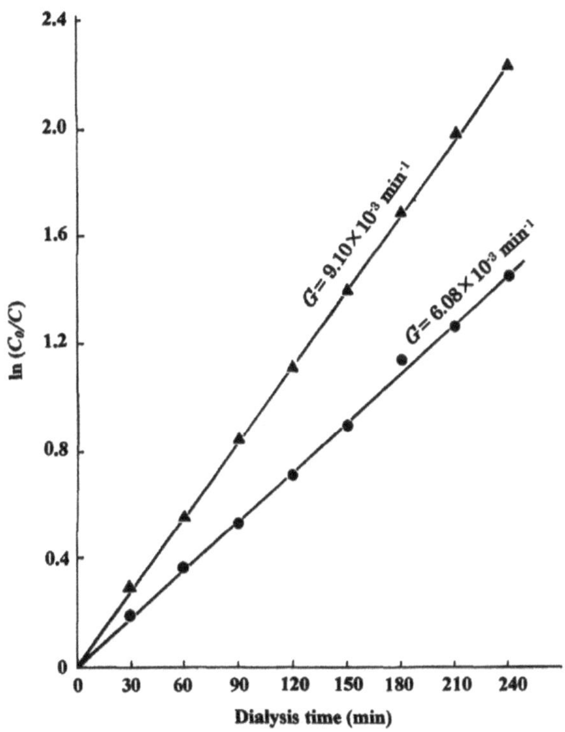

FIGURE 21.4 Copper concentration change of feed solution in Donnan dialysis. Feed solution, 10 dm^3, 0.001 N $CuSO_4$ (31.8 ppm Cu); Strip solution, 0.5 dm^3, 0.3 N Na_2SO_4. *Wen and Hamil (1981), p. 58, Fig. 1.*

21.3 Practice

21.3.1 Concentration of UO_2^{2+} Ions in a $UO_2(NO_3)_2$ Solution

Wallace (1967) concentrated uranyl ions from dilute solutions of uranyl nitrate applying DD. Figure 21.5 represents experimental unit consisting of two end plates, shown as shaded areas, and a series of membrane spacers (feeding cell and extracting cell), each separated from an adjacent one by a membrane. Figure 21.6 is a schematic of a membrane spacer. Each spacer was 3 × 3 inch square with a 2 × 2 inch area in the middle comprising the flow compartment. The membranes were AMF ion C-103C cation exchange membrane (American Machine & Foundry Co.). The membranes between the spacers are punched with two holes at opposite sides of the same edge to accommodate the flow between the spacers. After these preparations, a 0.01 M $UO_2(NO_3)_2$ and a 2.0 M HNO_3 solution were supplied to feeding and extracting cells, respectively in a single pass mode, and DD performance was evaluated. Here, $UO_2(NO_3)_2$ and HNO_3 correspond to respectively *AX* and *BX* in Section 21.2.3.

In this study, the cation concentration at the outlet of extracting cell C''_{out} is expressed by the following equation:

$$C''_{out} = \frac{Q'_{in}(C'_{in} - C'_{out})}{Q''_{out}} \tag{21.20}$$

where C is the concentration of the cation, Q is the volumetric flow, and subscripts in and out refer to the inlet and outlet of feeding cell and extracting cell, respectively (Figure 21.1).

Wallace (1967) suggests that if the feed stream flows too rapidly or the strip stream too slowly, not enough anions are introduced into the strip to satisfy the valence of the cations from the feed. When this happens, the product becomes saturated with the cation and some loss to the discharge must occur. Therefore, for complete recovery of the cation, the charge balance across the membrane requires:

$$z_- C''_{in} Q''_{in} \geq z_+ C'_{in} Q'_{in} \tag{21.21}$$

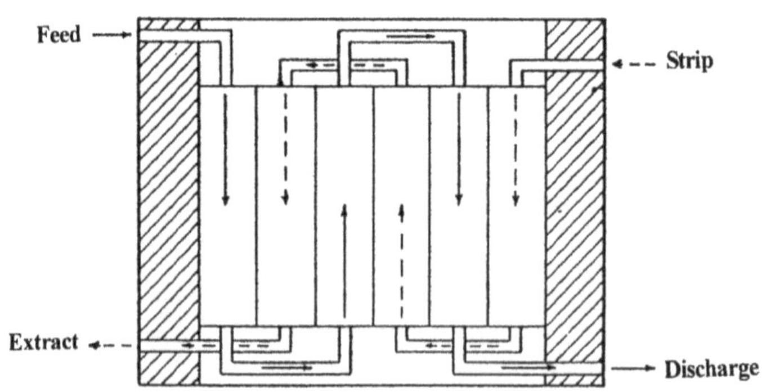

FIGURE 21.5 Donnan dialysis experimental unit. *Wallace (1967), p. 424, Fig. 1.*

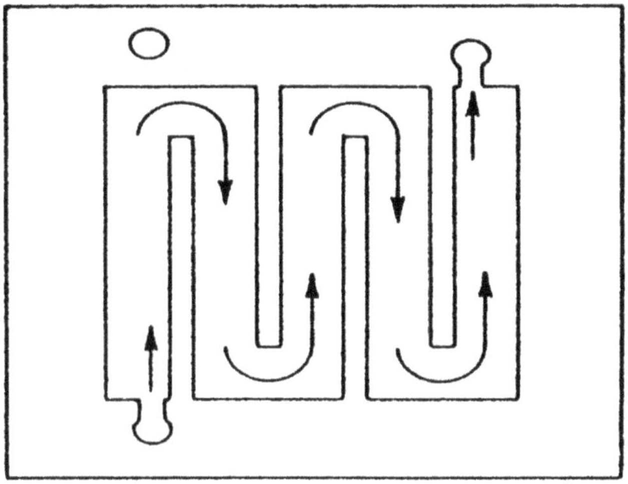

FIGURE 21.6 Membrane separator (feeding cell and extracting cell). *Wallace (1967), p. 424, Fig. 2.*

in which z_+ and z_- are the charge number of cations $(UO_2)^{2+}$ and anions $(NO_3)^-$, respectively and the unit of C'' in and C' in is M.

In cation concentration studies, it is convenient to define an operating parameter α:

$$\alpha \equiv \frac{z_- C''_{in} Q''_{in}}{z_+ C'_{in} Q'_{in}} \tag{21.22}$$

Based on the ion exchange kinetics (Section 21.2.2), the following ion exchange equilibrium reaction is proceeded at the solution/membrane interface in the feeding cell.

$$UO_2^{2+} + \overline{H^+} \underset{k_2}{\overset{k_1}{\rightleftarrows}} \overline{UO_2^{2+}} + H^+ \tag{21.23}$$

In order to establish the above equilibrium reaction, a large amount of H^+ ions (driving ions) must be supplied toward the membrane surface in the extracting cell by increasing $C''_{in,HNO_3} Q''_{in}$. On the other hand, in order to reduce sufficiently the amount of UO_2^{2+} ions (uranium loss) in the discharge (Raffinate) in the feeding cell, $U'_{in,UO_2} Q'_{in}$ can not be increased largely. These requirements are assumed to agree with the requirement of Eq. (21.21) to keep charge balance and satisfy the DD operation. Thus, α defined in Eq. (21.22) is understood by the following empirical formula:

$$\alpha \equiv \frac{C''_{in,HNO_3} Q''_{in}}{C'_{in,UO_2(NO_3)_2} Q'_{in}} \equiv \frac{z_- C''_{in} Q''_{in}}{z_+ C'_{in} Q'_{in}} \geq 1 \tag{21.24}$$

in which the unit of C''_{in,HNO_3} and $C'_{in,UO_2(NO_3)_2}$ is equivalent/ml

As the strip stream progresses through the cell, water is transferred by osmosis from feed stream into the strip stream, thereby decreasing the concentration in the strip and increasing its flow. Although the rate of osmosis is slow, the strip stream flow is also

Table 21.1 Concentration of Uranyl Nitrate with Nitric Acid

Flow				UO_2^{2+} concentration		H^+ concentration	
Feed Q'_{in} ml/min	Strip Q''_{in} ml/min	Extract Q''_{out} ml/min	α	Extract $C''_{out,UO2}$ M	$C'_{out,UO2}/C'_{in,UO2}$ %	Extract $C''_{out,H}$ M	Discharge $C'_{out,H}$ M
5.0	0.040	0.0995	0.80[a]	0.282	24.6	0.0126	0.0292
	0.050	0.112	1.0[a]	0.274	17.8	0.0297	0.0231
	0.060	0.142	1.2	0.273	7.46	0.0466	0.0244
	0.075	0.176	1.5	0.268	3.34	0.109	0.0270
	0.100	0.230	2.0	0.210	1.33	0.269	0.0292
	0.200	0.339	4.0	0.148	0.67	0.551	0.0315
4.0	0.060	0.158	1.5	0.220	1.41	0.0941	0.0274
	0.080	0.207	2.0	0.189	0.56	0.218	0.0293
	0.160	0.314	4.0	0.124	0.18	0.534	0.3420
3.0	0.036	0.111	1.2	0.189	1.76	0.0207	0.0266
	0.045	0.141	1.5	0.185	0.15	0.0726	0.0285
	0.060	0.181	2.0	0.155	0.09	0.188	0.0308
	0.160	0.344	5.3	0.084	<0.05	0.596	0.0401

Feed $C'_{in} = 0.01$ M $UO_2(NO_3)_2$(no HNO_3).Strip $C''_{in} = 2.0$ M HNO_3.
[a]Not at steady state.
Wallace (1967), p. 425, Table 1.

slow, so that considerable dilution can occur. Osmosis, therefore, limits the concentration of cations in the product.

The investigation proceeded by Wallace is presented in Table 21.1. In this experiment, UO_2^{2+} ions were concentrated from a 0.01M $UO_2(NO_3)_2$ feed with a 2 M HNO_3 strip. The feed flow Q'_{in} was kept constant within separate groups of the test, but α was varied. Each test was run for 7 h.

Steady state was attained in all tests, except the first two at a feed flow of 5 ml/min and α at 0.8 and 1.0. Uranium losses to the discharge $C'_{out,UO2}/C'_{in,UO2}$ decreased with increasing values of α at constant feed flow; the uranium concentration in the product (extract) $C''_{out,UO2}$ remained fairly constant up to $\alpha = 1.5$, then decreased with increasing α. Uranium loses $C'_{out,UO2}/C'_{in,UO2}$ decreased with decreasing feed flow Q'_{in} at constant α. The concentration of free acid $C''_{out,H}$ in the product increased with increasing α and approximated the value predicted from the excess acid added into the strip. The residual free acid in the discharge $C'_{out,H}$ was in all tests slightly higher than that predicted for a simple exchange of H^+ for UO_2^{2+} This indicates that a small amount of HNO_3 diffuses through the membrane. This diffusion is also indicated by the small increase in $C'_{out,H}$ as α is increased and feed flow Q'_{in} is decreased.

21.3.2 Fluoride Removal from Diluted Solutions

Too many or too few fluoride ions in drinking water are harmful to the consumer's health. The acceptable fluoride concentration is generally in the range of 0.5–1.5 mg/l Hichour et al. (1999) applied the DD with an anion exchange membrane for the

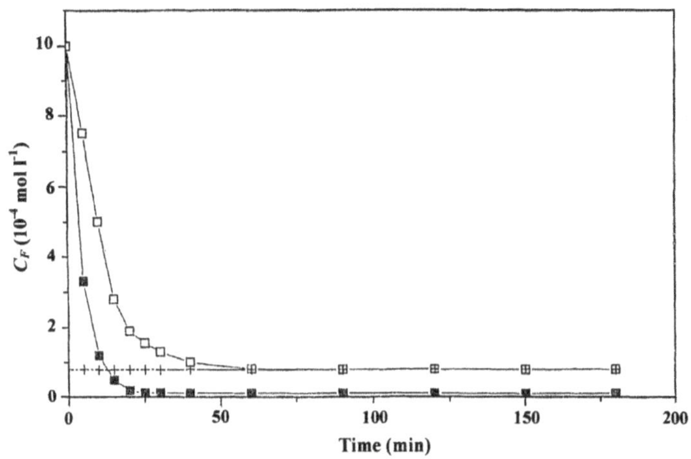

FIGURE 21.7 Time dependence of the feed concentration of the F^- ions. ■; 10^{-1} M NaCl, □; 5°×°10^{-2} M Na_2SO_4, +; Upper norm. *Hichour et al. (1999), p. 58, Fig. 4.*

defluorization of diluted NaF solutions. The commercially available anion exchange membranes tested are Selemion DSV (Asahi Glass) and Neosepta AFX, AFN, AMX, and ACS (Tokuyama Soda). After the preliminary experiments performed using a laboratory cell, a preindustrial pilot was composed of 11 cells (five feed cells and six receiver cells) separated by 10 DSV membranes having a total area of 1760 cm^2.

A 10^{-3} M (19 mg/l) NaF solution was supplied to the feeding cells. A 10^{-1} M NaCl solution or a 5×10^{-2} M Na_2SO_4 solution was supplied to extracting cells The solutions were circulated to each cell as a batch system, and the DD operation was performed. Figure 21.7 shows F^- ion concentration $C'_{in,F}$ changes in feeding cells with time. $C'_{in,F}$ becomes equal or lower than the upper norm ($=1.5$ mg/l $\fallingdotseq 0.8 \times 10^{-4}$ M) with sulfate or chloride ions, respectively.

21.3.3 Ammonium Ion Removal from Waste Water

The electrodialysis process of waste water for removing ammonium ions gives rise to organic fouling of ion exchange membranes (Fundamentals Section 11.5). To prevent such a trouble, Ajinomoto Co. developed DD process as follows (Takeuchi et al., 1987).

Figure 21.8 illustrates a DD unit (Effective membrane area: 0.5 dm^2) incorporated with a Selemion CMV cation exchange membrane (Asahi Glass Co.). A total of 400 ml of waste water before an activation sludge treatment process dissolving soluble organic materials and organic solid matter was fed into the feeding cell. A 2N HCl, 2N NaCl, or 2N KCl solution (strip) was supplied into the extracting cell. The solutions in both cells were stirred (300 rpm) using stirrers at room temperature. The ammonium state nitrogen concentration in the feeding cell was decreased with time, as shown in Figure 21.9, indicating the HCl strip to be most effective.

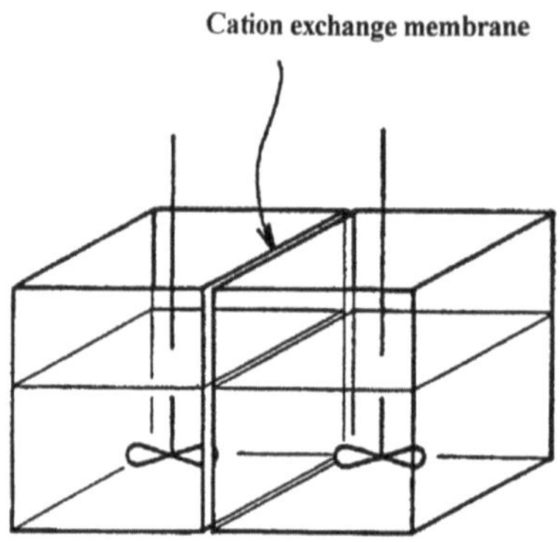

FIGURE 21.8 Donnan dialysis experimental unit. *Takeuchi et al. (1987), p. 533, Fig. 2.*

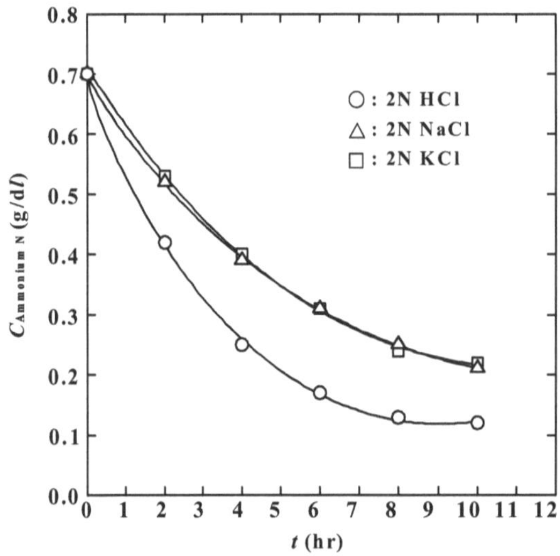

FIGURE 21.9 Changes of the ammonium state nitrogen concentration in the feeding cell with time. *Takeuchi et al. (1987), p. 532, from Table 4.*

References

Akretche, D.E., Kerdjoudj, H., 2000. Donnan dialysis of copper, gold and silver cyanides with various anion exchange membranes. Talanta 51, 281–289.

Cengeloglu, Y., Kir, E., Ersoz, M., Buyukerket, T., Gezgin, S., 2003. Recovery and concentration of metals from red mud by Donnan dialysis. Colloids Surf. A 223, 95–101.

Cwirko, E.H., Carbonell, R.G., 1990. A theoretical analysis of Donnan dialysis across charged porous membranes. J. Membr. Sci. 48, 155–179.

Davis, T.A., Wu, J.S., Baker, B.L., 1971. Use of the Donnan equilibrium principle to concentrate uranyl ions by ion-exchange membrane process. AIChE J. 17, 1006–1008.

Donnan, F.G., 1924. The theory of membrane equilibria. Chem. Rev. 1, 73–90.

Durmaz, F., Kara, H., Cengeloglu, Y., Ersoz, M., 2005. Fluoride removal by Donnan dialysis with anion exchange membranes. Desalination 177, 51–57.

Helfferich, F., 1962a. Ion Exchange. McGraw-Hill, p. 237.

Helfferich, F., 1962b. Ion Exchange. McGraw-Hill, p. 156.

Hichour, M., Persin, F., Molenat, J., Sandeaux, J., Gavach, C., 1999. Fluoride removal from diluted solutions by Donnan dialysis with anion-exchange membranes. Desalination 122, 53–62.

Hichour, M., Persin, F., Sandeax, J., Gavach, C., 2000. Fluoride removal from waters by Donnan dialysis. Sep. Pur. Tech. 18, 1–11.

Higa, M., Tanioka, A., Miyasaka, K., 1988. Simulation of the transport of ions against their concentration gradient across charged membranes. J. Membr. Sci. 37, 251–266.

Ktari, T., Auclair, B., 1987. Simplified flux equation and Donnan dialysis. J. Membr. Sci. 32, 251–265.

Ktari, T., Larchet, C., Auclair, B., 1993. Mass transfer characterization in Donnan dialysis. J. Membr. Sci. 84, 53–60.

Miyoshi, H., 1998. Diffusion coefficients of ions through ion exchange membrane in Donnan dialysis using ions of different valence. J. Membr. Sci. 141, 101–110.

Ruiz, T., Persin, F., Hichour, M., Sandeaux, J., 2003. Modelisation of fluoride removal in Donnan dialysis. J. Membr. Sci. 212, 113–121.

Sato, K., 2002. Effects of the feed solution concentration on the separation degree in Donnan dialysis for binary systems of amino acids. J. Membr. Sci. 196, 211–220.

Sudoh, M., Kamei, H., Nakamura, S., 1987. Donnan dialysis concentration of cupric ions. J. Chem. Eng. Jpn. 20, 34–40.

Takeuchi, H., Nagai, K., Ioh, H., Ozawa, M., 1987. Treatment of Waste Water Dissolving Ammonium Ions. JP S62-160189.

Tor, A., Cengelogue, Y., Ersoz, M., Arsla, G., 2004. Transport of chromium through cation-exchange membranes by Donnan dialysis in the presence of some metals of different valences. Desalination 170, 151–159.

Wallace, R.M., 1967. Concentration and separation of ions by Donnan membrane equilibrium. Ind. Eng. Chem. Process. Des. Dev. 6, 423–431.

Wen, C.P., Hamil, H.F., 1981. Metal counterion transport in Donnan dialysis. J. Membr. Sci. 8, 51–68.

Wisniewski, J., Rozanska, A., 2007. Donnan dialysis for hardness removal from water before electrodialytic desalination. Desalination 212, 251–260.

22

Fuel Cell

22.1 Overview of Technology

The fuel cell is a primary battery that was invented by Lord Glove in England in 1839. However, it was not attractive until General Electric Co. started investigating it in the latter half of the 1950s and a 1-kW battery was loaded in *Gemini 5* in 1965. During daytime, a space station generates electric power from solar energy and electrolyzes water using the electricity to obtain hydrogen and oxygen. During nighttime, the station creates water from hydrogen and oxygen using the fuel cell. This generation system has become widely applied in NASA space development projects because it does not discharge wastes, it is lightweight, and its generation efficiency is high.

Ion exchange membranes incorporated into the fuel cell in the first stage were inferior in chemical durability because they were based on polystyrene. However, perfluorinated Nafion membranes developed by Du Pont Co. exhibited excellent durability and were applied to fuel cells in the Biosatellite program in 1969. Ion exchange membrane technology in Japan was also developed with sodium chloride electrolysis for chlor-alkali production (Chapter 19). In the course of development research, Asahi Glass Co. and Asahi Chemical Co. produced perfluorinated Flemion and Aciplex membranes, respectively, with excellent endurance in an alkaline atmosphere.

Japan began investigating fuel cell technology in 1981 in the Moonlight program, developed by the Agency of International Science and Technology, Ministry International Trade and Industry. Research development advanced extensively in cooperation with many research organizations.

By the middle of the twenty-first century, it is estimated that fuel cells will be widely applied to electric traction for industrial tracks, delivery vehicles, and passenger cars; to cogeneration systems for households and buildings; to portable telephone power sources, and so forth.

22.2 Principle

Fuel cells are classified as phosphoric acid fuel cell, molten carbonate fuel cell (MCFC), solid oxide fuel cell, polymer electrolyte fuel cell (PEFC), and direct methanol fuel cell (DMFC). Ion exchange membranes are integrated into PEFC and DMFC.

The greatest merit of the PEFC is that it can be operated at temperatures between 60 and 100 °C. It consists of a fuel electrode, an air electrode, and an ion exchange

Ion Exchange Membranes. http://dx.doi.org/10.1016/B978-0-444-63319-4.00022-5

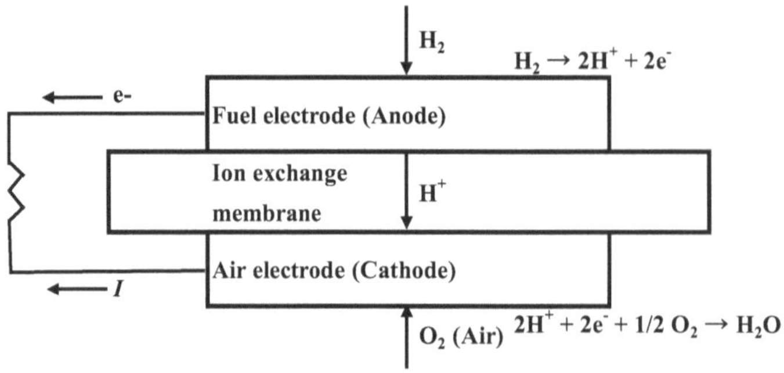

FIGURE 22.1 Polymer electrode fuel cell.

membrane, as illustrated in Fig. 22.1. In this system, water molecules are produced from hydrogen gas and oxygen gas by the following chemical reaction:

$$H_2 + 1/2O_2 \rightarrow H_2O \tag{22.1}$$

The relationship between the enthalpy change ΔH and Gibbs' free energy change ΔG in the reaction (22.1) is:

$$\Delta G = \Delta H - T\Delta S \tag{22.2}$$

in which T is the absolute temperature and ΔS is entropy change. ΔH and ΔG are evaluated as $\Delta H = -285.83$ kJ/mol and $\Delta G = -237.13$ kJ/mol. Accordingly, the theoretical maximum efficiency of the PEFC η is calculated as:

$$\eta = \frac{\Delta G}{\Delta H} \times 100 = \frac{-237.13}{-285.83} \times 100 = 82.9\% \tag{22.3}$$

The efficiency calculated above far exceeds the theoretical efficiency of a heat engine given by Carnot's cycle.

Electrochemical reactions on the electrodes in the PEFC in Fig. 22.1 are:

$$\text{Anode (Fuel electrode): } H_2 \rightarrow 2H^+ + 2e^- \tag{22.4}$$

$$\text{Cathode (Air electrode): } 1/2\,O_2 + 2H^+ + 2e^- \rightarrow H_2O\text{ (Liquid)} \tag{22.5}$$

Gibb's free energy change in the above reactions is:

$$-\Delta G = nFE_0 \tag{22.6}$$

in which F is the Faraday constant (96,485 C/mol). The theoretical electromotive force of the PEFC E_0 is:

$$E_0 = -\frac{-\Delta G}{nF} = \frac{237.13 \times 1000}{2 \times 96{,}485} = 1.229\text{ V} \tag{22.7}$$

The fuel (H_2) in the PEFC system is produced from natural gas (CH_4 + impurities), which is purified by desulfurization. The fuel is reformed by adding steam (steam methane reforming) as follows.

In the fuel reforming unit:

$$CH_4 + 2H_2O \rightarrow 4H_2 + CO_2 \tag{22.8}$$

$$CH_4 + H_2O \rightarrow 3H_2 + CO \tag{22.9}$$

In the CO treatment unit:

$$CO + H_2O \rightarrow H_2 + CO_2 \quad (22.10)$$

$$CO + 1/2\,O_2 \rightarrow CO_2 \quad (22.11)$$

Electrochemical reactions in the DMFC are:

$$\text{Anode (Fuel electrode): } CH_3OH + H_2O \rightarrow 6H^+ + CO_2 + 6e^- \quad (22.12)$$

$$\text{Cathode (Air electrode): } 3/2\,O_2 + 6H^+ + 6e^- \rightarrow 3H_2O\ (\text{Liquid}) \quad (22.13)$$

A direct methanol fuel cell can generate electricity directly from methanol without reforming. However, it has a limitation in that the fuel electrode reaction accompanies CO generation, which deteriorates catalyst (Pt) activity.

Under an applied electric current, the unit cell voltage in the PEFC decreases, as shown in Fig. 22.2, because of energy loss generated in the cell. This phenomenon is caused by (1) diffusion polarization, (2) resistance polarization, and (3) activation polarization, as explained below (New Sunshine Program Promotion Center, 1999).

1. Diffusion polarization

The electrochemical reaction in Eqns (22.4) and (22.5) brings about diffusion as a result of the occurrence of the concentration difference (concentration polarization) of each component. Part of the electromotive force of the PEFC is consumed by the concentration polarization.

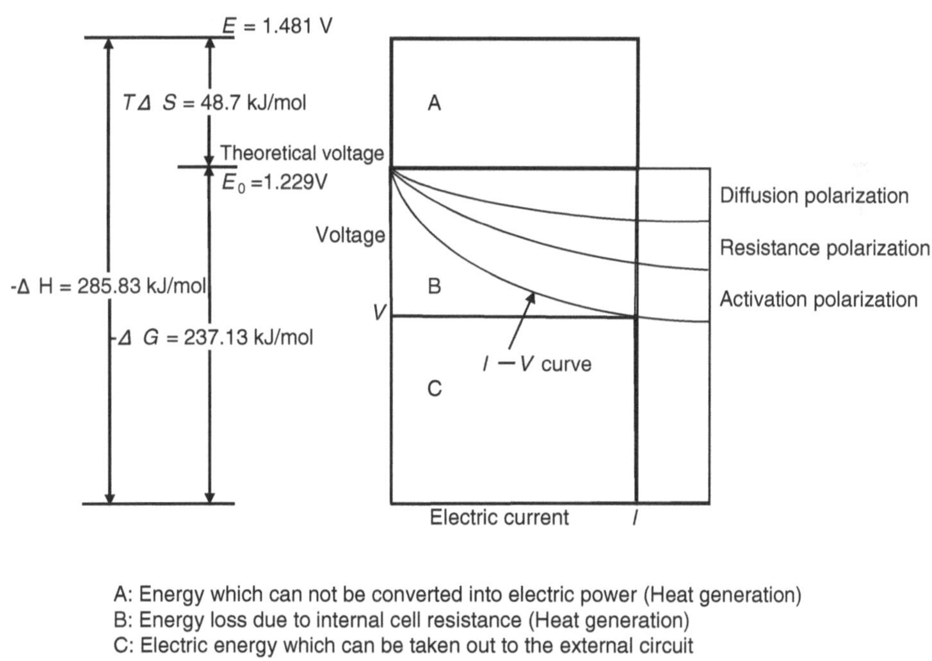

FIGURE 22.2 Performance and energy consumption in a fuel cell. *New Sunshine Program Promotion Center (1999), p. 14, Fig. 8.*

2. Resistance polarization

Electric resistance of the ion exchange membrane, electrode, and separator and the interface between the membrane and electrode consumes the electromotive force (*IR* loss).

3. Activation polarization

The oxidation/reduction reaction in Eqns (22.4) and (22.5) proceeds via the peak of activation energy. The reaction is accelerated by the potential shift (over voltage) established in the reaction system. A part of electromotive force in the PEFC is consumed to generate the over-voltage.

22.3 Parts of a Fuel Cell

22.3.1 Membrane Electrode Assembly and Cell Stack

The membrane electrode assembly (MEA) is composed of an ion exchange membrane (IEM), a cathode catalyst layer (CC), an anode catalyst layer (AC), and gas diffusion layers (GDL) placed between an anode separator (AS) and a cathode separator (CS), as illustrated in Fig. 22.3. The catalyst layers are formed on both surfaces of the membrane (Fundamentals Fig. 1.12).

Ducts are provided in the AS and CS for feeding fuel gas and air, respectively, into the electrodes. The ducts in the water separator (WS) are for cooling the assembly. The material of the separator is selected from sintered graphite carbon, compression-molded graphite/plastic mixed powders, press-formed expanded graphite sheets, etc.

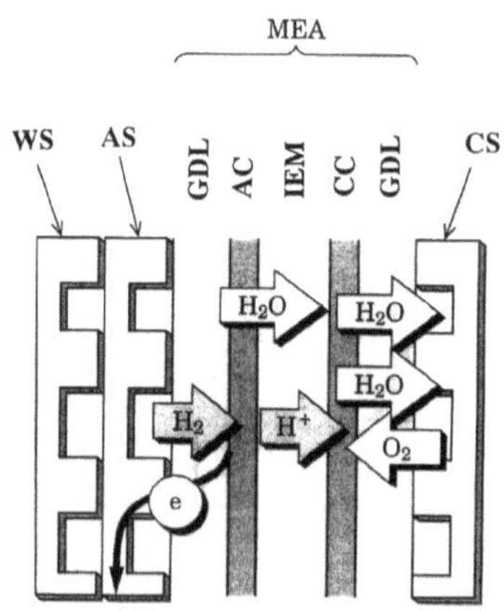

FIGURE 22.3 Unit cell arrangement in a PEFC. MEA, membrane electrode assembly; IEM, ion exchange membrane; GDL, gas diffusion layer; AC, anode catalyst layer; CC, cathode catalyst layer; AS, anode separator; CS, cathode separator; WS, water separator. *Fuel Cell Technology Working Committee (2002), p. 57, Fig. 2.2.*

The GDL allows the flow of gases H_2, air/oxygen, and product gases to pass through it. The main functions of the GDL are to:

1. keep a gas pathway from flow channels to the catalyst layer.
2. help removing by-produced water to the outside of the catalyst layer.
3. keep some water on the surface to increase membrane conductivity.
4. keep heat transfer during the cell operation.
5. provide enough mechanical strength to hold MEA from extension caused by water absorbance.

The GDL is commercially available in various forms such as carbon paper and woven carbon fabrics.

The cell stack is an array of MEA between terminal electrodes (current collectors) placed on both outsides of the array, as shown in Fig. 22.4. The cell stack is unified by fastening plates attached on both outsides of the terminal electrodes.

22.3.2 Ion Exchange Membrane

1. Functions of IEMs

Perfluorinated IEMs (Fundamentals Section 1.7.1) such as Nafion (Du Pont Co.), Flemion (Asahi Glass Co.), Aciplex (Asahi Chemical Co.), Dow (Dow Chemical Co.), and Hyflon (Solvay) are supplied to the PEFC (or DMFC) system. The membrane characteristics are generally, the thickness; 30–175 μm, and ion exchange capacity; 0.91–1.1 mEq/g. The functions of the IEM are to:

1. carry H^+ ions generated at the fuel electrode toward the air electrode.
2. prevent direct contact between H_2 and O_2.
3. prevent a short circuit current between both electrodes.

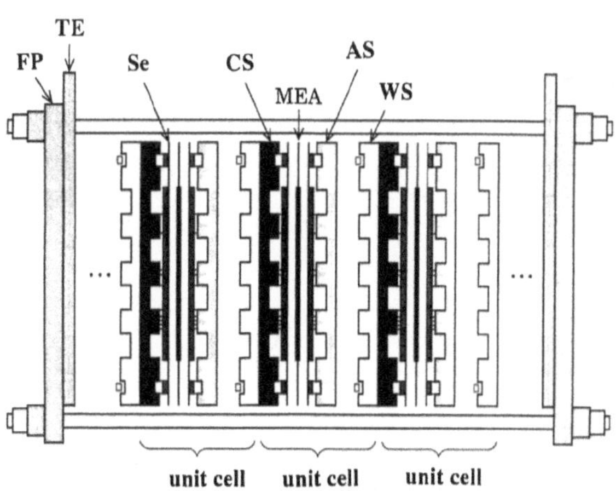

FIGURE 22.4 Cell stack arrangement in a PEFC. MEA, membrane electrode assembly; AS, anode separator; CS, cathode separator; WS, water separator; Se, seal; TE, terminal electrode; FP, fastening plate. *Fuel Cell Technology Working Committee (2002), p. 58, Fig. 2.3.*

2. Water content

The ion conductivity of the membrane is enhanced by water in the membrane, so it is necessary to increase the water content of the membrane by injecting moisture into the feeding fuel and air.

Figure 22.5 shows that the proton conductivity of Nafion membranes is strongly affected by the water content compared with the proton conductivity of sulfonated PEEK membranes (Gubler et al., 2004). To keep a suitable water content, it is important to control water transport in the assembly. Moisture injected into the fuel gas transports from the anode toward the cathode across the membrane with H^+ ions generated by Eqn. (22.4) (electro-osmosis). The H^+ ions are changed to H_2O by Eqn. (22.5) at the cathode, and the H_2O generated at the cathode diffuses from the cathode toward the anode (back diffusion) (Yasuda, 2000). If the feeding water or generated water in the assembly fills the fine pores in the electrodes, energy and voltage are increased by the increase in cell resistance. Accordingly, it becomes necessary to control the water supply to the system.

3. Micro-structure

The micro-structure of perfluorosulfonic acid membranes is an inverse micelle phase–separated structure, as shown in Fig. 22.6 (Gubler and Scherer, 2010) (cf. Fig. 19.6). In the hydrophobic domains, hydrophilic domains (ionic-cluster regions) with ion-conductive sites are connected and enhance ion conductivity owing to the cohesive structure of sulfonic acid groups, as shown in Fig. 22.5. Water transference coefficient $t_{H2O} = FJ_{H2O}/i$ (where F is the Faraday constant, J_{H2O} is the water flux, and i is the current

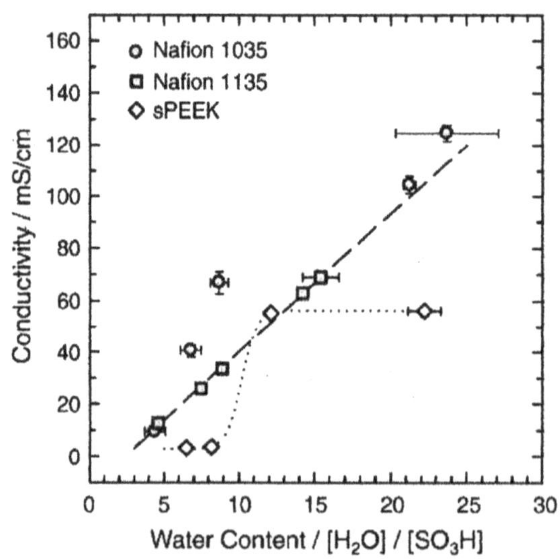

FIGURE 22.5 Proton conductivity of Nafion membranes and a sulfonated PEEK membrane as a function of water content. *Gubler and Scherer (2010), p. 1036, Fig. 3(b).*

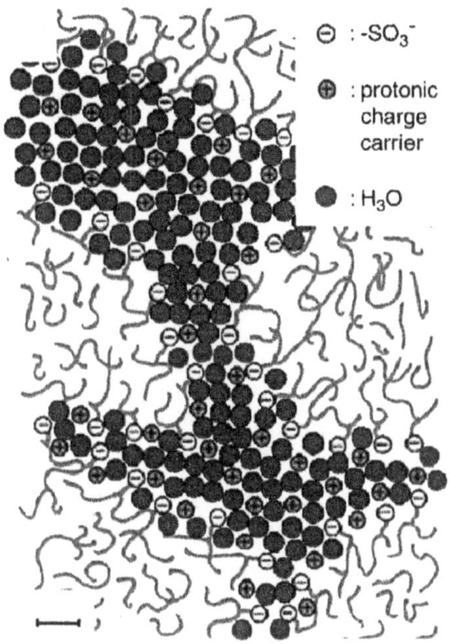

FIGURE 22.6 Nano-phase–separated structure of proton exchange membranes such as Nafion. *Gubler and Scherer (2010), p. 1036, Fig. 3(a).*

density) is increased with the water content (Okada et al., 1998). H^+ ions are assumed to be transferred in the hydrophilic domains with the Grotthuss mechanism (de Grotthuss, 1806). This phenomenon is strongly related to the water content.

4. Mechanical strength

Besides providing a proton transport medium, fuel cell membranes have to be mechanically robust. Fuel cells are mechanically compacted and prevent reactant leakage to the outside. Often, the membrane is a part of the gasket arrangement for sealing purposes. Therefore, the polymer has to be ductile, have high strength and fracture toughness, exhibit a low propensity to undergo creep deformation, and be dimensionally stable (Gubler and Scherer, 2010).

5. Desirable properties

The desirable characteristics of IEMs integrated into fuel cells are summarized as follows (Yoshitake, 2008):

1. High proton conductivity
2. Prevent a direct reaction between hydrogen and oxygen
3. Insulate between an anode and a cathode
4. High chemical and electrochemical stability
5. High mechanical strength and dimensional stability

6. High water permeability
7. Heat resistance
8. Ability to be joined to electrodes
9. Low cost
10. Easy handling

22.3.3 Catalyst

1. Anode catalyst

Catalysts reduce the energy consumption of the cell by decreasing the activation energy peak of the electrode reactions. The foremost exploited anode catalysis for PEFC (including DMFC) is still PtRu/C (Bockris, 1975), owing to its tolerance to poisoning of strongly adsorbed CO species on Pt. The PtRu catalyst is supported by graphitic carbon C to prevent corrosion. The promotional role of ruthenium in the PtRu/C is discussed based on the biofunction mechanism (Watanabe and Motoo, 1975) or the ligand effect mechanism (Frelink et al., 1995). An advanced anode catalyst such as PtRuRh/C has been developed and has been on the market (Kawaguchi et al., 2006).

2. Cathode catalyst

A typical cathode catalyst for PEFC is Pt/C because of its high activity toward oxygen reduction. However, the further enhancement of both catalytic activity and durability is inevitable for practical use. The over-potential for oxygen reduction needs to be reduced and the dissolution of platinum has to be minimized. The PtCo/C binary catalyst (Beard and Ross, 1990; Toda et al., 1999) has attracted attention owing to its enhanced specific activity and durability.

The membranes are subject to aggressive conditions because the interaction of H_2 and O_2 on the surface of the Pt catalyst leads to the formation of aggressive HO* and HOO* radical species, which attack the polymer and lead to chain scission and thus membrane degradation (LaConti et al., 2003). For this reason, perfluorinated IEMs described in Section 22.3.2 are applied to the fuel cell. Pt is expensive as the raw material of the fuel cells, so reduction of the Pt content is the main target in the development of technology.

22.4 Performance of Fuel Cells

The cell voltage change of a unit cell PEFC system (membrane area, 225 cm^2) being supplied pure hydrogen is exemplified in Fig. 22.7 (Mitsuta et al., 1997). The current density and operating pressure in this operation were, respectively, 250 mA/cm^2 and 1 ata. The decreasing rate of cell voltage was 4 mV/1000 h. Figure 22.8 shows the performance of a 20-cell unit operated at 367 mA/cm^2 and 6 ata (Washington, 2000). The cell voltage decreasing rate in this unit was 2.2 mV/1000 h.

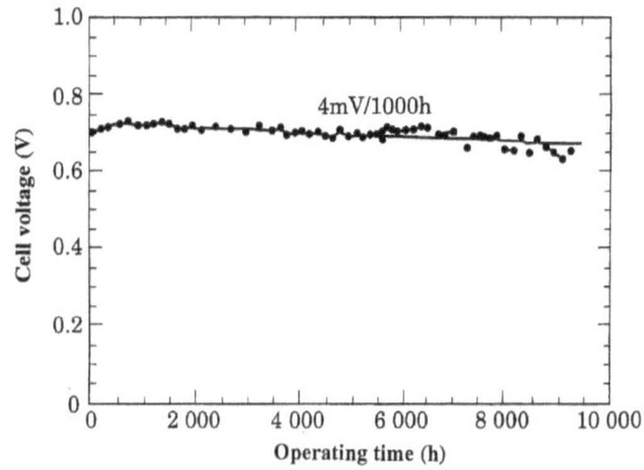

FIGURE 22.7 Cell voltage change of a unit cell PEFC system with time. *Mitsuta et al. (1997).*

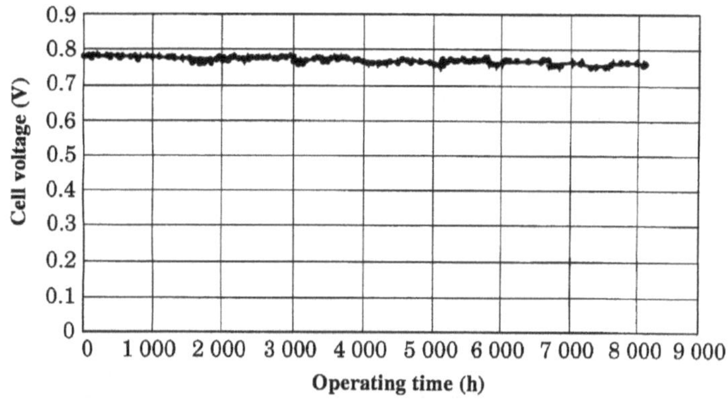

FIGURE 22.8 Performance of a 20-cell PEFC unit. *Washington (2000).*

The cell voltage decrease during the operation is caused by (1) a decrease in the effective reaction area of the catalyst owing to the increase in catalysis particle diameter or increase in moisture in the catalysis layer, (2) fouling of the membrane, and (3) an excessive or insufficient moisture injection to the electrodes.

22.5 Practice

22.5.1 Electrically Powered Car

Figure 22.9 exemplifies an electrically powered car system in which hydrogen and air are supplied to the PEFC and a motor is derived by generated electric power. The ultra-capacitor (electric double-layer capacitor) is incorporated to recover braking energy and increase driving force.

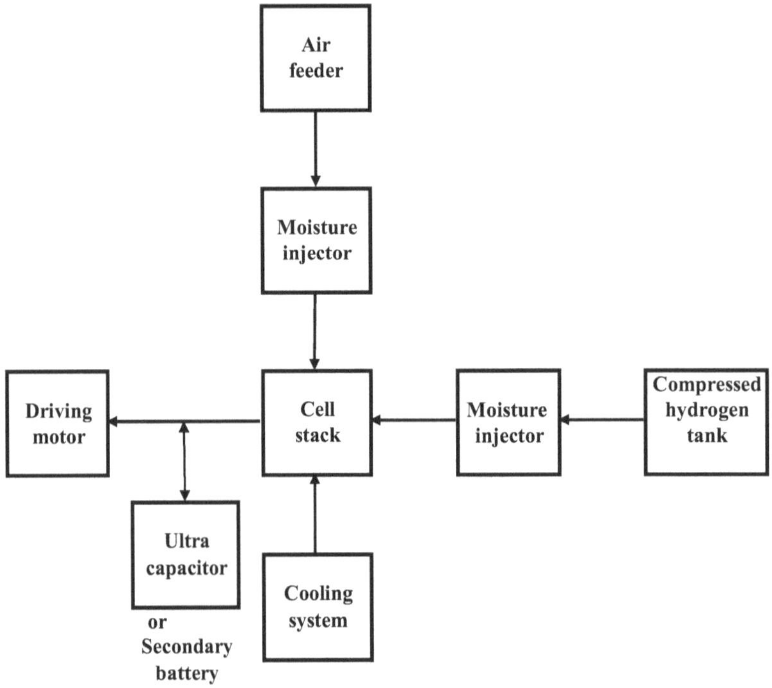

FIGURE 22.9 Polymer fuel cell system in an electrically powered car. *Fuel Cell Technology Working Committee (2002), p. 78, Fig. 2.36.*

In 1993, Ballad Power Systems, Inc., in Canada developed an electrically powered car (electric output, 120 kW; motor output, 80 kW; mileage, 160 km) by supplying compressed hydrogen and air to the PEFC. In 1997, DaimlerChrysler developed the NECAR three-passenger car (electric output, 50 kW; mileage, 400 km) refueled by methanol. In 1998, a fuel cell bus service seating 62 passengers started between Vancouver and Chicago refueled by compressed hydrogen.

A fuel cell car developed by Adam Opel Gmbh in Germany employing liquid hydrogen led marathon runners in the Sydney Olympics in 2000. Highway test driving was started in California with participation by NECAR 4a (DaimlerChrysler, compressed tank), Focus (Ford, compressed hydrogen), FCX-3V (Honda, compressed tank), and so forth, in November 2000.

The upper-limit price of an electrically powered fuel car is assumed to be 5–15% up of traditional cars, and the cost is estimated to be 5,000–10,000 yen/kW.

Advantages of IEMs incorporated with the cell stack are that they:

1. provide a compact and light-weight component at high maximum power density.
2. operate at normal temperature (20–100 °C).
3. correspond to load changes promptly.
4. are easy to integrate into the component.

Disadvantages are:

1. slow start operation. It takes several seconds until the cell stack works effectively.
2. malfunction owing to freezing of moisture at low temperatures below the freezing point.
3. need for a larger radiator to eliminate exhausted heat.
4. poorer responses to changes in load than batteries.

Hydrogen gases are supplied in hydrogen stations onsite (produced in hydrogen stations) or offsite (reformed in chemical plants and delivered to hydrogen stations). The raw materials of hydrogen gases are fossil fuels, and the fuel cell system does not cause zero emission. It is expected to generate hydrogen gases economically by means of water hydrolysis with natural energy use. Honda (2012) installed a solar hydrogen station (maximum output, 9 kW) that does not exhaust CO_2 in the hydrogen production, storage and supply process. The station produces 1.5 kg of hydrogen over 24 h.

22.5.2 Household Cogeneration

Figure 22.10 shows a household cogeneration system, consisting of (1) a fuel treating unit (FTU), (2) a blower (Blw), (3) a PEFC cell stack (Cs), (4) a heat recovering unit (HRU), and (5) an inverter (Inv).

In the fuel treating unit, town gas is converted into hydrogen including carbon dioxide and a small quantity of carbon monoxide. CO is selectively oxidized and converted to CO_2. Moisture is injected to enhance membrane conductivity. In the PEFC cell stack, a direct current and heat are generated from H_2 and O_2 supplied from the fuel treating unit and the blower, respectively. In the inverter, the direct current is inverted into an alternating current, which is consumed in the household for lighting, heating, etc. In the heat recovering unit, heat discharged from PEFC cell stack and fuel treating unit is converted into warm water with a temperature higher than 60 °C, which is also consumed in the household.

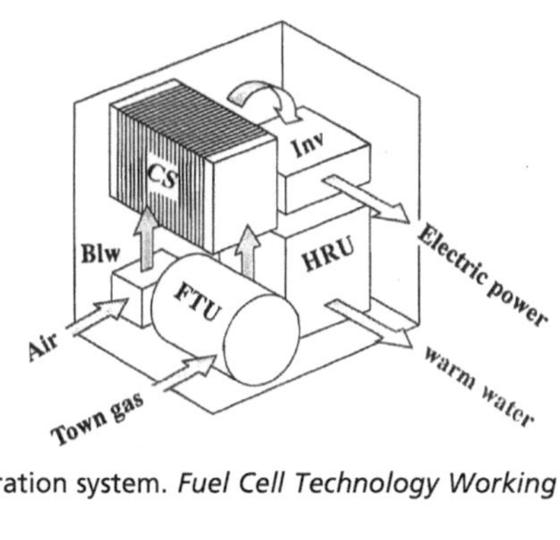

FIGURE 22.10 Household cogeneration system. *Fuel Cell Technology Working Committee (2002), p. 85, Fig. 2.43.*

The household cogeneration apparatuses are developed and commercialized by many electrical products firms and gas companies.

References

Beard, B.C., Ross, P.N., 1990. Structure of activity of Pt-Co alloys as oxygen reduction. J. Electrochem. Soc. 137, 3368–3374.

Bockris, J., Wroblowa, Ó.M., 1975. Electrocatalysis. J. Electroanal. Chem. 7, 428–451.

Frelink, T., Visscher, W., van Veen, J.A.R., 1995. On the role of Ru and Sn as promotors of methanol electro oxidation over Pt. Surf. Sci. 335, 353–360.

Fuel Cell Technology Working Committee, 2002. Polymer Electrolyte Fuel Cell. In: Institute of Electrical Engineers of Japan (Ed.), Fuel Cell Technology. Ohm Co., Tokyo. pp. 55–98.

de Grotthuss, C.J.T., 1806. Sur la decomposition de l'eau et des corps qu'elle tient en dissociation a l'aide de l'elecricite galvanique. Am. Chim. (Paris) 58, 54–73.

Gubler, L., Finsterwald, T., Scherer, G.G., 2004. PSI Electrochemistry Laboratory – Annual Report. Also available on the World Wide Web: http://ecl.web.psi.ch/SciRep.html, p. 111.

Gubler, L., Scherer, G.G., 2010. Trends for fuel cell membrane development. Desalination 250, 1034–1037.

Honda, 2012. http://honda.co.jp/news/2012/4120327.html.

Kawaguchi, T., Rachi, Y., Sugimoto, W., Murakami, Y., Takasu, Y., 2006. Performance of ternary PtRuRh/C electrocatalyst with varying Pt:Ru:Rh ratio for methanol electro-oxidation. J. Appl. Electrochem 36, 1117–1125.

LaConti, A.B., Hamdan, H., McDonald, R.C., 2003. In: Vielstich, W., Gasteiger, H.A., Lamm, A. (Eds.), Mechanisms of Membrane Degradation, Handbook of Fuel Cells – Fundamentals, Technology and Applications, vol. 3. John Wiley & Sons, pp. 647–662.

Mitsuta, K., et al., 1997. Development of an exterior manifold type PEFC module. In: The 4th FCDIC symposium, pp. 265–268.

New Sunshine Program Promotion Center, 1999. Polymer Electrolyte Fuel Cell. Agency of Industrial Science and Technology, Ministry of International Trade and Industry, Japan, pp. 13–14.

Okada, T., Xie, G., Gorseth, O., Kjelstrup, S., Nakamura, N., Arimura, T., 1998. Ion and water transport characteristics of Nafion membranes as electrolytes. Electrochim. Acta 43, 3741–3747.

Toda, T., Igarashi, H., Uchida, H., Watanabe, M., 1999. Enhancement of the electroreduction of oxygen on Pt alloys with Fe, Ni, and Co. J. Electrochem. Soc. 146, 3750–3756.

Washington, K., 2000. Development of a 250 kW class polymer electrolyte fuel cell Stack. Fuel Cell. Semin. pp. 468–472.

Watanabe, M., Motoo, S., 1975. Electrocalysis by ad-atoms, part II. Enhancement of the oxidation of methanol on platinum by ruthenium ad-atoms. J. Electroanal. Chem. 60, 267–273.

Yasuda, K., 2000. Development and Application of Polymer Electrolyte Fuel Cells, Lecture 2, Material Development in PEFC. NTS Co., Tokyo.

Yoshitake, Y., 2008. Recent development of parts in fuel cells – electrode catalyst and electrolytes. In: 37th Charged Membrane Colloquium, Research Group of Electrodialysis, Soc. Sea Water Science, Jpn. March, 21, Tokyo.

23

Redox Flow Battery

23.1 Overview of Technology

The term redox means reduction and oxidation, and refers to the oxidation–reduction reaction occurring between two chemical species on the inactive electrode surfaces of a battery. The redox flow battery is a secondary battery; it is termed redox flow because the chemical species are stored outside the battery and supplied to it by pumps.

An early patent of a titanium chloride flow battery was registered and granted in Germany in 1954; several metallic ions having different charges were applicable to the battery. However, the basic principles of redox flow technology were established by NASA in 1974 (Thaller, 1974), and the iron/chromium (Fe/Cr) system was initially selected (Hagedorn and Thaller, 1980). The Electrochemical Laboratory of Japan's Ministry of International Trade and Industry initiated the development research of the Moon Light project for the Fe/Cr system battery (Nozaki et al., 1984). However, the system did not lead to practical applications because of problems in the electrolysis solution and the battery stack: (1) occurrences of cross-contamination of Fe ions and Cr ions across ion exchange membranes; (2) breaking of charge balances between electrodes, caused by side reactions; and (3) low electromotive force (1 V) and output density. In response to these problems, development research was redirected toward a vanadium (V^{2+}/V^{3+} for an anode, V^{5+}/V^{4+} for a cathode) system.

Although the use of vanadium in batteries had been suggested earlier by Pissoort (1933), and Pelligri and Spaziante (1978), the first known successful demonstration and commercial development of the all-vanadium redox flow battery (VRB), employing vanadium in a solution of sulfuric acid, occurred at the University of New South Wales in Australia (Skyllas-Kazcos et al., 1986, 1988). The advantages of this system were: (1) high electromotive force (1.4 V); (2) high output density (several times that of a Fe/Cr system); (3) no deterioration in performance that could be attributed to contamination of ions across membranes. VRB technology is now developing widely throughout the world (International Flow Battery Forum).

23.2 Principle

Figure 23.1 shows the principle of a VRB consisting of solution tanks dissolving active electrolytes, pumps for circulating the electrolysis solution among the tanks and the cells, and pipelines. The electrolysis cell is separated from the anode cell and cathode cell by a cation exchange membrane. Alternating current transmitted from a power station is inverted to direct current through an A/D inverter, and the resulting charge is

Ion Exchange Membranes. http://dx.doi.org/10.1016/B978-0-444-63319-4.00023-7

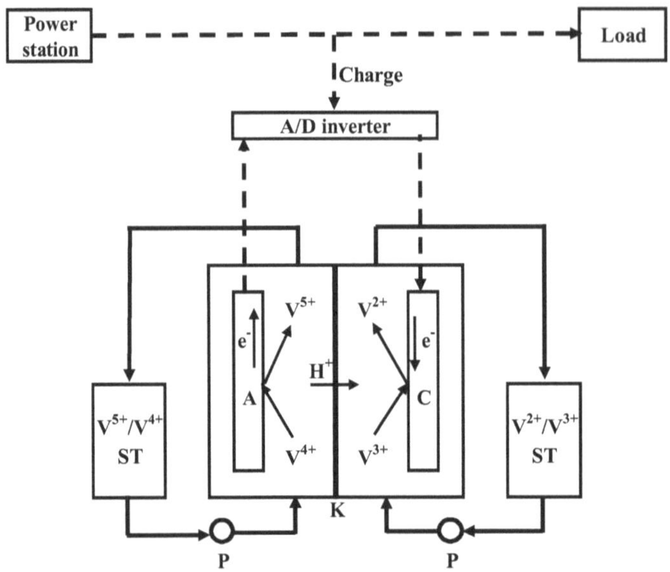

FIGURE 23.1 Principle of a vanadium redox flow battery (charging process).

transmitted into the battery. The electrochemical reaction in the vanadium battery is presented as follows.

On the anode:

$$VO^{2+}(\mathit{tetravalent}) + H_2O \underset{discharge}{\overset{charge}{\rightleftarrows}} VO_2^+(\mathit{pentavalent}) + 2H^+ + e^- : E^0 = 1.00V \tag{23.1}$$

On the cathode:

$$V^{3+}(\mathit{trivalent}) + e^- \underset{discharge}{\overset{charge}{\rightleftarrows}} V^{2+}(\mathit{divalent}) : E^0 = -0.26V \tag{23.2}$$

The forward reactions in Eqns (23.1) and (23.2) show the battery reaction in the charging process. In Eqn (23.1), tetravalent V^{4+} ions (VO^{2+}) are oxidized to pentavalent V^{5+} ions (VO_2^+) on the anode. In Eqn (23.2), trivalent V^{3+} ions are reduced to divalent V^{2+} ions on the cathode. The H^+ ions generated in the forward reaction in Eqn (23.1) pass through the cation exchange membrane to maintain the electric neutrality of the solution. The supplied electric power is accumulated in the cathode and anode solution tanks in the form of altered-valent vanadium ions.

Generated electromotive force E from the above electrochemical reactions is given using the following Nernst equation (Fundamentals Section 2.3.1):

$$E = E^0 + \frac{RT}{F}\ln\frac{[VO_2^+][V^{2+}]}{[VO^{2+}][V^{3+}]} \tag{23.3}$$

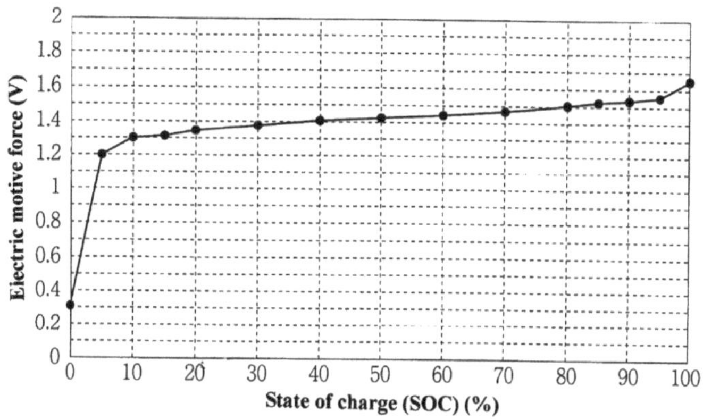

FIGURE 23.2 Electromotive force property of a vanadium redox flow battery (VRB). *Shigematsu (2011), p. 74, Fig. 3.8.*

where E^0 is the standard electromotive force, R is the gas constant, T is the absolute temperature, and F is the Faraday constant. For this equation, E^0 is computed as $E^0 = 1.00-(-0.26) = 1.26$ V from Eqns (23.1) and (23.2). The practical standard electromotive force E corresponds to the value at 50% state of charge (SOC), and it is $E \fallingdotseq 1.4$ V as shown in Figure 23.2. The battery is generally operated in a batch mode at SOC = 10–90%.

In order to generate practically available high electrical potential, the unit battery cells are arranged in series with bipolar plates. Figure 23.3 shows the VRB unit consisting of anodes, cathodes, cation exchange membranes, bipolar plates, and terminal electrodes. Electrolyte solutions dissolving active vanadium ions into sulfuric acid solutions are placed in the cathode and anode solution tanks, and electric power is charged and discharged through the electrochemical cycling reactions in Eqns (23.1) and (23.2). The direct current generated is inverted to alternating current through an inverter and supplied to external loads. Figure 23.4 (Shigematsu, 2011) exemplifies cell voltage and stack voltage generated in a 108-cell array operated at 360 A.

23.3 Advantages and Disadvantages of the Redox Flow Battery

23.3.1 Advantages

1. The electrochemical reaction is simple, stable for long periods, and responds instantaneously to load variation.
2. The system is designed to easily correspond to desired output and is easy to scale up.
3. The system is operated at room temperature.

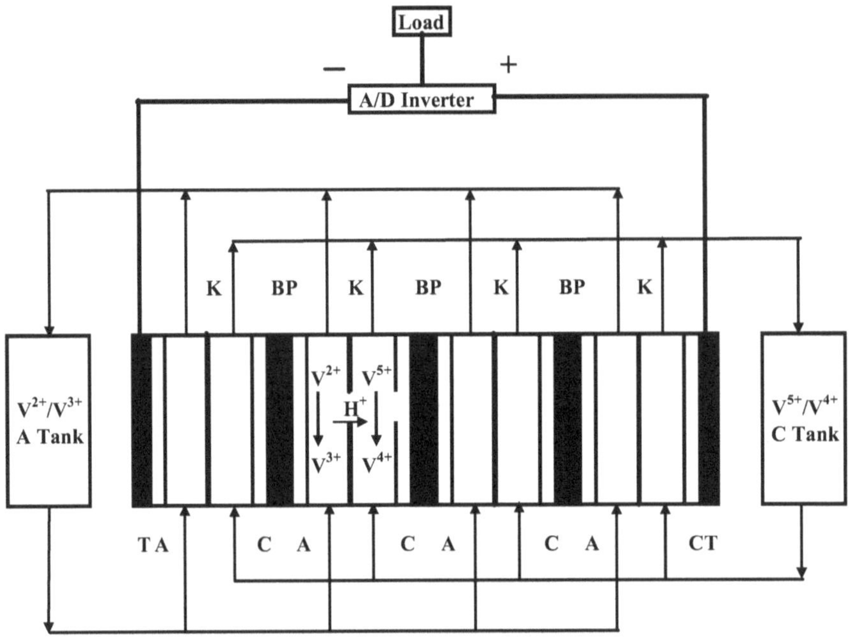

C: Cathode
A: Anode
K: Cation exchange membrane
T: Terminal electrode
BP: Bipolar plate
C Tank: Cathode solution tank
A Tank: Anode solution tank

FIGURE 23.3 Vanadium redox flow battery unit (discharging process). *Tanaka (2007), p. 511, Fig. 8.4.*

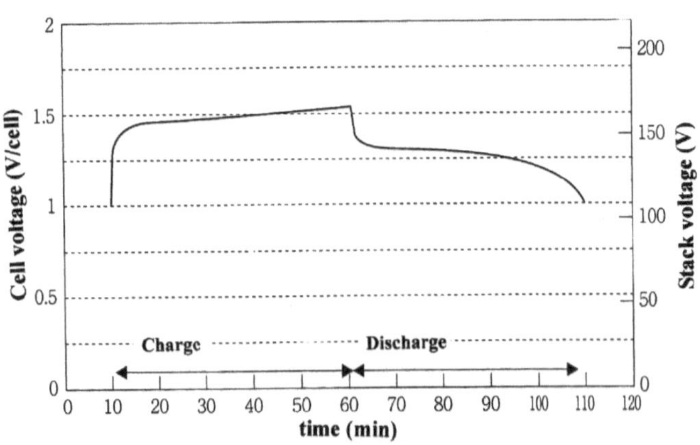

FIGURE 23.4 Voltage property of a vanadium redox flow battery (VRB). *Shigematsu (2011), p. 75, Fig. 3.9.*

23.3.2 Disadvantages

1. The volume of the tank is large, because solubility of the active electrolytes (vanadium) is limited.
2. Pump power is required for circulation of the electrolyte solution.
3. Electric current loss and leakage are generated through the electrolyte solution.

23.3.3 Remarks

1. Electric current leakage

Battery cells are arranged electrically in series, and an electrolyte solution is supplied in parallel to the cells through ducts. In this process, part of the electric current does not pass through the membranes and leaks through ducts (cf. Fundamentals Section 10.2). Electric current leakage increases with cell number, and it decreases the current efficiency of the process. Electric current leakage is reduced, for example, by elaborating pipeline design.

2. Sealing

Battery cells are sealed to prevent oxidation of the electrolyte solution due to contact between the solution and air. In this case, it is necessary to control volume changes of gases in the unit that are caused by temperature changes—sometimes explosion-proof valves are necessary.

3. Temperature control

The battery reaction is influenced by temperature, and to some extent the reaction is expected to proceed at a high temperature. However, temperature should be controlled in order to maintain the durability of unit materials.

23.4 Parts of the Redox Flow Battery

The battery system is composed of the following parts.

1. Battery cell

The battery cell illustrated in Fig. 23.3 should be designed to make oxidation–reduction reaction efficiency as high as possible and internal electric resistance as low as possible. The surface material in contact with the feeding electrolysis solution must be anti-acidic, because the solution includes sulfuric acid.

2. Electrodes

Electrodes are prepared using woven or nonwoven carbon fibers. The electrodes offer a place for oxidation–reduction reactions to occur, but do not react themselves. Accordingly, the surface area and solution penetrability should be as large as possible, and the fiber material should have affinity to the solution. Further, it is desirable to have the largest hydrogen and oxygen overpotential possible, in order to prevent side reactions from water splitting.

3. Bipolar plate

Many battery cells are arranged in series through bipolar plates as seen in Fig. 23.3. The functions of bipolar plates are to connect a cathode to an anode, and to prevent the mixing of the solutions. Bipolar plates are prepared with carbon resins (plastic carbon).

4. Flame

The above-mentioned parts are arranged in a flame prepared from polyvinyl chloride, polyethylene, etc.

5. Tank

Polyethylene tanks, rubber tanks, plastic lined metallic tanks, etc., are used.

6. Pump

Acid-resistant pumps must be used.

7. Ion exchange membrane

The ion exchange membrane is discussed in the following section.

23.5 Ion Exchange Membrane

H^+ ions generated on the cathode during the charging process, as expressed in Eqn (23.1), pass through cation exchange membranes (Fig. 23.1). Electrolyte ions passing through the membrane are mainly H^+ ions. The function of the cation exchange membrane is to permeate the H^+ ions and reject vanadium ion transport. The transfer of H^+ ions is accelerated by the Grotthuss mechanism for charge transfer (de Grotthuss, 1806). Vanadium ion transfer is caused by migration and diffusion in the sulfuric acid, which dissociates in the solution in two steps as follows:

$$H_2SO_4 \rightleftarrows H^+ + HSO_4^- \quad k_{a1} = 10^5$$
$$HSO_4^- \rightleftarrows H^+ + SO_4^{2-} \quad k_{a2} = 1.02 \times 10^{-2}$$

The first dissociation constant k_{al} is significantly larger than the second dissociation constant k_{a2}; thus, sulfuric acid ions are dissolved in the form of HSO_4^-. In the cation exchange membrane, HSO_4^- ions are rejected as a result of the Donnan exclusion (Fundamentals Section 2.2.1). In this circumstance, the H^+ ion transport number is estimated to be considerably high but not equal to 1. A small number of vanadium ions transferred by migration and diffusion cannot be ignored; the transport is related to the mobility of vanadium ions in the membrane. This phenomenon indicates that vanadium ions are not completely excluded from the membrane, thus allowing for vanadium ion crossover (Xi et al., 2007) that can cause self-discharge of the battery (Weidmann et al., 1998).

The H^+ ions pass through anion exchange membranes, and as with cation exchange membranes, this is due to the Grottuss mechanism. Thus, anion exchange membranes are also applicable to operating the VRB. The H^+ ion mobility in the anion exchange membrane is less than that in the cation exchange membrane, because of repulsive forces between H^+ ions and ion exchange groups ($-N^+(CH_3)_3$). As HSO_3^- ion transfer is not rejected by the Donnan exclusion, the H^+ ion transport number in the anion exchange membrane is less than that in the cation exchange membrane. However, vanadium ions are strongly excluded because of the Donnan exclusion, and as a result, vanadium ion crossover is negligible.

One must pay attention to water transfer across the membrane due to electroosmosis and volume osmosis. Electroosmosis is caused by water molecules hydrated to H^+ ions; the driving force of the electroosmosis is the potential gradient generated in the membrane (Fundamentals Section 2.8). In the ion exchange membrane, water molecules are transferred with H^+ ions, from the anode half-cell toward the cathode half-cell. This phenomenon induces an increase and decrease in the vanadium ion concentration in the anode half-cell and cathode half-cell, respectively. Thus, excessive water transfer between two half-cells can lead to precipitation of vanadium salt in the cell (Sukkar and Skyllas-Kazacos, 2003a,b). Volume osmosis is caused by the concentration (osmotic pressure) difference between the two half-cells (Fundamentals Section 2.7), and thus the phenomenon depends on the state of charge (SOC). Studies of cation exchange membranes found that the SOC of vanadium electrolytes influences the direction of volumetric transfer (Sukkar and Skyllas-Kazacos, 2003a,b). In the discharging process, the direction of volumetric transfer was from anode half-cell to cathode half-cell at an initial SOC from 100% to 50%. During further discharge, the direction of volumetric transfer reverses and occurs from cathode half-cell to anode half-cell until a SOC of 0% is attained.

Besides the above membrane characteristics, it is desirable that the electric resistance of the membrane be as low as possible. Membranes should have high chemical stability because they are exposed to the stressful oxidative effects of vanadium ions. A further desirable characteristic of the membrane is a low price. Consequently, it is necessary to design to minimize total cost while meeting technical requirements.

The VRB technology was developed vigorously in the research and development program carried out by the University of New South Wales. Parasuraman et al. (2013) performed a comprehensive historical review of materials research development and highlighted those findings that have contributed to VRB performance improvements. The following are descriptions of their review that are focused on the development of membrane technology.

During the early stages of the VRB research and development program at the University of New South Wales, the availability of low-cost, chemically stable commercial membranes proved to be a major obstacle to commercialization of the technology. Although good energy efficiency could be obtained with Selemion AMV, CMV, and DMV membranes (Asahi Glass Co.), their chemical stability in the highly oxidizing V (V) solution was limited, and significant chemical degradation was observed after only three months of cell operation (Mohamadi and Skyllas-Kazacos, 1996, 1997). Of the anion exchange membranes, Selemion AMV showed relatively poor stability in long-term charge–discharge cycling. In the case of New Selemion type 2 anion exchange membrane, however, area resistance and diffusivity values remained almost constant after six months of testing, while weight loss was negligible after two months of soaking in the V (V) solution. On the other hand, though Nafion membranes were found to have excellent chemical stability, their high price was commercially prohibitive, and their excessive swelling behavior led to serious water transfer problems during charge–discharge cycling (Hickner et al., 2004).

Membranes for the VRB have been extensively studied, particularly in recent years. Although many membranes provide promising results, meeting all the requirements is still a challenge requiring further development, in order to achieve the required cost structure for large-scale grid-connected applications. Excellent performance was achieved with anion exchange membranes by both Kashima-Kita Electric Power Corporation and Sumitomo Electric Industries in Japan between 1995 and 2005 (Shibata and Sato, 1999), but these tended to suffer from fouling by silica impurities in the vanadium electrolyte. Fouling could be eliminated by using high-purity vanadium pentaoxide for electrolyte preparation, but this in turn increased the cost of the system. Membranes for VRB, therefore, must be able to withstand fouling caused by impurities, so that low-purity vanadium oxide can be utilized in electrolyte production in order to reduce costs.

Although very expensive, Nafion has been one of the widely studied membranes for VRB. Several groups around the world have worked on modifying Nafion in order to reduce its permeability to vanadium ions (Xi et al., 2007, 2008; Schwenzer et al., 2011; Zeng et al., 2008; Vijayakumar et al., 2011).

Nonfluorinated membranes have received broad attention because of their low cost, low electrical resistance, and high ion selectivity. Sulfonated polyether ether ketone (SPEEK) has been one of the most widely studied. Several novel sandwich-type membranes involving SPEEK, polypropylene, and tungstophosphoric acid have been prepared and tested in VRB, and have produced promising results (Jia et al., 2010; Luo et al., 2008; Chen et al., 2010; Fujimoto et al., 2012).

Considerable work on anion exchange membranes in VRB has also been carried out by several research groups around the world (Hwang and Ohya, 1997; Qiu et al., 2007; Zhang et al., 2010; Xing et al., 2010). Although these membranes significantly reduce vanadium ion crossover resulting from the Donnan exclusion effect, the lower conductivity of protons across anion exchange membranes tends to increase resistivity and decrease the voltage efficiency of the battery.

23.6 Operation

Redox flow batteries are now operating in the following environments.

1. Effective electric power utilization

   In factories or buildings, electric power is charged at night time and discharged during the day time. Electricity charges are reduced by using power from the municipal grid at night when power rates are typically lower, and using power from VRBs during the day time, when municipal power rates are typically higher.

2. Measures against an interruption or instantaneous lowering of power supply

   VRBs are available to maintain stable power supply, thus avoiding inferior semi-finished goods production that can result from power interruptions or fluctuations.

3. Combination with natural energy utilization

   Output from natural energy devices such as solar generators and wind generators is not stable. The output can be stabilize with the additional of a redox flow battery operation.

4. Electric load leveling

   VRBs can adjust the supply of electricity in response to fluctuating demand levels, thus matching supply and demand levels in electrical environments that experience fluctuating loads.

5. Emergency uninterruptible power supply

   VRBs can be used for uninterruptible power supply in emergency lighting, hospital equipment, etc.

23.7 Practice

23.7.1 Kashima-Kita Electric Power Plant

Vanadium is a rare metal, and its production is maldistributed—the countries that produce it are not uniformly distributed with those countries having the greatest demand. As a result, it is easy to experience raw-materials shortages in nonproduction countries. This problem was solved by developing the technology to recover vanadium from soot discharged by thermal power plants. Kashima-Kita Electric Power Corporation (Ibaraki, Japan) developed the VRB system for utilizing vanadium recovered from thermal power plants. Kashima-Kita acquired a license for VRB technology from the University of New South Wales in 1992, and developed the VRB process to create a solar power generation system over four years, beginning in 1993 (Sato et al., 1998). The research and development was carried out in the New Sunshine Project managed by the Agency of Industrial Science and Technology (AIST).

In this research and development, meta vanadium acid ammonium (NH_4VO_3) was recovered from boiler-combustion soot and dissolved into high-temperature water. Then poly vanadic acid was generated, refined, and burned to produce high-purity vanadium pentoxide (V_2O_5), which was reduced with hydrogen to obtain vanadium trioxide (V_2O_3). The V_2O_5 and V_2O_3 were then dissolved into sulfuric acid that had been recovered from the gas-desulfurization equipment.

Polysulfone ion exchange membranes were found to show low oxidation resistance properties in the VRB. Kashima-Kita jointly developed polysulfone ion exchange membranes with Asahi Glass Co. Figure 23.5 shows that the power efficiency of polysulfone membranes is higher than that of polystirene and fluorine-based membranes. The oxidation-resistance property of the polysulfone membrane was largely improved, and the electric resistance lowered, as a result of decreased membrane thickness (Sato et al., 1998).

In September 1997, Kashima-Kita built a 200 kW × 4 h-rate battery after having built both 2 and 10 kW prototype units. The battery was interconnected to the company's power plant grid system and achieved 650-cycle continuous operation at the rated efficiency. Figure 23.6 exemplifies the performance of a charge–discharge cycle operation using the 10 kW prototype battery (Sato et al., 1998).

The specifications of a 200 kW battery are:

Basic specifications:
- Output: 200 kW, 4 h
- Current density: 80–100 mA/cm^2
- Electric power efficiency: 80%

Battery module:
- Module: eight stacks
- Stack: three substacks
- Substack: 21 cells
- Electrode area: 4000 cm^2
- Stack output: 25 kW

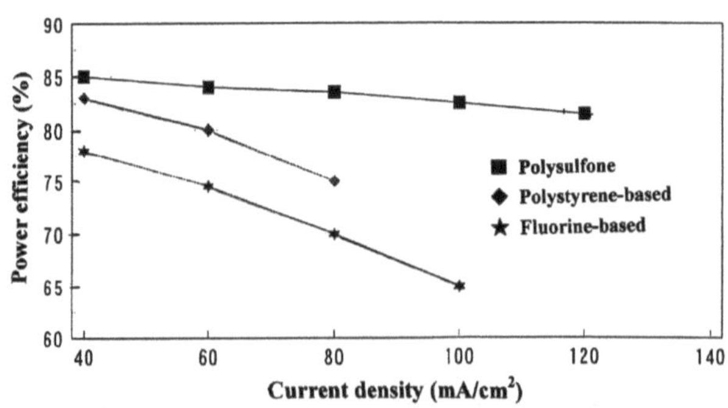

FIGURE 23.5 Power efficiency of ion exchange membranes. *Sato et al. (1998), p. 154, Fig. 23.5.*

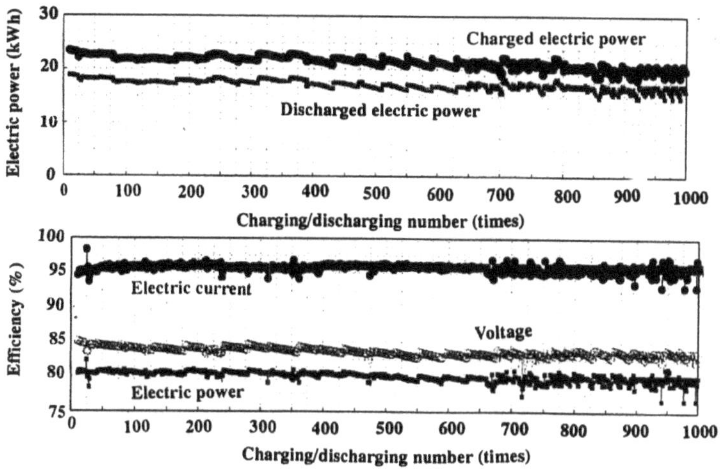

FIGURE 23.6 Charge–discharge operation of a 10 kW redox flow battery. Current density: 80 mA/cm^2, Charge–discharge time: 2.5 h cycle. *Sato et al. (1998), p. 155, Fig. 23.7.*

Electrolyte solution
 Volume: 22 m^3
 Concentration: 1.8 M

System cost for a 50 MW plant (8 h output) is estimated to be less than 200,000 yen/kW as follows, which is less than for a newly established thermal power plant (200,000–250,000 yen/kW).

Ion exchange membrane	26,000
Carbon resin electrode	19,000
Stack material	12,000
Electrolysis solution	36,000
Pipe, rack, tank	37,000
Meter, electrical system	9,000
Total	139,000 yen/kW

This success has proved that a high-efficiency heat power station system equipped with such an electricity storage battery, reusing byproducts of the power plant, can be commercially viable (Shibata and Sato, 1999).

23.7.2 Winbira Wind Power Plant

The redox flow battery system (rated output capacity AC: 4,000 kW–6,000 kWh; maximum output capacity: 6,000 kW) was installed in Electric Power Development Co., Winbira Wind power plant (Hokkaido, Japan; output capacity: 30,600 kW; wind turbine units: 19). An operation test was carried out to demonstrate output stabilization in the

Table 23.1 Specifications of the Wind Power Stabilization System (6 MW)

Item	Specifications
Connection voltage	Three-phase three-wire 6600 V ± 5%, 50 Hz ± 1 Hz
Rated AC power output	Inverter rated value: 1500 kVA × 4 bank
	Battery rated value: 1000 kW × 4 bank
Rated electric discharge capacity	6000 kWh
Stabilized output range	Less than ±6000 kW

Shigematsu (2011), p. 95, Table 3.4.

Table 23.2 Components of the Wind Power Stabilization System (6 MW)

Component	Quantity	Item
A/D inverter	4	1 unit/bank
Cell stack	96	24 units/bank
Heat exchanger	16	4 units/bank
Electrolyte solution tank	32	4 units/bank
Electrolytic solution circulation pump	32	4 units/bank
Battery control panel	4	1 unit/bank
Bank controller	1	1 unit/4 banks

Shigematsu (2011), p. 95, Table 3.5.

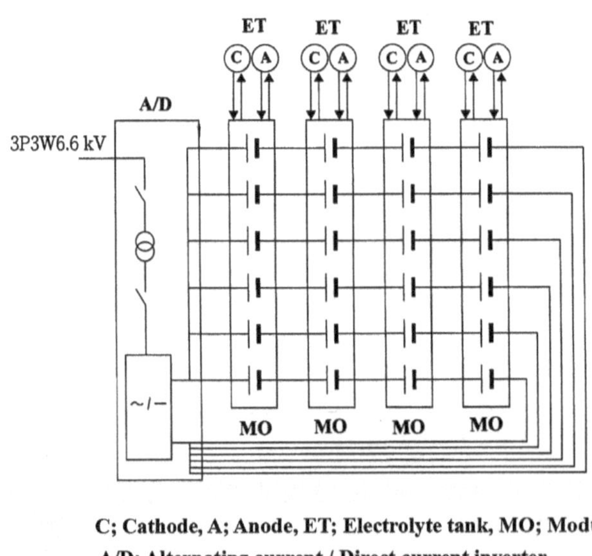

FIGURE 23.7 Main electric circuit for a bank. *Shigematsu (2011), p. 97, Fig. 3.44.*

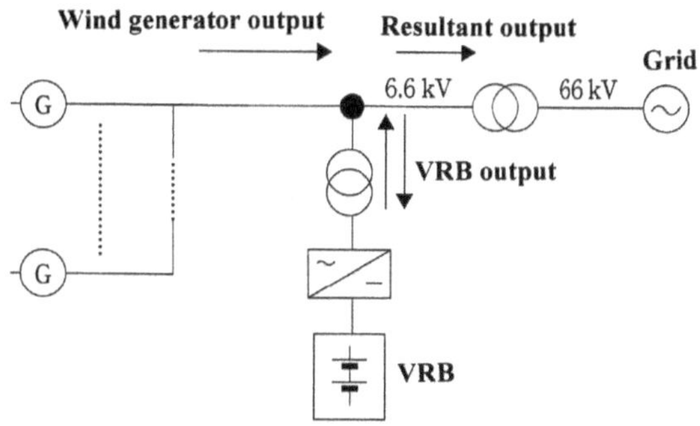

FIGURE 23.8 Connection diagram between generated electric power and an external grid. *Shigematsu (2011), p. 98, Fig. 3.45.*

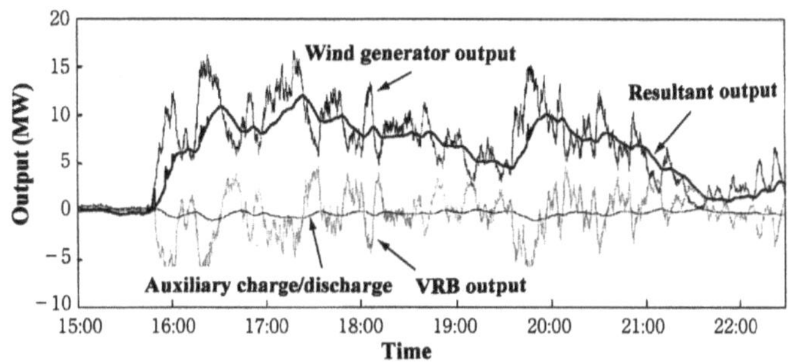

FIGURE 23.9 Electric power stabilization using the vanadium redox flow battery (VRB). *Shigematsu (2011), p. 100, Fig. 3.48.*

NEDO (National Economic Development Office) project beginning in 2003 (NEDO, 2010; Koshimizu et al., 2007; Yoshimoto et al., 2009).

Tables 23.1 and 23.2 show, respectively, the basic specifications of the battery system and instruments composing the system (Shigematsu, 2011). Figure 23.7 illustrates the main electric circuit for a bank. The resultant plant output is obtained by adding the wind generator output and the battery output, and is supplied to an external grid (Figure 23.8). The wind generator output usually fluctuates, and thus a low-pass filter is incorporated within the system to remove short-frequency components and minimize output fluctuations. Further, the battery system is operated with auxiliary charge and discharge to keep a fixed charge level. Figure 23.9 shows the changes of generator output, battery output, and resultant output over time. Auxiliary charge and discharge is also shown in the figure (Shigematsu, 2011).

References

Chen, D., Wang, S., Xiao, M., Meng, Y., 2010. Preparation and properties of sulfonated poly(fluorenyl ether ketone) membrane for vanadium redox flow battery application. J. Power Sources 195, 2089–2095.

Fujimoto, C., Kim, S., Stains, R., Wei, X., Li, L., Yang, Z.G., 2012. Vanadium redox flow battery efficiency and durability studies of sulfonated Diels Alder poly(phenylene)s. Electrochem. Commun. 20, 48–51.

de Grottuss, C.J.T., 1806. Sur la decomposition de l'eau et des corps qu'elle tient en dissolution a l'aide de l'electricite galvanique. Ann. Chim. Paris. 58, 54–73.

Hagedorn, N.H., Thaller, L.H., 1980. Redox Storage Systems for Solar Application. NASA TM-81464, National Aeronautics and Space Administration, U.S. Dept. of Energy.

Hickner, M.A., Ghassemi, H., Kim, Y.S., Einsla, B.R., McGrath, J.E., 2004. Alternative polymer systems for proton exchange membranes (PEMs). Chem. Rev. 104, 4587–4611.

Hwang, G.J., Ohya, H., 1997. Crosslinking of anion exchange membrane by accelerated electron radiation as a separator for the all-vanadium redox flow battery. J. Membr. Sci. 132, 55–61.

International Flow Battery Forum; IFBF, http://www.flowbatteryforum.com/.

Jia, C., Liu, J., Yan, C., 2010. A significantly improved membrane for vanadium redox flow battery. J. Power Sources 195, 4380–4383.

Koshimizu, G., Numata, T., Yoshimoto, K., Hasuike, H., Shibata, T., 2007. Project analysis of field test results for stabilization of 30.6 MW wind farm with energy storage. EESAT Conference.

Luo, Q., Zhang, H., Chen, J., You, D., Sun, C., Zhang, Y., 2008. Preparation and characterization of Nafion/SPEEK layered composite membrane and its application in vanadium redox flow battery. J. Membr. Sci. 325, 553–558.

Mohammadi, T., Skyllas-Kazakos, M., 1996. Modification of anion-exchange membranes for vanadium redox flow battery applications. J. Power Sources 63, 179–186.

Mohammadi, T., Skyllas-Kazakos, M., 1997. Evaluation of the chemical stability of some membranes in vanadium solution. J. Appl. Electrochem. 27, 153–160.

NEDO, 2010. http://www.nedo.go.jp/content/1000098129.pdf.

Nozaki, K., Kaneko, H., Negishi, A., Ozawa, T., 1984. In: Proc. Symp. Advances in Battery Materials, Proceedings, vol. 84-4. Electrochemical Society Inc, Princeton, NJ, p. 143.

Parasuraman, A., Lim, T.M., Menictas, C., Skyllas-Kazacos, M., 2013. Review of material research and development for vanadium redox flow battery applications. Electrochim. Acta 101, 27–40.

Pelligri, A., Spaziante, P. M., 1978. in GB 2030349.

Pissoort, P. A., 1933. in FR 754065.

Qiu, J.Y., Li, M.Y., Ni, J.F., Zhai, M.L., Peng, J., Xu, L., Zhou, H.H., Li, J.Q., Wei, G.S., 2007. Preparation of ETFE-based anion exchange membrane to reduce permeability of vanadium ions in vanadium redox battery. J. Membr. Sci. 297, 174–180.

Sato, K., Sawahama, M., Miyahayashi, M., Kageyama, Y., Nakajima, M., 1998. Development of new fuel and electric power storage battery. Soda Chlorine 49, 149–158.

Schwenzer, B., Kim, S., Vijayakumar, M., Yang, Z.G., Liu, J., 2011. Correlation of structural differences between Nafion/polyaniline and Nafion/polypyrrole composite membranes and observed transport properties. J. Membr. Sci. 372, 11–19.

Shibata, A., Sato, K., 1999. Development of vanadium redox flow battery for electrical energy storage. Power Eng. J. 13, 130–135.

Shigematsu, T., 2011. Redox flow battery. In: Ohta, K., Nozaki, K. (Eds.), Large-Scale Electric Power Storage Battery. Nikkan Kogyo Shimbun Publisher, Tokyo, pp. 63–112.

Skyllas-Kazacos, M., Rychcik, M., Robins, R., Fane, A.G., 1986. New all-vanadium redox cell. J. Electrochem. Soc. 133, 1057–1058.

Skyllas-Kazacos, Rychick, M., Robins, R., 1988. All-vanadium redox battery, U. S. 4,786,567.

Sukkar, T., Skyllas-Kazakos, 2003a. Water transfer behavior across cation exchange membranes in the vanadium redox battery. J. Membr. Sci. 222, 235–247.

Sukkar, T., Skyllas-Kazakos, 2003b. Modification of membranes using polyelectrolytes to improve water transfer properties in the vanadium redox battery. J. Membr. Sci. 222, 249–264.

Tanaka, Y., 2007. Ion Exchange Membranes: Fundamentals and Applications. Elsevier, Amsterdam.

Thaller, L.H., 1974. In: Proc. 9 Th Inter-Soc. Energy Conv. Eng. Conf. American Society of Mechanical Engineering, p. 924.

Vijayakumar, M., Bhuvaneswari, M.S., Nachimuthu, P., Schwenzer, B., Kim, S., Yang, Z., Liu, J., Graff, G. L., Thevuthasan, S., Hu, J., 2011. Spectroscopic investigation of the fouling process on Nafion membranes in vanadium redox flow batteries. J. Membr. Sci. 366, 325–334.

Weidmann, E., Heintz, E., Lichtenthaler, R.N., 1998. Sorption isotherm of vanadium with H_3O^+ ions in cation exchange membranes. J. Membr. Sci. 141, 207–213.

Xi, J., Wu, Z., Qiu, X., Chen, L., 2007. Nafion/SiO_2 hybrid membrane for vanadium redox flow battery. J. Power Sources 166, 531–536.

Xi, J., Wu, Z., Teng, X., Zho, Y., Chen, L., Qiu, X., 2008. Self-assembled polyelectrolyte multilayer modified Nafion membrane with suppressed vanadium ion crossover for vanadium redox flow batteries. J. Mater. Chem. 18, 1232–1238.

Xing, D.B., Zhang, S.H., Yin, C.X., Zhang, B.G., Jian, X.G., 2010. Effect of amination agent on the properties of quaternized poly(phthalazinone ether sulfone) anion exchange membrane for vanadium redox flow battery application. J. Membr. Sci. 354, 68–73.

Yoshimoto, K., Narahara, T., Koshimozu, G., 2009. Analysis of data obtained in demonstration test about battery energy storage system to mitigate output fluctuation of wind farm. IEEE PES CALGARY.

Zeng, J., Jiang, C., Wang, Y., Chen, J., Zhu, S., Zho, B., Wang, R., 2008. Studies on polypyrrole modified nafion membrane for vanadium redox flow battery applications. Electrochem. Commun. 10, 372–375.

Zhang, S.H., Yin, C.X., Xing, D.B., Yang, D.L., Jian, X.G., 2010. Preparation of chloromethylated/quaternized poly(phthalazinone ether ketone) anion exchange membrane materials for vanadium redox flow battery applications. J. Membr. Sci. 363, 243–249.

24 Companion Site

24.1 Introduction

The single-pass process is operated in a steady state and described by the stand-alone program integrated into the Excel spread sheet. The single-pass program consists of (1) constant current program, (2) constant voltage program, and (3) constant salt concentration program. Each program calculates the performance of the electrodialyzer. The computation is completed within about 10 min according to the procedure explained in Chapter 13, Section 13.15.

The feed-and-bleed process is also operated in a steady state and described by the stand-alone program integrated into the Excel spread sheet. The feed-and-bleed program calculates the performance of the electrodialyzer and circulation tank. The computation is carried out within about 15–25 min according to the steps explained in Chapter 14, Section 14.7.

The batch process is operated in a nonsteady state, thus it cannot be integrated into a stand-alone program. The computation is carried out according to the explanation in Chapter 15, Section 15.6 in the Excel spread sheet. The batch program computes the performance of the circulation tank. However, in order to operate the above computation, it is necessary to predetermine the 11 coefficients explained in Section 15.4 using the constant voltage single-pass program (companion site 2). The coefficients express the performance of the electrodialyzer.

24.2 Companion Site

The programs are integrated into the following companion sites and can be operated in the Internet web site.

Companion site 1
Constant current single-pass (continuous) program
Companion site 2
Constant voltage single-pass (continuous) program
Companion site 3
Constant salt concentration single-pass (continuous) program
Companion site 4
Constant voltage feed-and-bleed program

Ion Exchange Membranes. http://dx.doi.org/10.1016/B978-0-444-63319-4.00024-9

Companion site 5
Batch program for saline water desalination
Companion site 6
Batch program for drinking water production.

24.3 Web Site Address

http://booksite.elsevier.com/9780444633194.

Index

Note: Page numbers followed by "f" and "t" indicate figures and tables respectively.

N

T

Kolibri 360 GmbH
Ferdinand-Jühlke-Straße 7
99095 Erfurt, Deutschland
produktsicherheit@kolibri360.de